Acid Deposition

Environmental, Economic, and Policy Issues

Acid Deposition

Environmental, Economic, and Policy Issues

Edited by

Donald D. Adams

Center for Earth and Environmental Science
State University of New York at Plattsburgh
Plattsburgh, New York

and

Walter P. Page

Center for Business and Economics
State University of New York at Plattsburgh
Plattsburgh, New York

Plenum Press • New York and London

Library of Congress Cataloging in Publication Data

Main entry under title

Acid deposition.

   "Expanded version of the proceedings of a conference on acid deposition: environmental and economic impact, held June 10–12, 1983, in Plattsburgh, New
York"—T.p. verso.
   Includes bibliographical references and index.
   1. Acid deposition—Environmental aspects—United States—Congresses. 2. Acid
deposition—Environmental aspects—Government policy—United States—Congresses.
3. Environmental chemistry—Congresses. I. Adams, Donald D. II. Page, Walter P.
TD196.A25A244 1986   363.7'386   85-9456
ISBN 0-306-42062-7

Expanded version of the proceedings of a conference on Acid Deposition:
Environmental and Economic Impact, held June 10–12, 1983, in Plattsburgh,
New York

© 1985 Plenum Press, New York
A Division of Plenum Publishing Corporation
233 Spring Street, New York, N.Y. 10013

Printed in the United States of America

PREFACE

Concern about acid deposition, commonly referred
to as acid rain, as a widespread pollution problem with
severe ecological consequences has heightened public
awareness. Many authorities fear that acid deposition
may be the worst environmental crisis of our
industrialized society because of both the global
implications and possible widespread, irreversible
damage to lakes, soils, and forested ecosystems.
Neither state nor international boundaries are exempt
from the transport and deposition of airborne
pollutants resulting from local and distant emission
sources. The dilemma and debate will continue as long
as society requires fossil fuels for its energy needs
without regard to emission constraints.

This book started as a modest attempt to provide a
status report on atmospheric transport, the chemical
processes which produce acidifying agents, and
resultant ecological and economic consequences. The
materials in this book have been substantially revised
from those presented at the conference in 1983. It
became obvious that additional chapters were required
when sudden and profound changes occurring in European
forests were reported. It is felt that perhaps such
damages could be an early warning to forested
ecosystems in the northeastern United States and Canada
as well as other places throughout the world. Most
importantly, it is essential that gained scientific
knowledge be translated into required legislation - a
section on Policy Issues was incorporated to address
these concerns. It is hoped that the reader will
become informed and concerned enough to be involved in
this "global" debate.

Donald D. Adams
Walter P. Page

v

ACKNOWLEDGEMENTS

The success of the Conference on "The Environmental and Economic Effects of Acid Deposition" held at the State University of New York at Plattsburgh during the summer of 1983, where the basic foundations for this book were formulated, was due in large part to the enthusiasm and interest of many individuals at the Centers for Earth and Environmental Science, Business and Economics and Life-Long Learning. Special reference should be made to Professor Walter P. Page and especially Bette S. Brohel for their help in organizing the conference, and to Professors Donald J. Bogucki, James C. Dawson, Gerhard K. Gruendling, and John L. Malanchuk for chairing the sessions. Approximately half of the chapters in this book deal with topics presented at this conference. Financial support came from a donor wishing to remain anonymous, the Canadian Consulate in New York and over 80 participants from 13 states and 4 countries.

I wish to thank the numerous authors whose research and contributions brought this book to fruition. Dr. Volker Mohnen, Augustin Hogan, and other colleagues at the Atmospheric Science Research Center, Albany, NY, are acknowledged for providing editorial help with the Atmospheric Processes and Transport section. Mr. Gordon G. DeVries and staff at the SUNY-Plattsburgh Technical Assistance Center provided substantial help and access to computer facilities, while Mr. Kai-yan Yip developed the indexing program. The final editing and printing of the index was completed in Germany at the Christian-Albrechts-University in Kiel; Bernhard Kalhoff and Jens van Almelo provided considerable assistance. Financial support for book preparation came from the William H. Miner Agricultural Institute, Chazy, NY, and the Center for the Study of Canada at

SUNY-Plattsburgh.   Typing   was   done by Ms.   Yvette M. Pageau and Ms.   Theresa L.   Bosley.

        Lastly, I would like to thank Gerlinde   H.   Adams for her assistance during the editorial process and for indexing the entire book.

                                        Donald D. Adams, Editor

CONTENTS

POLICY ISSUES

ACID DEPOSITION: A NATIONAL PROBLEM

George R. Hendrey

Terrestrial and Aquatic Ecology Division, Department of
Applied Science, Brookhaven National Laboratory
Upton, New York 11973

ABSTRACT

   The deposition of excessive quantities of sulfur and nitrogen
from the atmosphere constitutes a problem encompassing all of the
United States east of the Mississippi River. It also occurs in
some areas of the western U.S. Calculations based on emission
inventories and simplifying assumptions indicate electric utilities
account for 66% of SO_2 emissions, 29% of NO_x emissions and about
half of precipitation acidity. Acidification of clearwater lakes
and streams is a widespread problem only in areas receiving rain
with an average acidity of pH $\leq$ 4.7. The dominant anion in such
waters is SO_4^{2-} and concentrations of aluminum derived from
watershed acidification may exceed 200 µg l^{-1}. Changes in
assemblages of aquatic biota become increasingly apparent as pH
decreases below 6.0, and elimination of fish from some waters has
been documented. The sensitivity of surface waters is controlled
by and represents an integration of biogeochemical processes in
their edaphic settings. Changes in surface water chemistry imply
changes in the terrestrial environment. Direct evidence of changes
in terrestrial environments is sparse. Nevertheless, observations
of forest dieback in the U.S. and abroad suggest that acid
deposition may contribute to the problem. Very few credible
studies are available which allow an evaluation of acid
precipitation effects on crops.

INTRODUCTION

Acid deposition is discussed on a daily basis by senators and congressmen, it has been the principal topic of presidential cabinet council meetings and it enters into discussions between the President and other heads of state (particularly with the Canadian Prime Minister). In less ethereal realms, the topic of acid deposition enters our homes in news programs and newspapers; children bring assignments relating to "acid rain" home from school; and the idea that our lakes and forests are "dying" contributes to a nagging fear that "they" are wrecking the world for the rest of us. Of course, we are "they." Acid deposition results from our own activities, particularly those that consume electricity from fossil fuel-fired power plants. Since the problem is an anthropogenic one, it is subject to control by man. Since the distribution of costs and benefits associated with both the creation of the pollutants causing acid deposition and the effects of acid deposition are neither equally distributed among the populace, nor evenly distributed over the landscape, solving the problem has become a highly political issue. All of this discussion has more than a trivial amount of fact behind it.

The problem, however, is how to evaluate just what the facts are. In this introductory chapter, a sketch of at least some of the facts will be provided as an overview of the topic with the effects area being emphasized. The following chapters present more detail on specific topics. Because of the importance of sulfate and nitrate in accounting for rain acidity, because they are primarily anthropogenic pollutants, and because of their importance to ecological processes following deposition, only these two materials will be followed in any detail.

PRECIPITATION AND DEPOSITION

The term "acid deposition" refers to deposition from the atmosphere of pollutants which when dissolved in water alter the solution concentrations of H^+ or acid anions. The most important of these are hydrogen ion (H^+), oxides of sulfur (SO_x), oxides of nitrogen (NO_x), strong acid anions (SO_4^{2-}, NO_3^-, Cl^-), ammonium (NH_4^+), and other cations (Ca^{2+}, Mg^{2+}, Na^+, K^+). These materials may be deposited in wet deposition (rain, fog, dew, snow, hail) or by dry deposition of gases (e.g., SO_2) or dry aerosols (e.g., ammonium sulfate).

Precipitation Acidity

Acid deposition occurs primarily in the eastern United States, but it has been observed in the West as well. For example, fog with hydrogen ion concentrations as high as 6000 microequivalents per liter (μeq l^{-1}), corresponding to a pH of 2.2, has been observed in southern California (Waldman et al., 1982). Over much of eastern North America all forms of precipitation on the average are quite acidic. The mean annual H^+ concentration averaged over all these forms of "wet" precipitation, and weighted by the volume of each precipitation event, ranged from 10 to 80 μeq l^{-1} (pH 5.0 to 4.1; Work Group 2, 1982). Individual rain episodes with H^+ concentrations around 1000 μeq l^{-1} (pH 3.0) are observed in the northeastern United States.

The H^+ in precipitation is associated with acid-forming anions. Stensland (1983) evaluated the associations among the various cations and anions observed in the annual average precipitation chemistry of widely separated atmospheric deposition monitoring sites in 1979. In the Adirondack Mountain region of New York, H^+ accounted for 67% of all cations. The strong acid anions sulfate (SO_4^{2-}) and nitrate (NO_3^-) comprised 60% and 34% respectively, of all anions (all values were reported in μeq l^{-1}). Together, SO_4^{2-} and NO_3^- accounted for 94% of all anions. In southwest Minnesota, H^+ made up less than one percent of cations; SO_4^{2-} and NO_3^- accounted for 54% and 29%, respectively, of all anions. Calcium and magnesium, together, made up 39% of the cations with ammonium contributing 42%, at the Minnesota site. In contrast, for the Adirondacks, calcium and magnesium were only 12% and ammonium was 12% of all cations. These comparisons are representative examples of the ranges of mean annual concentrations of ions observed in wet precipitation over the eastern United States.

Sources of SO_x and NO_x

The sources of sulfur found in acid deposition have been studied extensively. It is now well established that in eastern North America, and particularly in the northeastern U.S., natural sources of sulfur emissions (volcanoes, bogs, marshes, tide flats, other soils and vegetation) are insignificant, accounting for less than 1% of regional sulfur emissions (Robinson and Homoyla, 1983). Sources of NO_3^- and other oxides of nitrogen (NO_x) in the atmosphere which may contribute to the formation of NO_3^- are less well known, but Robinson and Homoyla (1983) concluded that "...perhaps only a few percent of the NO_x contribution to acid precipitation may be due to natural NO_x sources." Thus, it is reasonable to conclude that acid deposition is anthropogenic in origin. From evidence such as that presented above, it has been stated that over 90% of precipitation acidity in eastern North America may be the result of human activities (Work Group 2, 1982). This may be checked rather simply. A "back-of-the-envelope" calculation can be made using national pollutant emissions data for 1980. Of the 24.1 million metric tons (tonnes) of sulfur dioxide emitted in the U.S. in 1980,

electric utilities accounted for 66%, all other industrial sources
released 28%, and the transportation sector released a little over
3%. Of the 19.3 million tonnes of NO_x emitted in 1980, electric
utilities released 29%, all other industrial sources released 22%,
and the transportation sector released 44%. The approximate
contributions of each of these sources to precipitation acidity are
calculated in Table 1. Highway vehicles, which make up half of the
transportation sector, account for only 14% of precipitation
acidity on a national basis (Hendrey, 1983). However, this
fraction must certainly be higher in areas such as southern
California. These calculations assume that ground-level emissions
from automobiles are mixed into the atmosphere and transported in
the same way as emissions from smokestacks. While this may not be
exactly correct, the error introduced by this assumption is not
likely to be large (L. Newman, pers. comm.).

Deposition

    While the acidity of precipitation is of ecological concern,
the total deposition of both cations and anions associated with
acid deposition is also important to ecological processes. Other
papers in this volume will discuss these processes in greater
detail, but a general description will be presented here.

    Wet disposition of substances involves processes associated
with in-cloud droplet formation and scavenging. Aerosol particles
act as nuclei for water condensation. These small, wet particles
then grow by condensation and merge to form droplets. Gases such
as SO_2 and small particles may be incorporated during this process
as the droplets scavenge materials within the cloud. Below the

Table 1. Contribution (%) of electric utilities, other industries,
          and transportation to precipitation acidity. Calcu-
          lations based on data in Work Group 3-B (1982) and
          Environmental Protection Agency (1981).

	% of precip. acidity due to SO_x	% of SO_x emitted by source	% of precip. acidity due to NO_x	% of NO_x emitted by source	% contribution to precip. H^+
Electric Utilities	(60 x	66) +	(40 x	29) =	51.2
Other Industries	(60 x	28) +	(40 x	22) =	25.6
Transportation	(60 x	3) +	(40 x	44) =	19.4

Total contribution of these sources to precipitation acidity   96.2

cloud, falling droplets (or snowflakes) may also take up pollutants by interception and impaction processes. The amounts of wet materials deposited are reasonably well estimated by collecting precipitation and measuring its constituents. For a local area, averaging over period of a season or longer, a single properly operated monitoring site can provide sufficient information on wet deposition to determine the local inputs of substances resulting from this mechanism. For a larger region, such as a state, the Southeast or the Nation, depositional estimates will depend on one or more sampling networks. Spatial and temporal resolution of their patterns will depend on the area covered by the sampling networks, the density of the sampling stations over the area being monitored, the period of time (years of operation) over which the networks were operating, and the frequency with which samples are collected (Stensland, 1983). Currently, reasonable estimates of wet deposition are available.

The mean annual wet deposition of SO_4^{2-}, weighted by the amount of precipitation, ranged from 19 to 48 kilograms per hectare (kg ha^{-1}) in 1980 in the eastern United States. The corresponding values for nitrate were 12 to 33 kg ha^{-1} (Work Group 2, 1982).

While wet deposition is a rapid process, with most of the wet deposited pollutants arriving in a few major precipitation events, dry deposition is a slower but more continuous process. Dry deposition is also extremely difficult to actually measure in a way that is applicable to a region containing a variety of different surfaces, such as lakes, trees, buildings, soil, etc. (National Research Council, 1983). Seasonal factors are also important. Sulfate and H^+ concentrations are much higher in warm than cold season precipitation in the eastern United States (Stensland, 1983). Current best estimates hold that on a regional basis, wet and dry deposition are approximately equal (Galloway and Whelpdale, 1980; Shannon, 1981; Stensland, 1983). However, it can be expected that dry deposition is of greater importance in areas close to major emission zones (conurbations) than at sites remote from emissions. For a large region such as the Northeast or the Southeast, total deposition can be approximated by doubling wet deposition values.

EFFECTS OF ACID DEPOSITION TO AQUATIC ECOSYSTEMS

The link between acid deposition and environmental injury was first observed in Scandinavia in 1959. Widespread extinctions of fish populations in southern Norway now have been extensively documented. A preponderance of evidence indicates that surface water acidification due to acid deposition caused the loss of fisheries (Drabløs and Tollan, 1980). Since the early 1970s, when ecologists in the United States became aware of the Canadian and

Scandinavian acidification problems, increasing evidence became available that surface water acidification was occurring here as well.

Surface Water Acidification

    Lakes and streams in several areas of the world have been acidified in the last few decades. An extensive body of literature supports this statement and lengthy, critical reviews are available (Linthurst, 1983; Work Group I, 1983). Generally, waters which are sensitive to acidification are those with low alkalinity values (below 200 μeq 1^{-1}). This is because alkalinity is the principal form of buffering to acidic components in unacidified waters. This alkalinity is derived by weathering of minerals in the watershed. Thus, acidification of a lake by acid deposition is not simply the chemical titration of alkalinity from the water. Acidification will only occur when the supply of acids to waters flowing through a watershed into a lake plus those acids falling directly on the surface of a lake is sufficient to overcome both the rate at which alkalinity can be supplied by the watershed (including within the lake) and that which is stored within the lake.

    The importance of sulfur deposition on lake chemistry has been studied intensively at three lakes in the Adirondack Mountains by Galloway et al. (1983). Concentrations of sulfate in these waters were controlled primarily by deposition of anthropogenic sulfur. Sulfate concentrations in 1978-1980 were four to five times greater than historical values. This increased sulfate concentration must have had a strong influence on other aspects of chemistry. For example, from principals of conservation of mass and charge, either alkalinity had to decrease and/or base-forming cations must have increased along with sulfate. Decreasing alkalinity would reduce H^+ buffering resulting in a long-term decrease in pH or increased episodic fluctuations in pH. Increased concentrations of base cations could only arise from accelerated removal of these materials from watershed soils (Galloway et al., 1983).

    Acidified, oligotrophic, clear water lakes have only been reported in areas receiving precipitation with mean annual H^+ concentrations of pH 4.7 or less. Also, this acidification process appears to be a recent phenomenon, occurring in the last few decades. Acidification problems thus are coincident, both in space and in time, with acid deposition. However, the historical record of water chemistry is flawed in one or more ways. Because of this, for almost any water quality sampling station, it is difficult to present an unambiguous evaluation showing it to have been acidified. Other circumstantial evidence exists, however. Perhaps the most striking evidence is that in many lakes, such as some of these in the Adirondack Mountains from which fish have been eliminated, the waters are now so acidic (pH <5.0), and aluminum concentration so elevated (>200 µg l^{-1}) that fish cannot live in them. Obviously, the water chemistry has changed (Cronan and Schofield, 1979; Baker, 1983).

Aquatic Biota

Numerous studies indicated that increasing acidity is deleterious to aquatic organisms. Extensive reviews of this topic are presented elsewhere (Hendrey et al., 1980; Work Group I, 1983; Linthurst, 1983; Hendrey, 1984).

Even the most acidic natural waters have many kinds of organisms. Species of protozoans and insects are found at pH 2.0, rotifers and cladocerans at pH 3.0, and even some fish at pH 3.5 (Magnuson and Rahel, 1983). Acidified waters are not "dead," but loss of fisheries in areas receiving acid deposition is well documented (Baker, 1983; Baker and Schofield, 1985). Many other kinds of organisms have also been affected by acidification of surface waters.

Changes in assemblages of aquatic biota become increasingly apparent as pH decreases below 6.0. However, the extent to which biological changes occur is strongly influenced by other chemical concentrations. The biota of waters richer in divalent cations, in general, are less susceptible to injury at a particular pH level than are organisms in waters with lower cation concentrations; but the presence of aluminum, which increases with the addition of strong acid anions to a watershed, provides added stress.

Morphological features of a lake or stream can influence how the biotic community may respond to episodic acidification. For example, surges of strongly acidic water may occur during spring snowmelt because snow accumulation allows the temporary storage of pollutants in the snowpack. On melting, 50%-80% of the pollutant load may be released in the first thirty percent of meltwater (Johannessen and Henriksen, 1978). Lakes which are frozen in the winter generally retain their ice cover during the period when this polluted water is discharged. In lakes having diffuse inputs (as opposed to one or a few major tributaries) or during periods when snow melts rather slowly, this water will pass through the system just under the ice cover. Organisms can generally avoid this water, except those in the most shallow areas. In lakes with large tributaries or during rapid snow melting, the snowmelt water may mix to several meters depth thus exposing much or all of the littoral zone, and the organisms living there, to strongly acidic water (Hendrey et al., 1980). This is thought to be a major factor in stresses produced by acid deposition on aquatic biota (Hultberg, 1976).

In summary, the effects of surface water acidification on aquatic biota are observed at all trophic levels. Many species are lost and major processes, such as decomposition and primary and secondary production, are altered. Of greatest popular interest is the actual loss of fisheries (Work Group I, 1983).

The principal question now being addressed in the area of aquatic effects concerns estimating how many lakes and how many miles of streams have lost their normal fish population as a result of acidification. Such estimates would require both an inventory of water resources and either empirical correlations among deposition variables, watershed variables (size, soils, vegetation, etc.), and surface water chemistry or deterministic models. Such deterministic models would have to be based on both a detailed knowledge of the intervening processes and substantial amounts of field data for the specified sites. Both approaches might be required to address aquatic effects.

EFFECTS OF ACID DEPOSITION TO TERRESTRIAL ECOSYSTEMS

There are, as yet, no satisfactory answers concerning the importance which acid deposition may have in altering terrestrial systems. Evidence of terrestrial impacts is much more sparse than for aquatic systems. This may be because the chemistry of surface waters actually integrates all of the processes occurring in a watershed, thus reducing sampling heterogeneity and allowing spatial and temporal patterns to be more easily detected. Principal concerns in terrestrial environments, however, include the possibility that acid deposition might produce long-term changes in soil chemistry or alterations in forest or crop productivity. There is also concern for potential effects on structures (buildings, bridges, monuments) or painted surfaces (e.g., automobiles). All such injuries, if they occur, would lead to substantial losses in economic and aesthetic terms.

Soils

The characteristics of natural terrestrial ecosystems are largely determined by the properties of their soils. It has been pointed out above that watershed soils are a principal factor in controlling surface water alkalinity. Intimately linked with the process of alkalinity formation is the weathering of cations and other nutrients as well as the solubilization of organic substances which together provide the basic chemical constituents important to freshwater biota. It is also obvious that soil characteristics are a major factor in how man uses a particular area, i.e., for cultivation, range land, or forestry, etc.

In humid areas like the eastern and northwestern U.S., there is sufficient precipitation to cause the leaching of base-forming cations to be approximately equal to the rate of primary mineral weathering of these ions. Thus, soils become increasingly acidic over time due to entirely natural processes. The principal acid cations in soils are H^+ and aluminum (Al). As soil acidity increases, Al becomes increasingly soluble, but also adsorbs onto clay minerals in an equilibrium exchange with Al in the soil-water. Soil solution Al can hydrolyze to increase soil H^+ concentrations (Brady, 1974). In areas where sulfate adsorption in soils is low, such as in the Adirondacks, sulfur deposited from the atmosphere behaves more or less conservatively with sulfate passing through soils into lakes and streams being about equal to the total deposition of sulfur (Galloway et al., 1983). Because of a variety of geochemical processes, including cation exchange within the soils, some of the H^+ associated with acid rain is retained resulting in an increase in aluminum mobilization. Sulfur deposition, wet sulfate or dry SO_2, will have either an acidifying effect, or will leach base cations, or both, in such soils. In agricultural soils, agronomic practices of tilling, fertilization, and liming are judged to be far more important factors in altering soil chemistry than acid deposition.

In a broad review by McFee et al. (1983), it was concluded that additions of HNO_3 by acid deposition will generally be beneficial for forest growth since nitrogen is frequently a growth-limiting nutrient. Benefits from H_2SO_4 deposition were viewed as "minimal." Acid deposition might contribute to forest injury as a result of toxicity from increased soil solution aluminum concentrations. Soil microbiological processes (which are important for regulating soil nutrient status) may also be significantly influenced (McFee et al., 1983).

Forests

Currently, it is not known whether or not acid deposition is limiting forest growth. There is, however, strong circumstantial evidence that air pollutants are implicated in the well documented, large-scale forest decline in Europe. Spruce, fir, and beech forests at several locations in Germany are impacted. Evidence indicates that changes in soil chemistry, including high aluminum concentrations and calcium deficiency, may contribute to the problem in some areas (Johnson and Siccama, 1983; McLaughlin et al., 1983). On the other hand, similar declines in fir forests have been observed on calcarious soils; this would therefore rule out Al toxicity or calcium deficiency as contributors to the problem in those cases (McLaughlin et al., 1983).

Widespread declines in red spruce forests of New York, Vermont, and New Hampshire have been described by Johnson and

Siccama (1983). The decline is evidenced by foliar dieback and by
a very large reduction in basal area and density. At Camels Hump,
Vermont, between 1965 and 1979, a decrease in red spruce diameter
at breast height was 32%-84% across various age classes and was
observed over boreal, transition, and hardwood stand types. Camels
Hump was not the only forest affected for a wide variety of forests
covering a broad area have experienced substantial losses of
spruce. This problem, however, appears primarily in high elevation
stands. In the Green Mountains, Vermont, spruce mortality has also
ranged from about 15%-60%, while at Whiteface Mountain, New York,
about 20%-70% of the spruce showed dieback symptoms.

A variety of possible causes of the dieback and mortality were
examined by Johnson and Siccama (1983). The following did not seem
to be important factors: long-term climatic change, synchronous
forest development and mortality, natural thinning and breakup of
old stands, primary stress due to pathogens or atmospheric
oxidants. Studies of tree ring cores found a dramatic decrease in
increment size beginning in the mid-1960s and continuing to the
present. This was temporally unique; it did not occur previously.
This shift to abnormally small annual rings was widespread and
first appeared at about the same time as a major drought in the
early 1960s.

In the northeastern United States, middle to upper slopes
receive an estimated 2-3 Keq H^+ ha^{-1}yr^{-1} in precipitation. This is
three to four times the deposition measured in the lowlands. Such
high deposition rates are caused by high rates of precipitation and
cloud water interception. Similar values are estimated for high
elevation forests in the Southeast. Cloud moisture may be
especially important because, with an average pH of about 3.5, it
is much more acidic than rainfall. Furthermore, high elevation
vegetation is immersed in clouds for about one-fourth of each year.
Deposition of heavy metals, especially lead, is much greater in the
high elevation forests, and may be a contributor to the observed
forest reduction. On the other hand, there is some question as to
the importance of Al toxicity. Johnson and Siccama (1983)
suggested that multiple stresses may be responsible for the spruce
dieback and decline: "Drought followed by infection and secondary
organisms appears to be likely. Conceivably, acid deposition could
enhance drought stress or vice versa. The interaction of acid
deposition and secondary pathogens is also a possibility that
warrants investigation."

It is not known if the spruce decline is a harbinger of much
more widespread forest injuries which have yet to be discovered.
It is of great concern that this might be so, but hard data, other
than that reviewed here, have not been published.

Crops

Both dry and wet components of acid rain have been shown to cause injury to plants. Sulfur dioxide (SO_2) is known to be a potent phytotoxin in high concentrations. In locations near to emission sources, this is a well known problem which has led to devastation of plant communities in extreme cases. Plants considered to be sensitive to a three-hour exposure to SO_2 in concentrations ranging from 790-1570 µg m^{-3} (0.3-0.6 ppm) include ragweeds, legumes, blackberry, southern pines, red and black oaks, white ash, and sumac. Many other plants are also known to have various degrees of sensitivity (EPA, 1982).

In areas far removed from major emissions sources, SO_2 concentrations are generally low (0.5-5.0 µg m^{-3}) and are not likely to cause negative effects on vegetation. In fact, in some situations it may have a fertilizer effect (Noggle and Jones, 1979). Generally, nitrate can be expected to be beneficial to crops, but to the extent that it contributes to precipitation acidity it may also be harmful.

Many studies have examined the effects of various combinations of nitric and sulfuric acids as well as other substances commonly found in rain (EPA, 1982; Irving, 1983; Work Group 1, 1983). Most of these dealt with detection of morphological or physiological changes in greenhouse or "semi-field" conditions. These studies, while providing insight into mechanisms of injury, are not directly useful in estimating field crop loss (or gain) associated with acid deposition.

Although several true field studies of the vegetative effects of acid deposition have been published in the past few years, few of these can be used to provide estimates of a dose-response function applicable to field-grown crops. A major problem concerns the statistical designs used in field experiments. Evans and Thompson (1983), showed that field location is an important source of variation in field studies using soybean (this is a widely recognized fact in agronomic research). Using eight latin squares each with 4 x 4 plots, highly significant reductions in soybean yield were observed for simulated acidic rainfalls. For pH 4.1, 3.3, and 2.7 treatments, yields were 10.6%, 16.8%, and 23.9% less than that with a pH 5.6 treatment. The linear component of pH treatment differences was significant at p = 0.001. But when the latin squares were analyzed as separate experiments, erratic and in some cases contradictory conclusions could have been reached. The results reported by Evans and Thompson (1983) confirmed several previous similar field experiments conducted by Evans et al. (1983) and were very similar to results obtained in nearly identical experiments with the same cultivar (Amsoy) at the University of

Illinois in 1983 (W. Banwart, pers. comm.). These studies contrast
sharply with several others cited by Irving (1983), who stated:

> "...14 crop cultivars (nine species) studied, only one
> exhibited a consistently negative yield effect at acidity
> levels used (garden beet), three were negatively affected
> by at least one of the acidity levels used in the study
> ('So. Giant Curled' mustard green, 'Pioneer 3992' field
> corn and 'Amsoy' soybean), and six had higher yields from
> at least one acidity level ('Champion' and 'Cherry Bell'
> radish, 'Vernal' alfalfa, 'Alta' fescue, 'Beeson' soybean
> and 'Williams' soybean). The most frequent response
> reported to results from simulated acidic rain was 'no
> effect.'"

After reviewing the numerous studies cited by Irving (1983),
it is clear (as of June 1983) that only the experiments on soybean
cultivar "Amsoy" had a statistical design which included
estimations of field position effects, and these effects were found
to be quite significant. The consistency of the results with Amsoy
among five experiments by two different investigators at widely
separated sites is reasonably convincing. It is likely that
experiments which did not account for field position could not
separate variance due to this term from variance due to the
principal treatment, i.e., precipitation pH (Evans et al., 1984).
Unfortunately, other studies on field crops which can be relied on
to provide reasonable assessments of acid deposition effects are
lacking.

SUMMARY

Acid deposition is a consequence of the combustion of fossil
fuels, primarily in the process of generating electricity. While
it has been observed in the western United States, it is a major
problem over all of the eastern states, where precipitation acidity
averages less than pH 5.0; over most of the northeastern region
acidity averages less than pH 4.5. Wet and dry deposition
contribute about equally to total sulfur deposition when averaging
over the entire region, but dry deposition is thought to be more
important in areas close to major sources.

Surface water acidification is the most widely documented
environmental change associated with the deposition of strong
mineral acids. While a number of flaws exist in the historical
record of water quality data, the overall picture of surface water
acidification, particularly in the northeastern United States, is
convincing. Acid deposition is the only mechanism which has been
able to explain the spatial and temporal pattern of acidification.

Injuries to terrestrial environments due to acidic deposition have not been convincingly demonstrated. Evidence implicating acid deposition in the decline of forests, especially stands of spruce in higher elevation sites in the Northeast, is increasing. Changes in forest soils are a point of concern, but again detrimental impacts have not been clearly demonstrated. Although several crop plants appear to be sensitive to precipitation acidity, only one cultivar of soybeans has been shown to be injured by current levels of rain acidity. Few other credible field crops studies have been conducted.

REFERENCES

Baker, J.P., 1983, Fishes, pp. 5-76 to 5-137. In: The Acidic Deposition Phenomenon and Its Effects. Critical Assessment Review Papers. Vol. II, Effects Sciences. R.A. Linthurst, ed., Report No. EPA-600/8-83-016B, U.S.E.P.A., Washington, DC.

Baker, J.P. and Schofield, C.L., 1985, Acidification impacts on fish populations: A review, pp. 183 to 221. In: Acid Deposition - Environmental, Economic, and Policy Issues, D.D. Adams, ed., Plenum Publ., New York, NY.

Banwart, W.L. (personal communication). University of Illinois, Urbana, Ill.

Brady, N.C., 1974, The Nature and Properties of Soils. MacMillan, New York, NY. 672 pp.

Cronan, C.S. and Schofield, C.L., 1979, Aluminum leaching response to acid precipitation: effects on high elevation watersheds in the Northeast. Science 204:304.

Drabløs, D. and Tollan, A., eds., 1980, Ecological Impact of Acid Deposition, Proc. Intern. Conf., Sandefjord, Norway, SNSF Project, Ås-NLH, Norway. 383 pp.

EPA, 1981, National Air Pollutant Emission Estimates 1970-1979. EPA-450/4-81-010, U.S.E.P.A., Washington, DC.

EPA, 1982, Air Quality Criteria for Particulate Matter and Sulfur Oxides, Vol. III. Report No. EPA-600/8-82-029C, U.S.E.P.A., Washington, DC.

Evans, L.S. and Thompson, K.H., 1983, Comparison of experimental designs used to detect changes in yields of crops exposed to acidic precipitation. Agron. J. 76:81-84.

Evans, L.S., Hendrey, G.R., and Thompson, K.H., 1984, Comparison of statistical design and experimental protocols used to evaluate rain acidity effects on field-grown soybeans. J. Air Pollut. Control Assoc. 34:1107-1114.

Evans, L.S., Lewin, K.F., Patti, M.J., and Cunningham, E.A., 1983, Productivity of field-grown soybeans exposed to simulated acidic rain. New Phytol. 93:377-388.

Galloway, J.N. and Whelpdale, D.M., 1980, An atmospheric sulfur budget for eastern North America. Atmos. Environ. 14:409-417.

Galloway, J.N., Schofield, C.L., Peters, N.E., Hendrey, G.R., and
    Altwicker, E.R., 1983, Effect of atmospheric sulfur on the
    composition of three Adirondack lakes. Can. J. Fish. Aquat.
    Sci. 40:799-806.
Hendrey, G.R., 1983, Automobiles and acid rain. Science 222:8.
Hendrey, G.R., ed., 1984, Early Biotic Responses to Advancing Lake
    Acidification. Butterworth, Boston, MA. 173 pp.
Hendrey, G.R., Galloway, J.N., and Schofield, C.L., 1980, Temporal
    and spatial trends in the chemistry of acidified lakes under
    ice cover, pp. 266 to 267. In: Ecological Impact of Acid
    Deposition, D. Drabløs and A. Tollan, eds., Proceedings
    International Conference, Sandefjord, Norway, SNSF Project,
    Ås-NLH, Norway.
Hendrey, G.R., Yan, N.D., and Baumgartner, K.J., 1980, Response of
    freshwater plants and invertebrates to acidification, pp. 457
    to 465. In: Inland Waters and Lake Restoration, Proceedings
    International Conference. U.S.E.P.A., Washington, DC.
Hultberg, H., 1976, Thermally stratified acid water in late winter
    - a key factor inducing self-accelerating processes which
    increase acidification, pp. 503 to 517. In: Acid
    Precipitation and the Forest Ecosystem, Proceedings
    International Conference. Gen. Tech. Rep. NE-23, U.S.D.A.
    Forest Service, NE Forest Experiment Station, Upper Darby, PA.
Irving, P.M., 1983, Crops. pp. 3-42 to 3-62. In: The Acidic
    Deposition Phenomenon and Its Effects. Central Assessment
    Review Papers. Vol. II, Effects Sciences. R.A. Linthurst,
    ed., EPA-600/8-83-016B, U.S.E.P.A., Washington, DC.
Johannessen, M. and Henriksen, A., 1978, Chemistry of snowmelt
    water: Changes in concentration during melting. Water
    Resources Res. 14:615-619.
Johnson, A.H. and Siccama, T.G., 1983, Acid deposition and forest
    decline. Environ. Sci. Technol. 17:294A-305A.
Linthurst, R.A., ed., 1983, The Acidic Deposition Phenomenon and
    Its Effects. Critical Assessment Review Papers. Vol. II.
    Effects Sciences. Report No. EPA-600/8-83-016B, U.S.E.P.A.,
    Washington, DC.
Magnuson, J.J. and Rahel, R.J., 1983, Biota of naturally acidic
    waters, pp. 5-3 to 5-14. In: The Acidic Deposition
    Phenomenon and Its Effects. Critical Assessment Review
    Papers. Vol. II, Effects Sciences. R.A. Linthurst, ed.,
    Report No. EPA-600/8-83-016B, U.S.E.P.A., Washington, DC.
McFee, W.W., Adams, F., Cronan, C.S., Firestone, M.K., Foy, C.D.,
    Harter, R.D., and Johnson, D.W., 1983, The acidic deposition
    phenomenon and its effects. Effects on Soil Systems, pp. 2-1
    to 2-73. In: The Acidic Deposition Phenomenon and Its
    Effects. Critical Assessment Review Papers. Vol. II.
    Effects Sciences. R.A. Linthurst, ed., Report No.
    EPA-600/8-83-016B, U.S.E.P.A., Washington, DC.
McLaughlin, S.B., Raynal, D.J., and Johnson, A.H., 1983, Forests,
    pp. 3-27 to 3-88. In: The Acidic Deposition Phenomenon and

Its Effects. Critical Assessment Review Papers. Vol. II,
Effects Sciences. R.A. Linthurst, ed., Report No.
EPA-600/8-83-016B, U.S.E.P.A., Washington, DC.

National Research Council, 1983, Acid Deposition Atmospheric
Processes in Eastern North America. A Review of Current
Scientific Understanding. National Academy Press, Washington,
DC. 375 pp.

Newman, L. (personal communication). Brookhaven National
Laboratory, Upton, NY.

Noggle, J.C. and Jones, H.C., 1979, Accumulation of Atmospheric
Sulfur by Plants and Sulfur-Supplying Capacity of Soils.
Report No. EPA-600/7-79-109, U.S.E.P.A., Washington, DC.
49 pp.

Robinson, E. and Homoyla, J.B., 1983, The acidic deposition
phenomenon and its effects, natural and anthropogenic
emissions sources, pp. 2-1 to 2-108. In: The Acidic
Deposition Phenomenon and Its Effects. Critical Assessment
Review Papers. Vol. I, Atmospheric Sciences.
A.P. Altshuller, ed., Report No. EPA-600/8-83-016A,
U.S.E.P.A., Washington, DC.

Shannon, J.D., 1981, A model of regional long-term average sulfur
atmospheric pollution, surface removal, and net horizontal
flux. Atmos. Environ. 13:1155-1163.

Stensland, G.J., 1983, Wet deposition network data with
applications to selected problems, pp. 8-28 to 8-69. In: The
Acidic Deposition Phenomenon and Its Effects. Critical
Assessment Review Papers. Vol. I, Atmospheric Sciences.
A.P. Altshuller, ed., Report No. EPA-600/8-83-016A,
U.S.E.P.A., Washington, DC.

Waldman, J.M., Munger, J.W., Jacob, D.J., Flagan, R.C.,
Morgan, J.J., and Hoffman, M.R., 1982, Chemical composition of
acid fog. Science 218:677-680.

Work Group 1, 1983, Impact Assessment. United States - Canada
Memorandum of Intent on Transboundary Air Pollution. Final
Report. Office of Environmental Processes and Effects
Research (RD-682), U.S.E.P.A., Washington, DC.

Work Group 2, 1982, Atmospheric Sciences and Analysis. United
States - Canada Memorandum of Intent on Transboundary Air
Pollutants, Final Report. Report 2F. Office of Environmental
Processes and Effects Research (RD-682), U.S.E.P.A.,
Washington, DC.

Work Group 3B, 1982, Emissions, Costs, and Engineering Assessment.
United States - Canada Memorandum of Intent on Transboundary
Air Pollution, Office of Environmental Engineering and
Technology (RD-681), U.S.E.P.A., Washington, DC.

ATMOSPHERIC ACIDIFICATION CHEMISTRY: A REVIEW

Jack L. Durham and Kenneth L. Demerjian

U.S. Environmental Protection Agency
Environmental Sciences Research Laboratory
Research Triangle Park, NC

ABSTRACT

Atmospheric acidification is the result of the oxidation of
sulfur, nitrogen, and organic compounds to form their corresponding
acids.   The gas- and aqueous-phase pathways depend on the
production of oxidizing free radicals (HO, CH_3O_2) that react
directly with these compounds or produce molecular oxidants.   The
most important molecular oxidants are H_2O_2, organic peroxides, and
O_3.   Except for O_3-olefin reactions, these molecular oxidants are
not reactive in the gas phase, but they are highly reactive in the
aqueous phase with dissolved reductants (SO_2, HNO_2, organics).
Thus, the molecular oxidants may be generated in a photochemically
active region and transported long distances before reacting with
dissolved reductants in the aqueous phase, such as fogs, clouds,
and rain.   Changes in H_2SO_4 and HNO_3 atmospheric production rates
are expected to be a function of changes in the emission rates of
SO_2, NO_x, and VOC.   A reduction in emission rate of SO_2 will lead
to a reduction in gas-phase production of H_2SO_4.   However, if SO_2
concentration exceeds that of H_2O_2, reduction in SO_2 may not lead
to a reduction in aqueous-phase production of H_2SO_4.   Effective
control of H_2O_2 production is dependent on scientific advances to
determine its formation mechanisms and rates.

INTRODUCTION

Increases in environmental acidification result from the
oxidation of sulfur, nitrogen, and carbon compounds to form the
strong mineral acids H_2SO_4 and HNO_3 and organic acids (Work

17

Group 2, 1982). The principal paths of atmospheric acidification are reactions among pollutants in the gas phase, in or on particles, and in the aqueous phase. Ecosystem acidification occurs through the wet and dry deposition of these acids. Also, the wet and dry deposition of acidic precursors and oxidants may lead to acidifying reactions in terrestrial and aquatic ecosystems. Presently it is not possible to formulate a set of mathematical expressions to describe completely the source/receptor pathways, which are shown in Figure 1. Meteorological processes and complex chemical processes can be formulated, but there are important deficiencies in the knowledge of both areas. This review shall focus on the atmospheric chemistry of acidification. Conclusions and possible general control options shall be stated but with the recognition that the distribution of anthropogenic sources, contribution of natural sources, and meteorology are important and cannot be dismissed.

ACIDIFICATION PATHWAYS

The chemical paths leading to the acidification of the atmosphere depend on the production of oxidizing free radicals. The oxidizing free radicals may react directly with SO_2, NO_2, and organic compounds to yield H_2SO_4, HNO_3, and carboxylic acids, or they may react to produce molecular oxidants. The lifetime of the strongly oxidizing free radicals in the atmosphere is less than one second, so their transport distance in the atmosphere is less than several meters, which is insignificant. The most important molecular oxidants are hydrogen peroxide (H_2O_2), organic peroxides (CH_3OOH, etc.), ozone (O_3), and perhaps nitrogen dioxide (NO_2), nitrous acid (HNO_2), dinitrogen trioxide (N_2O_3). Ozone may react directly in the gas phase with olefinic organic pollutants to produce carboxylic acids, but in general, the molecular oxidant reactions with SO_2 and NO_2 are unimportant for atmospheric acidification. However, they are important for dissolved SO_2, HNO_2, and perhaps organics in the aqueous phase. Thus, the molecular oxidants may be generated in a photochemically active region and transported long distances (greater than 1000 km) before reacting with acid precursor gases in the aqueous phase (clouds, rain, or ecosystem) to produce the acids.

Photochemical Oxidation Cycle

Free radical attack on atmospheric volatile organic compounds (VOC) is initiated by a select group of compounds which are for the most part activated by sunlight. Formaldehyde (CH_2O) and HNO_2 photolyses show high potential as free radical initiators during the early morning sunrise period. The free radical attack causes the VOCs to decompose by various paths resulting in the production of the oxidizing free radicals HO (hydroxy), HO_2 (hydroperoxy), RO

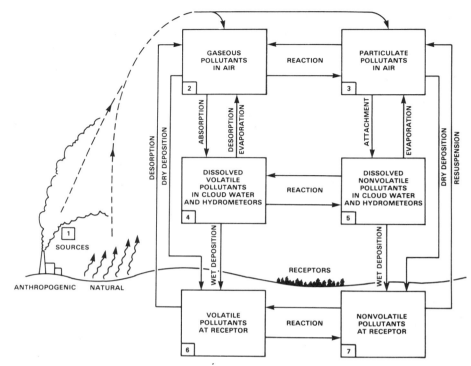

Figure 1.   Simplified schematic of pollutant chemical trans-
             formation, scavenging, and deposition pathways between
             sources and receptors. This diagram does not illustrate
             other important processes and factors affecting
             acidification, such as meteorological transport and
             source emission type and distribution.

(alkoxy), RO_2 (alkylperoxy) and partially oxidized products which
may also be photoactive radical-producing compounds.   The RO_2
radicals react with NO, converting it to NO_2, and in the process
produce HO and RO radicals.   RO radicals can be further oxidized,
forming additional RO_2 radicals and partially oxidized products,
thereby completing the inner cyclical loop reaction process
illustrated in Figure 2; or they may attack the VOC pool (this is
the major path for producing HO) present in the atmosphere, thereby
completing the outer loop reaction process.   The resultant effect
in either case is the conversion of NO to NO_2 with a commensurate
oxidation of reactive organic carbon to form the oxidation products
CO_2 and H_2O.   The complex mixture of organic compounds present in
the polluted atmosphere react with initiator radicals at different
rates dependent upon their molecular structure, the result being
varying yields of free radical species, O_3, NO_2, H_2O_2, peroxyacetyl
nitrate (PAN), and other partially oxidized organic products as a

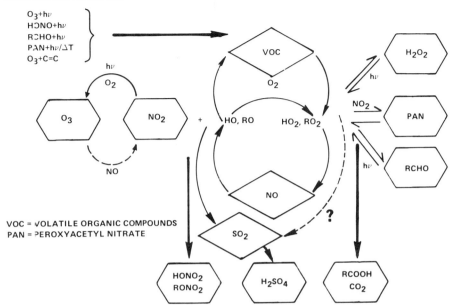

Figure 2.   Schematic of the polluted atmospheric photo-oxidation
            cycle. Pollutants emitted from anthropogenic sources
            are shown in diamond boxes; pollutants formed in the
            atmosphere are shown in hexagonal boxes. The reaction
            scheme represents gas-phase chemistry only.

function of VOC composition and [VOC] and [NO_x], i.e. the
concentrations of these two atmospheric components, respectively.
For a more detailed discussion of the oxidant chemistry of polluted
atmospheres, see Calvert (1983), Durham et al. (1982) and their
references.

        HO radical reactions seem to be the dominant gas-phase
mechanism by which carbon monoxide, hydrocarbons, nitrogen dioxide
and sulfur dioxide are consumed in the atmosphere (Niki et al.,
1972; Demerjian et al., 1974; Calvert et al., 1978). HO is a
highly reactive transient species, which contrary to its organic
free radical conterparts, shows limited variations in concentration
with changes in atmospheric [VOC] and [NO_x] levels, a result
readily explainable upon consideration of the free radical
production and consumption sources. In the case of the HO radical,
ambient conditions which enhance its production tend also to
consume the radical at an equivalent rate. The result is a faster
cycling in VOC-NO_x oxidation but very little perturbation in the HO
steady state concentration. Under similar conditions, an enhanced

production of organic free radicals, mainly RO_2, is not offset by an increased consumption, and this results in increased steady state concentrations. Hence, the concentration of organic free radicals is subject to variations due to changes in VOC and NO_x levels, in contrast to the HO concentration.

Gas-Phase Acidification Reactions

Calvert and Stockwell (1984) reviewed current rate constant data for SO_2 reactions with free radicals and concluded that the HO radical dominates the rate of SO_2 oxidation in the clean and polluted lower troposphere:

$$SO_2 + HO \longrightarrow H_2SO_4 \text{ (unbalanced)} \tag{1}$$

Typical average rates of SO_2 oxidation at mid-northern latitudes were estimated to be 16% per 24-hour period for summer and 3% per 24-hour period for winter conditions. The alkyl peroxyl radicals (especially CH_3O_2) and ozone-alkene (Crigree, 1957) intermediates may make important contributions to the gas-phase oxidation of SO_2 during periods of elevated $[NO_x]$ and [VOC].

NO_2 gas-phase oxidation in the lower troposphere is also likely dominated by HO radical reaction, shown as

$$NO_2 + HO \longrightarrow HNO_3 \tag{2}$$

For the same conditions as for SO_2, the NO_2 oxidation rate is 150% per 24-hour period for the summer and 25% per 24-hour period for the winter period (Calvert and Stockwell, 1984).

The known features of the gas-phase oxidation of SO_2 and NO_2 in the lower troposphere are presented in Table 1. The rates are bimolecular concentration functions. They are first-order in $[SO_2]$ or $[NO_2]$, but they may not be linear on a regional transport scale. This is described as:

$$d[H_2SO_4]/dt = -d[SO_2]/dt \neq \text{(constant)}[SO_2] \tag{3}$$

and

$$d[HNO_3]/dt = -d[NO_2]/dt \neq \text{(constant)}[NO_2]. \tag{4}$$

The non-linearity in the rates is due to: (a) the variation of [HO] as a function of sunlight intensity, and to some extent, [VOC], $[NO_x]$, and $[H_2O]$, and (b) the variations of $[CH_3O_2]$ and $[RO_2]$ are complex functions that depend strongly on sunlight intensity, the ratio $[VOC]/[NO_x]$, and [VOC]. Thus, the rate functions for the gas phase oxidation of SO_2 may be highly

Table 1.   Features of gas-phase SO_2 and NO_2 oxidation in the troposphere.

Reaction	Free Radical Oxidizer	Rate function[a]	Linear 24-hr Average Rate?[b]	Oxidizer Transported on Regional Scale?
(a)	HO	$k_a [SO_2] [HO]$	No	No
(b)	RO_2	$k_b [SO_2] [RO_2]$	No	No
(c)	HO	$k_c [NO_2] [HO]$	No	No

[a] Rate function: $-d[SO_2]/dt = k_a[SO_2][HO]$, etc.

[b] Overall rate of disappearance of SO_2 can be expressed as:

$$-d[SO_2]/dt = (constant)[SO_2]$$

variable. It has not been demonstrated that these functions may be represented by a 24-hour average rate calculation for transport/transformation/deposition modeling over transport time scales of several days. Because the atmospheric life-times of the free radicals HO and CH_3O_2 are less than one second, they cannot be transported for scale lengths greater than several meters.

The gas-phase oxidation of volatile organic compounds is known to occur through reactions with free radicals, and in the case of alkenes, with O_3. Other reaction paths are not established, but aqueous pathways may also be important.

Aqueous-Phase Acidification Reactions

The relative magnitudes of the contribution of gas-phase and aqueous-phase formation of H_2SO_4 are not established but expected to vary as a function of season. At southern latitudes for all seasons, it is likely that the gas-phase reaction,

$$SO_2 + HO \longrightarrow H_2SO_4 \text{ (unbalanced)}, \tag{1}$$

is an important acidification pathway. At southern and northern latitudes for all seasons, it is likely that the aqueous reactions are important and in the winter dominate the gas-phase pathway for northern latitudes. The aqueous-phase reactions thought to be the most important are (Work Group 2, 1982):

$$SO_{2(g)} + H_2O_{(liq)} \longleftrightarrow SO_2 \cdot H_2O \tag{5}$$

$$SO_2 \cdot H_2O \longleftrightarrow H^+ + HSO_3^- \tag{6}$$

$$HSO_3^- + H_2O_2 \longrightarrow H^+ + SO_4^{2-} + H_2O \tag{7}$$

$$HSO_3^- + O_3 \longrightarrow H^+ + SO_4^{2-} \quad \text{(unbalanced)} \tag{8}$$

$$HSO_3^- + NO_2^- \longrightarrow H^+ + SO_4^{2-} \quad \text{(unbalanced)} \tag{9}$$

$$HSO_3^- + O_2 \xrightarrow{\text{Mn(II)}} H^+ + SO_4^{2-} \quad \text{(unbalanced)} \tag{10}$$

$$HSO_3^- + O_2 \xrightarrow{\text{Fe(III)}} H^+ + SO_4^{2-} \quad \text{(unbalanced)} \tag{11}$$

$$HSO_3^- + O_2 \xrightarrow{\text{C(0)}} H^+ + SO_4^{2-} \quad \text{(unbalanced)} \tag{12}$$

At this time, the importance of the aqueous path of NO_2 oxidation to HNO_3 is not known, but it may be significant. The knowledge of aqueous oxidation rates of dissolved SO_2 is barely adequate for simple clean systems, inadequate for NO_x and N-oxy acid systems, and practically non-existent for complex sulfur dioxide/N-oxides/N-oxy acid/organic/catalyst/oxidizer atmospheric systems. Studies of these polluted air-aqueous systems are made difficult by the lack of high-purity reagents, the lack of sensitive instrumentation and methods to determine the reaction rates at ambient pollutant concentrations, and mass transfer rate limitations of practical bulk gas-liquid contact chemical reactors (Durham et al., 1982).

The dissolved SO_2 oxidation rate expressions for reactions thought to be important are shown in Table 2 for H_2O_2, O_3, NO_2, HNO_2, and catalysts. All of the rates are a function of the liquid water content (LWC) of the atmosphere, except for carbon, $C(0)$. These rates have been expressed in terms of the concentration of $SO_2 \cdot H_2O$, often referred to as "sulfurous acid," which is directly related by its Henry's Law constant (H_L) to the gas-phase concentration of SO_2:

$$[SO_{2(g)}] = H_L [SO_2 \cdot H_2O] \tag{13}$$

In similar fashion, the aqueous phase concentrations of H_2O_2, O_3, and HNO_2 are also directly related by their respective H_L constants to their gas-phase concentrations. Thus, a change in the gas-phase concentration of any of these species has a direct effect on the rate expression in Table 2. Since the concentrations of the dissolved oxidants and LWC are expected to be highly variable, none of the reaction rates are likely to be adequately represented over

Table 2.   Features of aqueous phase SO_2 oxidation in the lower
           troposphere

Reaction	Oxidizer	Rate function[a]	Linear 24-hr Average Rate?[b]	Oxidizer Transported on Regional Scale?
(d)	H_2O_2	$k_d[H_2O_2][SO_2 \cdot H_2O]$	No	Yes
(e)	O_3	$k_e[O_3][SO_2 \cdot H_2O]/[H^+]$	No	Yes
(f)	HNO_2	$k_f[HNO_2][SO_2 \cdot H_2O]/[H^+]^2$	No	Yes
(g)	$Mn(II)$	$k_g[Mn(II)]$	?	Yes
(h)	$Fe(III)$	$k_h[Fe(III)][SO_2 \cdot H_2O]/[H^+]$	?	Yes
(i)	$C(O)$	$(LWC)^{-1}k_i[C(O)]$	?	Yes

[a]Rate function: $-(LWC)^{-1}d[SO_2 \cdot H_2O]/dt = k_d[H_2O_2][SO_2 \cdot H_2O]$, etc.

[b]Overall rate of disappearance of SO_2 can be expressed as:
   $-d[SO_2 \cdot H_2O]/dt = (constant)[SO_2]$

a regional scale by a linear rate equation.   That is,

$$\text{rate of } H_2SO_4 \text{ formation} \neq K_a[SO_2]. \tag{14}$$

The oxidation rate of dissolved SO_2 by H_2O_2 possesses an
important feature.   When the rate is expressed formally as in
Table 2 in terms of $[SO_2 \cdot H_2O]$ instead of $[HSO_3^-]$, the aqueous rate
is dependent on the gas-phase concentrations of SO_2 and H_2O_2 and is
independent of $[H^+]$ over the pH range 2-6 (Martin and Damschen,
1981).   Thus, as the solution becomes more acidic, the rate of acid
formation does not decrease. Since the SO_2 oxidation rate by H_2O_2
is fast, the acidification of the aqueous phase can be considered
in terms of the "limiting" and the "excess" reactants.   Such
systems will continue to react until the "limiting" reactant is
totally consumed.   The concentration of the "excess" reactant does
not significantly influence the amount of acid that is formed.   In
terms of total conversion, the system is "linear" with respect to
the limiting reactant, and "non-linear" with respect to the
excess reactant.

The oxidation rate of dissolved SO_2 by O_3 may be expressed
formally as in Table 2 in terms of $[SO_2 \cdot H_2O]$ instead of $[HSO_3^-]$, and
it is dependent on the gas-phase concentrations of SO_2 and O_3 and
on $[H^+]^{-1}$.   Therefore, the rate will respond directly to changes in
gas-phase SO_2 and O_3, but more importantly, as the aqueous solution
acidity increases, the oxidation (acidification) rates decrease.
For pH less than about 6 and gas-phase $[O_3] = 50$ ppb, this reaction
is not as effective as the H_2O_2 oxidation reaction at gas-phase
concentrations greater than 1 ppb (Penkett et al., 1979; Durham et

al., 1984). In terms of total system conversion, the O_3 oxidation should be regarded as "non-linear." For pH values below about 5, the conversion in clouds and rain is unimportant for likely concentration ranges in regional air for SO_2 (5-20 ppb) and O_3 (40-80 ppb).

The oxidation of dissolved SO_2 is not important for the concentrations of NO and NO_2 usually found in the atmosphere (Lee and Schwartz, 1981). These gases are relatively insoluble in water. Their atmosphere concentrations are too low to form appreciable concentrations of reactive and soluble species such as dinitrogen trioxide (N_2O_3) and dinitrogen tetroxide (N_2O_4), and modeling of sub-cloud scavenging has demonstrated that they do not compete with H_2O_2 and O_3 in the acidification of raindrops (Durham et al., 1981). However, NO and NO_2 may be important aqueous-phase oxidants near their strong point sources.

Dissolved SO_2 may be oxidized by HNO_2, Mn(II), Fe(III), and C(O). However, modeling calculations (Durham et al., 1981; Durham et al., 1984) indicate that these reactions for a sub-cloud scavenging event are not competitive with those of H_2O_2 and O_3. Therefore, reactions 9 through 12 are likely to be important only in special circumstances, such as near point sources or, perhaps in the absence of H_2O_2 and O_3.

The major pathways of atmospheric acidification are summarized in Figure 3. In the gas phase, the main oxidants are the free radicals: HO, CH_3O_2, and perhaps the Crigee (1957) intermediates and other free radicals. These reactions require the establishment of the photochemical cycle shown in Figure 2. The photochemical cycle also leads to the formation of the molecular oxidants, especially H_2O_2, organic peroxides, and O_3. These oxidants are not effective in the gas phase for the oxidation of SO_2 but are reactive in the aqueous phase, such as atmospheric aqueous aerosols, cloud and fog droplets, raindrops, and surface waters.

ATMOSPHERIC OBSERVATIONS

A two-dimensional reactive scavenging model has been evaluated against the data of a single storm event for three SO_2 oxidation schemes: (a) aqueous-phase oxidation by dissolved O_3, (b) aqueous-phase oxidation by dissolved H_2O_2, and (c) pseudo first order aqueous oxidation at a rate proportional to dissolved S(IV) concentration (Easter and Hales, 1983). Aerosol scavenging was included in each scheme. Best agreement between the model and the experimental data was obtained for a combination of oxidation by H_2O_2 and O_3, with aerosol scavenging. It appears that H_2O_2 dominates the oxidation of SO_2, and O_3 does not contribute substantially to acidification until H_2O_2 has been consumed. This

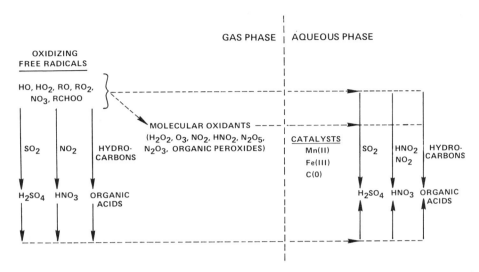

Figure 3.   Schematic of atmospheric acidification. The key step is
the formation of the oxidizing free radicals, which may
either oxidize directly SO_2, NO_2, and hydrocarbons to
their corresponding acids or may form the molecular
oxidants. The molecular oxidants may be transported
long distances, dissolve in the aqueous phase in fogs,
clouds or rain, and oxidize dissolved SO_2, HNO_2, NO_2 and
volatile organic compounds.

result is also consistent with reactive sub-cloud scavenging
modeling (Durham et al., 1984).

Summer and winter observations of cloud air and chemical
quality have been made at Whiteface Mountain (Camarota et al.,
1983; Galvin et al., 1983; McLaren et al., 1985). The summer time
observations indicated that SO_2 is rapidly and completely oxidized
in clouds, due to H_2O_2 aqueous concentrations in the range of 800
to 2000 ppb. The winter time observations are entirely different.
SO_2 is apparently not oxidized in the clouds at Whiteface, and H_2O_2
is also not present.

REDUCTION OF ATMOSPHERIC ACIDIFICATION

Effective options shown in Table 3 for the reduction of
atmospheric acidification must take into account the response of
H_2SO_4, HNO_3, and organic acid production as a function of SO_2, NO_x,
and VOC emission reductions. The task of formulating options is
difficult due to the lack of knowledge of the relative contribution
of gas-phase and of aqueous-phase pathways for the formation of

Table 3. Change in local H_2SO_4 atmospheric formation rates due to decrease in concentration of pollutants

Oxidizer		Decrease in Rate as a Function of Decrease in:	
		$[SO_2]$	[Oxidizer]
Gas Phase:			
Free radicals		linear	(impractical)
Aqueous Phase:			
H_2O_2	$<[SO_2]$	non-linear[a,b]	linear
	$>[SO_2]$	linear	non-linear[a,b]
O_3	$<[SO_2]$	non-linear[a,c]	non-linear[a,c]
	$>[SO_2]$	non-linear[a,c]	non-linear[a,c]

[a] The rate of H_2SO_4 formation is: $d[H_2SO_4] < $ (constant) $[SO_2]$
[b] Due to consumption of limiting reactant
[c] Due to H^+ inhibition

acids. The production of H_2SO_4 is the one pathway which is best understood, but the relative contributions of the major gas-phase reactions (HO oxidation of SO_2, etc.) has not been established. The rate of H_2SO_4 gas-phase production may be reduced by reducing the concentration of oxidizing free radicals (HO, CH_3O_2), or SO_2, or both. Reduction of [HO] is not practical; it is produced in the photochemical oxidant cycle and its concentration is not very sensitive to changes in precursor materials, such as NO_x and VOC. Also, this reaction is unimportant in locations where the photochemical oxidant cycle is not well established, such as latitudes greater than about 35°N from late fall to early spring.

The H_2O_2 oxidation of dissolved SO_2 in cloud and rain droplets is likely the dominating H_2SO_4 production reaction for all seasons and locations. Considered in terms of the "limiting reactant," H_2SO_4 aqueous-phase production will be responsive to emission controls of either SO_2 or H_2O_2, depending on which is present at the lower concentration. During non-photochemical conditions, (e.g., the northeastern U.S. in the winter months), the oxidants H_2O_2 and O_3 are not formed locally but may be transported from production regions in southerly latitudes. In such a case, oxidants would be the "limiting reactants" (which is consistent with observations at Whiteface Mountain), and control of SO_2 emissions to reduce H_2SO_4 aqueous-phase production will not be effective. In regions that are photochemically active, the oxidants may be present at concentrations that exceed those of SO_2. Effective reduction of H_2SO_4 production in that case will require SO_2 reduction.

Reduction in oxidant production (to reduce its contribution to aqueous-phase acidification) poses difficult problems. H_2O_2, and perhaps organic peroxides, are regarded as the main aqueous oxidizers of SO_2. However, their atmospheric formation mechanisms have not been established. If their production is related to O_3 production, the control of H_2O_2 poses difficult problems. It would be necessary to reduce VOC emissions (and perhaps NO_x emissions) in such a manner that [VOC]/[NO_x] is reduced in the photochemically active regions. However, the degree of source control to achieve reduction in H_2O_2 production cannot be presently specified; this is the object of a major research program (NAPAP, 1984). It is expected that reductions in acidic deposition to specific locations will require strategies that include SO_2, NO_x, and VOC emission reductions as a function of source location and season.

SUMMARY

Acidification of the atmosphere has been reviewed from the perspective of the influence of oxidizing agents. The photochemical oxidant cycle generates free radicals in the gaseous phase that may oxidize the reductants (sulfur dioxide, nitrogen dioxide, and organic compounds) to acids. Also, the free radicals react to produce stable molecular oxidants (hydrogen peroxide, ozone) that may be transported on a regional scale; these oxidants react with sulfur dioxide dissolved in cloud nuclei, rain, and surface waters to produce sulfuric acid.

The extent of the role of oxidants in limiting the acidification of the atmosphere is not known. Therefore, strategies to reduce only the concentrations of the reductants (sulfur dioxide and nitrogen oxides) may not produce the desired reduction in acid deposition at target areas. Effective options for reducing atmospheric acidification must take into account the response of sulfuric acid, nitric acid, and organic acid production as a function of emission reductions in sulfur dioxide, nitrogen oxides, and volatile organic compounds.

Acknowlegements - This review was performed as part of the National Acid Precipitation Assessment Program by the Environmental Protection Agency.

REFERENCES

Calvert, J.G. (Committee Chairman), 1983, Appendix A. The
    chemistry of acid formation, pp. 155 to 201. In: Acid
    Deposition: Atmospheric Processes in Eastern North America,
    National Research Council, National Academy Press, Washington,
    DC.
Calvert, J.G. and Stockwell, W.R., 1984, The mechanism and rates of
    the gas phase oxidations of sulfur dioxide and the nitrogen
    oxides in the atmosphere, pp. 1 to 62. In: Acid
    Precipitation--SO_2, NO, and NO_2 Oxidation Mechanisms:
    Atmospheric Considerations, J.G. Calvert, ed., Butterworth
    Publishers, Boston, MA.
Calvert, J.G., Su, F., Bottenheim, J.W., and Stausz, O.P., 1978,
    Mechanism of the homogeneous oxidation of sulfur dioxide in
    the troposphere. Atmos. Envir. 12:197-226.
Camarota, N.A., Kadlecek, J.A., McLaren, S.A., and Mohnen, V.A.,
    1983, Winter Cloud Study: ASRC Whiteface Mountain Field
    Station Project Report. Publication No. 971, Atmospheric
    Sciences Research Center, State University of New York,
    Albany, NY. 50 pp.
Criegee, R., 1957, The course of ozonization of unsaturated
    compounds. Record Chem. Progr. 18:111-120.
Demerjian, K.L., Kerr, J.A., and Calvert, J.G., 1974, Mechanisms of
    photochemical smog formation, pp. 1 to 262. In: Advances in
    Environmental Science and Technology, Vol. 4. J.N. Pitts and
    R.J. Metcalf, eds., John Wiley, New York, NY.
Durham, J.L., Barnes, H.M., and Overton, J.H., 1984, Acidification
    of rain by oxidation of dissolved sulfur dioxide and
    absorption of nitric acid, pp. 197 to 236. In: Acid
    Precipitation Series - Vol. 2, Chemistry of Particles, Fogs,
    and Rain, J.L. Durham, ed., Butterworth Publishers, Boston,
    MA.
Durham, J.L., Overton, J.H. Jr., and Aneja, V.P., 1981, Influence
    of gaseous nitric acid on sulfate production and acidity in
    rain. Atmos. Envir. 15:1059-1068.
Durham, J.L., Demerjian, K.L., Barnes, H.M., and Wilson, W.E.,
    1982, Chapter 2. Physical and chemical properties of sulfur
    oxides and particulate matter, pp. 1 to 100. In: Air Quality
    Criteria for Particulate Matter and Sulfur Oxides, Vol. I.
    Report No. EPA-600/8-82-029a, U.S.E.P.A., Office of Research
    and Development, Research Triangle Park, NC.
Easter, R.C. and Hales, J.M., 1983, Interpretations of the OSCAR
    data for reactive gas scavenging, pp. 649 to 662. In:
    Precipitation Scavenging, Dry Deposition, and Resuspension.
    Vol. I: Precipitation Scavenging. H.R. Pruppacher,
    R.G. Semonin, and W.G.N. Slinn, Coordinators, Elsevier, New
    York, NY.
Galvin, P.J., Mohnen, V.A., Kadlecek, J.A., Wilson, J.W., and
    Kelly, T.J., 1983, Cloud chemistry studies at the Whiteface

Mountain Field Station. Publication No. 973, Atmospheric
Sciences Research Center, State University of New York,
Albany, NY. 51 pp.

Lee, Y.N. and Schwartz, S.E., 1981, Evaluation of the rate of
uptake of nitric acid in liquid water. J. Geophys. Res.
86:11971-11983.

Martin, L.R. and Damschen, D.E., 1981, Aqueous oxidation of sulfur
dioxide by hydrogen peroxide at low pH. Atmos. Envir.
15:1615-1621.

McLaren, S.E., Kadlecek, J.A., and Mohnen, V.A., 1985, SO_2
oxidation in summertime cloud water at Whiteface Mountain,
pp. 31 to 48. In: Acid Deposition - Environmental, Economic,
and Policy Issues, D.D. Adams, ed., Plenum Publ., New York,
NY.

National Acid Precipitation Assessment Program (NAPAP), 1984,
Annual Report, 1983. Council on Environmental Quality,
Washington, DC. 74 pp.

Niki, H., Doby, E.E., and Weinstock, B., 1972, Mechanisms of smog
reactions. Adv. Chem. Ser. 113:16-57.

Penkett, S.A., Jones, B.M.R., Brice, K.A., and Eggleton, A.E.J.,
1979, The importance of atmospheric ozone and hydrogen
peroxide in oxidizing sulfur dioxide in cloud and rainwater.
Atmos. Envir. 13:123-137.

Work Group 2, 1982, U.S.-Canada Memorandum of Intent in
Transboundary Air Pollution (August 5, 1980), Report No. 2F.
U.S.E.P.A., Office of Research and Development, Washington,
DC. 224 pp.

SO_2 OXIDATION IN SUMMERTIME CLOUD WATER AT WHITEFACE MOUNTAIN

Scott E. McLaren, John A. Kadlecek, and Volker A. Mohnen

Atmospheric Sciences Research Center
State University of New York at Albany
Albany, NY 12222

ABSTRACT

The rate of within cloud aqueous oxidation of sulfur dioxide by hydrogen peroxide was calculated from existing laboratory data. For conditions typically encountered this rate was quite fast, at a few percent per minute, and was not effected by temperature, pH and liquid water content of the cloud. Summertime measurements at Whiteface Mountain of cloud water sulfate and H_2O_2 concentrations, pH, SO_2, cloud liquid water content, along with air mass trajectories and enhanced satellite imagery, were used to verify the calculated oxidation rate. Because of the very low measurements of H_2O_2 and high SO_2 concentrations within the clouds during a wintertime cloud water monitoring project for the 1983-1984 period, it was suspected that such an oxidation mechanism was not present.

INTRODUCTION

Before calculating the effects of any proposed SO_2 emissions control strategy, it is necessary to understand the important transport and transformation mechanisms. To this end, the Atmospheric Sciences Research Center maintains a laboratory on the summit of Whiteface Mountain in the Adirondack Mountains near Lake Placid, NY for the purpose of collecting cloud water and measuring its pH, conductivity, Cl^-, NO_3^-, SO_4^{2-}, Na^+, K^+, H_2O_2 (hydrogen peroxide), HSO_3^- (bisulfate), and HCHO (formaldahyde) concentrations. These measurements are supported by aerosol collection

31

studies, measurements of gas phase SO_2, NO_x, and O_3, along with monitoring of standard meteorological parameters.

The results of the measurements of H_2O_2 in cloud water and gas phase SO_2 at Whiteface Mountain will be discussed in this paper. These measurements permitted the establishment of a lower limit for the SO_2 oxidation rate; a limit consistent with a calculated SO_2 oxidation process in cloud considering only the parameters: temperature, liquid water content, pH, H_2O_2 and S(IV) concentrations. The high concentrations of hydrogen peroxide, $[H_2O_2]$, simplified the model by allowing only the reaction of SO_2 with H_2O_2 to be considered. The oxidation rate was proportional to both H_2O_2 and SO_2 concentrations; the more complicated effects of temperature, liquid water content and pH on the oxidation rate are also evaluated.

PREVIOUS STUDIES

The H_2O_2 concentrations found in cloud water at Whiteface Mountain during the summer were sufficiently high that it was considered the primary oxidizing agent for SO_2. Significant H_2O_2 concentrations persisted throughout cloudy periods thus insuring that this oxidation pathway was continuously available (Kadlecek et al., 1982). Therefore, the chemical reactions involved with H_2O_2 oxidation of SO_2, as described below, must be evaluated in greater detail.

Oxidation Rates

The aqueous phase oxidation of SO_2 by H_2O_2 in a cloud involves mass transfer of the SO_2(g) and H_2O_2(g) to the cloud drop, aqueous phase formation of H_2O_2, ionization of SO_2(aq), and reaction of S(IV)(aq) species with H_2O_2(aq) to form sulfuric acid (Schwartz, 1982). Schwartz and Freiberg (1981) concluded that under usual atmospheric conditions, SO_2 transfer to droplets will be oxidation rate controlled; not mass transfer limited. They assumed that no surfactants were present on the drop.[1]

---

[1]Recent work by Graedel et al. (1982) suggested that if insoluble organic compounds form a surface film on drops, a significant decrease in the mass transfer rate might occur. More work is needed to determine conditions for the existence of surface films and the effect they would have on mass transfer mechanisms in cloud drops. Calculations assume that if surface films are present on cloud drops, they do not hinder the mass transfer rate enough to prevent the gaseous and aqueous phases from attaining equilibrium.

Martin and Damschen (1981) studied the aqueous reaction of S(IV)(aq) with H$_2$O$_2$(aq) and obtained an expression for the rate of aqueous oxidation of S(IV) at 25°C over the pH range 0-5:

$$\frac{d[S(IV)]}{dt} = \frac{[5.2 \times 10^6] \, [H_2O_2] \, [HSO_3^-] \, [H^+]}{0.10 + [H^+]} \tag{1}$$

Penkett et al. (1979) found the activation energy for this reaction at pH = 4.6 to be 30.30 kJ/mol. Using this activation energy to incorporate a temperature-dependent term in equation 1 results in:

$$\frac{d[S(IV)]}{dt} = \frac{[5.2 \times 10^6] [H_2O_2][HSO_3^-][H^+]exp[-3655.3(\frac{1}{T} - \frac{1}{298})]}{0.10 + [H^+]} \tag{2}$$

This expression was used to calculate the aqueous oxidation rate of sulfur dioxide.

Liquid-vapor Equilibrium Calculations

Henry's law applies for dilute solutions in equilibrium with the gaseous phase, where the ratio of aqueous to gaseous phase concentrations of a solute is a constant for both H$_2$O$_2$ and SO$_2$:

$$\frac{[H_2O_2]}{P_{(H_2O_2)}} = H_{H_2O_2} \tag{3}$$

where $\log H_{H_2O_2} = \dfrac{3032.3}{T} - 5.317$ M atm^{-1} $\tag{4}$

and $\dfrac{[SO_2]_{aq}}{P_{SO_2}} = HSO_2$ $\tag{5}$

where $\log H_{SO_2} = \dfrac{1376.1}{T} - 4.521$ M atm^{-1} $\tag{6}$

Henry's law constants for H$_2$O$_2$ and SO$_2$ above were taken from Penkett et al. (1979) and Maahs (1982), respectively. When SO$_2$ dissolves in water it undergoes rapid ionization to form HSO$_3^-$ and SO$_3^{2-}$. The total S(IV)aq concentration becomes

$$[S(IV)] = [SO_2]_{aq} + [HSO_3^-] + [SO_3^{2-}] \tag{7}$$

which can be related to the dissolved SO$_2$ concentration by:

$$[S(IV)] = [SO_2]_{aq} [1 + \frac{k_1}{[H^+]} + \frac{(k_1) \, (k_2)}{[H^+]^2}] \tag{8}$$

where k_1 and k_2 are the first and second dissociation constants for sulfurous acid (Maahs, 1982):

$$\log k_1 = \frac{853.0}{T} - 4.740 \ M \quad \text{and} \quad \log k_2 = \frac{621.9}{T} - 9.278 \ M \qquad (9)$$

Henry's law allows the ratio of aqueous to gaseous concentrations of a solute to be calculated but does not provide information concerning the amount of solute in each phase. The molar ratio between the aqueous and gaseous phase of a solute (on a per unit total volume) is given by

$$\text{aqueous moles/gaseous moles} = L \ H \ R \ T \qquad (10)$$

where L is the liquid water content, H is the appropriate Henry's law constant, R is the universal gas constant and T is the absolute temperature.

RESULTS OF THE MODEL

Under actual atmospheric conditions, SO_2 will react with other oxidants or catalysts found in cloud water, and H_2O_2 can react with compounds other than SO_2. Furthermore, there is considerable evidence of both gas and aqueous phase production of H_2O_2 (Heikes et al., 1982; Cooper and Zika, 1983). Early in the life of a cloud, however, the reaction of H_2O_2 and SO_2 will so dominate other reactions that this mechanism alone can account for the observed rates.

For these model calculations, the different temperatures and liquid water contents were held constant. The initial acidity of the cloud was chosen by specifying a pre-cloud hydrogen ion concentration consistent with pH values in the low to mid 4 range. In a real cloud, this pre-cloud acid would be the net sum of soluble gaseous acids (HNO_3, HCl, etc.) and particulate acids (NH_4^+, HSO_4^-, H_2SO_4, etc.), modified by accumulated bases. The only new source of hydrogen ion considered in this model would result from the formation of sulfuric acid.

Given pre-set concentrations of SO_2, H_2O_2 and acid, a model cloud was allowed to form and chemical equilibrium distributions between aqueous and gaseous phases was calculated. Using these concentrations, the amounts of S(IV) and H_2O_2 that would react in a short time (1 sec.) were calculated and subtracted from the aqueous phase concentrations. The process of establishing equilibrium and reacting H_2O_2 and SO_2 was repeated until most of the SO_2 had been oxidized.

The modeled concentration profiles in a cloud containing $0.25 \ g \ m^{-3}$ of liquid water formed in air initially containing

2 ppbv SO_2, 2 ppbv H_2O_2, and 20 nM m^{-3} of acid at 10°C is provided in Figure 1. Except for vapor phase H_2O_2 (for which there are no reliable measurement techniques) concentrations were typical of those measured at Whiteface Mountain, and 2 ppbv for H_2O_2 was consistent with aqueous H_2O_2 measurements. S(IV) oxidized very quickly under these conditions; 70% reacted after 30 minutes and 82% after one hour.

The effect of temperature on the oxidation process is shown in Figure 2. The initial conditions were the same as in Figure 1 except for temperature, which was varied over the range of values normally seen during the summer at Whiteface Mountain. The overall oxidation rate decreased with increasing temperature because the decrease in solubility of SO_2 and H_2O_2 more than compensated for the increase in the aqueous phase oxidation rate. The difference in percent of original SO_2 remaining after one hour for the T = 10°C case was less than ten percent.

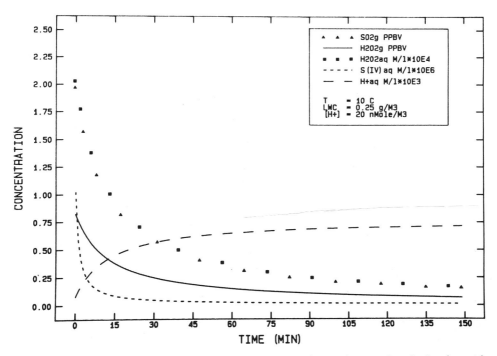

Figure 1. Calculated concentration profiles in a cloud during the oxidation of SO_2 by H_2O_2. Concentrations of SO_2 (g) and H_2O_2 (ag) followed the same profile.

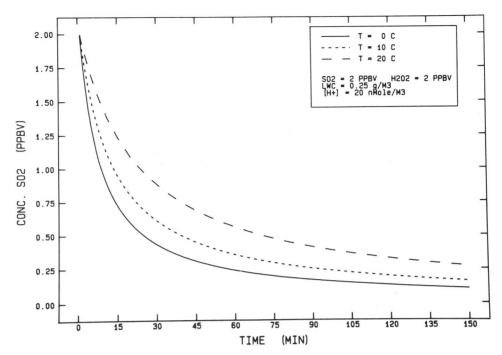

Figure 2. Effect of temperature on oxidation process.

     The influence of liquid water content on the oxidation process
is shown in Figure 3. The overall oxidation rate was faster with
increasing liquid water content. Even though a higher liquid water
content would cause the concentration of SO_2 and H_2O_2 to decrease,
thus reducing the aqueous reaction rate, the volume in which the
reaction occurs is greater, thereby causing the overall reaction
rate to increase. Over the most commonly observed ranges for
liquid water (0.1-0.5 g m^{-3}) the difference in original fraction of
SO_2 remaining after one hour was only a few percent.

     In Figures 2 and 3 only the SO_2(g) concentration is shown.
The majority of S(IV) in a cloud is present as gaseous SO_2 (98% at
pH 5 and > 99% at pH 3). The concentration of SO_2(g) shown in the
figures was thus a convenient indicator of the rate at which the
model was predicting the oxidation of S(IV).

     A similar study was performed to determine the effect of pre-
cloud acid concentrations on the oxidation process. By varying the
acid concentration from 1 to 50 nM m^{-3}, less than one percent
difference in SO_2 concentration was observed even after 150 mins.
This occurs because the rate constant is proportional to, and the

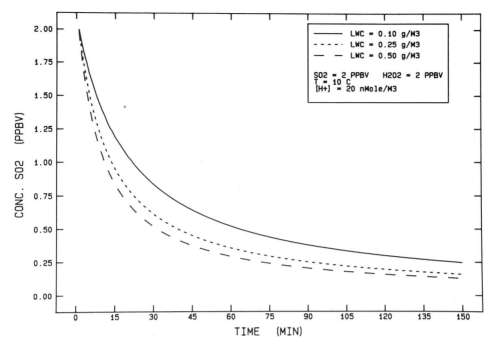

Figure   3.   Effect of cloud liquid water content on oxidation
              process.

solubility is inversely proportional to $[H^+]$, as shown in equations
1 and 8.

     In the calculations discussed so far, the initial concen-
trations of SO_2 and H_2O_2 were set at 2 ppbv. The percent of
limiting reagent remaining in cloud plotted as a function of time
is shown in Figure 4.   Since the aqueous reactions of S(IV) and
H_2O_2 are first order in both S(IV) and H_2O_2, this plot would be
applicable for either SO_2 or H_2O_2 as the limiting reagent. As seen
from Figure 4, when SO_2 or H_2O_2 is present in excess of the other,
the limiting reagent will be consumed fractionally faster than if
the two reagents are present in equimolar quantities.   Actual
atmospheres may have a ratio of excess to limiting reagent greater
than shown here, but this would only serve to increase the rate of
consumption of the limiting reagent. Under conditions leading to
the slowest oxidation rates conceivable at Whiteface (T = 20°C, LWC
= 0.1 g m⁻³, equimolar concentrations of SO_2 and H_2O_2), this model
predicts that about 3/4 of the limiting reagent (SO_2 or H_2O_2)
would be consumed in two hours.

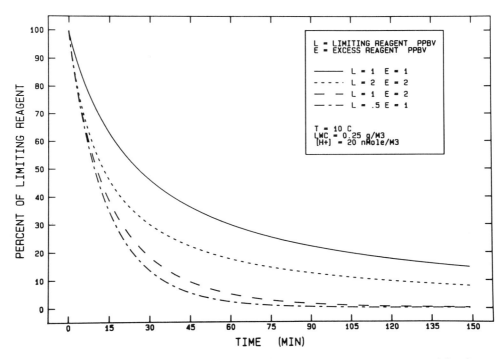

Figure   4.   Effect of charging ratio of SO_2 to H_2O_2 on oxidation
             process.

DISCUSSION OF FIELD STUDIES AT WHITEFACE MOUNTAIN

     The chemical composition of cloud water at the summit of
Whiteface Mountain has been monitored for three summers.   The
collector employed for these studies consists of 0.4 mm Teflon
threads strung over a polyethylene frame (Kadlecek et al., 1982).
The collector is mounted three meters above the summit laboratory
roof, where cloud water impacts on the threads and runs down the
collector into a polyethylene funnel and through a teflon tube into
the laboratory.   Collection rates are typically between 5 and
15 ml/min.

     The pH of the cloud water is monitored continuously using a
flow through pH cell.   This flow is briefly interrupted to collect
samples for other analyses (ion chromatography, HSO_3^-, and HCHO at
the rate of twice an hour; samples for H_2O_2 are collected 2 to
10 times per hour).   Analyses for H_2O_2 are performed at the summit
laboratory within one minute after sample collection.   Before
July 1982, the luminol reaction (Kok et al., 1978) was used for
detection; since then the enzyme technique developed by Lazrus et

al. (ms) has been employed. The procedure involves the reaction of H$_2$O$_2$ with p-hydroxyphenylacetic acid in the presence of the enzyme peroxidase to form a fluorescent dimer. The dimer is excited at 321 nm and the fluorescence measured at 401 nm. The fluorescence of the dimer is directly proportional to its concentration, which in turn is proportional to the H$_2$O$_2$ concentration. Measurements from the summer (June, July, and August) of 1983 are shown in Figure 5. Ninety percent of the samples had H$_2$O$_2$ concentrations of 200 ppbm[1] or greater, with an average concentration of about 700 ppbm. In every sample, H$_2$O$_2$ was found in easily measurable concentrations. Since H$_2$O$_2$ was observed throughout cloudy periods, the model (Figure 4) suggested that for summertime conditions, SO$_2$(g) could not persist for more than about an hour within the clouds.

The measurement of SO$_2$ was performed by Brookhaven National Laboratory using a modified flame photometer as described by Kelly et al. (in press). The instrument operated continuously, and

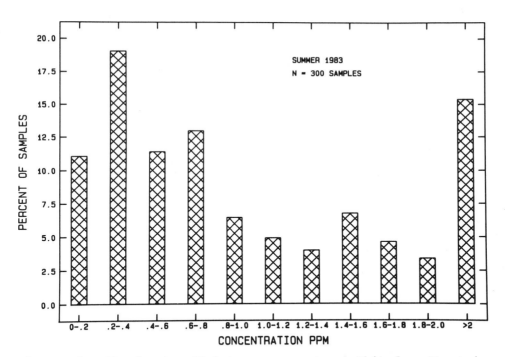

Figure 5.   Cloud water (H$_2$O$_2$) measurements at Whiteface Mountain during the summer of 1983.

---

[1]Parts per billion by mass

values were taken from the data record every 15 mins. Only one month of SO_2 measurements, however, were collected during the summer at Whiteface Mountain (July 1982). From these analyses, two of the events verify the rapid oxidation of SO_2.

On July 6, 1982, at approximately 0500 hrs EST, the summit of Whiteface was covered by clouds; this lasting until shortly after 0900 hrs that morning. As shown in Figure 6, SO_2 concentrations dropped to less than 0.2 ppbv (0.01 µM m^{-3}) with the onset of clouds and remained at this level until 0915 hrs when the clouds began to dissipate. The SO_2 concentration then returned to about 2.2 ppbv (0.1 µM m^{-3}) and remained there for several hours. Liquid water content (LWC) measurements during this event were obtained by a Knollenberg FSSP probe (Knollenberg, 1970). While the absolute values were uncertain (the Knollenberg Probe perhaps overestimated the LWC), the relative LWC during this event can be described. During about the first two hours the clouds were too light to provide a steady collection rate (<<0.1 g m^{-3}), but the rest of the event had approximately 0.4 g m^{-3} LWC. Aqueous cloud sulfate concentrations of about 0.12 µM m^{-3} were measured during this period. The sharp drop in $[SO_2]$, starting just before 1400 hrs EST and lasting for about one hour, is noteworthy. FSSP cloud droplet spectrum measurements showed a light cloud at Whiteface during this same interval of time. While the LWC history of this particular air mass was unknown, the consequence of the cloud reactions on $[SO_2]$ concentrations was obvious.

Using a filter pack sampler (T. Kelley, BNL) aerosol sulfate measurements were made during July 1982. These could not be interpreted if cloud water was collected on the filter, but SO_4^{2-} and SO_2 concentrations averaged over the same time interval as the filter sampler (Galvin et al., 1983) showed that generally most of the sulfur in cloud-free air arrived at Whiteface as SO_2, while most of the sulfur in cloudy air was present as sulfate. Most of the sulfate measured in cloud water comes from within cloud oxidation of SO_2 and not from nucleation about sulfate aerosols.

The initial concentration of H_2O_2 in a cloud was unknown because the cloud mass came to Whiteface Mountain in the Adirondacks after an indeterminate lifetime, during which much of the oxidation may have already occurred. While $[H_2O_2]$aq was measured at about one ppm, the actual concentration during early cloud formation would need to be considerably greater than the approximately 0.01 µM m^{-3} indicated by these liquid phase measurements. The model calculation was consistent with this interpretation. During this event $[H_2O_2]$ was not monitored, however, since every summer sample taken during 1982 and 1983 possessed concentratins capable of dominating SO_2 oxidation, there was no reason to presume otherwise during this event.

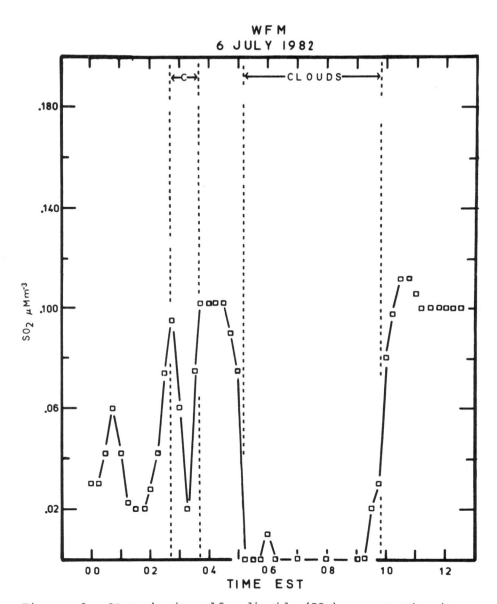

Figure   6.   Atmospheric sulfur dioxide (SO₂) measurments at
             Whiteface Mountain during the July 6, 1982 event.
             Clouds were present from 0240 to 0340 hrs. and again
             from 0500 to 0950 hrs. EST.

The mean transport layer trajectories[1] (Figure 7) suggested no major change in source region during this brief period. The "C" trajectory described most of the collection period. This trajectory showed a brief stagnation south of Lake Ontario, where it resided for about 40 hours after crossing the lake. This region received abundant sunshine the day before, enhancing photochemical formation of strong oxidants. Satellite imagery showed clouds that extended about 50 km upwind (southwest) of Whiteface during the period of sample collection. The wind speed measured at the summit during this time varied from about 15 m sec^{-1} at the beginning to about 8 m sec^{-1} at the end of the event. The time this air mass was in clouds was on the order of one hour. Calculations indicated that if the pre-cloud gas phase H_2O_2 concentration was at least of the same order as the SO_2 concentration, it could oxidize sulfur dioxide in this amount of time. Under these conditions, oxidation by ozone would be at least two orders of magnitude slower.

Most events displayed a more complicated behavior, involving major changes in regional influences. Shown in Figure 8 are distributions of $SO_2(g)$, $[H_2O_2]$ and $[SO_4^{2-}]$ (liquid) obtained during another event on July 25-26, 1982. Lines connecting some data points have been added to aid in following the sequence of point measurements. At about 1800 hrs EST, intermittent clouds began and at about 1915 hrs the cloud base lowered to below the summit and clouds persisted until about 0600 hrs the next morning, although evidence of breaking up started as early as 0515 hrs. The cloud layer was fairly thin (<1 km) since temperatures at the top of the clouds, derived from satellite images, were within about 2°C of the summit temperatures. Also evident in the infrared satellite pictures was a path in which there was little cloud formation, in an otherwise largely overcast region, extending from just upwind (west) of Whiteface to Lake Ontario. This channel coincided with the ARL trajectory during the first part of the event. During this period, the length of time this air mass could have spent within clouds as it approached Whiteface was relatively short (< 0.5 hour). This would account for the presence of both SO_2 and H_2O_2 in the cloud. As the trajectory becomes less precise, several hours back in time, the potential sources and cloud formations capable of affecting the oxidation of SO_2 cannot be accurately described.

When this channel filled in, sulfur dioxide concentrations dropped. However, since total sulfur in the air decreased as the

---

[1]Air mass trajectories were prepared using the NOAA/ARL atmospheric transport and dispersion model (Heffter, 1980).

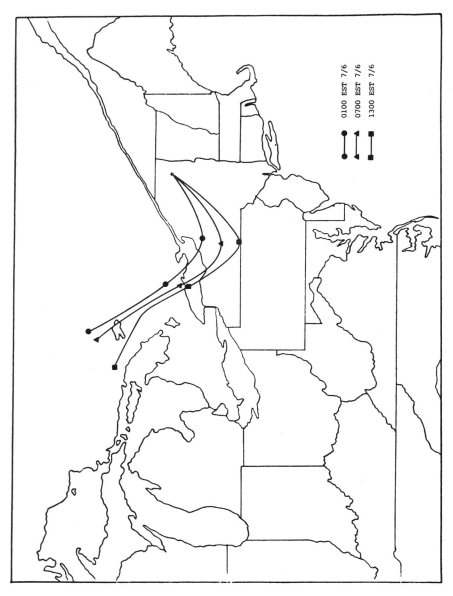

Figure 7. Trajectory arrival times and course predictions at Whiteface Mountain during July 6, 1982 event.

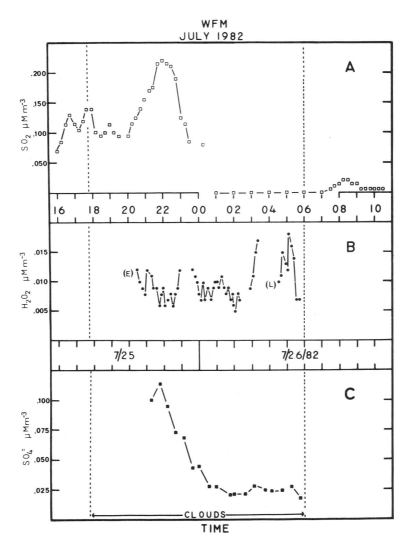

Figure 8. Measurements of (A) atmospheric sulfur dioxide
(μM m^{-3}), (B) aqueous H_2O_2 (μM m^{-3}), and (C) sulfate
(μM m^{-3}) at Whiteface Moutain during the
July 25-26, 1982 event (see text).

region of possible sources shifted (Figure 9), it was impossible to assign changes in [SO_2] concentrations to oxidation alone. During this event [H_2O_2] was measured by both the luminol and enzyme techniques. The two measuring techniques were very similar even during the rather complicated peaking at the middle of the event. Sampling times lasted about one to three minutes, which was short enough to display considerable, even microstructure, variability in strong oxidant concentrations. Since measurments of [H_2O_2] in summer clouds were always greater than 50 ppbm (for about 500 samples), the unique set of conditions during the first part of this event should be presented. SO_2 concentrations peaked (about 0.2 µM m^{-3}) at about an order of magnitude greater than [H_2O_2]. Color enhanced satellite photographs allowed for interpretation of concentrations resulting from a relatively short (< 0.5 hr) cloud history. Aerosol sulfate measurements during the six hours before the event showed an average concentration of 0.03 µM m^{-3} while the cloud water had sulfate concentrations about 0.1 µM m^{-3}. Air arriving at Whiteface before the event had high sulfur dioxide concentrations for the entire day, peaking at about twice the maximum concentration shown in Figure 8. If lower than normal H_2O_2(g) and higher than normal SO_2(g) concentrations are postulated, such that peroxide becomes the limiting reagent, then a cloud lifetime at Whiteface of substantially less than one hour would generate the concentration of SO_2 and H_2O_2 observed. The cloud would have oxidized some SO_2 in its short lifetime (about 0.7 µM m^{-3} - the difference between within cloud measurements and sulfate concentrations just before the cloud arrived). The high rate of oxidation would not continue as the clouds passed downwind of Whiteface unless there was a rapid H_2O_2 production mechanism. After the total sulfur burden began to drop, as the source region and cloud lifetime changed, the SO_2 chemistry returned to that typically seen during the rest of July 1982.

CONCLUSION

The concentrations of H_2O_2 in summertime cloud water collected at Whiteface Mountain were sufficiently high that it was the dominant oxidizing agent. Under most summertime conditions, the in cloud oxidation of SO_2 reduced this gas below the analytical detection limit (0.3 ppbv). This complete conversion of SO_2 to sulfuric acid will ensure a linear relationship between cloud water sulfate concentrations and the amount of SO_2 emitted into the air mass.

A cloud water monitoring project similar to that previously discussed was initiated during the 1983-84 winter season. Preliminary analysis of the data indicate that the efficient mechanism presented in this paper for summer SO_2 oxidation was no longer present. [H_2O_2] were generally less than 10 ppb (in

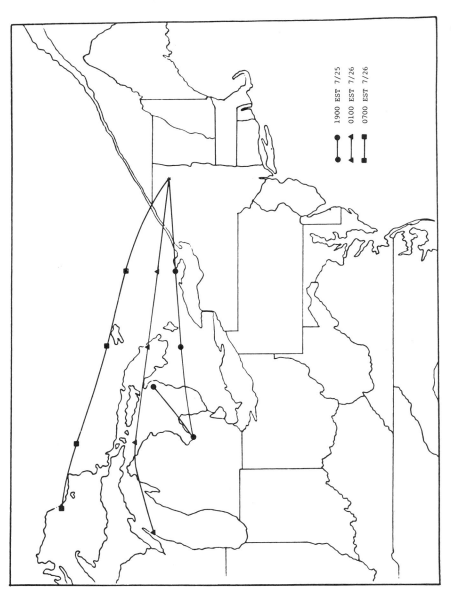

Figure 9. Trajectory arrival times and course predictions at Whiteface Mountain during the July 25-26, 1982 event.

contrast to the summer values of about 1 ppm), and substantial SO$_2$(g) concentrations (>10 ppb) were frequently observed, even under cloud conditions. Given this kind of change in the winter oxidation rate, oxidation mechanisms which could be ignored in the presence of H$_2$O$_2$ will need to be re-evaluated.

Acknowledgments - Support for this work was provided by the National Science Foundation (#ATM-821220201) and ESEERCO (New York State Power Pool - project no. 8210) and is gratefully acknowledged. H$_2$O$_2$ measurements were made with instrumentation on loan from the National Center for Atmospheric Research, Boulder, CO. Cloud droplet spectrum measurements were made by D. Wellman (NOAA/ARI-ERL). Hi Vol filter samples were analyzed by L. Husain during the July 25-26, 1982 event by the N.Y.S. Health Department. Without the untiring dedication of the Whiteface technical staff, Ann Kadlecek, Nancy Camarota and Roxana Jiminez, these measurements could never have been completed. We also acknowledge the invaluable help in meteorological analysis provided by Dr. Robert Pratt (McIdas, Satellite Photographs) and Dr. Jerre Wilson (ARL trajectories).

REFERENCES

Cooper, W.J. and Zika, R.G., 1983, Photochemical formation of hydrogen peroxide in surface and ground water exposed to sunlight. Science 220:711-712.
Galvin, P.J., Mohnen, V.A., Kadlecek, J.A., Wilson, J.W., and Kelly, T.J., 1983, Cloud chemistry studies at the Whiteface Mountain field station. ASRC-SUNYA Report No. 973, Atmos. Science Res. Center, SUNY, Albany, NY. 136 pp.
Graedel, T.E., Gill, P.S., and Weschler, C.J., 1982, Effects of organic surface films on the scavenging of atmospheric gases by rain drops and aerosol particles, pp. 417 to 430. In: Precipitation Scavenging, Dry Deposition, and Resuspension, H.R. Pruppacher, R.G. Semonin, W.G.N. Slinn, eds., Vol. 1, Proc. Fourth Intl. Conf., Santa Monica, CA.
Heffter, H.L., 1980, Air Resources Laboratories Atmospheric Transport and Dispersion Model. NOAA Technical Memorandum ARL-81, NOAA/Air Resources Laboratories, Silver Springs, MD.
Heikes, B.G., Lazrus, A.L., Kok, G.L., Kunen, S.M., Gandrud, B.G., Gitlin, S.N., and Sperry, P.D., 1982, Evidence of aqueous phase hydrogen peroxide synthesis in the troposphere. J. Geophys. Res. 87:3045-3051.
Kadlecek, J.A., McLaren, S.E., Camarota, N.A., Mohnen, V.A., and Wilson, J.W., 1982, Cloud water chemistry at Whiteface Moutain, pp. 103 to 114. In: Precipitation Scavenging, Dry Deposition, and Resuspension, H.R. Pruppacher, R.G. Semonin,

W.G.N. Slinn, eds., Vol. 1, Proc. Fourth Intl. Conf., Santa
    Monica, CA.
Kelly, T.J., Tanner, R.L., Newman, L., Galvin, P.J., and
    Kadlecek, J.A., in press, Trace gas and aerosol measurements
    at a remote site in the N.E. United States. Atmos. Envir.
    (Dec. 1984).
Knollenberg, G.R., 1970, The optical array. An alternative to
    scattering or extinction for airborne particle size
    determination. J. Appl. Met. 9:86-103.
Kok, G.L., Holler, T.P., Lopez, M.B., and Nachtrieb, H.A., 1978,
    Chemiluminescent methods for determination of H_2O_2 in ambient
    atmosphere. Envir. Science Tech. 12:1072-1078.
Lazrus, A.L., Kok, G.L., Gitlin, S.N., Lind, J.A., and
    McLaren, S.E., Manuscript. An automated fluorometric method
    for peroxides in atmospheric precipitation. Submitted to
    Anal. Chem.
Maahs, M.G., 1982, Sulfur-dioxide/water equilibria between 0° and
    50°C. An examination of data at low concentrations, pp. 187
    to 195. In: Heterogeneous Atmospheric Chemistry, D. Schryer,
    ed., Monograph 26, Amer. Geophysical Union, Washington, DC.
Martin, L.R. and Damschen, D.E., 1981, Aqueous oxidation of sulfur
    dioxide by hydrogen peroxide at low pH. Atmos. Envir.
    15:1615-1621.
Penkett, S.A., Jones, B.M.R., Brice, K.A., and Eggleston, A.E.J.,
    1979, The importance of atmospheric ozone and hydrogen
    peroxide in oxidizing sulphur dioxide in cloud and rainwater.
    Atmos. Envir. 13:123-137.
Schwartz, S.E., 1982, Gas aqueous reactions of sulfur and nitrogen
    oxides in liquid water clouds. Paper ENVI-J1, Acid Rain
    Symposium, Division of Environmental Chemistry. Amer. Chem.
    Soc., Las Vegas, NE.
Schwartz, S.E. and Freiberg, J.E., 1981, Mass-transport limitation
    to the rate of reaction in liquid droplets: Application to
    oxidation of SO_2 in aqueous solutions. Atmos. Envir.
    15:1129-1144.

SNOW AS AN INDICATOR OF ATMOSPHERIC FLUORIDE CONTAMINATION

Marcel Ouellet and H. Gerald Jones

INRS (Institut National de la recherche scientifique)
Universite du Quebec, C.P. 7500
Sainte-Foy (Quebec), Canada G1V 4C7

ABSTRACT

During the month of March 1978, 344 snow samples from the Saguenay - Lac Saint-Jean Region (Quebec) were analyzed for fluoride ion (F^-). Concentrations of individual samples ranged from <20 to 50,000 µg l^{-1} with a mean value of 925 µg l^{-1}. The major sources of this pollutant in this area are two aluminum smelters (Alma and Arvida) having an annual production of 74,000 and 435,000 tons. The approximate fallout distribution, defined as that area around the aluminum smelters containing snow showing fluoride levels >50 µg l^{-1}, was 3,000 km². Although the results of the present program concern short to medium range transport of fluoride within the atmosphere, it emphasizes the need for studies on the long range transport of fluoride, an important environmental pollutant. These studies should be carried out in conjunction with other anthropogenic pollutants using routine chemical analytical methods with a detection limit close to 1 µg l^{-1}.

INTRODUCTION

During the last decade, a considerable number of publications have appeared on the various aspects of environmental impacts due to atmospheric emissions of pollutants. As an example one can cite the large number of articles concerned with the acidification of surface waters caused by atmospheric emissions of sulfur and nitrogen oxides (Drabløs and Tollan, 1980; Ouellet and Jones, 1983). On the other hand, it appears that the impact of fluorides

49

on the environment has been relatively neglected. This would seem to be a certain paradox as fluoride contamination of the atmosphere is ranked as one of the major pollutant problems with sulfur and nitrogen oxides and ozone (Marier, 1972). Some of American and German coals used in thermal power-generating plants contain up to 500 mg/kg of fluorine and can give rise to atmospheric precipitation containing up to 14.1 mg l^{-1} of fluoride (Garber, 1970; Tourangeau et al., 1977). In the USA, the EPA (1972) estimated that American industries emitted over 150,000 t[1] of soluble fluorides annually into the atmosphere in the late sixties. A recent review (Galloway et al., 1980) carried out within the National Atmospheric Deposition Program and dealing specifically with toxic substances in atmospheric deposition did not include discussions concerning fluoride. In effect, most studies on fluoride have dealt with the various aspects of human, animal and plant health (Gabovich and Ovrutskiy, 1970; OMS, 1970; NAS, 1971; Rose and Marier, 1977; Waldbott et al., 1978).

The greater part of dissolved fluoride in surface waters originates from the fluoride content of atmospheric precipitation. This, in turn, depends on three main sources: 1) volcanic activity, 2) industrial emissions and 3) ocean spray. In Canada, 56% (8,000 t) of the annual fluoride atmospheric emissions (14,220 t) originates from the aluminum industry (Environment Canada, 1976).

Within the Province of Quebec, the contribution of this industry to atmospheric emissions of fluoride amounts to 6,870 t/yr (MEQ, 1979) which represents about 90% of the total. Cryolite (Na_3AlF_6) is used in the electrolytic processing of alumina as a fusing agent. At the fusion temperature of the metal some fluorides are volatilized and subsequently lost to the atmosphere.

SITE DESCRIPTION

Within the Saguenay - Lac Saint-Jean Region of Quebec, there are three (at Alma, Arvida, and Grande-Baie) aluminum smelters owned and operated by ALCAN with a potential production of aluminum close to 700,000 t/yr, while the one at Alma (Isle Maligne) produces 74,000 t per year. The two plants at Alma and Arvida have been in operation for several decades. The Grande-Baie smelter is the newest plant in Canada and emits substantially lesser amounts of atmospheric fluoride than the other two. It became operational in 1981 and has a maximum annual production capacity of 189,000 t. The energy needed to operate these three aluminum plants is produced in the Saguenay - Lac Saint-Jean catchment area by six

------

[1] Short ton, 2000 lbs.

hydro-electric stations developing a total power of 2,700 MW, which is about 10% of the total Quebec hydro-electric power presently generated by the provincial network (excluding Quebec-Labrador's 4,500 MW capacity).

METHODS

On March 26 and 27, 1978, 344 snow samples were collected within an area of 4,500 km² situated in the regional lowlands of the Saguenay River from the east shore of Lac Saint-Jean to Ha!Ha! Bay (Figure 1). Individual sites were on open ground well removed from the main roads in both urban and rural situations. Samples were taken with a small plastic scoop from the surface (0-15 cm) layer only, placed in plastic bags and kept frozen (-20°C) until melted for laboratory analysis. Meteorological conditions were assumed to have been quite uniform for the entire 45 x 100 km area before and during the short sampling period. The variability at one remote site was 20.5% from ten surface snow samples collected along a 500 m transect. More over, the fluoride content in composited 1-m long snow cores was compared to surface concentrations at ten different sites. The mean difference between these two data sets was 3 $\mu g \ l^{-1}$; there were no significant differences at the 0.5% probability level. The fluoride (F^-) content of each sample was estimated by means of an ionanalyzer (Orion Research, model 801A) coupled to a fluoride selective electrode (Orion, model 94-09; EPA, 1974). The detection limit of the method was 20 $\mu g \ l^{-1}$ with reproducibility between replicate samples being ± 2%.

RESULTS AND DISCUSSION

The fluoride content of the snow cover in March (Figure 1) reflected the fallout pattern from the Arvida and Alma smelters. Mean F^- content was 925 $\mu g \ l^{-1}$; a maximum of 23 mg l^{-1} was recorded close to the two plants in March at a site which during a preliminary sampling expedition in February had shown a maximum concentration of 50 mg l^{-1}. The fallout area (3,000 km²) has a west to east elongation as a result of westerly winds, which are dominant in this region, and the presence of high land (>200 m) immediately to the north and south of the sampling area.

These results may be partially correlated with those from the phytosociological studies of Leblanc et al. (1971, 1972); these authors demonstrated that communities of lichens and mosses were adversely affected in a northernly direction up to a distance of 12 km from the aluminum plants. In addition, fluorose in dairy cattle has been identified in this region since 1951, and approximately $3 million has been received by farmers from ALCAN as

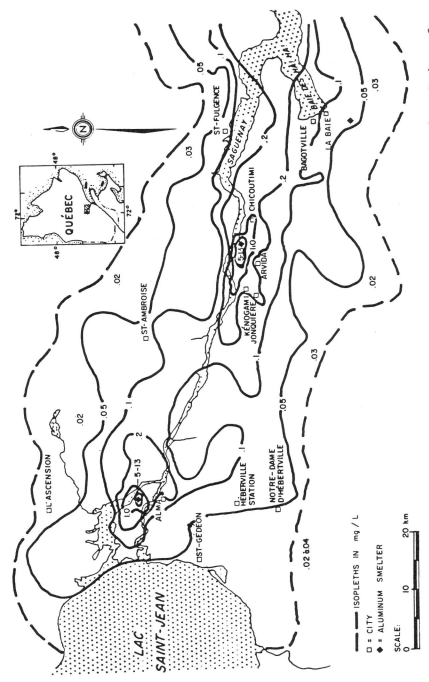

Figure 1. Fluoride distribution in snow of the Saguenay – Lac Saint-Jean Region, Quebec, for the month of March 1978.

compensation for the loss of production (MEQ, 1979). Suttie (1980) demonstrated that a gradual reduction of fluorides in pasture forage as well as a decrease in the incidence of fluorose in dairy cattle occurred in the area from 1970 to 1979. This was attributed to a reduction in emissions at two smelters. Further reductions were also observed during the labor strikes of 1976 and 1979.

Unfortunately, the global environmental impacts of these atmospheric fluoride emissions, which began in 1926, are still largely unknown. Preliminary dendrochronological studies are now being carried out in the area shown in Figure 1 in order to evaluate any adverse effects on tree growth. Studies have not been undertaken to estimate the threat to the health of the population either living or who have lived within the polluted area. It should be noted, however, that since 1972 the progressive installation of antipollution devices at both the Arvida and Alma plants has contributed substantially to the reduction of fluoride emissions. The 3-year old smelters of Grande-Baie seem to give even cleaner emissions (Bouchard et al., 1981; Paquet, 1983).

The fallout area as shown in Figure 1 is the typical result of a case study on the short range transport of an atmospheric pollutant of which the sources are easily identifiable. It cannot be compared, on a scale-wise basis, to the long-range atmospheric transport of acidic pollutants such as SO_4^{2-} and NO_3^- for which the emission sources are varied and the transport distances are of the order of thousands of kilometers.

The toxic nature of the fluoride ion does, however, raise two questions: 1) whether the various aluminum plants, brickworks and power stations in industrialized areas can contribute to the long-range transport of F^- on a continental scale and 2) whether this is of sufficiently high level to cause concern after such long-range transport. Very little knowledge on the environmental chemistry of fluorine is known, and data on the reservoirs and fluxes of this element in global cycles are very sparse. The absolute and relative abundances (Al = 1) of the elements associated mainly with the lithologic (Al, Fe, Ca; soil, rocks, coal), hydrologic (Ca, Cl; fresh and sea water) and atmospheric reservoirs (precipitation, air) in global chemical cycles is given in Table 1. Chlorine is enriched in the atmosphere and hydrosphere in comparison to the lithologic reservoirs. The main pathway of this element is atmospheric transport, precipitation and surface water transport between the land and sea masses.

Although there are no comparable data on the mean background content of fluoride in precipitation, it may be possible to derive an estimation by considering the respective origins and chemical characteristics of Cl^- and F^- aerosols in the atmospheric reservoir so that a subsequent evaluation of a mean washout ratio for

Table 1. Mean concentrations of aluminum, calcium, iron, chloride, and fluoride in the main reservoirs of the global chemical cycles.

RESERVOIRS	Al	Ca	Fe	Cl	F	Reference
Atmosphere* (ng m⁻³)	2×10^3 (1)	2.8×10^3 (1.4)	3.6×10^3 (1.8)	1.8×10^3 (0.9)	rural, 10 (0.005) urban, 400 (0.2)	1
Precipitations (µg l⁻¹)	10 (1)	5×10^2 (50)	15 (1.5)	5×10^2 (50)	rural, 2 (0.2) urban, 20 (2)	2, 4
Freshwater (µg l⁻¹)	3×10^2 (1)	1.5×10^3 (50)	5×10^2 (1.6)	7×10^3 (23.3)	1.5×10^2 (0.5)	1, 5
Seawater (µg l⁻¹)	2 (1)	412×10^3 (206×10^3)	2 (1)	19.4×10^6 (9.7×10^6)	1.35×10^3 (0.67×10^3)	1
Soil (mg kg⁻¹)	71×10^3 (1)	15×10^3 (0.21)	40×10^3 (0.56)	1×10^2 (0.001)	2×10^2 (0.003)	1
Rock (mg kg⁻¹)	82×10^3 (1)	41×10^3 (0.5)	41×10^3 (0.5)	13×10^2 (0.002)	9.5×10^2 (0.01)	1
Coal (mg kg⁻¹)	13×10^3 (1)	7.7×10^3 (0.6)	19.2×10^3 (1.5)	1.4×10^3 (0.11)	61 (0.005)	3

* northeastern North America
( ) geochemical ratio, element/aluminum (w/w)
1. Bowen (1979)
2. Tanaka et al. (1981)
3. Flanagan and Friedlander (1978)
4. Barnard and Nordstrom (1982)
5. Livingstone (1963)

fluoride ion may be made (the apparent mean washout ratio of chloride calculated from Table 1 is 2.78 x 10^5). If it is assumed that the source of fluoride in the atmosphere is sea water and that the phenomena giving rise to Cl^- aerosols also give rise to F^- aerosols, then one would expect from the respective concentration of these elements in sea water that the mean atmospheric background level of fluoride ion from the oceans would be 0.12 ng m^{-3}. This value is much closer to the mean value for Europe (1.5 ng m^{-3}; Bowen, 1979), which is strongly influenced by maritime winds, than that for the continental air masses of northeastern America (10-400 ng m^{-3}; Barnard and Nordstrom, 1982). This would suggest that the anthropogenic component of total fluorine loading in the atmosphere of the latter region is the major portion of the total.

It may also be possible that fluorine containing aerosols may not have any similarity to chlorine aerosols and the respective washout characteristics may be different. However, if fluoride is emitted as a submicron aerosol in an acidic environment, it will tend to associate more with acidic sulphate and nitrate aerosols than with clay particles where it could be fixed as a non-exchangeable form by the replacement of OH groups. The latter process would lead to short and medium range transport of fluoride around emission sources than transport over longer distances. In contrast, F^- associated either with submicron acidic aerosols (NH_4^+, SO_4^{2-}, H^+, NO_3^-) or in the vapor phase (HF) could be transported over long distances and, in the event of suitable precipitation episodes, could show high washout ratios of 10^4-10^6 (Scott, 1981). If such were the case, then the global background value of fluoride expected in rural precipitation (Table 1) from long range transport should range from 0.1 µg l^{-1} to 10 µg l^{-1} in eastern North America, which is in accordance with the mean value of 8.1 µg l^{-1} found by Barnard and Nordstrom (1982). Similarly, from the data for urban areas and assuming a maximum washout ratio of 10^6, it can be expected that the concentration of fluoride in polluted area precipitation would attain >400 µg l^{-1}. These projected values for the concentration of F^- in precipitation subsequent to either long range transport or regional pollution would require that its study with other aerosols of anthropogenic origin be carried out using routine analysis of the ion within the 1-10 µg l^{-1} range. It is not clear whether or not exposure of plant and animal populations to such levels of F^- in the atmosphere can have adverse effects on community structures. Even though hypothetical, the present slowing of growth and the die-back of the German as well as other forests (Ulrich et al., 1980; Bormann, 1982) might well be partly related to fluorine as well as other acidifying products.

CONCLUSION

     Fluoride ion analysis of snow samples from the Saguenay – Lac
Saint-Jean region of Quebec, Canada showed a highly polluted
fallout area around the two main aluminum smelters. Although data
are not available on background levels of F^- in eastern North
America, a consideration of fluoride concentrations in air masses
of this region and an estimation of the approximate range of values
for potential aerosol washout ratios suggest possible background
levels of 0.1 µg l^{-1} to 10 µg l^{-1} in rural precipitation for this
area. Actual F^- levels in the smelter fallout region averaged
925 µg l^{-1} in 1978 which would indicate that this region was
receiving extreme pollution loadings of this element.

Acknowledgements – The authors would like to express their deep
appreciation to J. Michaud, G. Bernier, and M. Bisson for their
assistance in sample handling and collection as well as for their
laboratory work. Financial support was provided by the Natural
Sciences and Engineering Research Council of Canada, the Quebec
Ministry of Education and Environment Quebec.

REFERENCES

Barnard, W.R. and Nordstrom, D.K., 1982, Fluoride in precipitation.
     II: Implication for the geochemical cycling of fluoride.
     Atmospheric Environment 16:105-111.
Bormann, F.H., 1982, The effects of air pollution on the New
     England Landscape. Ambio 11:338-346.
Bouchard, L., Theberge, D., Riverin, J., and Lavoie, E., 1981,
     Etude environnmentale relative a l'implantation de l'Alcan,
     Ville de la Baie. Les Presses de l'Universite du Quebec a
     Chicoutimi, Chicoutimi, Quebec. 308 p.
Bowen, A.J.M., 1979, The atmosphere, pp. 1 to 29. In:
     A.J.M. Bowen, ed., Environmental Chemistry of the Elements.
     Academic Press, New York, NY.
Drabløs, D. and Tollan, A., eds., 1980, Ecological Impact of Acid
     Precipitation. Proceedings International Conference,
     Sandefjord, Norway. SNSF Project, Ås-NLH, Norway. 383 pp.
Environment Canada, 1976, National Inventory of Sources and
     Emissions of Fluoride (1972). Report APCD 75-7, Air Pollution
     Control Directorate, Ottawa, Ontario, Canada. 31 pp.
Environmental Protection Agency, 1972, Engineering and Cost
     Effectiveness Study of Fluoride Emissions Control, Vol. 1.
     Report No. SN16893.000, Contract EHSD71-74, Office of Air
     Programs, Washington, DC. 404 pp.
Environmental Protection Agency, 1974, Manual of methods for
     Chemical Analysis of Water and Wastes. Report EPA-625A6-74-
     003, U.S.E.P.A., Office of Technology Transfer, Washington,
     DC. 298 pp.

Flanagan, R.C. and Friedlander, S.K., 1978, Particle formation in pulverized coal combustion - A review, pp. 25 to 60. In: D.T. Shaw, ed., Recent Developments in Aerosol Science. Wiley Interscience, New York, NY.

Gabovich, R.D. and Ovrutskiy, G.D., 1970, Fluorine in Stomatology and Hygiene (Traduit de la langue russe en 1977 par the American National Institute of Dental Research). Publ No. (NIH)78-785, U.S. Dept. Health, Education, Welfare, Bethesda, MD. 1028 pp.

Galloway, J.N., Eisenreich, S.J., and Scott, B.C., eds., 1980, Toxic substances in atmospheric deposition: A review and assessment. Proceedings of a workshop of the National Atmospheric Deposition Program, Report No. 560/5-80-001, U.S.E.P.A., Washington, DC. 146 pp.

Garber, K., 1970, Fluoride in rainwater and vegetation. Fluoride 3:22-26.

Leblanc, F., Comeau, G., and Roa, D.M., 1971, Fluoride injury symptoms in epiphytic lichen and mosses. Can. J. Bot. 49: 1691-1698.

Leblanc, F., Rao, D.N., and Comeau, G., 1972, Indices of atmospheric purity and fluoride pollution pattern in Arivda, Quebec. Can. J. Bot. 50:991-998.

Livingstone, D.A., 1963, Chemical Composition of Rivers and Lakes. Professional Paper 440-0, U.S. Geological Survey, Washington, DC. 64 pp.

Marier, J.R., 1972, The ecological aspect of fluoride. Fluoride 5: 92-97.

MEQ (ministere de l'Environnement du Quebec), 1979, Les fluorures, la fluoration et la qualite de l' environnement. Rapport prepare pour le ministre de l'Environnement par le Comite aviseur sur la fluoration des eaux de consommation. Government du Quebec, Quebec, Canada. 210 pp.

National Academy of Sciences, 1971, Fluorides. Committee on Biological Effects of Air Pollution, Division of Medical Sciences, NAS, Washington, DC. 295 pp.

OMS (Organisation mondiale de la sante), 1970, Fluorides and human health. Monograph Series 59, OMS, Geneva, Switzerland. 364 pp.

Ouellet, M., and Jones, H.G., 1983, Paleolimnological evidence for the long-range atmospheric transport of acidic pollutants and heavy metals into the province of Quebec, eastern Canada. Can. J. Earth Sci. 20:23-36.

Paquet, J.-L., 1983, L'industrie de l'aluminum et l'agriculture. Agriculture 40:63-68.

Rose, D. and Marier, J.R., 1977, Environmental fluorides 1977. Publication No. 16081, National Research Council Canada, Ottawa, Ontario, Canada. 151 pp.

Scott, B.C., 1981, Modeling of Atmospheric Wet Deposition in Atmospheric Pollutants in Natural Waters, pp. 3 to 21. In: S.J. Eisenreich, ed., Ann Arbor Science, Ann Arbor, MI.

Suttie, J.W., 1980, Performance of a dairy cattle herd in close
    proximity to an industrial fluoride-emitting source.
    Abstract, Annual meeting of the Air Pollution Control Assoc.,
    Montreal, Quebec, Canada.
Tanaka, S., Darzi, M., and Winchester, J.W., 1981, Elemental
    analysis of soluble and insoluble fractions of rain and
    surface waters by particle induced X-ray emission. Environ.
    Sci. Tech. 15:354-357.
Tourangeau, P.C., Gordon, C.C., and Carlson, C.E., 1977, Fluoride
    emissions of coal-fired power plants and their impact upon
    plant and animal species. Fluoride 10:47-62.
Ulrich, B., Mayer, R., and Khanna, P.Y., 1980, Chemical changes due
    to acid precipitation in a loess-derived soil in Central
    Europe. Soil Sci. 130:193-199.
Waldbott, G.L., Burgstahler, A.W., and McKinney, H.L., 1978,
    Fluoridation: the great dilemma. Coronoda Press Inc.,
    Lawrence, KS. 423 pp.

METHODS FOR DIAGNOSING THE SOURCES OF ACID DEPOSITION

Perry J. Samson

Department of Atmospheric and Oceanic Science
University of Michigan
Ann Arbor, MI   48109

ABSTRACT

A regional-scale sulfur transport and deposition model has been developed and applied to several receptor regions in eastern North America. Model estimates of the fraction of source region emissions of sulfur dioxide reaching receptor areas (source-receptor relationships) as ambient SO_2 or SO_4^{2-} and/or depositing there by precipitation or dryfall have been computed for nine potentially sensitive receptor regions in eastern North America. This modeling has illustrated that the source regions with the highest natural potential to contribute are those which are closest to the receptor region. The potential for a source region to contribute multiplied by its emissions density has yielded estimates of actual culpability. The source contributions are generally dominated by high-emission areas in the United States and Canadian provinces nearest the receptor.

INTRODUCTION

Our ability to forecast the environmental improvements to be gained as a result of reductions in sulfur dioxide emissions is limited by our ability to mathematically represent physical and chemical processes in the atmosphere. In recent years, a number of investigators have developed acid (actually sulfur) deposition models which attempt to incorporate the emissions, transport, dispersion, chemical transformation, and deposition of pollutants over regional scales (>500 km) where standard plume equations are no longer applicable. This paper presents the results of model estimates of source-receptor relationships calculated for nine

59

sensitive receptor areas for the United States/Canada Memorandum of
Intent (MOI, 1982) using the University of Michigan ACID model.
There is evidence that the natural potential for mass transfer of
pollutants to a receptor over the course of a year is highest for
sources closest to the receptor.

The Atmospheric Contributions to Interregional Deposition
(ACID) model (Samson and Small, 1984) is similar in design to other
Lagrangian-type transport and deposition models (Bhumralker et al.,
1981; Shannon, 1981; Voldner, et al., 1981; Eliassen and Saltbones,
1983), including first-order, linear chemical conversion and
deposition processes. The ACID model, however, is unique in that
it is receptor-oriented and is designed to assess the potential of
upwind regions for contribution to the sulfur deposition at
specific receptors. That is, the model calculates the potential
deposition or concentration per unit of emissions (e.g., kilograms
per hectare of sulfur wet deposition per kilograms of sulfur
emissions) based on the observed meteorological conditions and
assumed chemical transformation processes. This allows an analysis
of the variability in source-receptor relationships which is apart
from any variations in emission strength. The variability in the
natural potential quantifies the confidence with which an emissions
reduction program will result in measurable decreases in sulfur
deposition.

MODEL ASSUMPTIONS

The determination of the potential of upwind areas to contribute to the deposition at a receptor can be expressed in terms
of probability. For example, the natural potential for contribution to wet deposition is the probability of the sulfur species,
sulfur dioxide (SO_2) and sulfate (SO_4^{2-}), being transferred to the
receptor multiplied by the probability that, once there, the
material will be deposited. Let $S_g(\vec{x},t)$ be the probability of
sulfur as SO_2 arriving at a receptor located at x at time t, and
$S_p(\vec{x},t)$ represent the probability for sulfur as SO_4^{2-}. Then, in
terms of probability (Lamb and Seinfeld, 1973; Cass, 1981)

$$S_g(\vec{x},t) = \int_{-\infty}^{\infty} \int_{-\infty}^{\infty} \int_{t-\tau}^{t} Q(\vec{x},t|\vec{x}',t') \, D_g(\vec{x}',t') \, R(t|t') \, d\vec{x}' \, dt' \qquad (1)$$

where $Q(\vec{x},t|\vec{x}',t')$ is the mass transfer function defining the
probability of an air parcel located at x' at time t' arriving at a
receptor located at x at time t, $D_g(\vec{x}',t')$ is the probability of
sulfur as SO_2 not being deposited upwind at $(\vec{x}',t')$, and $R(t|t')$ is
the probability of sulfur as SO_2 not being reacted to SO_4^{2-} from
time t' to time t. The equation is integrated from t' to t over
the domain from t-τ assuming the probability of mass transfer
beyond time τ to be zero.

The mass transfer function can be approximated from the computation of upwind (backward) trajectories. The axis of the computed trajectory is assumed to represent the highest probability at a given time upwind of contributing to the receptor. The spatial distribution of the potential mass transfer function away from the axis of the trajectory will depend upon the degree of vertical mixing, coupled with the magnitude of the wind velocity shear. As a first approximation, it has been assumed that the "puff" of potential mass transfer is normally distributed about the trajectory with a standard deviation which is increasing linearly in time upwind (Draxler and Taylor, 1982; Samson, 1980; Samson and Moody, 1980).

Thus far, this model has been specifically applied to the transport, transformation, and deposition of sulfur species. Consequently, the probability of chemical transformation of SO_2 to SO_4^{2-} species has been included as a linear, first-order process. This is accomplished by assuming linear transformation rates which vary both diurnally and annually from maximum rates in June during daylight hours (when photochemistry is greatest) to minimum rates during the nighttime period. The rate of change of S_g with time is described by

$$dS_g/dt = -\kappa_t(t)S_g(t) \qquad (2)$$

$$dS_p/dt = +\kappa_t(t)S_g(t) \qquad (3)$$

so the probability of not being converted can be computed as

$$R(t|t') = 1 - \int_{t-t'}^{t} \kappa_t(t') \, S_g(t') \, dt' \qquad (4)$$

This approximation is a gross simplification of a very complex system which involves both gas-phase transformation processes and in-cloud transformation. The justification for such a parameterization is based on field studies conducted in episodes of high ambient SO_4^{2-} concentrations (Gillani et al., 1978). Its extension to all meteorological conditions is open to debate, but it has been utilized in these models for lack of more complete information about the variations in the transformation process.

The dry deposition of SO_2 and SO_4^{2-} assumes linear rate coefficients over three-hour periods. The dry deposition velocities vary diurnally and seasonally, having minimal values of 0.15 cm sec^{-1} for SO_2 and 0.05 cm sec^{-1} for SO_4^{2-} at nighttime regardless of season. The highest values are assumed to occur in the afternoon, with 0.8 cm sec^{-1} and 0.3 cm sec^{-1} assumed wintertime (January) SO_2 and SO_4^{2-} dry deposition velocities, respectively, increasing to 1.3 cm sec^{-1} and 0.45 cm sec^{-1} for summertime (July) calculations.

Rates for other months are linearly interpolated between these values. The height of the layer through which the deposition is occurring is assumed to be the monthly average mixing height computed along the trajectory, determined as the bottom of the first nonsurface based stable ($\partial\theta/\partial z \geq 0.005°C/m$) layer which meets the criteria

$$\theta_{TOP} - \theta_{BASE} \geq 2°C. \tag{5}$$

The wet scavenging of sulfur compounds along the trajectories and at the receptor is also assumed to be linearly proportional to the species concentrations. The rate coefficients for wet deposition of SO_2, k_w, and ambient sulfate, κ_w, are determined by

$$k_w = \omega(SO_2)R/h \tag{6}$$

$$\kappa_w = 23.3 \times 10^4 R^{+0.625}/h \tag{7}$$

where $\omega(SO_2)$ is the washout ratio for SO_2, R is the precipitation rate (mm hr^{-1}), and h is the mixing height (mm). The washout ratio is defined as the ratio of the aqueous concentration of pollutant to the ambient concentration in the atmosphere. The value of $\omega(SO_2)$ was chosen as a median number from the range described by Barrie (1981)

$$\omega(SO_2) = 5 \times 10^4, \tag{8}$$

and the coefficient for κ_w was derived from a formulation (Scott, 1978; McNaughton, 1984) for clouds containing ice nuclei.

It is further assumed that all scavenged sulfur dioxide arrives at the surface as SO_4^{2-} through in-cloud and/or in-precipitation conversion. This removal mechanism does not distinguish between scavenging within the cloud (rainout) and scavenging beneath the cloud (washout), but rather lumps the two together as net removers of the sulfur mass in the transporting layer.

RESULTS

The ACID model was applied to several receptor location in the eastern United States and Canada as part of the studies for the United States/Canadian Memorandum of Intent (MOI, 1982). The model was run for the year 1978 using observed meteorological and emissions data. A detailed description of the model results for the Adirondack Mountains is contained in Samson and Small (1984). This paper summarizes the UMACID results for all the sites included in the MOI study.

In general, the potential of upwind regions to contribute to annual ambient sulfate concentrations was highest for regions nearest the receptors. The locations of the nine MOI receptors are listed in Table 1, while the states and provinces with the highest potential for contributing to each of these receptor locations based on estimates of the ACID model are given in Table 2.

The potential contribution of various states and provinces to sulfur dry and wet deposition, respectively, at the nine MOI receptors are listed in Tables 3 and 4. Again, inspection reveals that the source regions having the highest possibility of contributing to the annual deposition are those which are closest to the receptors.

This finding illustrates that the magnitude of dispersion created by the meandering of trajectories over the course of a year is sufficient to negate the fact that time is required for conversion of SO_2 to SO_4^{2-}. The slow conversion process would be expected to result in a higher probability for contribution by sources far upwind of the receptor to ambient sulfate concentrations.

Table 1. Name and location of nine targeted sensitive areas used in United States/Canada Memorandum of Intent (MOI)

NAME	Acronym	Latitude	Longitude
Boundary Waters, Minnesota	BDW	49°00'	93°00'
Algoma, Ontario	ALG	46°30'	84°00'
Muskoka, Ontario	MUS	45°00'	79°30'
Quebec City, Quebec	QUE	47°00'	72°00'
Southern Nova Sciota	SNS	44°00'	66°00'
Vermont/New Hampshire	VNH	45°00'	72°00'
Adirondacks, New York	ADR	44°00'	74°00'
Western Pennsylvania	WPA	41°00'	78°00'
Southern Appalachia	SAP	35°00'	84°00'

Table 2. Potential contribution of states and provinces to the annual average concentration of ambient sulfate at the nine MOI targeted sensitive receptors. Estimates have been made using the University of Michigan ACID model with 1978 sulfur dioxide emissions and 1978 meteorology. Units are micrograms per cubic meter of ambient sulfate per teragram (10^{12} grams) of sulfur emissions

Region	\multicolumn{9}{c}{Receptors}								
	BDW	ALG	MUS	QUE	SNS	VNH	ADR	WPA	SAP
Ontario	1.38	2.37	7.65	4.32	2.20	3.91	5.41	3.19	0.42
Quebec	0.01	2.11	0.45	2.77	1.86	2.58	1.36	0.23	0.02
Maritime	0.00	0.02	0.02	0.20	2.24	0.38	0.19	0.02	0.00
Manitoba	1.03	0.28	0.14	0.05	0.02	0.04	0.07	0.06	0.05
Ohio	0.00	0.10	0.38	0.38	0.28	0.43	0.69	2.14	0.59
Illinois	0.05	0.32	0.41	0.21	0.13	0.19	0.31	0.81	0.71
Pennsylvania	0.00	0.04	0.17	0.30	0.40	0.48	0.68	2.67	0.30
Indiana	0.01	0.16	0.33	0.22	0.14	0.21	0.36	1.05	1.17
Kentucky	0.01	0.09	0.18	0.14	0.10	0.15	0.25	0.82	1.57
Michigan	0.07	0.62	1.20	0.53	0.29	0.48	0.72	1.37	0.28
Tennessee	0.00	0.06	0.11	0.08	0.07	0.10	0.16	0.56	2.54
Missouri	0.03	0.19	0.19	0.09	0.06	0.09	0.16	0.46	0.81
W. Virginia	0.00	0.03	0.12	0.20	0.32	0.36	0.48	1.61	0.65
New York	0.00	0.05	0.24	0.42	0.67	0.73	1.06	0.64	0.09
Alabama	0.00	0.03	0.06	0.04	0.04	0.05	0.08	0.33	1.91
Wisconsin/Iowa	0.19	0.70	0.64	0.27	0.14	0.23	0.35	0.71	0.32
Minnesota	0.90	0.66	0.42	0.18	0.07	0.16	0.25	0.33	0.17
Virginia/North Carolina	0.00	0.01	0.04	0.07	0.20	0.17	0.19	0.47	0.76
Georgia/South Carolina	0.00	0.01	0.04	0.03	0.05	0.06	0.07	0.27	2.25
Maryland/New Jersey	0.00	0.01	0.07	0.16	0.46	0.36	0.39	0.52	0.18
New England	0.00	0.07	0.20	0.84	3.69	2.99	1.61	0.33	0.06
Other	0.37	0.21	0.17	0.06	0.04	0.07	0.12	0.34	0.99

Table 3. Potential contribution of states and provinces to the annual sulfur dry deposition at the nine MOI targeted sensitive receptors. Estimates have been made using the University of Michigan ACID model with 1978 sulfur dioxide emissions and 1978 meteorology. Units are kilograms of sulfur dry deposited per hectare per teragram (10^{12} grams) of sulfur emissions

Region	Dry Deposition at Receptors								
	BDW	ALG	MUS	QUE	SNS	VNH	ADR	WPA	SAP
Ontario	3.74	4.31	20.70	7.35	2.68	6.21	10.33	4.91	0.38
Quebec	0.01	0.12	0.82	7.60	3.10	5.90	2.44	0.29	0.02
Maritime	0.00	0.02	0.04	0.38	6.73	0.74	0.28	0.03	0.00
Manitoba	2.44	0.25	0.13	0.04	0.01	0.04	0.06	0.06	0.05
Ohio	0.00	0.16	0.65	0.46	0.32	0.54	1.04	5.07	0.70
Illinois	0.04	0.45	0.58	0.22	0.13	0.20	0.35	1.08	0.94
Pennsylvania	0.00	0.05	0.34	0.40	0.47	0.70	1.19	7.90	0.32
Indiana	0.01	0.22	0.52	0.24	0.15	0.24	0.46	1.59	1.80
Kentucky	0.00	0.11	0.26	0.15	0.11	0.16	0.31	1.21	2.82
Michigan	0.09	1.30	2.44	0.63	0.31	0.57	1.01	2.24	0.28
Tennessee	0.00	0.07	0.15	0.09	0.07	0.11	0.20	0.77	6.08
Missouri	0.03	0.22	0.24	0.09	0.06	0.09	0.17	0.53	1.15
W. Virginia	0.00	0.04	0.22	0.25	0.43	0.50	0.75	3.75	0.87
New York	0.00	0.06	0.48	0.66	0.98	1.34	2.38	1.35	0.09
Alabama	0.00	0.03	0.08	0.04	0.03	0.06	0.10	0.44	4.11
Wisconsin/Iowa	0.22	1.26	0.88	0.28	0.13	0.24	0.39	0.88	0.36
Minnesota	1.80	0.96	0.46	0.17	0.07	0.14	0.24	0.35	0.18
Virginia/North Carolina	0.00	0.01	0.07	0.08	0.21	0.21	0.28	0.83	1.22
Georgia/South Carolina	0.00	0.01	0.06	0.03	0.05	0.07	0.09	0.37	5.82
Maryland/New Jersey	0.00	0.01	0.14	0.22	0.57	0.58	0.67	1.12	0.19
New England	0.00	0.07	0.34	1.59	8.46	7.64	3.54	0.53	0.06
Other	0.56	0.23	0.17	0.05	0.04	0.07	0.13	0.35	1.32

Table 4. Potential contribution of states and provinces to the annual sulfur wet deposition at the nine MOI targeted sensitive receptors. Estimates have been made using the University of Michigan ACID model with 1978 sulfur dioxide emissions and 1978 meteorology. Units are kilograms of sulfur wet deposited per hectare per teragram (10^{12} grams) of sulfur emissions

Region	Wet Deposition at Receptors								
	BDW	ALG	MUS	QUE	SNS	VNH	ADR	WPA	SAP
Ontario	0.61	1.14	4.36	1.75	0.54	0.67	1.21	0.74	0.09
Quebec	0.01	0.03	0.22	1.41	0.58	0.35	0.29	0.10	0.00
Maritime	0.00	0.00	0.03	0.11	0.97	0.17	0.12	0.02	0.00
Manitoba	0.52	0.14	0.03	0.02	0.00	0.00	0.01	0.01	0.02
Ohio	0.00	0.09	0.81	0.14	0.11	0.17	0.62	1.50	0.20
Illinois	0.01	0.11	0.51	0.06	0.03	0.05	0.20	0.47	0.18
Pennsylvania	0.00	0.02	0.46	0.11	0.17	0.28	0.51	1.82	0.15
Indiana	0.00	0.08	0.49	0.08	0.04	0.09	0.34	0.93	0.30
Kentucky	0.00	0.04	0.32	0.05	0.04	0.09	0.28	0.98	0.54
Michigan	0.02	0.48	1.43	0.17	0.07	0.10	0.30	0.37	0.05
Tennessee	0.00	0.02	0.23	0.03	0.03	0.07	0.18	0.79	1.26
Missouri	0.01	0.05	0.32	0.03	0.02	0.04	0.13	0.41	0.23
W. Virginia	0.00	0.01	0.34	0.08	0.14	0.21	0.43	1.48	0.33
New York	0.00	0.02	0.43	0.18	0.26	0.38	0.67	0.36	0.04
Alabama	0.00	0.01	0.15	0.01	0.02	0.04	0.09	0.48	1.85
Wisconsin/Iowa	0.06	0.33	0.43	0.08	0.03	0.04	0.12	0.21	0.07
Minnesota	0.31	0.31	0.16	0.06	0.01	0.01	0.05	0.07	0.05
Virginia/North Carolina	0.00	0.00	0.14	0.02	0.08	0.17	0.16	0.71	0.64
Georgia/South Carolina	0.00	0.01	0.10	0.01	0.02	0.05	0.07	0.36	3.80
Maryland/New Jersey	0.00	0.00	0.22	0.04	0.18	0.35	0.23	0.73	0.13
New England	0.00	0.00	0.28	0.41	1.74	0.55	1.28	0.35	0.03
Other	0.12	0.07	0.24	0.01	0.02	0.03	0.08	0.36	0.97

When the potential fields listed in Tables 2 to 4 are multiplied by the SO$_2$ emissions density for 1978, an estimate is
obtained for the actual contribution of source regions to the
predicted concentration and deposition which occurred over the
year. The percent of estimated contribution of upwind states and
provinces to the total (dry plus wet) sulfur deposition at the
Adirondack Mountain receptor is shown in Figure 1. The states and
provinces estimated to contribute the most to the total sulfur
deposition include Ohio, Ontario, Pennsylvania, and New York, in
that order.

Similar estimates for the other eight receptor locations have
been calculated. The contribution of upwind states and provinces
to total sulfur deposition at the Muskoka, Ontario, receptor is
shown in Figure 2. The major contribution regions include Ontario
itself and the upwind states of Ohio and Michigan. While the
natural potential for Michigan is considerably higher than that of
Ohio (Tables 3 and 4), the larger emissions strength in Ohio
accounts for its relative importance at this receptor. Algoma,
Ontario, located near Sault St. Marie, Michigan, also receives a
substantial fraction of its total sulfur deposition from the State
of Michigan as shown in Figure 3, with other significant contributions from the combination of Wisconsin/Iowa and Illinois.

The estimated culpability for total sulfur deposition at
Boundary Waters, Minnesota; southern Appalachia; western Pennsylvania; Vermont/New Hampshire; Quebec City; and southern Nova
Scotia; respectively, are shown in Figures 4 through 9. The probable source for Boundary Waters deposition appears to be from

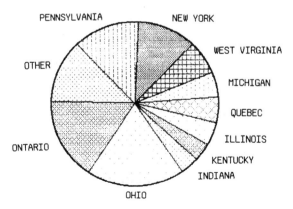

Figure 1.   The percent of contribution to total sulfur deposition
            at the Adirondack receptor estimated for 1978 using
            the University of Michigan ACID model

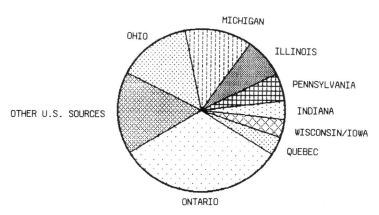

Figure 2.   The estimated percent of contribution to total 1978
            sulfur deposition at the Muskoka, Ontario, receptor
            as derived using the University of Michigan ACID model

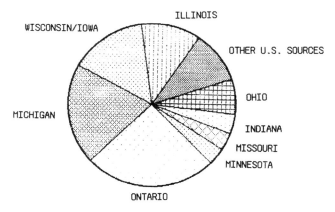

Figure 3.   The estimated percent of contribution to total 1978
            sulfur deposition at the Algoma, Ontario, receptor
            as derived using the University of Michigan ACID model

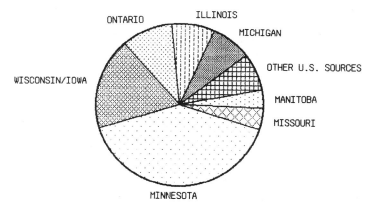

Figure 4.   The estimated percent of contribution to total 1978
            sulfur deposition at the Boundary Waters canoe area of
            Minnesota receptor as derived using the University of
            Michigan ACID model

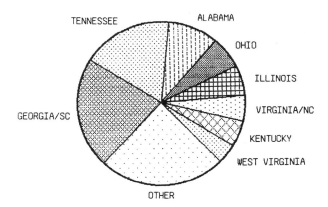

Figure 5.   The estimated percent of contribution to total 1978
            sulfur deposition at the southern Appalachian Moun-
            tains receptor as derived using the University of
            Michigan ACID model

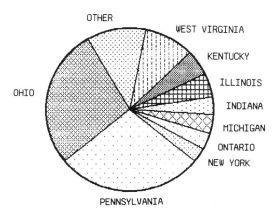

Figure 6. The estimated percent of contribution to total 1978
          sulfur deposition at the western Pennsylvania receptor
          as derived using the University of Michigan ACID model

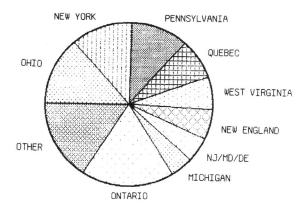

Figure 7. The estimated percent of contribution to total 1978
          sulfur deposition at northern Vermont/New Hampshire
          receptor as derived using the University of Michigan
          ACID model

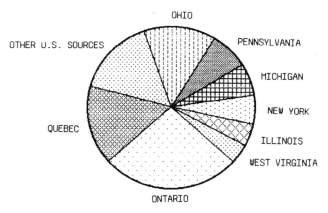

Figure 8. The estimated percent of contribution to total 1978
           sulfur deposition at the Quebec City, Province of
           Quebec receptor as derived using the University of
           Michigan ACID model

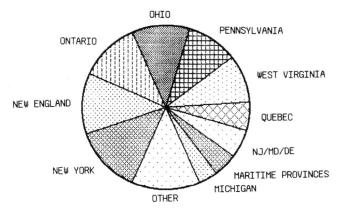

Figure 9. The estimated percent of contribution to total 1978
           sulfur deposition at the southern Nova Scotia receptor
           as derived using the University of Michigan ACID model

within the State of Minnesota, but the results from this site were questionable due to a lack of reliable precipitation data in this area. Therefore, the results in Figure 4 must be viewed with caution. On the other hand, the source data for southern Appalachia, shown in Figure 5, represents the most complete set of input trajectories and precipitation data of all the sites analyzed. It is somewhat surprising to note the relatively high contribution from the southern states of Georgia, South Carolina, and Alabama to this site (only parts of Florida were included in this model computation and, thus, its contribution has not been tabulated). Contributions from the midwestern United States make up only a small fraction of the estimated deposition. In contrast to this, the major contributors to deposition in western Pennsylvania, illustrated in Figure 6, are (not surprisingly) the two closest states with the largest amount of emissions, Pennsylvania and Ohio.

Contributions to the northern Vermont/New Hampshire region (Figure 7) include a significant fraction from Canadian sources. Sources in the states of Ohio, New York, and Pennsylvania are also estimated to contribute significant percentages of the total sulfur deposition at this site. The Canadian contribution increases to the north at the Quebec City, Province of Quebec receptor (Figure 8). Emissions in Ontario and Quebec are estimated to be the dominant contributors to southern Nova Scotia (Figure 9). Additions of sulfur from the industrial east coast, including New York and the New England States, is significantly higher for southern Nova Scotia than at the other sites studies.

CONCLUSIONS

Current acid deposition modeling efforts have focused on estimating source-receptor relationships. Performed computer analyses given in this report show that:

° the source regions which have the greatest potential for contributing to ambient sulfate concentrations on an annual basis (for calculated annual total deposition) are those areas closest to the receptor, and

° the source regions which are estimated to have the highest actual influence on ambient sulfate concentrations and sulfur deposition are those nearby with the highest emissions.

The first finding illustrates the dominance of atmospheric dispersion on an annual basis over the influences of atmospheric chemistry in the spatial distribution of culpability. This finding also could have some important implications on the choice of location for future industrial growth. Given the higher potential of states and provinces near ecologically sensitive area, increased emissions in these areas pose a higher risk to the ecosystem than do similar emission increases in regions farther removed.

The second finding illustrates that the variance in the natural potential contribution field is far smaller than the variance in the emissions field (i.e., the ratio of the potential of New York State to contribute to Adirondack sulfur deposition as compared to its sulfur emissions strength is less than the ratio of Ohio's potential as compared to its emission strength). Consequently, while nearby sources have the highest potential to contribute to sulfur deposition, in reality the large more distant sources dominate the current estimates of culpability.

REFERENCES

Barrie, L.A., 1981, The prediction of rain acidity and SO_2 scavenging in eastern North America. Atmos. Envir. 15:31-41.

Benkovitz, C.M., 1982, Compilation of an inventory of anthropogenic emissions in the United States and Canada. Atmos. Envir. 16:1551-1564.

Bhumralker, C.M., Mancuso, R.L., Wolf, D.E., Johnson, W.B., and Prankrath, J., 1981, Regional air pollution model for calculating short-term (daily) patterns and transfrontier exchanges of airborne sulfur in Europe. Tellus 33:142-161.

Cass, G.R., 1981, Sulfate air quality control strategy design. Atmos. Envir. 15:1227-1249.

Draxler, R.R. and Taylor, A.D., 1982, Horizontal dispersion parameters for long-range transport modeling. J. Appl. Meteor. 21:367-372.

Eliassen, A. and Saltbones, J., 1983, Modeling of long-range transport of sulphur over Europe: a two-year model run and some model experiments. Atmos. Envir. 17:1457-1473.

Gillani, N.V., Husar, R.B., Husar, J.D., Patterson, D.E., and Wilson, W.E., 1978, Project MISTT: Kinetics of particulate sulfur formation in a power plant plume out to 300 km. Atmos. Envir. 12:589-598.

Heffter, J.L., 1980, Air Resources Laboratories Atmospheric Transport and Dispersion Model (ARL-ATAD). NOAA Tech. Memo. ERL ARL-81, National Oceanic Atmospheric Administration, Rockville, MD. 17 pp.

Lamb, R.G. and Seinfeld, J.H., 1973, Mathematical modeling of urban air pollution-general theory. Envir. Sci. Tech. 7:253-261.

McNaughton, D.J., 1984, A second look at a theoretical approach for sulfate aerosol scavenging, pp. 483 to 495. In: APCA Specialty Conference on the Meteorology of Acid Deposition, P.J. Samson, ed., Air Pollution Control Assoc., Pittsburgh, PA.

Samson, P.J., 1980, Trajectory analysis of summertime sulfate
    concentrations in the northeastern United States. J. Appl.
    Meteor. 19:1382-1394.
Samson, P.J. and Moody, J.L., 1981, Trajectories and two-
    dimensional probability fields, pp.43 to 54. In: Air
    Pollution Modeling and Its Application I, C. DeWispelaere,
    ed., Plenum Press, New York, NY.
Samson, P.J. and Small, M.J., 1984, Atmospheric trajectory models
    for diagnosing the sources of acid precipitation, pp. 1 to 24.
    In: Modeling of Total Acid Precipitation Impacts, J.L.
    Schnoor, ed., Butterworth Publ., Boston, MA.
Scott, B.C., 1978, Parameterization of sulfate removal by precip-
    itation. J. Appl. Meteor. 17:1375-1389.
Shannon, J.D., 1981, A model of regional long-term average sulfur
    atmospheric pollution, surface removal, and net horizontal
    flux. Atmos. Envir. 15:689-701.
U.S./Canada Memorandum of Intent, 1982, Regional Modeling Subgroup
    Report. Report 2F-M, U.S.E.P.A., Washington, DC.
Voldner, E.C., Olson, M.P., Oikawa, K., and Loiselle, M., 1981,
    Comparison between measured and computer concentrations of
    sulfur compounds in eastern North America. J. Geophys. Res.
    86:5339-5346.

THE USE OF LONG RANGE TRANSPORT MODELS IN DETERMINING EMISSION CONTROL STRATEGIES FOR ACID DEPOSITION

Roderick W. Shaw

Atmospheric Environment Service
Environment Canada
4905 Dufferin Street
Downsview, Ontario, Canada   M3H 5T4

ABSTRACT

Long range transport (LRT) models attempt to simulate over large regions the atmospheric pathway between emission sources and acid deposition. The output of these models can be expressed as source-receptor matrices in which acid deposition at a given sensitive receptor point is the sum of contributions from individual source regions. The largest element in the matrix belongs to the source region with the largest contribution to deposition per unit emission. Therefore, reducing emissions in these source regions with the largest matrix elements will minimize the amount of sulphur control that is needed to achieve a desired reduction in acid deposition. Based on this principle, an optimized emission reduction scheme applied to eastern North America indicated that it was most effective to confine emission reductions to the Ohio River Valley, northern Appalachia, the lower Great Lakes and St. Lawrence River Valley and the smelters at Sudbury.

INTRODUCTION

Scientists use models to represent processes that may be difficult to duplicate in reality. In particular, air pollution meteorologists use mathematical models to simulate atmospheric processes affecting the fate of pollutants emitted into the air. There has been increasing concern during the past twenty years about the long range transport of acidifying pollutants through the

75

atmosphere and their subsequent deposition to sensitive receiving surfaces that may be several hundred or as much as several thousand kilometers from the original source of the pollutant. As a result, long range transport (LRT) models are being developed and tested with an attempt to incorporate not only atmospheric transport and diffusion (processes which were the primary concern of short range models) but also the atmospheric chemistry and deposition affecting the behavior of acidifying pollutants.

Although some experiments in the mitigation of acid deposition, i.e. the liming of sensitive fresh water bodies, are being carried out in Nova Scotia (Watt, 1980) and in Sweden (Dickson, 1979), it appears that a more fundamental approach is limiting emissions of acidifying pollutants, namely the oxides of sulphur and nitrogen, at their sources. Abatement measures for these pollutants are expensive (for example: $100 to $4000 per tonne for sulphur dioxide). It is important that these emission reductions occur at sources such that acid deposition at sensitive ecosystems will be reduced to a specified value with the least amount of pollutant removal or, better still, at the least cost.

Because LRT models attempt to determine the strength of the atmospheric link between sources and receptors, they are a potentially useful tool in determining in an optimum manner where emission reductions should take place. This paper will briefly describe some work that has been done within Environment Canada in using LRT models to formulate emission control strategies. Examples of results shall also be given. Although nitrates may in the long term be included in control strategies, most work thus far has concentrated on sulphur oxides because: 1) they are believed to contribute two-thirds of the acidity in precipitation, and 2) potentially harmful ion exchange in soils may result from the deposition of sulphate. The method that will be described minimizes the amount of sulphur removed from emissions. Methodologies that are still being developed will go one step further and attempt to minimize costs.

LONG RANGE TRANSPORT MODELS AND SOURCE-RECEPTOR MATRICES

On 5 August 1980, the governments of the United States and of Canada signed a Memorandum of Intent on Transboundary Pollution. Under the Memorandum, bilateral technical work groups were established. The Modelling Work Group examined eight available LRT models and attempted to define their applicability for assessing the environmental results of changes in annual emission rates of sulphur dioxide in eastern North America.

Space does not permit a detailed description of these models and how they simulate the physical and chemical processes during long range atmospheric transport; this may be found in the final report of the Modelling Work Group (MOI, 1982). Suffice it to say that the models use input fields of emission rates over North America and meteorological data such as wind velocity and precipitation to predict monthly, seasonal and annual fields of concentrations and deposition of sulphur dioxide and sulphates. An example of such a predicted field for 1977 is shown in Figure 1, where the isopleths show annual rates of deposition of sulphur by precipitation (wet deposition) as predicted by the Advanced Statistical Trajectory Regional Air Pollution (ASTRAP) model as developed by Shannon (1981). Many aquatic effects have been indexed to the annual wet deposition rate of sulphur. Furthermore, a target value of 20 kilograms sulphate per hectare per year has been proposed by some workers to protect moderately sensitive fresh water bodies (MOI, 1983). Therefore, annual wet deposition has received the greatest interest so far and will be used in this paper. However, as the role of dry deposition is better understood, target values for total (wet plus dry) deposition may be established in the future, as was proposed at the 1982 Stockholm Conference on Acidification of the Environment. If and when such targets are accepted, emission control strategies may then be based upon total deposition.

The pattern of wet deposition in Figure 1 is elongated in a southwest-northeast direction with the area of maximum deposition (3.5 times that of the proposed target of 20 kg SO_4 ha^{-1} yr^{-1}) located just to the northeast of the strong emitting areas in the Ohio valley, northern Appalachia, and the lower Great Lakes area. The region of significant deposition extends, however, as far as the sensitive receptor areas in the Adirondacks and the Atlantic provinces of Canada.

For purposes of comparison with Figure 1, observed wet sulphate deposition for 1980 is illustrated in Figure 2 (from MOI, 1982). The predicted pattern in Figure 1 has been obviously smoothed, but despite the difference in the period (1977 vs. 1980) the predicted pattern reproduced the observed pattern reasonably well, including the orientation from southwest to northeast and the location and magnitude of the maximum deposition. Many models can predict wet sulphur deposition within a factor of two of the observed values in the areas of high deposition, a finding which gives some confidence in using models for control strategies on a broad geographical basis.

Long range transport models are generally time-consuming and expensive to run because they must process large amounts of emission and meterological data at short time intervals over long periods (a year, for example) to make a prediction. To simplify matters, the output of the LRT models can be conveniently expressed as an "atmospheric transfer matrix" that links emissions with

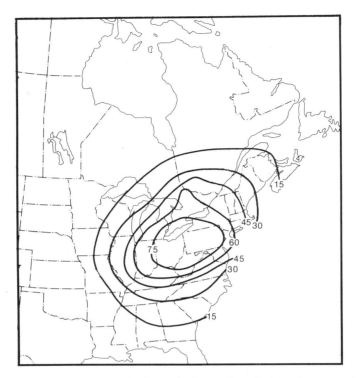

Figure 1.   Annual rates of deposition of sulphur by precipitation
            (kg SO_4 ha^{-1} y^{-1}) as predicted by the Advanced
            Statistical Trajectory Regional Air Pollution (ASTRAP)
            model (developed by Shannon, 1981).

predicted concentrations and/or depositions at specified receptors.
The MOI modellers have divided North America into 40 source regions
and 9 sensitive receptor points, as shown in Figure 3.   The source
regions are generally based upon political units while the
sensitive receptor points, shown in Figure 3, are locations where
considerable monitoring and research has taken place.

     The "atmospheric transfer matrix" developed in this paper is
shown in Table 1.   The elements in the 40 x 9 matrix indicate the
amount of deposition (kilograms sulphur per hectare per year; kg S
$ha^{-1}y^{-1}$) at each of the 9 receptor points, per teragram sulphur
emission (10^{12} g S or 10^6 tonnes S) at one of the 40 source
regions, as predicted by another model considered by the MOI Work
Group (the Monte Carlo Model; Patterson et al., 1981).   For
example, one teragram of emission per year at source No. 6
(southwestern Ontario) will lead to 2.95 kg S $ha^{-1}y^{-1}$ deposition at
Muskoka (receptor No. 3), while the same emission in Michigan would
lead to almost the same deposition (2.19 kg S $ha^{-1}y^{-1}$).   Not

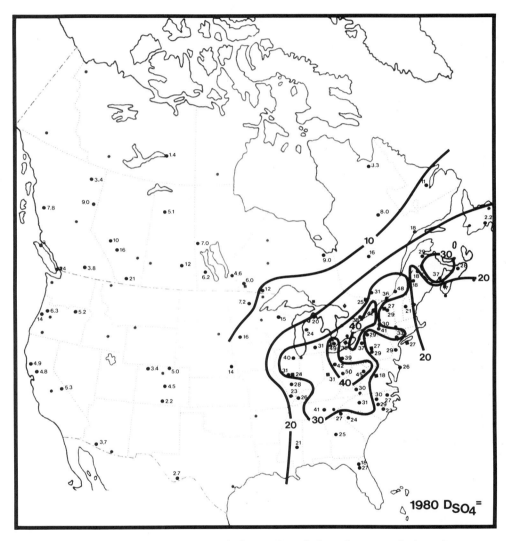

Figure 2.   Observed deposition of sulphur by precipitation
            (kg SO$_4$ ha^{-1} y^{-1}) in 1980 (from MOI, 1982).

Figure 3. The 40 source regions and 9 sensitive receptor points as
           designated by the modellers under the U.S.-Canada
           Memorandum of Intent on Transboundary Air Pollution
           (MOI, 1982).

Table 1.   Atmospheric transfer matrix showing wet deposition in
           1978 at each of 9 receptor points for each teragram (10^{12}
           g or 10^6 tonnes) of sulphur emitted in each of the 40
           source regions, as estimated by the Monte Carlo model
           (Patterson et al., 1981).   For locations of sources and
           receptors see Figure 2.   Units are kg S ha^{-1}y^{-1}.

SOURCE REGION	STATE/ PROV CODE	RECEPTOR								
		BOUNDARY WATERS (1)	ALGOMA ONTARIO (2)	MUSKOKA ONTARIO (3)	QUEBEC (4)	NOVA SCOTIA (5)	VT/ NH (6)	ADIRON- DACKS (7)	PENN (8)	SMOKEY MTS. (9)
1	N. Manitoba	0.00	0.00	0.00	0.00	0.00	0.00	0.00	0.00	0.00
2	S. Manitoba	0.86	0.28	0.18	0.15	0.05	0.19	0.13	0.07	0.01
3	N.W. Ontario	2.30	1.23	0.56	0.28	0.10	0.24	0.26	0.22	0.03
4	N.E. Ontario	0.03	1.97	1.66	1.23	0.23	0.89	0.83	0.27	0.12
5	Sudbury	0.00	1.26	2.76	1.31	0.29	0.80	1.22	0.24	0.16
6	S.W. Ontario	0.01	0.41	2.95	1.18	0.30	0.83	1.42	0.82	0.12
7	S.E. Ontario	0.00	0.21	1.30	3.06	0.43	2.11	1.43	0.28	0.05
8	St. Lawrence	0.00	0.02	0.21	4.16	0.56	2.45	1.26	0.26	0.01
9	N. Cent. Quebec	0.00	0.48	0.93	1.79	0.32	0.76	0.78	0.26	0.04
10	Gaspe, Quebec	0.00	0.00	0.05	0.42	0.52	0.14	0.06	0.00	0.00
11	New Brunswick	0.00	0.00	0.03	0.41	1.49	0.57	0.23	0.00	0.00
12	Nova Scotia	0.00	0.00	0.01	0.24	1.88	0.21	0.06	0.00	0.00
13	Newfoundland	0.00	0.00	0.00	0.02	0.02	0.00	0.00	0.00	0.00
14	Sask., Alta	0.00	0.00	0.00	0.00	0.00	0.00	0.00	0.00	0.00
15	Brit. Columbia	0.00	0.00	0.00	0.00	0.00	0.00	0.00	0.00	0.00
16	Ohio	0.00	0.13	1.38	1.00	0.33	0.48	1.20	2.13	0.15
17	Illinois	0.04	1.06	0.82	0.36	0.15	0.32	0.47	0.55	0.28
18	Pennsylvania	0.00	0.02	0.81	1.09	0.44	0.74	1.73	4.53	0.07
19	Indiana	0.00	0.26	0.74	0.63	0.18	0.40	0.71	1.05	0.28
20	Kentucky	0.00	0.12	0.78	0.42	0.10	0.33	0.76	0.91	0.45
21	Michigan	0.07	1.24	2.19	0.55	0.25	0.52	0.91	0.98	0.10
22	Tennessee	0.00	0.06	0.60	0.19	0.05	0.17	0.33	0.62	1.70
23	Missouri	0.04	0.64	0.48	0.25	0.04	0.14	0.27	0.42	0.26
24	West Virginia	0.00	0.01	0.83	1.14	0.33	0.71	1.40	1.89	0.24
25	New York	0.00	0.04	0.73	1.48	0.54	1.44	2.49	0.88	0.05
26	Alabama	0.00	0.02	0.30	0.09	0.02	0.08	0.11	0.19	1.47
27	WI/IO	0.38	2.06	0.77	0.36	0.13	0.31	0.39	0.55	0.09
28	Minnesota	2.78	0.80	0.31	0.20	0.06	0.14	0.25	0.26	0.13
29	VA/NC	0.00	0.00	0.26	0.45	0.21	0.50	0.76	0.78	0.28
30	Florida	0.00	0.00	0.04	0.07	0.01	0.05	0.06	0.13	0.71
31	GA/SC	0.00	0.00	0.19	0.17	0.04	0.10	0.11	0.41	3.36
32	MD/DL/NJ/DC	0.00	0.00	0.24	0.69	0.44	0.82	1.10	1.33	0.02
33	AK/LA/MS	0.00	0.09	0.25	0.07	0.02	0.07	0.20	0.19	0.38
34	MA/CT/RI	0.00	0.00	0.06	0.61	0.88	0.95	1.43	0.22	0.00
35	Maine	0.00	0.00	0.03	0.72	0.99	1.10	0.46	0.02	0.00
36	VT/NH	0.00	0.00	0.09	1.51	0.85	4.07	2.63	0.32	0.00
37	NE/ND/SD/MT/WY	0.68	0.34	0.12	0.06	0.04	0.06	0.12	0.09	0.02
38	OK/KA/TX/CO/NM	0.12	0.25	0.14	0.06	0.02	0.04	0.11	0.13	0.10
39	WA/ID/OR	0.00	0.00	0.00	0.00	0.00	0.00	0.00	0.00	0.00
40	CA/NV/UT/AZ	0.00	0.00	0.00	0.00	0.00	0.00	0.00	0.00	0.00

surprisingly, because of their proximity, sources in southwestern
Ontario, Sudbury and Michigan have the strongest atmospheric
linkage to Muskoka. In contrast, other source regions (e.g.
Nos. 13 to 15, 39 and 40) have no atmospheric linkage to Muskoka
(according to the Monte Carlo model) as indicated by the zero
values for these elements.

To predict deposition at the receptor points, the transfer
elements need to be multiplied by the appropriate emission values
in each of the 40 source areas. An example of the resulting
"absolute deposition matrix" is shown in Table 2, where the largest
contributors in an absolute sense to wet deposition at Muskoka are
Ohio and Sudbury (sources No. 16 and 5). Although the atmospheric
linkage of Ohio to Muskoka is not as strong as that of southwestern
Ontario, the sheer magnitude of Ohio's emissions makes it the
largest contributor. The matrix elements in Table 2 show only
anthropogenic contributions to deposition. Based upon the analysis
of Whelpdale (1978), a background deposition of 2 kg S ha^{-1}y^{-1} is
added for anthropogenic and natural sources not accounted for in
the emission column on the left hand side of Table 2.

It should be noted that the concept of a transfer matrix
implies that the deposition at a given location is the sum of
independent contributions from the 40 source regions, and that each
contribution is proportional to the rate of emission in the
respective source region. Due to the complexities of the
atmospheric chemistry, the above implications are not exactly true.
Shaw and Young (1983) examined the effect of non-linear atmospheric
chemistry on the above assumptions of proportionality and
independence and concluded that non-linearity does not seriously
affect the qualitative use of matrices in selecting where emission
reductions should take place. Nevertheless, more investigations of
this issue are required using updated atmospheric chemistry models.

Optimization Methodology

The deposition field that will result from a given emission
field is given in Table 2. The effects of changing the emission
field can be easily estimated by re-multiplying the appropriate
transfer elements by the new emission values. However, in
formulating emission control strategies, there is an opposite
problem: what emission field will produce a desired deposition
field? By trial and error, many emission fields could be used to
generate the desired deposition field. However, because emission
abatement measures are so expensive, it is very important that the
maximum emission field (i.e. reductions minimized) be utilized.
Linear programming is one technique that will produce a minimized
emission field. This paper will describe another technique that
has been developed to estimate an optimized emission field to meet
a set of target depositions.

Table 2. Annual values of wet deposition (kg S ha^{-1}y^{-1}) for 1978 obtained by multiplying the atmospheric transfer matrix in Table 1 by the 40 values in the "EMISSION" column in Table 2 (expressed as teragram, 10^{12} g or 10^{6} tons). Depositions are also summed from Canadian sources ("CAN" row), U.S. sources ("USA" row) and in "TOTAL." To the anthropogenic "TOTAL" deposition must be added a background deposition of approximately 2 kg S ha^{-1}y^{-1} from anthropogenic and natural sources not included in the atmospheric transfer matrix. See Table 1 for the receptor code.

SOURCE REGION	EMISSION Tg Sy^{-1}	1	2	3	4	5	6	7	8	9
1	0.232	0.00	0.00	0.00	0.00	0.00	0.00	0.00	0.00	0.00
2	0.013	0.01	0.00	0.00	0.00	0.00	0.00	0.00	0.00	0.00
3	0.012	0.03	0.01	0.01	0.00	0.00	0.00	0.00	0.00	0.00
4	0.081	0.00	0.16	0.13	0.10	0.02	0.07	0.07	0.02	0.01
5	0.470	0.00	0.59	1.30	0.62	0.14	0.38	0.57	0.11	0.03
6	0.325	0.00	0.13	0.96	0.38	0.10	0.27	0.46	0.27	0.04
7	0.018	0.00	0.00	0.02	0.06	0.01	0.04	0.03	0.01	0.00
8	0.234	0.00	0.00	0.05	0.97	0.13	0.57	0.29	0.06	0.00
9	0.280	0.00	0.13	0.26	0.50	0.09	0.21	0.22	0.07	0.01
10	0.065	0.00	0.00	0.00	0.03	0.03	0.01	0.00	0.00	0.00
11	0.108	0.00	0.00	0.00	0.04	0.16	0.06	0.02	0.00	0.00
12	0.112	0.00	0.00	0.00	0.03	0.21	0.02	0.01	0.00	0.00
13	0.030	0.00	0.00	0.00	0.00	0.00	0.00	0.00	0.00	0.00
14	0.299	0.00	0.00	0.00	0.00	0.00	0.00	0.00	0.00	0.00
15	0.097	0.00	0.00	0.00	0.00	0.00	0.00	0.00	0.00	0.00
16	1.199	0.00	0.16	1.65	1.20	0.40	0.58	1.44	2.55	0.18
17	0.668	0.03	0.71	0.55	0.24	0.10	0.21	0.31	0.37	0.19
18	0.906	0.00	0.02	0.73	0.99	0.40	0.67	1.57	4.10	0.06
19	0.909	0.00	0.24	0.67	0.57	0.16	0.36	0.65	0.95	0.25
20	0.497	0.00	0.06	0.39	0.21	0.05	0.16	0.38	0.45	0.22
21	0.411	0.03	0.51	0.90	0.23	0.10	0.21	0.37	0.40	0.04
22	0.488	0.00	0.03	0.29	0.09	0.02	0.08	0.16	0.30	0.83
23	0.590	0.02	0.38	0.28	0.15	0.02	0.08	0.16	0.25	0.15
24	0.495	0.00	0.00	0.41	0.56	0.16	0.35	0.69	0.94	0.12
25	0.432	0.00	0.02	0.32	0.64	0.23	0.62	1.08	0.38	0.02
26	0.354	0.00	0.01	0.11	0.03	0.01	0.03	0.04	0.07	0.52
27	0.422	0.16	0.87	0.32	0.15	0.05	0.13	0.16	0.23	0.04
28	0.119	0.33	0.10	0.04	0.02	0.01	0.02	0.03	0.03	0.02
29	0.437	0.00	0.00	0.11	0.20	0.09	0.22	0.33	0.34	0.12
30	0.497	0.00	0.00	0.02	0.03	0.00	0.02	0.03	0.06	0.35
31	0.527	0.00	0.00	0.10	0.09	0.02	0.05	0.06	0.22	1.77
32	0.336	0.00	0.00	0.08	0.23	0.15	0.28	0.37	0.45	0.01
33	0.315	0.00	0.03	0.08	0.02	0.01	0.02	0.06	0.06	0.12
34	0.196	0.00	0.00	0.01	0.12	0.17	0.19	0.28	0.04	0.00
35	0.043	0.00	0.00	0.00	0.03	0.04	0.05	0.02	0.00	0.00
36	0.045	0.00	0.00	0.00	0.07	0.04	0.18	0.12	0.01	0.00
37	0.276	0.19	0.09	0.03	0.02	0.01	0.02	0.03	0.02	0.01
38	0.934	0.11	0.23	0.13	0.06	0.02	0.04	0.10	0.12	0.09
39	0.543	0.00	0.00	0.00	0.00	0.00	0.00	0.00	0.00	0.00
40	0.764	0.00	0.00	0.00	0.00	0.00	0.00	0.00	0.00	0.00
CDA	2.376	.04	1.05	2.74	2.73	.89	1.64	1.68	.54	.14
USA	12.403	.87	3.44	7.24	5.95	2.28	4.58	8.45	12.36	5.12
TOTAL	14.779	.91	4.49	9.98	8.68	3.17	6.22	10.13	12.91	5.26

Several optimization methods to minimize emission reductions were described in Young and Shaw (1983). To reduce wet deposition in an optimal manner at one of the nine receptor points, this method applies emission reductions first at the source area which has the largest transfer element in Table 1 for that receptor, i.e. the source with the strongest atmospheric link. When emissions at that source area are reduced to the minimum allowable value, the method then selects a source region with the next largest element for abatement, and so on, until either the target deposition has been reached on emissions at the source have been reduced to the minimum allowable values.

Using Tables 1 and 2 as examples, suppose a reduction in wet deposition to 7 kg S ha^{-1}y^{-1} (anthropogenic component = 5 kg S ha^{-1}y^{-1}) at all receptor points is considered. As shown in Table 2, receptor 8 (western Pennsylvania) has the largest deposition and, therefore, efforts will initially be made to reduce deposition there. From Table 1, the source with the largest element in the column for receptor 8 (the strongest atmospheric linkage) is source 18 (not surprisingly, Pennsylvania). Emissions at source 18 are reduced in successive steps of 0.01 teragrams per year (10,000 tonnes sulphur); depositions are re-computed using the transfer matrix until one of three things occurs:

(i)   Emissions at the source undergoing abatement are reduced to the smallest allowable value determined from a technological and/or socioeconomic basis. If this occurs, emissions at the source with the next largest transfer element (i.e. the next strongest atmospheric linkage to the receptor) are then reduced, and so on including sources with ever-diminishing transfer elements. Emissions are never reduced at sources with zero transfer elements (i.e. no atmospheric linkage) as that would have no effect upon deposition at the receptor in question.

(ii)   Deposition at receptor 8 falls below that of another reception point. If this occurs, the other receptor point becomes the new target and the transfer elements in the transfer matrix are then scanned for the largest element.

(iii) Deposition at receptor 8 meets the target. If this occurs, optimization is then carried out for other receptor points where the target deposition has not yet been met.

The above emission reductions, re-calculation of depositions, and testing are carried on until either the deposition at all receptor points is reduced to the appropriate target value or the emissions at all pertinent source areas are reduced to the minimum allowable value. In either case, the optimization calculation is complete.

The above optimization calculation can be carried out for several receptors, as was described above, or for a single receptor point. In the latter case, the elements in only one column of the atmospheric transfer element are used for establishing the sequence of source areas where emission reductions should take place.

Sample Results

The above technique was applied to eastern North America, using the atmospheric transfer matrix for wet sulfur deposition for 1978 meteorology, as estimated by the Monte Carlo model. The minimum allowable emissions for the source areas in Canada were estimated using information about emission reductions at: 1) the large non-ferrous smelters in Manitoba, Ontario and Quebec; 2) thermal power stations in Ontario, New Brunswick and Nova Scotia; and 3) fuel oil desulfurization in Quebec.

In the United States, about 80% of SO_2 emissions in 1980 came from thermal generating stations and non-utility fuel combustion. Information is not available about possible abatement actions in the United States; therefore, an initial assumption was made that a maximum emission reduction of 70% is possible in each of the U.S. source regions through an average abatement of 90% at thermal generating stations and a reduction of the sulphur content of fuels burned in commercial and residential furnaces.

It is the experience of the author that the optimal emission field depends greatly upon the values that are assumed for the minimum allowable emission in each source region. Furthermore, testing of the MOI models is still being carried out and it is not yet known which model or models is best suited for optimization calculations. For these reasons, the results described below should be considered as preliminary examples of optimized emission reductions rather than definite recommendations for emission reductions in eastern North America. The latter must await improved input data and further testing of the models.

a)   Emission reductions to reduce deposition in the Adirondacks to 20 kg sulfate $ha^{-1}y^{-1}$

Optimized emission reductions to bring wet deposition in the Adirondacks to the above target value are shown in Figure 4. In this and the next four figures, the receptor points at which depositions are being reduced to the target value are shown as heavy rather than open circles. Emission reductions in Figure 3 are concentrated in the Great Lakes Basin, the Ohio River Valley, the Eastern Seaboard, and in Ontario and Quebec. These are the source areas, according to the atmospheric transfer matrix being used, which are most closely linked with the Adirondacks. The

10⁶ TONNES SULPHUR REMOVED PER YEAR
CANADA = 0.80 (34%), USA = 3.28 (26%) TOTAL = 4.08 (25%)
EQUIVALENT UNIFORM REDUCTION NEEDED = 7.99 (54%)

Figure 4.   Optimized percentage emission reductions in the 40
source regions, as estimated by the Monte Carlo model,
to reduce annual wet deposition in the Adirondacks
(shown as a heavy dot) to 20 kg SO_4 $ha^{-1}y^{-1}$.

optimization requires a 34% emission reduction in eastern Canada and a 26% reduction in the United States. The weighted average emission reduction in eastern North America would be 28% which is much less than the 54% needed for a uniform emission reduction achieving the same result.

b)   Emission reductions to reduce deposition at all United States receptor points to 20 kg sulfate ha^{-1}y^{-1}

The area where emission reductions are required to take place in Figure 5 is shifted to the west of that in Figure 4. Less control is required in Canada and more in the United States than in case a) above. Protecting other U.S. receptor points such as Western Pennsylvania requires emission reductions in source areas such as Indiana, Illinois, Iowa, and Wisconsin which replace emission reductions in Northern Ontario and Quebec.

c)   Emission reductions to reduce deposition at Muskoka to 20 kg sulfate ha^{-1}y^{-1}

Emission reductions in both countries are required, as shown in Figure 6. In contrast to case b) above, emission reductions are not required along the Eastern Seaboard, New England or New York, but slightly more is required in the Midwest due to the location of the receptor point.

d)   Emission reductions to reduce deposition at all Canadian receptor points to 20 kg sulfate ha^{-1}y^{-1}

The emission reduction pattern in Figure 7 is very similar to that in case c) above because measures to reduce deposition at Muskoka will contribute considerably to deposition reductions at the other Canadian receptor points. It should be noted, however, that more control is required in the province of Quebec to reduce deposition at the Montmorency receptor point.

e)   Emission reductions to reduce depositions at all nine North American reception points

The pattern of emission reductions needed to reduce depositions at all nine receptor points, as shown in Figure 8, is very similar to that needed for reduction at the U.S. receptor points because those measures which reduce deposition in the U.S. will also be effective in reducing deposition in Canada. More emission reduction in Canada (18% as compared to 8% in Figure 5) is needed.

In this case, an overall reduction of 37% is needed in North America to reduce deposition to the target value at all nine receptor points. Almost twice as much, or 64%, would be

10⁶ TONNES SULPHUR REMOVED PER YEAR
CANADA = 0.20 (8%), USA = 5.18 (42%) TOTAL = 5.38 (36%)
EQUIVALENT UNIFORM REDUCTION NEEDED = 9.47 (64%)

Figure 5.   As in Figure 3, but to reduce annual wet deposition at
            U.S. receptor points (heavy dots) to 20 kg SO₄ ha⁻¹y⁻¹.

10^6 TONNES SULPHUR REMOVED PER YEAR
CANADA = 0.69 (29%), USA = 3.45 (28%) TOTAL = 4.14 (28%)
EQUIVALENT UNIFORM REDUCTION NEEDED = 7.84 (53%)

Figure 6.   As in Figure 3 but to reduce wet deposition at Muskoka
            (heavy dot) to 20 kg SO$_4$ ha^{-1}y^{-1}.

10⁶ TONNES SULPHUR REMOVED PER YEAR
CANADA = 0.80 (34%), USA = 3.45 (28%) TOTAL = 4.25 (29%)
EQUIVALENT UNIFORM REDUCTION NEEDED = 7.84 (53%)

Figure 7. As in Figure 3, but for all Canadian receptor points.

10^6 TONNES SULPHUR REMOVED PER YEAR
CANADA = 0.43 (18%), USA = 4.98 (40%) TOTAL = 5.41 (37%)
EQUIVALENT UNIFORM REDUCTION = 9.47 (64%)

Figure 8. As in Figure 3, but for all nine receptor points in
          North America.

required for uniform emission reductions. It can be seen, therefore, that optimizing emission reductions could lead, at an assumed cost of $1,000 per tonne of sulphur, to a saving of approximately 4 billion dollars per year.

If the maximum allowable emission reductions were less than those assumed in the above cases, the area requiring emission reductions would be larger than those shown in Figures 4-8. If the allowable emission reductions were too small, it is unlikely that the target depositions could be met.

MINIMIZING THE COST OF EMISSION REDUCTIONS

The technique that was described above is designed to minimize the amount of sulphur removed from emissions to reduce deposition a given amount. It will tend to reduce the cost of emission reductions but not explicitly minimize them. Work is now underway in developing and testing a second generation optimization method which attempts to minimize control costs.

The method combines the atmospheric linkages as expressed by the source-receptor matrix and control costs, expressed in dollars per tonne of sulphur removed. The source receptor matrix expresses, after differentiation:

$$t_{ij} = \frac{\text{reduction in wet deposition at receptor } j}{\text{reduction in emission at source } i}$$

A cost factor is introduced:

$$c_i = \frac{\text{cost of control at source } i}{\text{unit of emission reduction at source } i}$$

By division, one can combine atmospheric and cost factors to produce a cost factor based upon reduction in deposition:

$$e_{ij} = \frac{\text{cost of control at source } i}{\text{reduction in deposition at receptor } j}$$

Values of e_{ij} are being produced for each receptor point and for each possible control step at smelters and utilities in Canada and the United States. To reduce deposition at a given receptor point, the control steps can be ranked in order of increasing values of e_{ij} such that the first control steps to be selected would be those involving the least cost per unit reduction in deposition. The cost of control can be minimized in this way. The

results of using this technique will be the subject of future
reports.

SUMMARY

     Several optimization methods have been developed to select a
sulphur emission control strategy that minimizes the amount of
sulphur removed to achieve a given reduction in the wet deposition
of sulphur.   The methods use the atmospheric linkages between
sources and receptors, as estimated by the source-receptor matrices
produced by long range transport models.   The emission source areas
are ranked in such a way that those sources causing the greatest
deposition per unit emission are abated first.

     The method appears to be robust and gives optimized fields of
emission reduction.   For example, the method estimates that to
reduce wet deposition to 20 kg sulphate per hectare per year at the
sensitive receptor points in North America, the amount of sulphur
removed from emissions will be minimized if emission reductions are
confined to the central part of the continent including the Ohio
River Valley, northern Appalachia, the lower Great Lakes and St.
Lawrence River Valley and the smelters at Sudbury.   If sufficient
emission reductions are attained in the foregoing source regions,
then little or no emission reductions are required in other areas
which are less strongly linked by atmospheric transport processes
to the receptor points.

     Optimization can potentially lead to considerable savings.   An
across-the-board emission reduction to reduce deposition at all
receptor points to 20 kg SO_4 $ha^{-1}y^{-1}$ would require approximately
9.5 million tonnes of sulphur to be removed per year from North
American emissions; an optimized emission reduction would amount to
only about 5.4 million tonnes removal per year.   At an average
removal cost (capital and operating) of $1000 per tonne, the saving
would amount to about 4 billion dollars per year.   A second-
generation method designed to minimize the cost, rather than the
amount, of removal is being developed and tested.

REFERENCES

Dickson, W., 1979, Experience from small scale liming in Sweden,
     pp. 303-304.   In: Sulphur Emissions Environ., Proc. Int.
     Symp. Sulphur Emissions into the Environment, Society Chemical
     Industries, London, U.K.
MOI, 1982, Final Report of Work Group II on Atmospheric Sciences
     and Analysis, United States - Canada Memorandum of Intent on
     Transboundary Air Pollution.   Available from LRTAP Liaison
     Office, Atmospheric Environment Service, 4905 Dufferin Street,
     Downsview, Ontario, Canada.

MOI, 1983, Final Report of Work Group I on Impact Assessment,
    United States - Canada Memorandum of Intent on Transboundary
    Air Pollution. Available from LRTAP Liaison Office,
    Atmospheric Environment Service, 4905 Dufferin Street,
    Downsview, Ontario, Canada.

Patterson, D.E., Husar, R.B., Wilson, W.E., and Smith, L.F., 1981,
    Monte Carlo simulation of daily regional sulphur distri-
    butions: Comparison with SURE sulfate data, and visual range
    observations during August 1977. J. Appl. Meteorol. 20:70-86.

Shannon, J., 1981, A model of regional long-term average sulphur
    atmospheric pollution, surface removal and wet horizontal
    flux. Atmos. Environ. 15:689-701.

Shaw, R.W. and Young, J.W.S., 1983, An investigation of the
    assumptions of linear chemistry and superposition in LRTAP
    models. Atmos. Environ. 17:2221-2229.

Watt, W., 1980, A proposal for Atlantic salmon habitat restoration
    by addition of lime to acidified rivers. Internal report,
    Fisheries and Oceans Canada, Halifax, Nova Scotia, Canada.
    6 pp.

Whelpdale, D.M., 1978, Atmospheric pathways of sulphur compounds,
    MARC Report No. 8, Chelsea College, University of London,
    London, U.K. 39 pp.

Young, J.W.S. and Shaw, R.W., 1983, A science-based strategy for
    the control of acidic deposition in North America. Proc. VI
    World Congress on Air Quality, Paris, France. 4 pp.

THE SEDIMENTARY RECORD OF ATMOSPHERIC POLLUTION IN

JERSEYFIELD LAKE, ADIRONDACK MOUNTAINS, NEW YORK

Stephen A. Norton

Department of Geological Sciences
University of Maine at Orono
Orono, ME 04469

ABSTRACT

Jerseyfield Lake is a clear water oligotrophic lake of 176 hectare with a watershed of 1,936 hectare. Maximum depth is 28 meters. The present pH is about 4.8-4.9. Portions of the watershed were cut in the 1920's, 1939-1941, ca. 1955-1961, and since 1981, concurrently with the emplacement of a circum-lake road. The chemistry of a sediment core from Jerseyfield indicates nearly steady state conditions over the period 1600-1850. The only systematic change in sediment chemistry was a persistent decline in MnO content. The cause of this is unknown. Pb, Zn, Cu, and V concentrations and deposition rates started to increase about 1850, 1900, 1925, and 1950, respectively. These changes are interpreted to be caused by increased atmospheric deposition of these metals from polluted air masses. Disturbance of the watershed, indicated by increased deposition of all major elements, occurred from ca. 1910 to 1970. This is followed by a decrease in deposition rates to 1982. Acidification of the lake water starting about 1950 is suggested by a decrease in the Zn and CaO deposition rates. The sharp reversal from Zn accumulation to Zn depletion suggests a rapid change in lake pH. The decline in MnO accelerated slightly starting about 1850. None of these changes appear to be synchronous with increased deposition of sediments, presumably related to concurrent lumbering in the watershed. The trends for changing deposition rates for Pb, Zn, Cu, V, Mn, and Ca appear to be reflective of atmospheric deposition of acids and metals.

95

INTRODUCTION

The Adirondack Mountains of New York are the most frequently cited area in the United States for which there is evidence suggesting a negative impact on the sports fisheries by regional atmospheric deposition (Baker and Schofield, 1984). Historic water quality data (Schofield, 1977; Pfeiffer and Festa, 1980), of questionable reliability (Kramer and Tessier, 1982), suggested lake acidification. Lake acidification due to atmospheric deposition of acidic and acidifying compounds is the most plausible cause of these observations. Natural acidification, changing vegetation, and climate change occur on a time scale which is too long to explain the apparent acidification and loss of fisheries. An alternative explanation for lake acidification is changing land use. However, land use changes elsewhere in the world have not produced acidification of clear water lakes, except where acidic deposition occurs.

This research was designed to document the chronology of atmospheric deposition of acids and metals and their effects on the chemistry of lake sediments in the Adirondack Mountains. The impact of changing land use on acidification as evidenced by lake sediment chemistry was also assessed.

METHODS

Jerseyfield Lake was one of ten lakes studied in this project. The lake was selected as a candidate by Schofield (pers. comm.) based on its pH (4.9-5.0) and loss of fisheries in the late 1950's and early 1960's (pers. comm. with local landowners and fishermen). Jerseyfield is a clear water (color $\leq$ 10 Pt-Co units) oligotrophic lake. The area of the lake is 176 hectare; the upstream drainage basin has an area of 1,936 hectare. Vegetation in the watershed consists of mixed hard and softwoods. Heavy cutting occurred in the 1920's. Selective cutting has occurred in 1939-1941 and ca. 1955-1961. Forest cutting has occurred since 1981, concurrently with the emplacement of a circum-lake road. Only a few percent of the drainage basin have been disturbed recently. In two successive years in the mid 1960's, the lake was limed from a boat with six tons of $CaCO_3$ to raise the pH to nearly 7 and save the fisheries. The pH returned to strongly acidic conditions within one year due to flushing. Several year-round residences are present near the lake. The lake level is regulated by a dam at the outlet. It raises the water a maximum of 2 m.

The maximum depth of Jerseyfield is about 28 m. A sediment core was collected in a water depth of 21 m, about 200 meters south of the deepest part of the relatively flatbottomed area, using a 6.5 cm diameter stationary piston corer (Davis and Doyle, 1969). Sediment was extruded upward from the core tube and sectioned in the field, in 0.5 cm increments from 0 to 20 cm and in 1.0 cm increments from 20 to 40 cm. Sediment was stored in whirl bags in the dark at 4°C until processed.

Water content of the sediment was determined from weight loss after oven drying at 105°C for 24 hours. Volatile organic matter was determined by weight loss of dried sediment by heating at 550°C for three hours in a muffle furnace. Pb-210 chronology was determined on an aliquot of oven dried sediment following the method of Eakins and Morrison (1978). Age calculations were made assuming the constant rate of supply (CRS) model of Appleby and Oldfield (1978). The constant input concentration model (CIC) age/depth relationship (Krishnaswami et al., 1971) was used for extrapolation below the depth of applicability for the CRS model. In the Jerseyfield core, this splice was made at a depth of 11.5 cm, or about 1880. A 0.1000 g aliquot of ashed sediment was acid solubilized (total digestion) using the method of Buckley and Cranston (1971). Al, Ti, Fe, Mn, K, Na, Ca, Mg, Zn, and Cu were analyzed by flame atomic absorption spectrophotometry (AAS) using a Perkin-Elmer model 703. Pb and V were determined by flameless AAS (HGA 2200). Standards were made up by combining solutions of the elements, in the same matrix, in proportion to expected concentrations. SiO_2 was not measured but is reported as the difference, i.e., 100% minus the other measured elements. This results in a slight overestimation because sulfur and phosphorus, which may comprise as much as one to two percent of the sediment, were not measured. Deposition rates (DR) of constituents were calculated with a knowledge of water content of sediment intervals, organic content, age of interval boundaries, and concentration of constituents within an interval.

RESULTS AND DISCUSSION

The Pb-210 profile (Figure 1) approximates an exponential function with increasing age; this, combined with relatively invariant chemistry through the lower two-thirds of the core, suggested comparative stability of the sedimentary record (e.g. Fe, Pb, Zn, etc.). However, inspection of deposition rates reveals otherwise. The chronology of the core (Table 1) indicated a nearly linear increase in age with depth in the sediments.

The chemistry of the core is shown in Figure 1; the units are concentration in the sediment (wet, dry, ashed) and thus are not independent. Variations in one parameter (e.g. FeO) in the ashed sediment can cause significant reciprocal changes in all other parameters if the change is independent and of large absolute magnitude. Thus, the FeO enrichment for surface sediments (from ca. 4-16%) caused an approximately 12 percent reduction in a number of dependent variables (e.g. TiO_2, Al_2O_3, Na_2O, K_2O, and MgO).

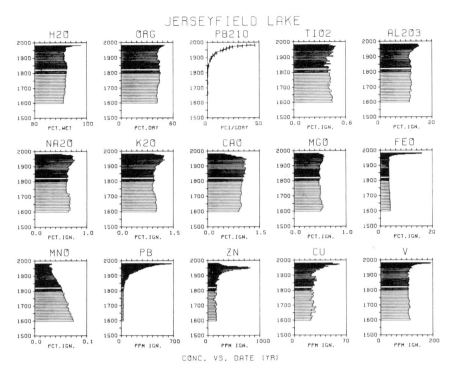

Figure 1.   Sediment chemistry of a core from Jerseyfield Lake,
             Adirondack Mountains, N.Y. Depth has been translated
             into age using Pb-210 chronology.

Analysis of deposition rates (DR) is a more meaningful way of
interpreting events from sediment chemistry. The TiO_2 DR
(Figure 2) declined slowly from about 1600 A.D. (ca. 23 μg $cm^{-2}y^{-1}$)
to slightly less than 20 μg $cm^{-2}y^{-1}$ and then increased to values
above this background, starting between 1910 and 1920, and peaking
about 1970 at 31 μg $cm^{-2}y^{-1}$. This is an increase of 50% for the
TiO_2 DR. Values then declined to background by 1982. However,
recent DR values were very sensitive to surface sediment Pb-210
activity; these must be extrapolated. TiO_2, Al_2O_3, Na_2O, K_2O and
MgO behaved in a parallel fashion (Figure 1) with respect to
concentration, and thus with respect to deposition rates. The
sharp increase in the deposition rates for TiO_2, Al_2O_3, Na_2O, K_2O
and MgO during the period 1910-1920 to about 1970 is interpreted to
be related to deforestation, accelerated erosion and subsequently
higher deposition, with a later return to near normalcy.

       Iron (FeO) behaved independently of all other elements
(Figures 1 and 3) as it does in slowly sedimenting oligotrophic
lakes. The long term decline in the DR from 1600 to
1900 paralleled that of TiO_2. However, it is suspected that the

Table 1. Chronology of sediment intervals from Jerseyfield Lake,
          Adirondack Mountains, NY, located at 43°18' N Lat. and
          74°46' W Long.

Sediment Interval (cm)		Age of bottom of interval	Sediment Interval (cm)		Age of bottom of interval
0.00	0.50	1980	15.00	15.50	1843
0.50	1.00	1977	15.50	16.00	1839
1.00	1.50	1975	16.00	16.50	1834
1.50	2.00	1972	16.50	17.00	1830
2.00	2.50	1968	17.00	17.50	1825
2.50	3.00	1964	17.50	18.00	1814
3.00	3.50	1960	18.00	18.50	1810
3.50	4.00	1956	18.50	19.00	1805
4.00	4.50	1952	19.00	19.50	1801
4.50	5.00	1948	19.50	20.00	1796
5.00	5.50	1943	20.00	21.00	1787
5.50	6.00	1939	21.00	22.00	1777
6.00	6.50	1933	22.00	23.00	1768
6.50	7.00	1928	23.00	24.00	1758
7.00	7.50	1923	24.00	25.00	1749
7.50	8.00	1918	25.00	26.00	1739
8.00	8.50	1913	26.00	27.00	1729
8.50	9.00	1908	27.00	28.00	1720
9.00	9.50	1902	28.00	29.00	1710
9.50	10.00	1897	29.00	30.00	1700
10.00	10.50	1891	30.00	31.00	1690
10.50	11.00	1885	31.00	32.00	1680
11.00	11.50	1880	32.00	33.00	1671
11.50	12.00	1875	33.00	34.00	1661
12.00	12.50	1870	34.00	35.00	1650
12.50	13.00	1865	35.00	36.00	1640
13.00	13.50	1861	36.00	37.00	1629
13.50	14.00	1856	37.00	38.00	1619
14.00	14.50	1852	38.00	39.00	1608
14.50	15.00	1847	39.00	40.00	1597

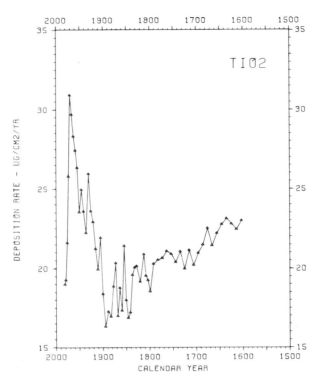

Figure 2.   TiO$_2$ deposition rate for a core from
            Jerseyfield Lake, Adirondack Moun-
            tains, N.Y.

five to six fold increase in the DR toward the surface was due to
sediment diagenesis: solution of Fe-phases in the reducing
(low Eh) portions of the sediment, upward migration of Fe^{2+} in the
interstitial waters, and precipitation of Fe^{2+}, as FeO(OH) or
Fe(OH)$_3$, at or near the sediment-water interface under oxidizing
conditions (Norton, 1974). This type of feature may migrate upward
keeping pace with the sediment-water interface or it may be buried
(Norton and Hess, 1980). We observed that MnO deposition rates
(Figure 4) commonly mimic FeO for the same reason in most
sedimentary environments but this is not the case for Jerseyfield
sediments. This is particularly striking in the zone of FeO
enrichment. Under higher Eh conditions Fe is less mobile than Mn,
suggesting the possibility of geochemical separation of the
elements during acidification.

Deposition rates for background lead (Pb) were about
0.15 µg cm^{-2}y^{-1} but increased, starting in the mid-1800's, to a
maximum of 3.5 µg cm^{-2}y^{-1} in 1979 followed by a slight decline
(Figure 5). Lead has two sources: from the bedrock and soils of
the drainage basin and from the atmosphere. Whereas the flux of

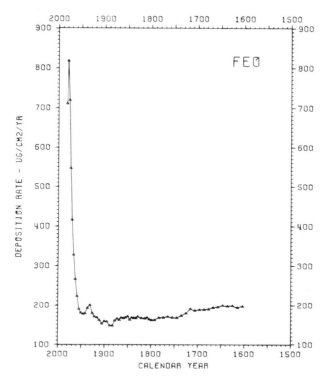

Figure 3.   FeO deposition rate for a core from
            Jerseyfield Lake, Adirondack Moun-
            tains, N.Y.

inorganic debris from the drainage basin (e.g. TiO_2) does not vary
by more than 50%, most (~95%) of the variation for Pb in the recent
sediment is interpreted to be atmospheric in origin.   Nearly the
entire profile since 1850 is dominated by atmospheric additions of
Pb into the lake, either directly or indirectly.   This pattern, its
magnitude and timing are ubiquitous in the northeastern U.S.
(Norton and Kahl, 1983).   A slight downturn in the last few years
is also common and may be a reflection of reduced emissions of Pb
to the atmosphere due to reduced use of leaded gasoline and lower
fossil fuel consumption.

The zinc (Zn) DR profile (Figure 6) suggested nearly steady
state conditions for Zn deposition until the late 1800's followed
by a five fold increase (to 4.5 µg $cm^{-2}y^{-1}$) in the DR.   This peak
occurred about 1950 followed by a decline to background values in
the more recent sediments.   This decrease is ubiquitous in all
lakes that we have studied in the northeast with average pH less
than about 5.5.   Background (pre-1850) values for Zn were high in
Jerseyfield sediments compared to other lakes in the northeastern
U.S. which resulted in a significant atmospheric signal/background

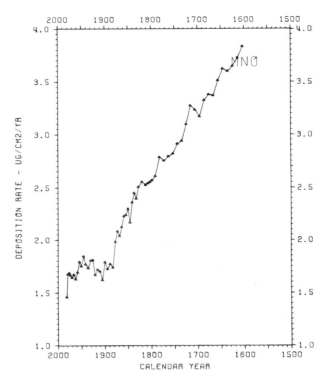

Figure 4.   MnO deposition rate for a core from
            Jerseyfield Lake, Adirondack Moun-
            tains, N.Y.

signal detection limit (Norton and Kahl, unpub. data).   The
magnitude and timing of the declines for rates of Zn deposition
vary from lake to lake (in acidified systems) in the Adirondacks.
However, circumneutral lakes (e.g. Sagamore Lake or Panther Lake)
in the same area do not exhibit these declines.   It is speculated
that a decrease in the pH of atmospheric precipitation and lake
water has resulted in either leaching of the Zn from recently
deposited sediment (Kahl and Norton, 1983), leaching of Zn from
particulates while they are in the water column, reduced uptake of
Zn by sediments due to a tendency toward desorption of metals, or
solubilization of metals from detritus in the terrestrial part of
the ecosystem with no subsequent adsorption of Zn during transport
and deposition (Hanson et al., 1982).   Increased SO_4^{2-} availability
in lake water probably increased SO_4^{2-} reduction in sediments, with
production of H_2S.   This would precipitate divalent heavy metals
(e.g. Zn) resulting in their enrichment within the sediment and

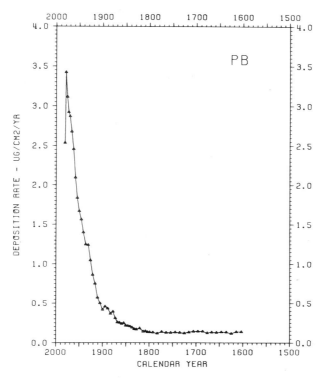

Figure 5.   Pb deposition rate for a core from
Jerseyfield Lake, Adirondack Mountains, N.Y.

setting up diffusion gradients into the sediment. This was
observed in "peeper" (Hesslein, 1976) studies by Carignan (pers.
comm.). While the Zn peak (Figures 1 and 6) may be caused partly
be post-depositional precipitation, the decline appears to be more
closely controlled by Zn mobilization. It also appears to be
approximately at the same time as the beginning of the decline in
fish populations (early 1950's according to anecdotal evidence).

The calcium (CaO) DR profile (Figure 7) showed a sharp decline
of about 65% over the period 1960-1983. This decline is earlier
and greater than that for TiO_2 (Figure 2). The CaO DR did not
increase appreciably during the period of accelerated deposition of
TiO_2. This suggested that CaO deposition, relative to TiO_2,
decreased in the period 1910-present. Progressive acidification of
the ecosystem, with processes operating which are similar to those
controlling Zn (except for SO_4^{2-} reduction and ZnS precipitation)
has probably caused this problem. A reduction in CaO relative
deposition rates has been observed in nearly all acidic (pH <5.5)
lakes in New England and New York (Hanson et al., 1982).

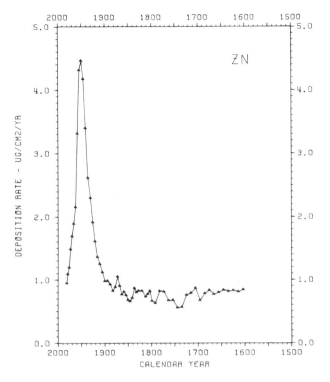

Figure 6.   Zn deposition rate for a core from
            Jerseyfield Lake, Adirondack Moun-
            tains, N.Y.

     Deposition rates for vanadium (V) were similar to those of
TiO$_2$ until about 1950 and then increased to values 2X background,
dropping slightly in the post-1975 sediment (Figure 8).
Atmospheric V is associated with the burning of oil. Its DR
profile is clearly out of phase with watershed disturbances and
increases in Pb and Zn. Its slight decline (as well as part of the
Zn and Pb decline) in the post-1975 sediment may be due to the
mini-recession and decline in emissions (Husar, pers. comm.) and/or
partly an artifact of dating techniques for the most recent
sediments near the sediment water interface. For example, the
Pb-210 activity in the top of the core (0.0-0.5 cm, 0.5-1.0 cm,
Figure 1) was estimated for dating purposes because of insufficient
sediment sample mass. An overestimation of the activity will
result in low deposition rates for the upper cm of sediment, and
vice versa. However, the Pb DR (aerosol component) decreased only
slightly in the post-1980 sediment; this change was close to the
decrease in aerosol Pb at other locations in the northeastern
United States.

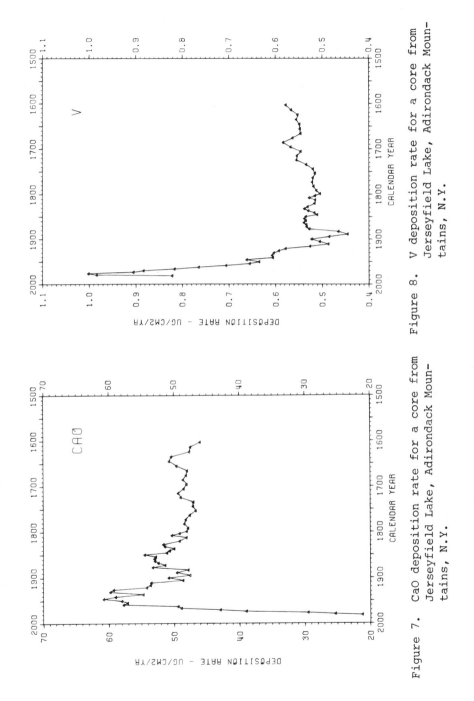

Figure 8.   V deposition rate for a core from Jerseyfield Lake, Adirondack Mountains, N.Y.

Figure 7.   CaO deposition rate for a core from Jerseyfield Lake, Adirondack Mountains, N.Y.

SUMMARY

   The chemistry of sediment from Jerseyfield Lake in the southwestern part of the Adirondack Mountains of New York indicated
nearly steady state conditions over the period 1600-1850 A.D.
During this time the only systematic change in sediment chemistry
was a persistent decline in MnO content, whose cause is unknown.
Lead, Zn, and Cu concentrations and deposition rates started to
increase in about 1850, 1900, and 1925, respectively. These
changes are interpreted as reflecting increased atmospheric
deposition of these metals from polluted air masses. Increasing
consumption of coal as well as smelting of sulfide ores has led to
increased metal emissions. Disturbance of the watershed, indicated
by increased lake deposition rates of all major elements, occurred
from 1910 to 1970, followed by a decrease in deposition rates from
1970 to 1982. Starting in 1950, the V DR increased relative to the
other major elements. This corresponded approximately in time to
sharply increased consumption of oil, the major source of
atmospheric V emissions.

   Acidification of the lake water is suggested by a decrease in
the Zn deposition rate, starting about 1950. The sharp reversal
from Zn accumulation to depletion, in spite of a continued increase
in atmospheric deposition of this element may indicate a rapid
change in lake pH. Deposition rates for CaO also declined sharply,
starting about 1950. This is still happening for both Ca and Zn
suggesting continued acidification of the lake. The decline in MnO
deposition rates started to accelerate slightly about 1850. These
trends only occur in lakes with pH values less than 5.5 and located
in areas receiving acidic precipitation. These changes do not
appear synchronous with accelerated deposition of sediments,
presumably related to be concurrent with lumbering in the
watershed. Consequently, the trends for changing deposition rates
for Pb, Zn, Cu, V, Mn, and Ca appear to be reflective of
atmospheric deposition of acids and metals.

Acknowledgments - The research was sponsored by U.S.E.P.A. Contract
APP-0119-1982. Scott Denning and Myron Buck (undergraduates in the
Department of Geological Sciences) and Geneva Blake, J. Steve Kahl,
and Marilyn Morrison (research associates) capably assisted in all
phases of the field and laboratory work.

REFERENCES

Appleby, P.G. and Oldfield, F., 1978, The calculation of lead-210
     dates assuming a constant rate of supply of unsupported Pb-210
     to the sediment. Catena 5:1-8.

Buckley, D.E. and Cranston, R.E., 1971, Atomic absorption analyses of 18 elements from a single decomposition of aluminosilicate. Chem. Geol. 7:273-284.

Carignan, R., personal communication, Inst. Nat. Res. Sci.-Eau, University of Quebec, Sainte-Foy, Quebec, Canada.

Davis, R.B. and Doyle, R.W., 1969, A piston corer for upper sediment in lakes. Limnol. Oceanogr. 14:643-648.

Eakins, J.D. and Morrison, R.T., 1978, A new procedure for the determination of lead-210 in lake and marine sediments. Intern. Appl. Rad. Isot. 29:531-536.

Hanson, D.W., Norton, S.A., and Williams, J.S., 1982, Modern and paleolimnological evidence for accelerated leaching and metal accumulation in soils in New England, caused by atmospheric deposition. Water, Air, Soil Pollu. 18:227-239.

Hesslein, R.H., 1976, An in situ sampler for close interval pore water studies. Limnol. Oceanogr. 21:912-914.

Husar, R., personal communication, Washington University, St. Louis, MO.

Kahl, J.S. and Norton, S.A., 1983, Metal input and mobilization in two acid-stressed lake watersheds in Maine. Project Rept. A-035ME, Land and Water Resources Center, Univ. of Maine, Orono, ME. 70 pp.

Kramer, J.R. and Tessier, A., 1982, Acidification of aquatic systems: a critique of chemical approaches. Envir. Sci. Tech. 16:606A-615A.

Krishnaswami, S., Lal, D., Martin, J.M., and Maybeck, M., 1971, Geochronology of lake sediments. Earth Planet. Sci. Lett. 11:407-414.

Norton, S.A., 1974, Postglacial iron-rich crusts in hemipelagic deep-sea sediment: Discussion. Geol. Soc. Am. Bull. 85:159-160.

Norton, S.A. and Hess, C.T., 1980, Atmospheric deposition in Norway during the last 300 years as recorded in SNSF lake sediments: 1. Sediment dating and chemical stratigraphy, pp. 268 to 269. In: Ecological Impact of Acid Precipitation, D. Drabløs and A. Tollan, eds., Proceedings International Conference, Sandefjord, Norway. SNSF Project, Ås-NLH, Norway.

Norton, S.A. and Kahl, J.S., 1983, Deposition of atmospheric Pb and Zn in New England lake sediment, pp. 128 to 131. In: Heavy Metals Environ., Proc. Intern. Conf., Vol. 1, Heidelberg, Federal Republic of Germany.

Pfeiffer, M.H. and Festa, P.J., 1980, Acidity status of lakes in the Adirondack region of New York in relation to fish resources. Dept. Environ. Conservation, Albany, NY. 36 pp.

Schofield, C.L., personal communication, Cornell University, Ithaca, NY.

Schofield, C.L., 1977, Acidification of Adirondack Lakes by atmospheric precipitation: Extent and magnitude of the problem. Final Rept. for Project F-28-R, U.S.F.W.S., Albany, NY. 11 pp.

TRANSFORMATION OF NITRIC, SULFURIC, AND ORGANIC ACIDS ON THE

BICKFORD RESERVOIR WATERSHED

Harold F. Hemond and Keith N. Eshleman

Department of Civil Engineering and M.I.T. Energy
Laboratory, Massachusetts Institute of Technology
Cambridge, MA   02139

ABSTRACT

The acid-base balance of Bickford Reservoir watershed in central Massachusetts is dominated by atmospheric mineral acid additions, mineral weathering, sulfur and nitrogen transport and transformation, and endogenous organic acids. Bickford surface water nitrate concentrations rarely exceed 1-2 $\mu eq\ l^{-1}$; annual neutralization of nitric acid in precipitation exceeds 97%, and no nitric acid peak has been observed. In contrast, the annual watershed sulfate budget indicates that twice as much sulfate leaves in streamwater than is measured in bulk collectors; the Bickford sulfate deficit of 608 $eq\ ha^{-1}\ yr^{-1}$ is higher than any value found in the literature. Charge balance analysis of water samples has inferred the presence and dominance of organic acids (up to 100 $\mu eq\ l^{-1}$) in one Bickford tributary during the growing season. Long-term assessment of the effects of acid deposition on surface water alkalinity should consider, in addition to mineral weathering, possible biotic nitrogen and sulfur transformations and organic acid production within a receptor watershed.

INTRODUCTION

A common approach to the modeling of ecosystem response to acid deposition is to consider that mineral acidity is neutralized by the addition of basic cations derived from minerals or cation exchange. Models, based on maps, equations or nomograms, incorporate sensitivity indices based on watershed bedrock mineralogy, surficial mineralogy, soil cation exchange capacity, or carbonate and exchangeable base content (Henriksen, 1979; Cowell et al.,

1980; McFee, 1980). However, because anions were not considered in weathering models, they cannot predict the effects of anion trans- formations on the sensitivity of a system to nitric, sulfuric, and other acidic additions. Only a few relatively complex numerical simulation models (Booty, 1980; Christophersen et al., 1982; Chen et al., 1983) consider acid neutralization through the removal of acidic anions. Data from Bickford watershed showed, however, that the biogeochemistry of anions within a watershed may be as impor- tant as the biogeochemistry of basic cations in determining how sensitive an ecosystem is to deposition of various acids.

STUDY SITE

    Bickford Reservoir is a small, softwater water supply reser- voir located near Fitchburg in central Massachusetts (latitude 42°29'07" N and longitude 71°55'57"W). The watershed is underlain by granites and schists and covered with a variable thickness of glacial till. The reservoir has an area of 65 ha, a mean depth of 6 m, and a watershed area of about 1800 hectares (Figure 1). The acid-base balance of the watershed is largely determined by atmos- pheric acid sources, mineral weathering, sulfur and nitrogen transformations, and endogenous organic acids. The Bickford watershed receives bulk atmospheric deposition containing approxi- mately 20 µeq l^{-1} of nitrate and 40 µeq l^{-1} of sulfate, with a pH typically about 4.2. Ionic chemistry of surface waters from the Bickford watershed varies seasonally, with alkalinities and pH values lowest in early spring. Annual weighted mean water chem- istry data from the Bickford Reservoir spillway and two inflowing brooks are shown in Table 1.

METHODS

    Concentrations of all major inorganic ions were measured in samples from the reservoir and tributaries at biweekly intervals beginning July 1981 and ending June 1983. In addition, nitrate, chloride, and sulfate were analyzed by ion chromatography, ammonium by phenol-hypochlorite, alkalinity by acidimetric titration using Gran plot end-point determinations, and major metals by atomic absorption or emission spectrophotometry. Anion and cation balances for reservoir surface water were consistently within 10% of the charge balance, but significant cation excesses in tribu- taries at certain times of the year are interpreted as due to unmeasured organic anions. Streams were gaged using Stevens type F stage recorders in heated stilling wells. Rating curves were established using a Marsh-McBirney electromagnetic current meter. Groundwater inflow was estimated by difference. The standard error of individual streamflow measurements was three percent for flows less than or equal to the maximum discharge for which current meter

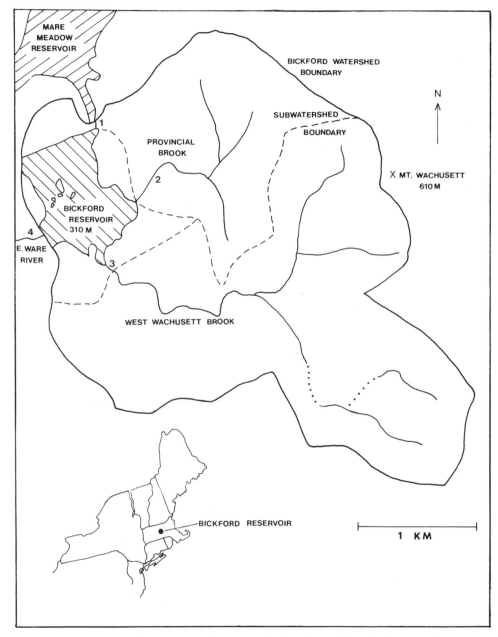

Figure 1. Location map and watershed of Bickford Reservoir near Fitchburg, Massachusetts

Table 1. Major ion chemistry of Bickford surface waters (μeq l^{-1});
          data are given as weighted means for the July 1981 to
          June 1983 period

Ion	Bickford Reservoir	West Wachusett Brook	Provencial Brook
Ca^{2+}	111.6	87.8	95.0
Mg^{2+}	36.5	33.9	31.4
K^+	14.4	10.5	11.8
Na^+	123.3	106.5	83.9
NH_4^+	0.8	0.6	0.4
Cl^-	114.7	97.9	64.9
SO_4^{2-}	145.6	145.8	118.7
NO_3^-	3.5	1.2	0.8
Titration Alkalinity	9.9	-14.2	1.0

calibration of the gages was performed. Over two years, surface inflow contributed about 92% of the water budget for the reservoir. Direct precipitation contributed 6% of the water input, and the groundwater contribution was 2% ± 1% of the total budget. A detailed discussion of the water budgets of the reservoir and watershed is presented in Eshleman and Hemond (1983).

Bulk precipitation (predominantly wet) chemistry was obtained from samples collected in 15 cm diameter glass funnel collectors with vapor traps (Likens et al., 1977) or in polyethylene buckets during winter months. Mass balances for the reservoir and watershed were computed using daily streamflows and ionic concentrations obtained by linear interpolation between measurements. Possible bias in chemical flux calculations due to systematic relationships between chemistry and streamflow is believed to be small. In sequential sampling of storm runoff, streamflow chemistry was observed to vary only moderately (factor of less than two) while streamflows varied over several orders of magnitude. Major ion concentrations did not change by more than a factor of two during any week-long period of intensive (hourly to daily) sampling, even during weeks when major storms occurred. Thus, it is felt that the major chemical fluxes are not in substantial error. The calculated chloride mass balance for the reservoir was within ±1% for the two years of sampling, which is further evidence that chemical and water budgets are not substantially in error.

RESULTS AND DISCUSSION

The annual and monthly major ion (and alkalinity) budgets for the reservoir and watershed were used to assess the relative importance of various geochemical and biochemical processes on surface water alkalinity. Because the titration alkalinity flux is equivalent to the strong base less strong acid flux, the role of each major anion in the alkalinity budget could be quantified.

Nitrate

   Biogeochemical mass balances for the reservoir and for the watersheds of the tributary streams (Table 2) indicated that removal of nitrate on the watershed was nearly complete throughout the two-year period. Nitric acid was consumed with high efficiency even during those months when plant uptake was believed to be at a minimum. The lowest monthly removal efficiency found in 24 months of data was 83% in February of 1983. Nitrate retention produced at least 250 eq ha^{-1} yr^{-1} of alkalinity, second in importance only to weathering and exchange of calcium ion, which generated 665 eq ha^{-1} yr^{-1} of alkalinity. Neglecting dry nitrate deposition, removal of nitrate from bulk deposition contributed 20% of the total watershed alkalinity production; the true contribution was certainly larger because dry nitrate deposition was underestimated by bulk collectors. Nitric acid removal occured throughout the winter and spring, during which times sulfuric acid output from the Bickford watershed exceeds input and caused a decline in pH and alkalinity of the upland tributaries. Because sulfate levels were particularly high during the spring, and streams were consequently more acidic, the absence of a springtime nitrate peak at Bickford was of considerable significance. Its absence greatly mitigated the spring acid pulse which might otherwise have been more severe.

   The nitrate levels observed at Bickford were exceptionally low compared to observations at other northeastern sites (Figure 2). The low nitrate levels in surface waters imply nearly complete neutralization of nitric acid deposition. From March through May, Bickford tributary waters had a weighted nitrate concentration of less than 1 µeq l^{-1}. By contrast, stream waters at the Hubbard Brook Experimental Forest (HBEF) in New Hampshire during the spring period were characterized by long term average nitrate concentrations on the order of 40 µeq l^{-1} (Likens et al., 1977). Corresponding differences in pH between the two sites (4.8 at Bickford

Table 2. Annual mass balances for nitrate at different temperate zone watersheds

System	Input (eq ha[-1])	Output (eq ha[-1])	Net Flowthrough (Losses in %)
Bickford Watershed			
W. Wachusett Brook	320	9.9	3
Provencial Brook	320	8.5	3
Reservoir	825	582	71
Hubbard Brook NH, USA[a]	316	277	88
Adirondack Park, NY, USA[b]			
Sagamore Lake	390	210	54
Woods Lake	475	160	66
Panther Lake	460	370	20
Three Swedish basins[c]	118	10.6	9
Birkenes, Norway[d]	530	80	15
Kejimkujik Lake, N.S., Can.[e]	349	5	1
Harp Lake, Ontario, Can.[f]	370	47	13

[a]Likens et al., 1977
[b]Calculated from data in Figure 2 of Galloway et al., 1980
[c]Andersson-Calles and Eriksson, 1979
[d]Wright and Johannessen, 1980
[e]Kerekes, 1980
[f]Dillon et al., 1982

versus 4.5 at HBEF) can be more than accounted for by the differences in nitrate concentration.

Sulfate

In contrast to the quantitative removal of nitric acid by the Bickford watershed at all times of the year, transformation of sulfuric acid appeared significantly more complex. Bickford Reservoir and watershed received an average annual sulfate load of 579 eq ha[-1] yr[-1] from bulk deposition during the two years reported here. The mean sulfate concentration in bulk precipitation was 40.3 μeq l[-1].

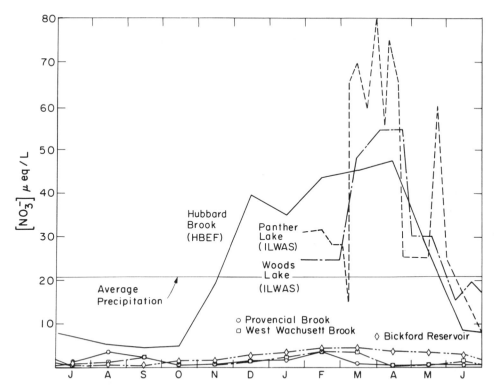

Figure 2.   Nitrate concentrations in surface waters at Bickford
            compared with precipitation and with other experimental
            watersheds in the northeastern United States

The annual measured loss of sulfate from the Bickford watershed was 1186 eq ha^{-1} yr^{-1} during this period, indicating a net
source of 608 eq ha^{-1} yr^{-1} of sulfate. Positive net watershed
output of sulfate did not occur continuously throughout the year
(Figure 3). Apparent watershed consumption of sulfate occurred
during the summer and fall, but apparent production began in
November and continued throughout the winter and spring. The same
pattern was repeated during both years of the study.

This "crossover pattern" of apparent sulfate production during
part of the year and apparent consumption at other times was
observed on other watersheds (Likens et al., 1977; Christophersen
and Wright, 1981). Christophersen and Wright (op.cit.) attributed
the crossover sulfate pattern at Birkenes, Norway to hydrologic and
meteorological characteristics of the temperate zone, with accumu-

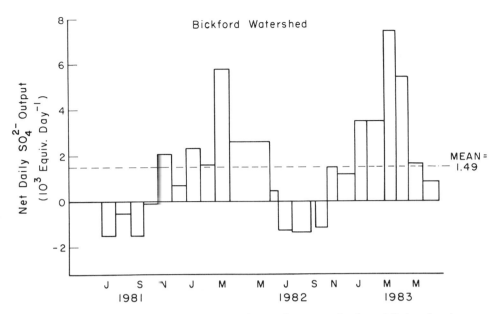

Figure 3.   Net apparent production of watershed sulfate during
two years at Bickford Reservoirs.   Dry deposition in
excess of what was collected in a bulk collector was
not included in the sulfate budget calculations

lation in the winter due to snowpack formation, and net accumulation in summer by evapotranspiration.   However, the Bickford and
HBEF watersheds did not accumulate sulfate in winter, despite the
formation of snowpacks at these sites.

Sulfate concentrations in streamwater at Bickford were not
entirely explained by water balance and snowpack arguments (Figure 4).   Sulfate concentrations in streamwater were relatively low
during the summer (120 µeq l^{-1}) when evaporation would be expected
to increase their concentrations.   Sulfate levels increased during
the fall, reaching a peak of 150-175 µeq l^{-1} during the winter
months.

The annual average sulfate budget for the Bickford watershed
was similar to sulfate budgets determined for other temperate zone
forested watersheds Table 3).   Eight of the ten sulfate budgets in
Table 3 show the watersheds to be apparent sources of sulfate; one
watershed (Woods Lake) appeared as a net sink, while sulfate was
apparently conservative at Rawson Lake.   All budgets in Table 3
were calculated from bulk (predominantly wet) atmospheric sulfate

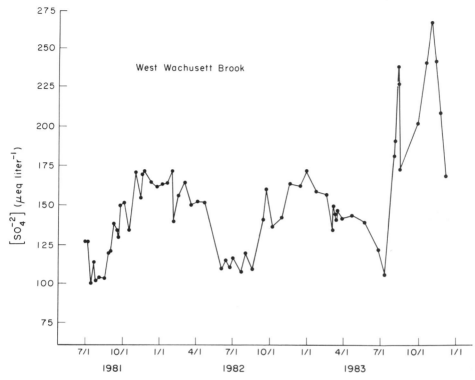

Figure 4.   Seasonal pattern of sulfate concentrations in Bickford
             tributary streamwater (West Wachusett Brook) during the
             thirty-month sampling period.

inputs and sulfate outputs from surface water runoff; estimates of
dry sulfate deposition are excluded.

     Comparisons of the net output to input ratios for the ten
watersheds in Table 3 indicated that annual net watershed sulfate
losses varied between 0 and 350 percent of bulk sulfate additions.
The highest ratios were found for the Sudbury, Ontario, watersheds
(Dillon et al., 1982); these were all within 50 km of the Inco.
Ltd. smelter at Copper Cliff.   The lowest ratios corresponded to
the Adirondack sites (Galloway et al., 1980) and the Rawson Lake
watershed in rural Ontario, Canada.   Intermediate ratios were found
for Bickford, Birkenes, and HBEF watersheds.

     A comparison of the calculated sulfate deficits (which equal
the magnitude of unknown sources) showed that the Bickford
watershed has the highest value (608 eq ha^{-1} yr^{-1}).   Data from
Dillon et al. (1982) may be biased on the low side, since their
winter/spring values of the third year were not included in their

Table 3.   Watershed sulfate budgets for various temperate zone
           watersheds

Watershed Site/Location	Dates	Sulfate Deficit (eq ha^{-1} yr^{-1})[1]	Output/Input[1]
Bickford watershed, MA, U.S.A.	1981-83	608	2.0
Hubbard Brook, NH, U.S.A.[a]	1963-74	319	1.4
Birkenes, Norway[b]	1972-78	606	1.6
Adirondack Park, NY U.S.A.[c]			
Sagamore Lake	1978-79	290	1.3
Panther Lake	1978-79	400	1.4
Woods Lake	1976-79	-40	1.0
Within 50 km of Inco Smelter[d] Sudbury, Ontario, Canada			
Middle Lake	1977-79	537	3.5
Clearwater Lake	1977-79	271	2.4
Nelson Lake	1977-79	291	2.6
Rawson Lake[e] Ontario, Canada	1972	9	1.0

[a]Likens et al., 1977
[b]Christophersen and Wright, 1981
[c]Estimated from Figures 2 and 3 in Galloway et al., 1980
[d]Dillon et al., 1982
[e]Schindler et al., 1976

2.5-year study. The apparent excess of output over input at Bickford and the other watersheds could be attributed to weathering of sulfur containing minerals on the watershed, or to some unmeasured atmospheric sulfate source. Customarily, researchers assumed that weathering of sulfur bearing minerals on these watersheds is small relative to total deposition (Likens et. al, 1977); if no other sources or sinks exist, the mass balances can be used as estimates of unmeasured dry sulfur deposition. However, sulfur is a macronutrient for higher plants, and sulfate is reduced to sulfide by sulfate-reducing bacteria under anaerobic conditions (e.g., wetland soils). Accordingly, the net annual sulfur budget may reflect an algebraic sum of weathering and dry deposition sources plus sinks associated with uptake and sulfate reduction.

Therefore, the relative importance of dry sulfur deposition to forested watersheds, the rate of weathering of sulfidic minerals, and the rates of assimilatory and dissimilatory sulfate reduction cannot be accurately stated without further study of each of these processes.

Organic Acids

Analysis of charge balance data from the Bickford watershed indicated that unmeasured chemical species seasonally represented a major component of the ionic charge of tributary waters. The sum of measured cation concentrations, especially in the summer and fall in Provencial Brook, significantly exceeded the sum of inorganic anions (Figure 5). The waters were, of course, electroneutral, so the only possibilities were the presence of unmeasured negatively charged anions or an overestimation of positively charged species. The unmeasured species are believed to be organic acids. Several lines of reasoning substantiate this conclusion.

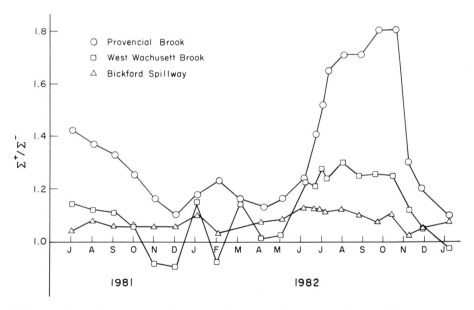

Figure 5.   Seasonal pattern of charge balance in surface waters at Bickford Reservoir. Peak of 1.8 for cation to anion concentration ratios observed for Provencial Brook corresponded to 100 μeq l^{-1} of anions

Exhaustive checks of the analytical procedures, including comparisons of cation analysis by ion chromatography and atomic absorption
spectrophotometry, indicated the presence of unmeasured negatively
charged species. The magnitude of this unmeasured charge appeared
to be correlated with the visual color of the waters. In addition,
irradiation by ultraviolet light (Cronan et. al, 1978) under
aerated conditions both destroyed the color and brought the
measured charge balance to a 1-to-1 ratio between cations and
anions.

Organic species actually dominated the charge balance of
Provencial Brook during the summer; as much as 100 µeq l^{-1} of
organic acids have been inferred (30% of the total charge). It is
estimated that on an annual basis, organic acid production neutralized as much as 20 of the total alkalinity generated by the
Provencial Brook subwatershed. An appreciable fraction of the
acidic functional groups associated with these organic acids have
apparent pKa's near or below 3.5 (the end point used in the Gran
plot titrations) because the titration alkalinity increased

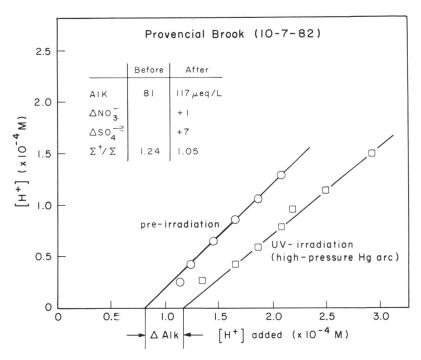

Figure 6.   Gran plot titrations of a tributary water before and
            after UV irradiation showing the possible presence of
            labile organic acids having pKa's sufficiently low as
            not to be completely titrated at pH 3.5

following the presumed oxidation of these organic compounds by ultraviolet light (Figure 6). The acidity due to natural organic acids thus appeared to significantly affect the pH of tributary waters to Bickford on a seasonal basis. Because the highest concentrations of organic acids were present during the periods of low flow, effects on the reservoir were somewhat mitigated. However, the fate of these organic acids (and, for that matter, their precise source) is not known. It is suspected that these acids may be destroyed by chemical and biotic oxidation processes in the water column as water moves through the reservoir. Nevertheless, the acids do substantially affect the chemical environment of the tributaries. More work is needed to elucidate the sources, fate, and significance of natural dissolved organic material, especially in soft, dilute waters.

CONCLUSIONS

    Based on the concepts of mass balance and alkalinity of solutions, the following conclusions have been drawn concerning the biogeochemistry of acidic anions at Bickford watershed:

º   Nitric acid removal provides at least 20% of the total alkalinity produced on an annual basis at Bickford. Nearly complete removal by the watershed underscores the fact that models of surface water alkalinity need to consider nitrogen biogeochemistry.

º   The Bickford annual sulfate budget demonstrates a need to better understand the relative roles of wet and dry sulfur deposition, sulfide weathering, and watershed sulfate reduction. While the data indicate an appreciable dry sulfur source, a biotic or abiotic removal process may be operative on a seasonal basis.

º   In a largely neutral system such as Bickford (where acid inputs are almost exactly titrated by watershed reations), organic anions can become a significant source of acidity. This is the case in one tributary which is dominated by organic acids during the summer. On an annual basis, the quantity of organic acids produced actually exceeds the total amount of nitric acid deposition to the subwatershed.

Acknowledgments - The authors gratefully acknowledge the support of Consolidated Edison Co. of New York, New England Electric Power Service Co., American Electric Power Co., and Northeast Utilities Service Co. in the conduct of this research.

REFERENCES

Andersson-Calles, U.M. and Eriksson, E., 1979, Mass balance of
    dissolved inorganic substances in three representative basins
    in Sweden. Nordic Hydrology 10:99-114.
Booty, W.G., 1980, Watershed Acidification Model. Environmental
    Geochemistry Report 1980/4, Dept. of Geology, McMaster Univer-
    sity, Hamilton, Ontario, Canada. 35 pp.
Chen, C.W., Gherini, S.A., Dean, J.D., and Hudson, R.J.M., 1983,
    Overview of the integrated lake-watershed acidification study
    (ILWAS), pp. 1-1 to 1-31. In: Proceedings of the ILWAS
    Annual Review Conference. Report No. EA-2827, Project 1109-5,
    Electric Power Research Institute, Washington, DC.
Christophersen, N. and Wright, R.F., 1981, Sulfate budget and a
    model for sulfate concentrations in stream water at Birkenes,
    a small forested catchment in southernmost Norway. Water
    Resources Res. 17:377-389.
Christophersen, N., Seip, H.M., and Wright, R.F., 1982, A model
    for streamwater chemistry at Birkenes, Norway. Water
    Resources Res. 18:977-996.
Cowell, D.W., Lucas, A.E., and Rubec, C.D.A., 1980, The develop-
    ment of an ecological sensitivity rating for acid precipita-
    tion impact assessment. Working Paper No. 10, Lands Direc-
    torate, Environment Canada, Ottawa, Canada. 42 pp.
Cronan, C.S., Reiners, W.A., Reynolds, R.C., and Lang, G.E., 1978,
    Forest floor leaching: Contributions from mineral organic and
    carbonic acids in New Hampshire Subalpine forests. Science
    200:309-311.
Dillon, P.J., Jeffries, D.S., and Scheider, W.A., 1982, The use of
    calibrated lakes and watersheds for estimating atmospheric
    deposition near a large point source. Water, Air, Soil Pollu.
    18:241-258.
Eshleman, K.N. and Hemond, H.F., 1983, Acid deposition effects on
    a Massachusetts watershed. I. Water Balance. Unpublished
    manuscript, Massachusetts Institute of Technology, Cambridge,
    MA. 28 pp.
Galloway, J.N., Schofield, C.L., Hendrey, G.R., Altwicker, E.R.,
    and Troutman, D.E., 1980, An analysis of lake acidification
    using annual budgets, pp. 254 to 255. In: Ecological Impact
    of Acid Precipitation, D. Drabløs and A. Tollan, eds.,
    Proceedings International Conference, Sandefjord, Norway.
    SNSF Project, Ås-NLH, Norway.
Henriksen, A., 1979, A simple approach for identifying and meas-
    uring acidification of freshwater. Nature 278:542-545.
Kerekes, J.J., 1980, Preliminary characterization of three lake
    basins sensitive to acid precipitation in Nova Scotia, Canada,
    pp. 232 to 233. In: Ecological Impact of Acid Precipitation,
    D. Drabløs and A. Tollan, eds., Proceedings International
    Conference, Sandefjord, Norway. SNSF Project, Ås-NLH,
    Norway.

Likens, G.E., Bormann, F.H., Pierce, R.S., Eaton, J.S., and John-
    son, N.M., 1977, Biogeochemistry of a Forested Ecosystem.
    Springer-Verlag, New York, NY. 146 pp.
McFee, W.W., 1980, Sensitivity of soil to acid precipitation.
    U.S.E.P.A. 600/3-80-013, Environment Research Laboratory, Cor-
    vallis, OR. 186 pp.
Schindler, D.W., Newbury, R.W., Beaty, K.G., and Campbell, P.,
    1976, Natural water and chemical budgets for a small Precam-
    brian lake basin in central Canada. J. Fish. Res. Board Can.
    33:2526-2543.
Wright, R.W. and Johannessen, M., 1980, Input-output budgets of
    major ions at gauged catchments in Norway, pp. 250 to 253.
    In: Ecological Impact of Acid Precipitation, D. Drabløs and
    A. Tollan, eds., Proceedings International Conference,
    Sandefjord, Norway. SNSF Project, Ås-NLH, Norway.

THE EFFECTS OF ACID PRECIPITATION ON GROUND WATER QUALITY

David M. Nielsen[a], Gillian L. Yeates[a] and
Catherine M. Ferry[b]

[a]IEP Inc., Consulting Environmental Scientists
Worthington, OH and
[b]Battelle Memorial Institute
Columbus, OH

ABSTRACT

The phenomenon of acid precipitation detrimentally affects aquatic ecosystems to a presently unknown degree. Complex geochemical processes in soils are important factors in determining the quality of percolate which infiltrates down to ground water systems. Elevated base cation and trace metal concentrations in ground water may result in corrosive or toxic water quality conditions. Subsequent discharge of affected ground water to surface water bodies results in detrimental effects to water quality and ambient aquatic life.

INTRODUCTION

The phenomenon of excesses of strong acids in precipitation over several regions of the United States, Canada, Sweden, and Norway is well documented in the literature. Whereas pure rain-water in equilibrium with atmospheric CO_2 should have a pH value of between 5.6 and 5.7, many individual precipitation events in these areas have a measured pH of less than 4.0. In the northeastern United States, the annual acidity value averages a pH of about 4.0, while pH values as low as 2.1 have been recorded for individual precipitation events (Likens, 1972; Cogbill and Likens, 1974; Likens and Borman, 1974; Likens, 1976). More recent information indicates that in the southeastern and far western regions of the United States, pH values of between 3.0 and 4.0 are routinely observed during individual events.

The occurrence of acids in precipitation is linked to increasing levels of anthropogenic gaseous atmospheric pollutants, particularly sulfur and nitrogen oxides. Sulfur dioxide, the most common oxide of sulfur generated by man, can be converted chemically in the atmosphere to sulfuric acid; nitrogen oxides, formed primarily during the combustion of fossil fuels, can be converted to nitric acid. Subsequent to being formed in the atmosphere, these acids fall to the earth in the form of wet and dry acid deposition, commonly referred to as acid precipitation.

As attention becomes increasingly focused on the occurrence of acid precipitation, the question of what impact this phenomenon has on the environment needs to be addressed. A growing body of evidence suggests that acid precipitation may be responsible for substantial adverse effects (Figure 1). Large portions of the eastern United States (particularly the Northeast) have been identified as being sensitive to the long-term environmental effects of acid precipitation (Glass et al., 1984). This has been determined through information on soils (McFee, 1980), bedrock geology (Kramer, 1978; Norton, 1980) and surface water (Norton et al., 1980). These effects may not only be the result of a cumulative build-up from years of exposure to acid precipitation but may also result from peak activity episodes. Some of the environmental consequences of the massive deposition of acid rain which can be quantified on the basis of extensive studies include:

o   deterioration of man-made materials such as statues and monuments, stone facings on buildings, metal structures, and painted surfaces on homes and automobiles;

o   possible reductions in forest productivity, damage to agricultural crops, and increased stress on other vegetative systems;

o   acidification and demineralization of soils and resultant reductions in soil fertility and possibly irreversible changes in soil geochemistry;

o   mobilization and leaching of toxic heavy metals and other cations from soils into ground and surface waters and resultant contamination of drinking water supplies; and

o   acidification of fresh water lakes and reservoirs and resultant damage to fish and other aquatic organisms.

Acidification of soils and their resultant demineralization as well as leaching and mobilization of heavy metals are effects which represent direct avenues for potential ground water contamination. Consequently, these areas are discussed in greater detail with regard to ground water quality and subsequent impacts on surface waters.

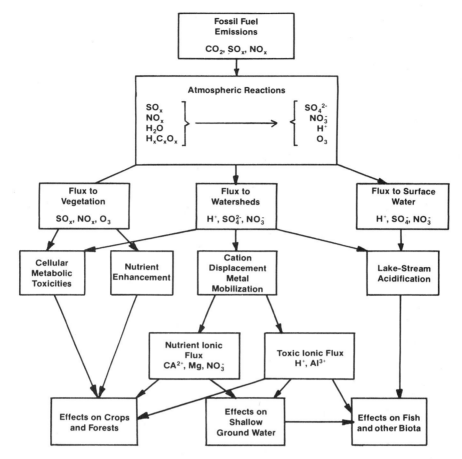

Figure 1. Effects of acid precipitation on the environment.

EFFECTS ON SOILS

Acid precipitation is documented to have profound effects on the chemistry of soils and the water which percolates through soils (Wiklander, 1974; Norton, 1975; Hutchinson, 1980; McFee, 1980; Wiklander, 1980). The introduction of acid precipitation to soils may acidify the soils and accelerate the natural weathering process resulting in increased leaching of essential soil nutrients (base cations) and enhanced mobility of trace metals. Water that infiltrates through soils transports base cations and trace metals to ground water and results in their increased concentrations.

Soils with calcium or other carbonates are not readily susceptible to acidification because buffering by carbonates keeps the soil pH, as well as the base saturation percentage (i.e., the total content of basic cations), relatively constant. Norton (1975) estimated that it may take as many as several hundred years before any effects from acid precipitation may be noted in calcareous soils. For noncalcareous soils, the soil texture and amount of organic debris affect the degree of acidification that occurs. A greater amount of silicate clay and organic matter gives soils a higher cation exchange capacity and, hence, a higher buffering capacity.

According to Gorham and McFee (1980), topography has a strong influence on soil acidification. Thin soil layers and greater precipitation typical of the headlands of lakes and streams render these areas susceptible to the effects of acid precipitation. At lower places in the watershed, thicker soils provide a higher chance of alkaline additions to ground water systems. This gradient, from upland soils which are rather thin and exhibit lower pH to the lowland soils which are thicker and have higher pH values, is known as a soil catena (Krug and Frink, 1983).

The cation exchange capacity of a soil describes the ability of the soil to store reserves of nutrient cations. The presence of negatively charged sites on colloidal clay and humus particles is the basis for cation exchange. The amount of this excess negative charge depends on the silicate clay mineral content, soil pH, and the amount of organic matter found in the soil. The natural tendency toward electrical neutrality causes adsorption of cations by clay particles found in the soil. The adsorbed cations create a positive ionic aura around the negative sites, and cation exchange equilibria are established between the solution of percolating water with soil particles and their charged surfaces. In noncalcareous soils, the cation exchange process results in increased leaching of bivalent and univalent cations (Ca^{2+}, Mg^{2+}, K^+, Na^+) from the soil, with a corresponding decrease in what is normally an already low base saturation percentage (Table 1).

The concentration of hydronium ions (H^+ or H_3O^+) contributed by acid precipitation is the acidifying factor in soils. As concentrations of H^+ increase, competition between hydronium ions and base cations for the negatively charged exchange sites on soil particles increases according to the law of mass action. Bache (1980) indicated that leaching of base cations results from three possible geochemical processes (Ca^{2+} is usually the main cation lost, as it is the predominant cation in soils, but Mg^{2+}, Al^+, K^+, and Na^+ undergo similar reactions):

Solution of free carbonates

$$H_3O^+ + CaCO_3 \longrightarrow Ca^{2+} + HCO_3^- + H_2O \qquad (1)$$

Table 1.   Sensitivity[1] of soils to acid precipitation (adapted from
           Glass et al., 1982).

	Nonsensitive	Slightly sensitive	Sensitive
Cation exchange capacity (CEC) meq/100 g	Any value	>15.4 $6.2 \leq CEC \leq 15.4$	<6.2
Other relevant conditions	Free carbonates present or subject to frequent flooding	Free carbonates absent; not subject to frequent flooding	Free carbonates absent; not subject to frequent flooding

[1]Sensitivity is predicted from the chemical characteristics of the top 25 cm of soil and cation input.

Replacement of base cations by H^+ at weak acid exchange sites

$$(R-O)_2Ca + 2H^+ \longrightarrow 2ROH + Ca^{2+} \tag{2}$$

Replacement of base cations at permanent exchange sites

$$\begin{smallmatrix}-O\diagdown\\-O\diagup\end{smallmatrix}Al-OH + H_3O^+ \longrightarrow \begin{smallmatrix}-O\diagdown\\-O\diagup\end{smallmatrix}Al^+ + 2H_2O \tag{3}$$

     The displacement of cations (equation 3) at permanent exchange
sites is not likely to occur to any great extent because the
exchange of aluminum ions is a more naturally preferred process.
Thus, absorption of H^+ and desorption of base cations (decrease of
base saturation) occurs.   Reuss (1978), in modeling ion loss from
soils as a function of soil properties and distribution of
rainfall, determined that significant acidification and depletion
of bases could occur over a period of several decades if rainfall
is consistently acidified to a pH of 4.0 (with sulfuric acid
predominating).   Wiklander (1974) noted that the soils that
released the most basic cations in response to acid precipitation
were the already slightly acid soils, i.e., podzols, which already
have considerable cation exchange capacity.   Experimental leaching
of a podzolic soil in field lysimeters by simulated rain acidified
with sulfuric acid (Abrahamsen et al., 1976) revealed a distinct
increase in losses of calcium, magnesium, potassium, and aluminum
from the soil with each unit decrease in pH.

     In addition to accelerating the natural tendency of soils to
lose basic cations, acid precipitation accelerates the loss of

trace metals from soils. Norton et al. (1980) pointed out that levels of trace metals, particularly heavy metals, are intimately affected by acidic precipitation. Their mobility in many cases is greatly altered. One of the most dramatic effects of acid precipitation on soil is greatly increased aluminum mobilization. The mobilization of aluminum ions in soil occurs in the pH range of 3.5 to 5.5, and is the main soil buffering process in this range. The geochemical buffering action which occurs is (Bache, 1980):

$$Al^{3+} \longleftrightarrow Al(OH)^{2+} \longleftrightarrow Al_n OH_m^{(3n-m)+} \longleftrightarrow Al(OH) \longleftrightarrow Al(OH)_4^- \quad (4)$$

pH: 3.5   5   6.5   8

As soil pH drops, production of soluble aluminum ions causes the displacement of more calcium. This is the reason why an acid mineral soil is sometimes termed an "aluminum soil." Probably the most important soil acidifying process is when Al^{3+} is exchanged for Ca^{2+} and Mg^{2+} cations in subsoils (Bache, 1980):

$$3 \text{ Soil } (Ca,Mg) + 2 Al^{3+} \longleftrightarrow 3 (Ca^{2+}, Mg^{2+}) + 2 \text{ Soil Al} \quad (5)$$

Gorham and McFee (1980) noted that elevated aluminum concentrations are not common in precipitation and concluded that high concentrations of aluminum found in strongly acidified southern Norwegian lakes must be due to the weathering of soil aluminosilicates and the subsequent displacement of aluminum ions in solution as a result of cation exchange processes. Soil chemistry data from the White Mountains region in New Hampshire show that aluminum concentrations in soils are abnormally high and continue to increase with depth (Cronan and Schofield, 1979). These authors determined that aluminum is the first or second most abundant cation, on a charge-equivalent basis, throughout the soil solution of the entire soil profile and that the high concentrations of dissolved aluminum are accounted for entirely by the low pH of the sulfuric acid-dominated soil solution. Norton (1975) noted that within the normal range of acidity of soils of humid temperate regions (which includes much of the eastern United States), the solubility of aluminum increases dramatically below pH 5.0. As most podzolic soils contain aluminum which can be removed by acid leaching (Abrahamsen et al., 1976), the potential for environmental damage from continued acid deposition is considerable.

If aluminum is mobilized in acidic soil environments, other metals with pH-sensitive solubility will also be mobilized; these are particularly pronounced in podzolic soils. Both experimental and field studies suggest that manganese, for example, is leached from the A horizon of soils (Hanson, 1980). It has also been demonstrated that leafy plants (such as lettuce) grown in soils exposed to acid precipitation experienced increased uptake of cadmium, indicating enhanced mobility of that element (Abrahamsen et al., 1976). The retention time of heavy metals in soils under various acidic precipitation regimes was considered in artificial leaching experiments conducted by Tyler (1978). The metals considered in this study -- lead, zinc, and manganese -- differed markedly in their retention times, but these differences decreased dramatically with increasing acidity of the percolate. Lead showed a 10 percent retention time of 70-90 years when leached with an acid solution of pH 4.2, while a stronger acid leach of pH 2.8 decreased the 10 percent retention time to only 20 years. Zinc and manganese showed even more highly reduced retention times compared with lead.

A number of studies correctly point out the fact that even normal precipitation, which is a weak solution of carbonic acid, causes soil acidification and some leaching of basic cations, irrespective of soil pH. However, while acidification and base leaching from soils are, to an extent, naturally occurring processes, a dramatic increase in rainfall acidity can be expected to accelerate the rates of these processes. The extent of this problem is currently unknown. Because natural soil acidification, though very slow, is generally accompanied by deleterious side effects, any acceleration of the process is viewed as undesirable.

EFFECTS ON GROUND WATER

Precipitation is introduced to the ground water system as a result of infiltration through soils, a phenomenon known as recharge. In the hydrologic cycle (Figure 2),

$$\text{Ground Water Recharge} = \text{Precipitation} - (\text{Evapotranspiration} + \text{Runoff}).$$

This water balance equation demonstrates how the ground water system is replenished. Any precipitation which reaches the earth may evaporate to the atmosphere, transpire to the atmosphere through plant metabolism, travel overland (runoff) to a surface water body, or percolate through soils to become ground water.

In areas sensitive to acid precipitation, especially where hard crystalline bedrock and thin or non-buffered soils occur, the chemistry of ground water should strongly reflect the chemistry of atmospheric precipitation. This has been demonstrated by Pearson and Fisher (1971) who found that the chemistry of ground water in the shallow sand and gravel aquifer on Long Island, New York, is largely determined by the chemistry of precipitation. On the other hand, where the soil mantle is thicker its texture and mineralogy become more significant, and ground water chemistry, partly because of a longer residence time and a longer time for reactivity, should be strongly influenced by soil chemistry.

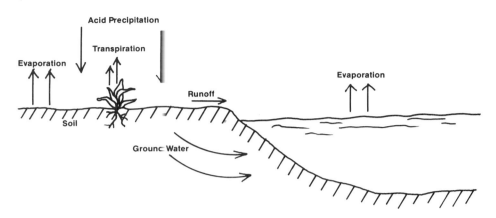

Figure 2. Acid precipitation in the hydrologic cycle.

        A study of acidified ground water in southwestern Sweden
(Hultberg and Wenblad, 1980) yielded pH data depicting trends in
the effects of acid precipitation on ground water (Figures 3-5).
An initial regional well water quality survey was performed on both
private shallow dug wells (1-10 m deep) and drilled wells (20-50 m
deep).   This survey was done in areas susceptible to acidification
because of the presence of shallow crystalline bedrock near acid
lakes.   The results of this preliminary study showed that more than
half of the wells surveyed in some areas had a pH of less than 6.0.
The effect of soil type on ground water acidification was apparent;
ground water underlying sandy soils incapable of neutralizing
acidic recharge yielded lower pH (median pH = 5.1) than the ground
water underlying soils with a high clay content (median pH = 6.1).
Long-term effects noted in this study include decreasing
bicarbonate (HCO_3^-) concentrations, shown as:

$$CaCO_3 + H_2CO_3 \longrightarrow Ca^{2+} + 2HCO_3^- \tag{6}$$

Hultberg and Wenblad (1980) concluded that in 10 to 20 years the
alkalinity (in the form of bicarbonate), which acts as a buffer to
this ground water, will be reduced to nearly zero and the pH will
drop dramatically.   Another study of ground water acidification in
Norway (Henriksen and Kirkhusmo, 1982) showed that ground water
found in some sensitive areas is in the pH range of 5.0-5.5 and
that the soil in these areas has exhausted almost all of its
bicarbonate buffering capacity.

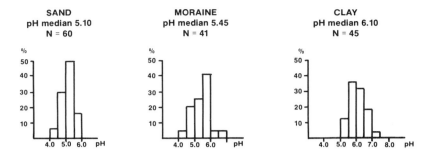

Figure 3.   Effects of overlying soil types on ground water pH
            (after Hultberg and Wenblad, 1980).

Expected effects of acid precipitation on ground water, drawn from these and other studies, especially with relationship to soils include:

°   Decreasing pH of ground water rendered sensitive by soil thickness and bedrock type;

°   Increasing levels of concentration of basic cations (Ca^{2+}, Mg^{2+}, Na^+, K^+) leached from soils; and

°   Increasing levels of dissolved aluminum, iron, manganese, zinc, lead and other trace metals, the mobility of which is increased by lower pH.

In a study of the Sudbury, Ontario, area, Hutchinson (1980) determined that metallic contaminants, deposited by either precipitation or dry deposition onto the soil surfaces, are being leached at toxic levels from the soil into ground water and lakes through the action of acid precipitation. This same region was studied by Gorham (1976), who determined that area precipitation, while it did deposit some metals directly into lakes, was also not solely responsible for the high levels of concentrations of nickel, copper, and cobalt found in nearby lake waters. He also cited enhanced acid leaching of metals from Sudbury area soils as the source of much of the problem. The quality of ground water is thus endangered to some degree by acid precipitation. Where these effects have already been fostered, the health of humans and animals who drink from springs or wells whose water is contaminated may also be endangered.

While the pH of drinking water in itself has no known adverse effect on human health, there are several problems connected with low pH levels in ground water. Corrosion, for example, is associated with pH levels below 6.5. In Sweden, both dug and drilled wells have yielded ground water with pH levels in the range of 4.0 to 6.0 (Hultberg and Wenblad, 1980; Cowling, 1982).

Corrosion releases toxic metals, such as lead, zinc, copper, and cadmium from well casing, pipes, and plumbing fixtures. Such effects have already been noted in systems drawing water from acidified surface reservoirs in New York State (Johnson, pers. comm.). In addition to corrosion effects, pH levels of less than 4.0 cause ground water to have a sour taste when used as drinking water.

Although increases in basic cations, such as calcium, magnesium, sodium, and potassium, in ground water with lowered pH have been documented (Figure 4), higher concentrations are not likely to produce a significant adverse effect on ground water quality. These elements are already relatively common in varying concentrations in most ground water and, for the most part, are harmless to humans in all but extremely high concentrations. The magnitude of increases of these constituents in ground water is not likely to be of a degree to cause public concern.

In contrast to the relatively innocuous nature of most basic cations in ground water, many trace metals exhibit deleterious effects on human health even at very low concentrations. Iron, manganese, copper and zinc are exceptions, though these exhibit other objectionable characteristics in water supplies. The rocks and minerals which are sources of the more harmful trace metals are only slightly soluble and only rarely occur in large amounts under normal weathering conditions. Abnormally high concentrations of trace metals of natural origin in ground water are thus rare. However, where acid precipitation may be leaching increasingly high levels of these trace metals from soils, ground water quality may be threatened.

While iron produces no adverse human health effects, high levels of dissolved iron impart an unattractive appearance and taste to drinking water. Concentrations in excess of 0.3 mg l^{-1} cause staining of laundry and may make water objectionable for processing food, making beverages, dyeing, bleaching, and processing many other items. If iron concentrations in ground water exceed 0.5 mg l^{-1}, wells and well screens are likely to become encrusted, and if concentrations exceed 1.0 mg l^{-1} the water becomes unpalatable. The Federal recommended limit (Lehr et al., 1980) for iron in drinking water, 0.3 mg l^{-1}, is not based on physiological reactions but on aesthetic and taste considerations, namely the staining characteristics and the metallic taste.

Manganese does not appear to have a toxicological significance in drinking water, at least in concentrations typical of natural waters. As with iron, the Federal recommended limit for manganese in drinking water, 0.05 mg l^{-1}, is based largely on aesthetic and taste considerations Upon oxidation, manganese in excess of 0.2 mg l^{-1} tends to precipitate and form noxious deposits on foods

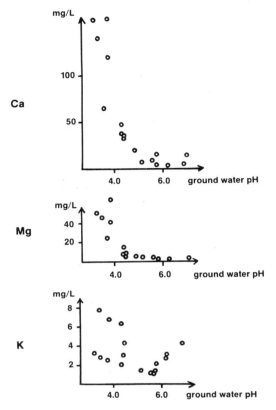

Figure 4.   Effects of pH on basic cation content in
            ground water (after Hultberg and Wenblad,   1980).

during cooking and black stains on laundry and plumbing fixtures.
Concentrations greater than 0.5 mg l^{-1} may impart a metallic taste
to both foods and water.

Federal drinking water standards impose a limit of 1.0 mg l^{-1}
of copper, not because it is a health hazard but because it can
impart an undesirable taste. For the most part, copper is readily
passed from the body, although it does tend to accumulate in the
liver.

Only at very high concentrations does zinc have any known
adverse physiological effects, though it may be undesirable in
drinking water for several other reasons. Zinc produces an
astringent taste above 5.0 mg l^{-1} and may cause water to appear
milky or, upon boiling, to seem to have a greasy surface scum.
Very high concentrations of zinc are associated with nausea and

fainting. Federal drinking water standards impose a limit of
5.0 mg l^{-1}, largely because of taste considerations.

Many other trace elements do occur at very low concentrations
in ground water (generally less than 1 mg l^{-1}) and, in fact, some
are essential to human well-being. However, in many cases, there
is only a small safety range between health requirement and
toxicity. Health problems related to an excessive intake of trace
elements are not usually widespread, but the symptoms of the
diseases they may cause are sometimes so subtle or commonplace that
they are not easy to diagnose. It is because of this potential
danger to health that federal drinking water standards include
several trace elements which have been documented to be leached
from soils by acid precipitation. These include cadmium, chromium,
lead, and mercury.

Cadmium in drinking water, even in relatively low
concentrations, causes nausea and vomiting. It is bioaccumulative,
collecting in the liver, kidneys, pancreas, and thyroid, and is a
recognized carcinogen. For these reasons, federal standards list a
maximum limit of 0.01 mg l^{-1} of cadmium in drinking water.

Chromium causes nausea, and may cause ulcers after long-term
exposure. It is not bioaccumulative but may be carcinogenic in its
hexavalent form (it is considered harmless in its trivalent form).
Because of this, federal standards limit hexavalent chromium in
drinking water supplies to a maximum of 0.5 mg l^{-1}.

Lead is a well-known toxic metal that is bioaccumulative.
Though it is not a known carcinogen, lead causes constipation, loss
of appetite, anemia, abdominal pain, and eventually paralysis. The
federal limit on lead in drinking water is 0.05 mg l^{-1}.

Mercury is highly toxic to man, causing gingivitis,
stomatitis, tremors, chest pains, and coughing. The most common
form of mercury in solution is organic methyl mercury, a highly
toxic compound. Because of the toxic nature of mercury, the
federal government has established a limit for public water
supplies of 0.002 mg l^{-1} of mercury.

The eventual accumulation of aluminum in ground water is one
of the most important consequences of its mobilization within soils
(Figure 5). Despite the fact that aluminum has been demonstrated
to be toxic to both terrestrial plants and fish at levels as low as
0.2 mg l^{-1} (Schofield, 1976), neither the U.S. Environmental
Protection Agency's Water Quality Criteria (U.S.E.P.A., 1972) nor
the Interim Primary Drinking Water Regulations (U.S.E.P.A., 1976)
deal with the health effects of minimum levels of aluminum
allowable in water supplies. Recent reports from Sweden suggest
that high levels of aluminum in drinking water are responsible for

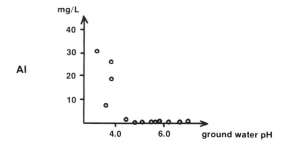

Figure 5. Effect of pH on aluminum content in ground
water (after Hultberg and Wenblad, 1980).

several toxic effects (specifically mental disorders) on humans.
Results from New York and New Hampshire demonstrated that
comparatively high concentrations of dissolved aluminum do occur in
ground water as a result of mineral weathering by acid
precipitation (Cronan and Schofield, 1979). In light of this,
aluminum should perhaps be recognized as another potentially
harmful constituent of water supplies.

EFFECTS ON SURFACE WATER

     Because ground water discharges to surface water as part of
the hydrologic cycle, the chemistry of ground water entering a lake
or stream will have an effect on the surface water chemistry of
that water body. Results of a study in Norway conducted by
Henriksen and Kirkhusmo (1982) demonstrated that ground water was
acidified and chemically affected by acid precipitation, and that
surface water bodies into which the acidified ground water
discharged were even more acidified. The extent of change in lake
acidity was mainly determined by the buffering capacity of the
surrounding soil and the geology and size of the lake's watershed.
Lakes most susceptible to acidification were: 1) those located
downwind from major air pollution sources; 2) those with watersheds
underlain by hard, insoluble bedrock with thin, sandy soil; and
3) those which have a low watershed-to-lake-surface ratio. Lakes
whose watersheds were underlain with soluble bedrock (i.e.,
limestone, dolomite, or marble), or whose soils were thick and of a
calcareous and/or clayey nature were relatively resistant to
acidification.

     Many lakes in the northeastern United States are highly
sensitive to acid precipitation. The watersheds of these lakes
are, for the most part, underlain by granite, quartzite, or other
crystalline bedrock which are highly resistant to chemical
weathering and contain few minerals which would contribute to

buffering. Many lakes in New York and New Hampshire exhibit a pH of less than 5.0 (Cronan and Schofield, 1979), while lakes in similar granitic terrain in Norway and in the Minnesota-Wisconsin area, which are not receiving substantial inputs of acid precipitation, exhibit an inherently slight to moderate acidity of between pH 6.5 and 5.5 (Wright and Henriksen, 1978; Glass, 1981). The water pH of the latter lakes is apparently controlled by the natural acidity of rainfall. In contrast, lakes in carbonate terrains have pH levels of at least 6.5 regardless of loadings of atmospheric acid (Wright and Henriksen, 1978).

Many lakes in the northeastern U.S. and southeastern Canada have experienced dramatic depletions of fish and other aquatic organisms as a result of becoming acidified (Beamish, 1976; Cronan and Schofield, 1979). In some cases lakes have become totally devoid of fish. A 1976 survey of 219 lakes in the Adirondack region of New York demonstrated that over 50% of the lakes had a pH which imperiled fish survival (pH less than 5.0) and that 82 lakes were in fact fishless (Schofield, 1976). The mean pH of these lakes, measured over the period 1969-1975, was 4.8. By 1979, the number of fishless lakes in the same area had grown to nearly 200 (U.S. Environmental Protection Agency, 1979). In contrast, pH measurements made in Adirondack lakes between 1929 and 1937 showed that only 4% of these lakes had a pH of less than 5.0 or were devoid of fish (Schofield, 1976). The mean pH of lakes during this period was 6.5.

The continued influx of acid precipitation to poorly buffered watersheds in the northeastern U.S. has not only lowered the pH of lakes, but has also drastically altered the chemistry of these lakes. Sulfate has replaced bicarbonate as the major anion, the hydrogen ion has become a major cation, and aluminum is present at levels 10-50 times that of unacidified lakes (Schofield, 1976). Similar observations have been made in Norway by Wright et al. (1976) and Wright and Henriksen (1978). This chemical change has, in turn, produced an even more hostile environment for fish and other aquatic organisms. Field and laboratory experiments in the Adirondacks demonstrated that, in acidified waters, toxic conditions for fish may be produced by dissolved inorganic aluminum even at lake pH values that are not physically harmful to the fish (Schofield, 1976).

The presence of aluminum in lake waters indicates a source other than precipitation, since precipitation, whether acidic or not, contains very little aluminum (Hutchinson, 1980). Limnological studies in the Adirondack lakes region indicated that much of the aluminum may be transported from soils in the surrounding watershed. Wright et al. (1976) also cited the leaching of aluminum from watershed soils as the cause of unusually high aluminum concentrations in Norwegian lakes.

As noted earlier (equation 3), aluminum becomes soluble and mobile in soils with a solution pH of 5.0 or less (Norton, 1975). Leaching transports this free aluminum to ground water which, in turn, discharges to acidified lakes where comparatively high concentrations of aluminum remain in solution. Concentrations as high as 0.75 mg l^{-1} have been reported in spring water feeding high-elevation lakes in the White Mountains region, New Hampshire (Cronan and Schofield, 1979). Schofield (1976) documented that aluminum concentrations of 0.2 mg l^{-1} or higher resulted in fish mortality.

The shift from an alkaline to an acidic lake also increases the load of mercury and other heavy metals which can be taken up by fish -- metals which can either be toxic to the fish or render them unfit for human consumption. A clear correlation exists between acid levels in lakes and mercury levels of its fish -- the more acid the lake, the more mercury in the fish (U.S. Environmental Protection Agency, 1979). Other toxic metals, such as lead and cadmium, may also be found to have such relationships with lake acidity and fish accumulation.

The acidification of surface waters not only affects fish and other aquatic organisms, but it may also affect man. As surface drinking water reservoirs acidify, concurrent increases in metal concentrations may cause water supplies to fail in meeting state and federal public water supply standards for water quality. Indications of such conditions have already been seen in New York, where water from at least one reservoir has been acidified to the level where contact with household plumbing systems and solder joints has resulted in concentrations exceeding maximum levels recommended by the New York State Department of Health for lead (Johnson, per. comm.).

CONCLUSIONS

As understanding of the phenomenon of acid precipitation expands, it becomes increasingly evident that interaction with various components of the environment is an exceedingly complex issue involving several distinct yet related problems. It is also evident that research is needed to better define the extent of the effects of acid precipitation. Clearly, assessing the impact of acid precipitation on ground water is an important factor in determining the gross impact of acid precipitation on the environment. The migration of base cations and, more importantly, trace metals from soil to ground water is detrimental both to the soil and to ground water quality. Given the behavior of contaminants in ground water systems, it is possible that potentially harmful substances leached from soils into ground water may render the ground water in some areas unfit for human consumption without extensive treatment.

140   D. M. NIELSEN ET AL.

The direct and indirect consequences of contaminated ground water on humans and on the environment pose a serious problem that needs to be addressed. The close relationships between soil chemistry and ground water quality and the intimate interrelationship between ground water and surface water demonstrates the need for determining the cumulative effects of acid precipitation through interdisciplinary research efforts.

REFERENCES

Abrahamsen, G., Bjor U., Horntuedt, R., and Treate, B., 1976, Effects of Acid Precipitation on Coniferous Forest; Research Report. SNSF Project, Part 2, Ås-NLH, Norway. 36 pp.
Bache, B.W., 1980, The acidification of soils, pp. 183 to 201. In: Precipitation on Terrestrial Ecosystems, T.C. Hutchinson and M. Havas, eds., Plenum Press, New York, NY.
Beamish, R.J., 1976, Acidification of lakes in Canada by acid precipitation and resulting effects on fishes. Water, Air, Soil Pollu. 6:511-514.
Cogbill, C.V. and Likens, G.E., 1974, Acid precipitation in the northeastern United States. Water Resources Res. 10:1133-1137.
Cowling, E.B., 1982, Acid precipitation in historical perspective. Environ. Sci. Technol. 16:110A-123A.
Cronan, C.S. and Schofield, C.L., 1979, Aluminum leaching response to acid precipitation: Effects on high-elevation watersheds in the northeast. Science 204:204-306.
Glass, G.E., 1981, Susceptibility of aquatic and terrestrial resources of Minnesota, Wisconsin, and Michigan to impacts of acid precipitation, pp. 112 to 113. In: Ecological Impact of Acid Precipitation, D. Drabløs and A. Tollan, eds., Proceedings International Conference, Sandefjord, Norway. SNSF Project, Ås-NLH, Norway.
Glass, N.R., Arnold, D.E., Galloway, J.N., Hendrey, G.R., Lee, J.J., McFee, W.W., Norton, S.A., Powers, C.F., Rambo, D.L., and Schofield, C.L., 1982, Effects of acid precipitation. Envir. Science Tech. 16:162A-169A.
Gorham, E., 1976, Acid precipitation and its influence upon aquatic ecosystems. Water, Air, Soil Pollu. 6:457.
Gorham, E. and McFee, W.W., 1980, Effects of acid deposition upon outputs from terrestrial aquatic ecosystems, pp. 465 to 480. In: Effects of Acid Precipitation on Terrestrial Ecosystems, T.C. Hutchinson and M. Havas, eds., Plenum Press, New York, NY.
Hanson, D.W., 1980, Acidic precipitation-induced changes in subalpine fir forest organic soil layers. M.S. thesis, University of Maine, Orono, ME. 124 pp.
Henriksen, A. and Kirkhusmo, L.A., 1982, Acidification of groundwater in Norway. Nordic Hydrology 13:183-192.

Hultberg, H. and Wenblad, A., 1980, Acid groundwater in
    southwestern Sweden, pp. 220 to 221. In: Ecological Impact
    of Acid Precipitation, D. Drabløs and A. Tollan, eds.,
    Proceedings International Conference, Sandefjord, Norway.
    SNSF Project, Ås-NLH, Norway.
Hutchinson, T.C., 1980, Effects of acid leaching on cation loss
    from soils, pp. 481 to 497. In: Proceedings of the NATO
    Conference on the Effects of Acid Precipitation on Vegetation
    and Soils. Plenum Press, New York, NY.
Johnson, R.L., personal communication, Dept. Environmental
    Conservation, Albany, NY.
Kramer, J.R., 1975, Geochemical and lithological factors in acid
    precipitation, pp. 611 to 618. In: Proceedings of the First
    International Symposium on Acid Precipitation and the Forest
    Ecosystems, Tech. Report No. NE-23, U.S. Forest Service, Upper
    Darby, PA.
Krug, E.C. and Frink, C.R., 1983, Acid rain on acid soil: A new
    perspective. Science. 22:520-525 pp.
Lehr, J.H., Gass, T.E., Pettyjohn, W.A., DeMarre, J., 1980,
    Domestic Water Treatment. McGraw-Hill Inc., New York, NY.
    264 pp.
Likens, G.E., 1972, The Chemistry of Precipitation in the Central
    Finger Lakes Region. Technical Report No. 50, Cornell
    University Water Resources Center, Ithaca, New York, NY.
    32 pp.
Likens, G.E., 1976, Acid precipitation. Chem. Engineering News.
    54:29-44.
Likens, G.E. and Bormann, F.H., 1974, Acid rain: A serious
    regional environmental problem. Science 184:1176-1179.
McFee, W.W., 1980, Sensitivity of Soil Regions to Acid
    Precipitation. Tech. Report 600/3-80-013, U.S.E.P.A.,
    Corvallis, OR. 12 pp.
Norton, S.A., 1975, Changes in Chemical Processes in Soils Caused
    by Acid Precipitation, pp. 711 to 724. In: Proceedings of
    the First International Symposium on Acid Precipitation and
    the Forest Ecosystem. Tech. Report No. NE-23, U.S. Forest
    Service, Upper Darby, PA.
Norton, S.A., 1980, Geologic factors controlling the sensitivity of
    aquatic ecosystems to acidic precipitation, pp. 521 to 530.
    In: Atmospheric Sulfur Deposition: Environmental Impact and
    Health Effects, D.S. Shriner, C.R. Richmond and S.E. Lindberg,
    eds., Ann Arbor Science Publ., Ann Arbor, MI.
Norton, S.A., Hanson, D.W., and Campana, R.J., 1980, The impact of
    acidic precipitation and heavy metals on soils in relation to
    forest ecosystems. Paper presented at the International
    Symposium on Effects of Air Pollutants on Mediterranean and
    Temperate Forest Ecosystems, Riverside, CA.
Pearson, F.J., and Fisher, D.W., 1971, Chemical Composition of
    Atmospheric Precipitation in the Northeastern United States.

U.S. Geological Survey Water Supply Paper 1535-P, U.S.G.S.,
    Washington, DC. 23 pp.
Reuss, J.O., 1978, Simulation of Nutrient Loss From Soils Due to
    Rainfall Acidity. Tech. Report No. 600/3-78-053, U.S.E.P.A.,
    Corvallis, OR. 4 pp.
Schofield, C.L., 1976, Acid precipitation: Effects on fish. Ambio
    5:228-230.
Tyler, G.T., 1978, Leaching rates of heavy metal ions in forest
    soil. Water, Air, Soil Pollu. 9:137-151.
U.S. Environmental Protection Agency, 1972, Water Quality Criteria.
    Committee on Water Quality Criteria, Technical Report No.
    EPA-R/3-73-033, U.S.E.P.A., Washington, DC. 128 pp.
U.S. Environmental Protection Agency, 1976, National Interim
    Primary Drinking Water Regulations. Technical Report No.
    EPA-570/9-76-003, U.S.E.P.A., Office of Water Supply,
    Washington, DC. 159 pp.
U.S. Environmental Protection Agency, 1979, Research Summary: Acid
    Rain. Office of Research and Development, U.S.E.P.A.,
    Washington, DC. 23 pp.
Wiklander, L., 1974, Leaching of plant nutrients in soils. Acta.
    Agric. Scand. 24:349-357.
Wiklander, L., 1980, The sensitivity of soils to acid precipi-
    tation, pp. 553 to 567. In: Effects of Acid Precipitation on
    Terrestrial Ecosystems, T.C. Hutchinson and M. Havas, eds.,
    Plenum Press, New York, NY.
Wright, R.F., Torsten, D., Gjessing, E.T., Hendrey, G.R.,
    Henricksen, A., Johannsen, M., and Muniz, I.P., 1976, Impact
    of acid precipitation on freshwater ecosystems in Norway.
    Water, Air, Soil Pollu. 6:483-499.
Wright, R.F. and Henriksen, A., 1978, Chemistry of small Norwegian
    lakes, with special reference to acid precipitation. Limnol.
    Oceanogr. 23:487-498.

ALKALINITY AND TRACE METAL CONTENT OF DRINKING WATER IN AREAS OF

NEW YORK STATE SUSCEPTIBLE TO ACIDIC DEPOSITION

G. Wolfgang Fuhs[a], Rolf A. Olsen, and Anthony Bucciffero

Wadsworth Center for Laboratories and Research
New York State Department of Health
Albany, New York   12201

ABSTRACT

Waters serving as sources of water supply in the central and northwestern Adirondack region were examined for alkalinity, pH, conductivity, and several major ions and trace metals. All waters from surface waters or shallow aquifers were corrosive. Elevated background concentrations of copper and lead were found sporadically but not at levels of significance to human health. Problems of copper dissolution and metallic taste existed in copper plumbing systems and were substantiated in the analyses. In about 10% of the individual systems with soldered copper pipes, volumes of first-flush water, ranging from 250 ml to one liter, contained amounts of lead equalling or exceeding the allowable daily intake from water for this element. Individual systems with lead service lines were not encountered in this study. Without appropriate treatment, such systems would pose a definitive health hazard with source waters of the type encountered.

INTRODUCTION

Ground and surface waters in many areas in the northeastern United States and eastern Canada are considered susceptible to acidic airborne pollutants, as local bedrock provides little leachable alkali. In New York State, the Adirondack and Catskill mountains have been considered particularly susceptible. Studies of lakes in the Adirondack Mountains suggested that elevation is

---

[a]Calif. Dept. Health Services, 2151 Berkeley Way, Berkeley, CA
  94704

another important factor in surface water acidification (Schofield, 1976). The apparent effect of altitude, however, is mostly explained by lack of adequate soil contact by infiltrating rain-water at higher elevations.

Local soils, however, vary not only in thickness, but also in their buffering capacity. Those of glacial origin may contain more leachable alkali than soils of local origin. Yet if acid deposi-tion is severe and of long duration, their buffering capacity may also become exhausted (McFee et al., 1976). It follows that acidification may occur in any shallow aquifer which is located in noncalcareous soil or not immediately underlain by calcareous material.

In the Adirondack Region of New York, the bedrock geology is generally known except for certain areas with extensive glacial and alluvial deposits. Soils, however, are not well characterized and generally are not mapped. The lack of this type of information is one of the major obstacles in assessing the long-term effect of acidic deposition in the area.

Drinking water in the Adirondack Region is mostly from individual supplies. Most common are dug or drilled wells, which are often located underneath or immediately adjacent to the residences they serve. Some residences are supplied by small streams or hillside springs connected to residences by gravity-fed delivery lines, sometimes several hundred meters long. Lakeshore cottages are often supplied by local lake water without treatment, although this practice is strongly discouraged by regulatory agencies. Municipal water supplies utilize large springs, lakes, reservoirs, or, less often, groundwater. The effects of acid deposition on the quality of these source waters are likely to vary with the local geologic and soil conditions and the extent of soil contact.

Most larger lakes in the Adirondacks are situated at lower elevations and generally have sufficiently large and varied water-sheds to receive runoff of varying alkalinity. Productive lakes also generate alkalinity from the reduction and incorporation into biomass of carbon dioxide and nitrates. Therefore, summer pH values in many of these lakes bear little evidence of acid runoff (Wood, 1978; Fuhs et al., 1982). In addition, lakes with populated shores receive septic-tank leachate which contains alkalinity, mainly from household detergents. A comparison of drinking water and domestic wastewater composition at Lake George, New York suggested that the septic discharge contribution was on the order of three milliequivalents (meq) of alkalinity per liter of dis-charge, or a daily rate of one meq per person (Fuhs, 1972). The depression of pH values in runoff during snowmelt can be signifi-cant (Jeffries et al., 1979; Pfeiffer and Festa, 1980). This effect is certain to be noticeable in many shallow groundwaters as well, although New York State Dept. of Health officials are not aware that this has been documented.

In this report, the results of several surveys of source waters for public and individual water supplies in the Adirondack mountains of New York is presented. The purpose was not to quantify acid deposition effects but to assess, by conventional methods, the present quality of waters in the area.

METHODS

Temperature, pH, and total alkalinity were determined immediately at each site. The thermometer was calibrated against a NBS-certified meter. Sample pH was determined with an Orion Model 401 specific-ion meter with Corning No. 746020 glass and Corning No. 476002 reference electrodes after calibration with buffers at pH 4.0 and 7.0. The sample was swirled lightly without equilibration with ambient air. Total alkalinity was determined by titration with methyl purple (for public water supplies) or by titration to pH 4.5 (for individual supplies) with 0.0200 N H_2SO_4 from a microburet with 0.05 ml divisions.

Calcium was usually titrated in the field with EDTA by using murexide (calcium purpurate) as an indicator. Color change was judged from the titration of standards. In the survey of Lewis and St. Lawrence counties, however, calcium was determined with the other metals by atomic absorption spectrometry.

Field conductivity measurements (k_t) either were made at the local temperature (t) and converted to conductivity at 25°C (k_{25}) by an empirical equation:

$$k_{25} = k_t [1 + 0.016 (25 - t) + 0.000555 (25-t)^2]   \quad (1)$$

derived from values in Golterman (1969) or, for the survey of Lewis and St. Lawrence counties, were determined in the laboratory at 25°C. Langelier's Index (LI), which Department of Health field staff utilize as a measure of corrosivity, was calculated as described in Standard Methods (American Public Health Association, 1975). Based on an analysis of tributary data for Lake George, New York (Fuhs, 1972):

$$\text{total dissolved solids (in mg } l^{-1}) \approx 0.7 \times k_{25}.   \quad (2)$$

Separate samples for magnesium, alkali, and trace metals were collected, preserved with Ultrex nitric acid, and transported to the Albany Department of Health Laboratory for analyses by atomic absorption (American Public Health Association, 1975). Lead was measured by the graphite furnace method and mercury by flameless atomic absorption. Quality control was employed with regard to all constituents, the bottles, and the use of nitric acid. Chloride and sulfate were determined in unpreserved samples by the automated ferricyanide and methyl thymol blue methods, respectively (American Public Health Association, 1975).

RESULTS

Public Water Supplies

    The surveys of public water supplies in the central Adirondacks were conducted in October 1978 (Figure 1). Sources were located at elevations up to 2050 ft (700 m) above mean sea level. Six groundwater sources and 23 surface waters were examined. Where possible, samples were collected from the source before treatment. One well, however, could be sampled only after chlorination. In another case, the system had to be sampled at a nearby residence, and the Old Forge, NY, system was sampled at a residence near the far end of the distribution system. Water in residences were sampled after the tap had been left running for at least one minute.

    The results are shown in Table 1. The pH range of the surface waters was from 5.7 to 7.6. Due to the low buffering of these waters, chlorination affected the pH; at Newton Falls, the pH decreased from 6.2 to 5.8 upon chlorination. Similar effects were observed at the Saranac Lake, NY, and Blue Mountain Lake, NY, systems (Table 1). Groundwater sources had pH values between 6.2 and 8.6. Total alkalinity ranged from 2.0 to 67 mg l^{-1} as $CaCO_3$, and calcium ranged from 3.2 to 33.2 mg l^{-1}.

    All waters were corrosive. The LI is primarily intended to indicate undersaturation with calcium carbonate. Copper dissolution was seen in the water supply at Camp Adirondack, better known as the site of the Olympic Village for the 1980 Olympic Winter Games. The source had less than 0.05 mg l^{-1} copper whereas the cold and hot water lines in a residence using this supply contained water with 0.44 and 0.67 mg Cu l^{-1}, respectively. A sample from the Star Lake Subdivision showed an unacceptably high copper concentration of 2.6 mg l^{-1}, probably the result of corrosive action (LI = −4.0) on the copper plumbing in the residence where the sample was taken. Mercury and cadmium data are not shown, as all results were below the limit of detection (0.0004 and 0.002 mg l^{-1}, respectively).

Individual Water Supplies

    Individual water supplies were surveyed in response to

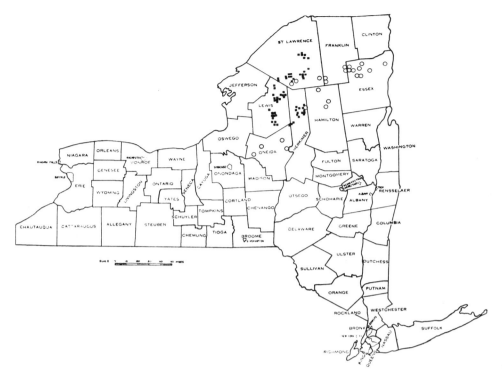

Figure 1.   Map of New York State showing locations of public water
            supplies (O) and individual water systems (■) samples.

concerns by local citizens in areas where acid deposition would
most likely affect water quality (Figure 1). The results, therefore, are not a statistically valid cross section for the region;
they are the result of a search for worst-case conditions.

       The Eagle Bay, NY, area of Herkimer County in the west-central
Adirondacks was selected specifically as was the western and
northwestern slopes of the Adirondacks. This study area borders on
the "marble belt" of northwestern New York and includes sites with
complex and unknown bedrock geology underlying glacial and alluvial
deposits. At all sites, the geologic map (New York State Museum
and Science Service, 1970) indicated an absence of limestone
formations.

       The wells and springs in the Eagle Bay area (elevation
600-700 m) are all located in a geologically similar area, but

Table 1. Adirondack public water supply sources, ranked by alkalinity, October 1978

Water Supply*	Water Supply Source	Latitude	Longitude	Alk. (mg l^{-1})	Temp. (°C)	pH	Cond. 25C ($\mu S/cm^{-1}$)	Calcium (mg l^{-1})	L.I.	Copper (mg l^{-1})	Zinc (mg l^{-1})	Lead (mg l^{-1})	Aluminum (mg l^{-1})
Tupper Lake (V)	Simond Pond	44°09'40"	74°26'30"	1	12.0	5.7	31	7.2	-4.7	<0.05	0.08	<0.01	0.17
Newton Falls (V)	Raquette River – Hotel	44°12'38"	74°59'24"	2	15.7	5.8	40	4.0	-5.2	0.34	0.05	0.01	0.18
		44°11'75"		3	11.0	7.1		3.6	-3.9	1.00			0.13
Old Forge W.D.	Private Home	43°44'28"	74°27'28"	4				2.1					0.16
Newton Falls	Raquette River	44°12'48"	74°59'24"	4	15.7	6.2	35	4.0	-4.5	0.09	0.05	0.01	0.20
Indian Lake (V)	Lake Abakanee	43°47'08"	74°14'27"	4	12.7	6.8	30	3.6	-4.1	<0.05	<0.05	0.01	N/A
Blue Mtn. Lake (V)	after treatment	43°51'46"	74°26'25"	4	16.4	6.2	50	3.2	-4.4	N/A	N/A	N/A	0.08
Whitney Park	Little Tupper Lake	44°02'54"	74°35'10"	5	12.4	6.9	30	4.8	-4.4	0.09	0.05	0.01	0.14
Saranac Lake (V)	McKenzie Pond	44°19'32"	74°05'20"	5	14.0	6.4	35	4.8	-4.2	0.23	0.07	0.01	0.12
Star Lake (V)	after treatment	44°19'23"	74°07'54"	N/A	N/A	N/A	N/A	N/A	N/A	0.17	<0.05	0.01	0.15
Star Lake	Star Lake	44°09'32"	75°02'14"	6	15.2	6.5	41	4.0	-4.0	<0.05	<0.05	<0.01	0.15
Blue Mtn. Lake (V)	Blue Mountain Lake	43°51'56"	74°26'25"	6	13.7	7.0	44	4.4	-3.6	<0.05	<0.05	<0.01	0.18
Lake Placid (V)	Lake Placid	44°17'52"	73°58'44"	7	14.1	6.9	39	5.2	-3.5	<0.05	<0.05	<0.01	0.16
Long Lake (V)	Long Lake	43°57'24"	74°23'21"	7	7.8	7.2	39	31.6	-3.4	<0.05	<0.05	<0.01	0.08
Keene	Upper Cascade Lake	44°13'31"	73°52'26"	11	11.0	7.4	78	18.4	-2.7	<0.05	<0.05	<0.01	0.08
Keene	Lower Cascade Lake	44°14'13"	73°51'47"	12	10.2	7.4	91	8.0	-3.5	<0.05	<0.05	<0.01	0.08
Camp Adirondack	Haystack Mtn./Saymore Mtn. Res.	44°11'46"	74°05'37"	14	9.0	7.2	56	7.2	-2.9	<0.05	<0.05	<0.01	0.09
Camp Adirondack	Residence, hot water tap	44°15'16"	74°05'37"	N/A	N/A	N/A	N/A	N/A	N/A	0.67	0.18	0.01	0.14
Camp Adirondack	Residence, cold water tap	44°15'16"	74°05'37"	N/A	N/A	N/A	N/A	N/A	N/A	0.44	0.19	0.01	0.12
Tupper Lake (V)	Cranberry Pond	44°11'07"	74°28'16"	16	8.9	7.0	64	4.0	-3.1	<0.05	0.07	<0.01	0.16
Wilmington (V)	White Brook Reservoir	44°23'37"	73°50'15"	18	10.1	6.6	61	7.2	-3.4	<0.05	<0.05	<0.01	<0.05
Jay (T)	Book Reservoir	44°20'51"	73°42'31"	19	10.2	7.6	124	16.8	-1.9	<0.05	<0.05	<0.01	0.10
Ausable Acres	Signor Cistern	44°25'41"	73°45'08"	20	12.9	6.5	70	19.2	-3.4	<0.05	<0.05	<0.01	0.07
Ray Brook	Well 25 ft.	44°17'32"	74°04'13"	28	8.0	6.8	89	10.4	-2.9	<0.05	<0.05	<0.01	0.08
Keene (T)	Spring	44°14'28"	73°46'21"	40	9.4	7.8	114	8.0	-1.5	<0.05	0.13	<0.01	0.10
Piercefield (V)	Raquette River	44°13'53"	74°33'33"	42	13.2	7.6	80	3.2	-2.3	0.05	0.05	0.01	0.11
Star Lake Subdiv.	Well 200 ft. (Residence)	44°09'21"	75°05'08"	48	15.0	6.2	401	33.2	-2.6	2.60	0.05	0.01	0.09
Wilmington Notch Campsite	Well 300 ft.	44°21'01"	73°51'39"	48	11.3	8.6	151	17.2	-0.6	<0.05	0.07	<0.01	0.06
Conifer (V)	Brock	44°12'58"	74°36'14"	50	10.4	7.1	91	7.2	-2.4	<0.05	0.14	<0.01	0.13
Ausable Acres	Kitzbuhl Well	44°23'25"	73°44'35"	67	13.2	8.5	154	6.8	-0.2	<0.05	0.05	<0.01	0.12
Twitchell Lake Water Supply	Well – Residence	43°50'58"	74°53'41"	N/A	N/A	N/A	N/A	1.4	N/A	1.40	1.00	0.13	1.40

Except where indicated, untreated waters were analyzed. Note two entries for Blue Mountain Lake.

*(V) = Village; (T) = Township

N/A: Data not available

their soil cover and well depths vary. Seven springs and shallow
wells (0-10 m) had the following average composition:

Cations	μeq l^{-1}	Ion %	Anions	μeq l^{-1}	Ion %
Calcium	178	42	Sulfate	167	39
Sodium	84	20	Bicarbonate	101	24
Aluminum	58	14	Chloride	86	20
Magnesium	44	10	Nitrate	73	17
Copper	25	6			
Potassium	17	4			
Zinc	17	4			

Ammonia was measured but amounts were negligible. Nitrate was 21
to 73 μeq l^{-1} in six sources and 129 μeq l^{-1} in one well, possibly
from a nearby septic leachfield. Contamination of this source is
also indicated by a pH of 7.95. The other sources were pH 4.95 to
5.71.

Conductivity averaged 54 μS cm^{-1} (25°C) with little variation.
Copper concentrations of 1.4 to 2.1 mg l^{-1} were found in three
upland sources and 0.03 to 0.05 mg l^{-1} in the others. None of this
water had been in direct contact with metal piping. Zinc concentrations were not determined in all wells, but were significant
where measured (0.1 to 1.9 mg l^{-1}). Severe corrosion of copper
piping was noted in the system with pH 4.95 water, and a taste
problem in first-flush water existed in one other system with
copper piping.

The depth of the wells affected water quality greatly. Depth
was positively correlated with conductivity, pH, alkalinity, and
magnesium concentration (Figure 2). Calcium concentration and
depth would undoubtedly have shown a significant correlation in a
larger sample. Sulfate concentrations did not change with depth,
nor did the concentrations of chloride, potassium, or sodium. For
sodium and chloride, contamination from sources near ground level
could not always be ruled out.

Aluminum concentrations decreased sharply with depth and then
leveled off (Figure 2). This pattern indicates dissolution near
the surface and possibly immobilization in the deeper layers. In
five upland sources there were slightly elevated lead concentrations not attributable to metal piping (0.08 to 0.16 mg l^{-1}).
The others were below 0.05 or undetectable at 0.01 mg l^{-1}.

A significant improvement in water quality was achieved by one
property owner who backfilled a well with limestone. A plastic
pipe led from the spring housing to a reservoir and from there to a
house, which had copper piping. Measurements were taken from a tap
in the basement of the house when the line was in use and well

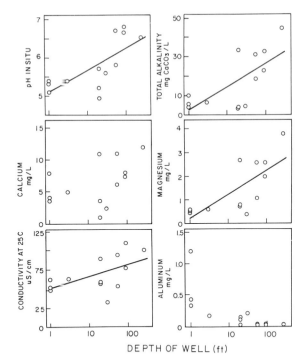

Figure 2   Water chemistry in wells in the
              Eagle Bay area, Herkimer County,
              New York

flushed.   Control samples were from an abandoned spring near the
source.   As a result of limestone treatment, the pH value increased
from 5.3 to 7.9, conductivity from 35.5 to 105 μS cm^{-1} (at
10–13°C), alkalinity and calcium from 101 to 1110 and from 210 to
1000 μeq l^{-1}, respectively.   Aluminum decreased from 1.2 to less
than 0.05 mg l^{-1} and was probably deposited in the reservoir as a
hydroxide.   The system did not contain measurable lead or copper.

     In Lewis and St. Lawrence counties, 44 individual supplies
were sampled in October 1979, selected on the basis of their likely
location on acidic bedrock.   Where acidic conditions and metal
plumbing existed, information about the construction of the well
was recorded.   Samples for determination of trace metals were
collected from flushed lines and also, where available, from lines
where water had been left standing for a period of time.   All metal
plumbing was copper; the type of solder was unknown.   Plumbing or
transmission lines made entirely of lead were not observed during
this survey.

Alkalinity in this group of supplies ranged from 70 to 3700 µeq l^{-1}, or 3.5 to 184 mg $CaCO_3$ l^{-1} (Figure 3). They followed a lognormal distribution with a geometric mean of 520 µeq l^{-1} (26.7 mg $CaCO_3$ l^{-1}) and an error factor of 2.70. Conductivity (at 25°C) ranged from 40 to 651 µS cm^{-1}, and pH from 5.4 to 8.4. Both were distributed as shown in Figure 4. Calculations showed, as the distributions (Figures 4b,c) suggest, that both conductivity and pH were correlated with alkalinity. A rather close correlation also existed between conductivity and alkalinity for a majority of the samples (Figure 4c). All samples, except one from the Eagle Bay area and most of public water supplies with one exception, fell on the same regression line ($r = 0.941$):

$$\text{conductivity (µS } cm^{-1}) = 30.23 + 2.33 \times \text{T.A. (mg } l^{-1}), \qquad (3)$$

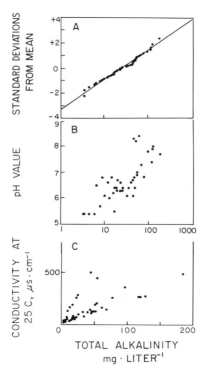

Figure 3. Water chemistry in individual water supplies in parts of Lewis and St. Lawrence counties, New York: (a) Normal probability plot of total alkalinity; (b) pH values; and (c) conductivity versus total alkalinity.

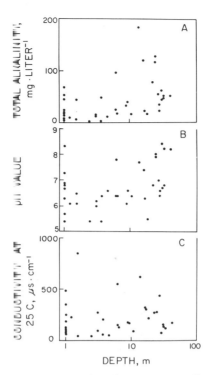

Figure 4. Water chemistry as a function of depth of the source
in individual water supplies
in parts of Lewis and St. Lawrence counties, New York.

where T.A. is total alkalinity. The outliers seem to form another
population of data through which a regression line could be drawn,
but since some of the additional conductivity may be from contamination it is not certain that such a regression would be meaningful.

Water quality in this group of samples was much more variable
than in the Eagle Bay area, and there was no consistent correlation
between alkalinity and depth (Figure 4a). Possible explanations
for this variability are: (1) failure to accurately locate sources
on acidic rock, or lack of accuracy in the geologic data; (2) variable thickness and leachable alkali content; (3) variable contamination from septic tank systems discharging into soil nearby or,
in the case of a spring or stream source, further upstream; or
(4) variable contamination with road salt. The actual correlations
of pH and conductivity versus alkalinity are not given because the
values may not have significance beyond this data set.

Concentrations of total calcium, aluminum, copper, and lead are given in Table 2 in the order of increasing alkalinity. Several supplies were sampled before and after flushing of a line, or else one sample was taken from a portion of the system which had not been in use for some hours or days. Only in three locations did residents indicate taste or corrosion problems. One individual, a plumbing contractor, was aware of a need for early replacement of copper piping in homes in the area.

A positive correlation was found between calcium and alkalinity and an inverse correlation between aluminum and alkalinity, except for one elevated value for aluminum which was related to a low pH of 5.5. High copper concentrations were found in some systems, both in flushed lines and in first-flush water. The few readings for total lead above the detection limit were in low-alkalinity waters, as would be expected.

During part of the survey, several samples for lead analysis of first-flush water were taken in 1-liter bottles and others in 250-ml bottles. Lower values were consistently found in the larger samples. This suggested that the lead content of the samples was affected by exposure to solder surfaces, which are concentrated near the tap or, by mixing of water from flushed and unflushed sections of pipe. The lead content of first-flush water in eleven systems was, therefore, examined further at places where the first survey had shown detectable lead (Table 2, asterisks). The residents agreed to cooperate by not flushing the lines in the morning

Table 2. Point-of-use water in individual supplies with copper plumbing in Lewis and St. Lawrence Counties

Site	Depth Total (m)	Alk. (mg l^{-1})	Temp. (°C)	pH	Cond, 25C (µS cm^{-1})	Al (mg l^{-1})	Pb (mg l^{-1})	Cu (mg l^{-1})	Ca (mg l^{-1})	Notes
1*	3	3.5	8.0	5.4	28.7	0.35	< 0.01	< 0.05	2.1	
2*	4	3.5	14.2	5.4	37.7	0.22	< 0.01	0.07	2.9	Flushed
2	4	3.5	14.0	5.4	37.4	0.08	< 0.01	2.1	3.1	Unflushed
3*	0	4.5	15.0	5.4	41.4	0.30	< 0.01	0.09	3.4	Corrosion
4*	0	6.0	15.9	5.4	47.0	0.23	< 0.01	0.42	5.1	
5*	2	6.5	17.2	6.5	29.0	0.27	< 0.01	0.07	3.6	Flushed
5	2	6.5	17.2	6.5	29.0	0.23	0.07	1.20	3.5	14 hrs
6	0	8.0	13.4	5.7	78.0	0.05	< 0.01	6.60	7.7	
7	0	9.0	19.0	6.8	57.0	0.09	< 0.01	< 0.05	3.8	
8*	1	10.0	20.2	6.1	137.0	< 0.05	0.03	1.60	16.0	Flushed
8	1	10.0	20.2	6.1	137.0	0.06	0.04	0.61	16.0	7 days
9*	5	12.0	18.2	6.6	36.0	< 0.05	< 0.01	0.09	3.3	
10*	3	13.0	11.0	6.0	170.0	0.10	0.02	0.10	N/A	Flushed
10	3	14.0	11.0	6.5	170.0	0.09	0.02	0.27	N/A	Unflushed
11	0	14.5	17.5	6.8	47.0	0.08	0.04	1.00	5.7	
12*	3	16.0	16.1	6.2	59.0	< 0.05	< 0.01	1.00	7.1	Flushed
12	3	16.0	16.1	6.2	59.0	< 0.05	1.50	0.31	6.7	15 days
13	12	17.0	17.0	6.4	68.7	< 0.05	< 0.01	0.17	7.3	
14	20	17.0	13.6	5.5	193.0	0.34	< 0.01	1.10	11.0	
15*	7	18.0	13.1	6.4	87.4	0.05	0.01	0.14	6.6	
16	0	20.0	16.1	6.7	61.0	< 0.05	< 0.01	0.52	9.8	
17	31	23.0	16.0	6.6	76.0	< 0.05	< 0.01	0.09	8.9	
18	0	25.0	16.6	6.3	286.0	0.05	< 0.01	0.55	18.0	
19	7	26.0	17.0	6.4	103.0	< 0.05	< 0.01	0.07	13.0	
20*	9	33.0	15.0	6.1	125.0	< 0.05	< 0.01	0.52	17.0	Taste
20	9	33.0	15.0	6.1	125.0	< 0.05	0.02	4.70	14.0	48 hrs
21	33	44.0	13.5	6.7	146.0	0.06	< 0.01	0.26	20.0	
22	0	69.0	15.9	7.3	149.0	< 0.05	< 0.01	0.16	20.0	
23	24	78.0	11.5	6.8	260.0	< 0.05	< 0.01	0.10	33.0	
24	27	128.0	17.1	8.0	278.0	< 0.05	< 0.01	0.16	44.0	

* Supply included in follow-up survey (see text for details).

Unflushed in last column means no use of tap overnight or longer; if known, residence times are given.

until the field chemist arrived. The system was then sampled,
usually at the kitchen tap, by first collecting four 250-ml
portions, then one 1-liter sample, and finally one sample after the
water had run for at least one minute.

The total lead concentrations in first-flush water decreased
from sample to sample in the order the samples were taken, except
in two cases where lead was undetectable even in the first portion.
The four highest concentrations, all in the first 250-ml portions,
were 0.39, 0.22, 0.18, and 0.085 mg l^{-1}. In every case, the second
liter and the sample from the flushed line contained either less
than 0.05 mg l^{-1} or no detectable lead (one suspicious result was
discarded).

A normal probability plot is given in Figure 5 of the total
lead concentrations in each of the eleven systems, i.e., the lead
concentrations in the first 250-ml portion, in the first full liter
collected in 250-ml portions, and in the first full liter less the
first 250-ml portion of water. The first two sets of data fall on
nearly straight lines indicating a lognormal distribution, but the
distribution of the third data set is skewed. The slope of each
regression line is the inverse of the geometric standard deviation
of the distribution, so that the increases of the slopes correspond
to decreases in the standard deviations.

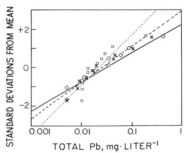

Figure 5.   Normal probability plots of
            total lead in first-flush
            water in selected individual
            water systems in Lewis and
            St. Lawrence counties, New York.
            Solid line (◊), concentration,
            per liter, in first 250 ml;
            dashed line (X), concentration
            in first liter; dotted line (□),
            concentration, per liter, in
            remaining 750 ml following
            first 250 ml

A fit to the lognormal distribution suggests a certain consistency of the data. In a much larger survey of lead concentrations in first-flush water throughout a large institutional water system (over three hundred 100-ml samples, unpublished data), the results were lognormally distributed with virtually the same geometric standard deviation as was found here, but with a lower log mean (the lower mean concentration for 100-ml portions indicates a much lower lead content in the other system).

Human Exposure

In these surveys, exposure to trace metals was investigated only for individual water systems. The public systems are subject to routine regulatory monitoring of finished water throughout the distribution system, and violations of the drinking water standard are investigated and corrected.

Copper - In natural waters not in contact with copper piping, copper concentrations were sometimes elevated but remained below the drinking water standard (1 mg l^{-1}) and the taste threshold (2 mg l^{-1}). Water in contact with copper plumbing occasionally exceeded these concentrations, particularly first-flush water. Toxic effects from copper occur only at concentrations above 2 mg l^{-1}. Personnel in local clinics, when polled and alerted to the possibility of copper poisoning, said that they had not heard of such cases in their areas.

Lead - Concentrations of lead in natural waters not in contact with metal plumbing occasionally approached the drinking water standard of 0.05 mg l^{-1}. These occurrences, however, were sporadic and without human health significance because of the slight safety margin built into the standard. In metal plumbing which contains lead piping or solder, however, lead concentrations in first-flush water reached levels of concern.

The question arises whether consistent use of first-flush water can cause lead intake to exceed the allowable daily intake. The World Health Organization (1972) recommended a maximum intake of inorganic lead from all sources of 3 mg per week for an adult. Kehoe (1966) recommended a maximum intake of 600 µg per day. An ad-hoc committee, appointed by the U.S. Public Health Service, established a permissible intake of 300 µg lead per day for an infant (King, 1971, cited from U.S.E.P.A., 1976).

The allowable intake from drinking water, considering possible contributions to total intake from other sources, is 100 µg per day, corresponding to two liters daily consumption at the drinking water standard of 0.05 mg Pb l^{-1} (U.S.E.P.A., 1976). Assuming a lead concentration after flushing of 0.01 mg l^{-1}, this maximum

allowable exposure is reached at concentrations in the first 250,
500, and 1000 ml of first-flush water of 0.330, 0.170, and
0.090 mg Pb l^{-1}, respectively. Each of these concentrations was
reached or exceeded in two of the 11 samples in our survey. The
lognormal distribution of values suggests that these concentrations
are reached or exceeded in 5, 19, and 14%, respectively, of first-
flush water samples from a group of similar water systems.

Results of a larger statistical sample, from the larger water
system mentioned before, show a low incidence of higher concentrations, as would be predicted by extrapolation from data provided in
the present study. It is believed, however, that the lead surface
exposed at solder joints in small residential piping systems will
set a practical limit to the lead concentration so that much higher
concentrations may occur only after very long residence times. If
the first 250 ml are discarded, the lead concentrations in the next
750 ml do not follow a lognormal distribution (Figure 5). Slightly
elevated lead concentrations are effectively lowered and stabilized
(increase in slope) whereas high concentrations are barely reduced
and remain quite variable (reduction in slope). This suggests both
a clustering of soldered connections near the tap and within long
piping systems from which high concentrations are cleared only
after 1000 ml of water have been passed. A 250-ml sample
corresponds to 2 m (6 ft) of 12.5 mm (0.5 in) i.d. pipe, and a
1000-ml sample corresponds to 8 m (approximately 25 ft)--dimensions
which are typical of small residential systems.

DISCUSSION

This work was undertaken to obtain information on raw drinking
water quality as it currently exists in an area of New York State
considered susceptible to acidification. The study area is
generally the same as in two previous studies by the Department of
Health Laboratory, which dealt with the quality of lake water and
stream runoff in the central Adirondack Mountains (Wood, 1978, Fuhs
et al., 1982).

The present study was not intended to quantify acid rain
effects nor to present information indicating a decrease of alkalinity with time. The presence of sulfate as the predominant anion
does suggest, however, an influence of acid deposition on water
chemistry. Calculations showed that the groundwaters were super-
saturated with carbon dioxide, and acidity from this source is
included in our measurements. Natural weak acids may have been
present, although none of the waters were colored. The suggestion
that much acidity in areas receiving acid deposition is due to
natural acids has merit but does not argue against the role of
airborne mineral acidity. On the contrary, natural acids are
formed, persist and accumulate where natural alkalinity is lacking.

Well-known examples are peat bogs, tree litter (particularly on acid soil), and lichens growing on tree bark or acidic rock surfaces (Abrahamsen et al., 1979). Under conditions where the leaching of alkalinity from soil and other substrates is increased by acid deposition, the production of natural acids also is likely to increase.

Soil contact is important in the neutralization of acid waters. Shallow wells and hillside springs typically produce water of pH 5 to 5.5, whereas even in the central Adirondacks 30 m deep wells produce neutral or near-neutral water. The use of limestone barriers to neutralize acid runoff was examined by Grahn and Hultberg (1975), Pearson and McDonnell (1975, 1976), and Barton and Vatanatham (1976). Department field staff recommend the use of such barriers in acid streams and shallow wells used as sources of small water supplies.

In artificial leaching experiments with a leucogranitic gneiss from the Cranberry Lake area of the Adirondack mountains, rapid leaching of 90% of total copper, nearly 50% of the total mercury, and 20 to 30% of the total lead, zinc, and cadmium content of the rock was observed (Fuhs et al., 1982). Etch pits in scanning electron micrographs suggested the selective dissolution of trace-metal-rich components such as hornblende and biotite. The rock matrix, which dissolves more slowly. It releases, in decreasing order, magnesium, calcium, potassium, aluminum, iron, manganese, and sodium. Such elevated copper concentrations in waters not in contact with metal plumbing have been observed, but the concentrations were not high enough to be of human health concern. The same applies to sporadic findings of somewhat elevated lead concentrations. Lead in this area is mostly derived from atmospheric deposition and is retained in the watersheds (Swanson and Johnson, 1980; Troutman and Peters, 1982) and lake sediments where it is mostly associated with organic matter (Galloway and Likens, 1979; Heit et al., 1981; Davis et al., 1982; Fuhs et al., 1982).

Aluminum in Adirondack waters is most likely derived from bedrock. In the New Jersey pine barrens, however, atmospheric additions and watershed losses are balanced (Budd et al., 1981). In each case, the concentrations vary inversely with the pH of the water (Fuhs et al., 1982). The toxicity of free aluminum to fish is well established; the toxicity to humans of dietary aluminum is not. A draft standard of 0.01 mg l^{-1} was, however, adopted by the Association for the Advancement of Medical Instrumentation (AAMI) for aluminum in hemodialysis water. This standard was endorsed by the U.S. Department of Health and Human Services (U.S.D.H.H.S., 1981).[1] This concentration is greatly exceeded in many acid waters

---

[1]The U.S. Food and Drug Administration has also unofficially endorsed this standard (telecon with an unnamed official).

in the Adirondacks and elsewhere. Limestone treatment is effective, as is reverse osmosis, in lowering the concentration of
soluble (dialyzable) aluminum. Total aluminum in drinking water
from public supplies which use alum in water treatment may occasionally reach or exceed 0.01 mg l^{-1}. This aluminum generally is not
in dialyzable form but contributes to dietary intake. It is most
likely aluminum hydroxide floc which can be easily removed by
physical means, e.g., membrane filtration. Neither the FDA recommended standard nor a recent survey of drinking waters by the U.S.
Environmental Protection Agency (Miller et al., 1984) made a
distinction between these forms of aluminum.

The corrosion potential of acid waters, particularly with
regard to copper and lead, has been studied and documented for many
years. Aggressivity toward copper and lead are clearly evident in
almost all waters included in this study. In copper plumbing with
lead solder, the use of first-flush water for drinking will provide
a dose of lead equalling or exceeding the allowable daily intake
from water in a significant percentage of homes which are served by
individual water systems. Intake which does not significantly
exceed the maximum allowable level is not associated with a
measurable risk to human health, but the safety margin provided by
the standard in the case of lead is small, and it cannot be ruled
out that intake in individual situations could be higher.

As there is no benefit associated with the ingestion of lead
at any level, intake should be reduced as much as possible.
Frequency distributions permit estimation of the dose to the
population, but are irrelevant for individuals. Therefore, it has
been standard practice for field staff to advise residents to avoid
the intake of first-flush water from metal piping in situations
where untreated, corrosive waters are used. Specifically, if water
remains in the system overnight, the first liter (or quart) of
water should be discarded. Use of tin-antimony solder, which is
quite inert, will prevent lead contamination; but contamination
with copper can still be a problem, particularly in hot-water
lines.

Piping systems made entirely of lead pose a definite health
hazard with acid and even neutral waters unless plumbosolvency is
controlled in some appropriate manner. In three rural homes west
of the Catskill mountains (Otsego and Delaware counties), where
long service lines of lead were used to supply water of pH 5.2 and
5.8 from wells or hillside springs, lead concentrations in first-
flush water were on the order of 5 mg l^{-1}, whereas in the flushed
lines, lead was at or near the drinking water standard of
0.05 mg l^{-1} (New York State Department of Health, 1981). The
drinking water in these homes was identified as the source of lead
after a U.S. government-funded health survey (Child Health Assur-

ance Program) had shown elevated blood lead levels (40 to 57 μg dcl^{-1}) in children of these households. Similar blood lead concentrations were found in the adults. Field staff involved with these cases have reported that backfilling of the sources with limestone has been attempted with apparently good results.

Personnel operating most public water supply systems with lead service connections are familiar with the requirement to control lead solubility and to carry out a monitoring program. U.S. Environmental Protection Agency and state regulatory agencies are placing increased emphasis on corrosion control in all public water supplies in susceptible areas. In addition, there is clearly a need for a program to determine long-term trends in the alkalinity and ionic composition of source waters and to obtain such data in a quality-controlled manner, coordinated with other measurements of acid deposition and its effects.

CONCLUSIONS

The chemical composition of drinking water in several regions of New York State sensitive to acid precipitation is described in this paper. Sources of public water supply, both groundwater and surface water, were generally slightly acid and corrosive. Chlorine additions slightly increased corrosivity. Shallow ground water serving as a source in many individual residences typically showed pH values near five and negligible alkalinity.

Increasing pH and electrolyte content was found in source waters which had been in substantial contact with mineralized soils. Dissolved copper concentrations in first-flush water from households containing copper plumbing were seen to reach levels exceeding the U.S. drinking water standard and taste threshold. Dissolved lead was also leached from lead-containing solder joints, and first-flush waters could reach unacceptable levels. This problem was evident in 10-15 percent of the households within the study area. Even higher levels of lead have been observed in individual water supplies which have service lines made entirely of lead and served by hillside springs or shallow wells. Even though it is not known what percentage of the acidity is directly attributable to acid deposition, it is clear that these source waters are susceptible to further acidification.

The problem of heavy metal contamination of source waters in New York State is manageable mostly because of the lack of deposits with readily leachable toxic metals in the affected areas. However, corrosivity of the water and the potential for leaching of heavy metals, such as copper and lead, require continued attention by the regulatory agencies and affected residents.

Acknowledgments - Staff of the Department of Health's District
Offices in Utica, Saranac Lake, Watertown, Amsterdam, and Massena
were most helpful as were the many local residents who cooperated
with our program. Special thanks are due to Mr. Kenneth Graulich
of the Utica Office for establishing contacts with many local
residents and to James Decker, P.E., of the Albany Area Office for
helpful discussions. Trace metals analyses were carried out by
staff at the Center's Chemical Sciences Laboratory.

REFERENCES

Abrahamsen, G., Stuanes, A., and Bjor, K., 1979, Interaction
    between simulated rain and barren rock surface. Water, Air,
    Soil Pollu. 11:191-200.
American Public Health Association, 1975, Standard Methods for the
    Analysis of Water and Wastewater, 14th Ed., American Public
    Health Association, Washington, DC. 1193 pp.
Barton, P. and Vatanacham, T., 1976, Kinetics of limestone neutral-
    ization of acid waters. Envir. Sci. Tech. 10:262-266.
Budd, W.W., Johnson, A.H., Huss, J.B., and Turner, R.S., 1981,
    Aluminum in precipitation, streams, and shallow groundwater in
    the New Jersey Pine Barrens. Water Resources Res. 17:1179-
    1183.
Davis, A.D., Galloway, J.N., and Nordstrom, D.K., 1982, Lake
    acidification: Its effect on lead in the sediment of two
    Adirondack lakes. Limnol. Oceanogr. 27:163-167.
Fuhs, G.W., 1972, The chemistry of streams tributary to Lake
    George, New York. Environmental Health Report No. 1, New York
    State Department of Health, Albany, NY. 53 pp.
Fuhs, G.W., 1979, A contribution to the assessment of human health
    effects of acid precipitation, pp. 113 to 116. In: Proceed-
    ings of the Action Seminar on Acid Precipitation, Toronto,
    Ontario. ASAP Organizing Committee, Toronto, Ontario, Canada.
Fuhs, G.W., Reddy, M.M., and Parekh, P.P., 1982, Distribution of
    mercury and 14 other elements in remote watersheds in the
    Adirondack mountains, pp. 63 to 88. In: Energy and Environ-
    mental Chemistry, Vol. 2, Acid Rain, L.H. Keith, ed., Ann
    Arbor Science, Ann Arbor, MI.
Galloway, J.N. and Likens, G.E., 1979, Atmospheric enhancement of
    metal deposition in Adirondack lake sediments. Limnol.
    Oceanogr. 24:427-433.
Golterman, H.L., 1969, Methods for Chemical Analysis of Fresh
    Waters (IBP Handbook No. 8). Blackwell Publ., Oxford, U.K.
    172 pp.
Grahn, O. and Hultberg, H., 1975, The neutralizing capacity of
    12 different lime products used for adjustment of acid water.
    Vatten 2:120-132.

Heit, M., Tan, Y., Klusek, C., and Burke, J.C., 1981, Anthropogenic trace elements and polycyclic aromatic hydrocarbon levels in sediment cores from two lakes in the Adirondack acid lake region. Water, Air, Soil Pollu. 15:441-464.

Jeffries, D.S., Cox, C.M., and Dillon, P.J., 1979, Depression of pH in lakes and streams in central Ontario during snowmelt. J. Fish. Res. Board Canada 36:640-646.

Kehoe, R.A., 1966, Under what circumstances is ingestion of lead dangerous?, pp. 51 to 58. In: Symposium on Environmental Lead Contamination. Publication No. 1440, Public Health Service, U.S. Department Health, Education, and Welfare, Washington, DC.

McFee, W.W., Kelly, J.M., and Beck, R.H., 1976, Acid precipitation effects on soil pH and base saturation of exchange sites, pp. 725 to 736. In: Proceedings of the First International Symposium on Acid Precipitation and the Forest Ecosystem, L.S. Dochinger and T.A. Seliga, eds., Gen. Tech. Rep. NE-23, USDA Forest Service, Northeastern Forest Experiment Station, Upper Darby, PA.

Miller, R.G., Kopfler, F.C., Kelly, K.C., Stober, J.A., and Ulmer, N.S., 1984, The occurrence of aluminum in drinking water. J. Amer. Works Assoc. 76:84-91.

New York State Department of Health, 1981, Lead poisoning related to private drinking water sources. Technical Memorandum, 23 October 1981, N.Y. Dept. Health, Albany, NY. 9 pp.

New York State Museum and Science Service, 1970, Geologic Map of New York, Adirondack Sheet, Map and Chart Series No. 15, New York State Education Department, State Univ. New York, Albany, NY.

Pearson, F.H. and McDonnell, A.J., 1975, 1976. Limestone barriers to neutralize acid streams. J. Envir. Engng. Div. ASCE 101:425-441 and 102:690.

Pfeiffer, M.H. and Festa, P.J., 1980, Acidity status of lakes in the Adirondack Region of New York in relation to fish resources. FW-P168 (10/80), New York State Dept. Environmental Conservation, Albany, NY. 36 pp.

Schofield, C.L., 1976, Dynamics and management of Adirondack fish populations. Final Report for Project F-28-R, New York State Dept. Environmental Conservation, Albany, NY. 11 pp.

Swanson, K.A. and Johnson, A.H., 1980, Trace metal budgets for a forested watershed in the New Jersey Pine Barrens. Water Resources Res. 16:373-376.

Troutman, D.E. and Peters, N.E., 1982, Deposition and transport of heavy metals in 3 lake basins affected by acid precipitation in the Adirondack Mountains, New York, pp. 33 to 61. In: Energy and Environmental Chemistry, Vol. 2, Acid Rain, L.H. Keith, ed., Ann Arbor Sci. Publ., Ann Arbor, MI.

U.S. Department of Health and Human Services, 1981. Bureau of Health Standards and Quality, Letter No. 226 (June 25, 1981) with attachments. New York, NY. and U.S. Department of Health and Human Services, Health Care Fianancing Administration, Region II Office, New York, NY. 9 pp.

United States Environmental Protection Agency, 1976, Quality Criteria for Water. U.S.E.P.A., Washington, DC. 256 pp.

Wood, L.W., 1978, Limnology of remote lakes in the Adirondack region of New York with emphasis on acidification problems. Environmental Health Report No. 4, New York State Department of Health, Albany, NY. 65 pp.

Wood, L.W. and Fuhs, G.W., 1979, An evaluation of the eutrophica-tion process in Lake George, New York based on historical and 1978 limnological data. Environmental Health Report No. 5, New York State Department of Health, Albany, NY. 73 pp.

World Health Organization, 1972, Lead, pp. 16 to 20. In: Expert Committee on Food Additives, 16th Report, World Health Organ-ization, Geneva, Switzerland.

EFFECTS OF ACIDIFICATION ON THE PRIMARY PRODUCERS OF SOFTWATER LAKES

Charles W. Boylen

Fresh Water Institute and
Department of Biology
Rensselaer Polytechnic Institute
Troy, NY

ABSTRACT

Increased acidification of the aquatic environment has brought about alterations in community diversity and distribution, shifts in species dominance, and changes in the levels of biomass and productivity. Phytoplankton biomass decrease is concomitant with increased water clarity. Benthic mats of Sphagnum or bluegreen algae often develop, effectively sealing sediments from overlying water. A relatively few species of macrophytes dominate the littoral zone which increases in depth due to improved water clarity. As pH decreases, the emergence of a more simplified ecosystem is observed; this is even more vulnerable to changing environmental conditions than the one which existed previously.

INTRODUCTION

The primary producers occupy a unique position within the aquatic environment. Through photosynthetic activity, they supply virtually all the autotrophically-derived organic carbon which, in the form of a host of taxonomically varied organisms, selectively becomes food for grazers. These, in turn, support higher trophic levels. The producers mobilize nutrients from the sediment as well as the water column which are then made available to the food chain directly by consumption or indirectly by excretion of dissolved organic carbon. Nutrients captured in plant biomass are recycled through senescence and decomposition, thus supplying a major component of the organic carbon which supports the detritus-based food chains.

This overview of the effects of acidification on the primary producers is not intended to be an exhaustive review of the literature. Our knowledge to date will be discussed in terms of the difficulties in interpreting this information in light of the overall complexity of the aquatic ecosystem. It will attempt to deal with the misconceptions, seeming contradictions and significance of many of the research data found in the literature.

DISCUSSION

Even though a description of the photosynthetic biota found in various acidic aquatic habitats will be briefly discussed, primary focus will be on clearwater, low alkalinity ecosystems. An examination will be made of each producer community in turn with a discussion of what is known about the effects of acidification upon the phytoplankton, the periphyton and benthic algae, and the macrophytes or higher aquatic plants.

Types of Acidic Aquatic Habitats

Inland waters exhibit a wide variety of pH values most often governed by the watershed geology and terrestrial biome in which they exist. Each is distinct as to water chemistry and the types of biota it will support. Naturally occurring acidic waters can be categorized into three groups:

°   inorganic acidotrophic waters associated with geothermal areas or the oxidation of exposed pyrite from mining tailings (pH range 1.0 - 3.5)

°   brownwater lakes and streams associated with peatlands and cypress swamps (pH range 3.5 - 5.0).

°   low alkalinity softwater oligotrophic waters occurring in watersheds of minimal buffering capacity (pH range 4.5 to circumneutral).

Documentation of acidic environments has been known since the 1860's (Hutchinson, 1957). Geothermal areas can be found worldwide. In North America, they are confined primarily to the Yellowstone Park region of Wyoming. Lignite burns are localized in areas of northern Canada (Sheath et al., 1982) whereas acid mine drainage is widespread in North America associated with coal mining operations in Pennsylvania and West Virginia and the midwest (Kentucky, Indiana, and Illinois and metal ore mining in western portions of the U.S. and Canada. The high acidity arises from sulfuric acid produced from the microbial oxidation of hydrogen sulfide (H_2S) and pyrite (FeS_2). In its extreme, surface water contamination by sulfate in streams or impoundments can lower the pH to 2.0 or less. In streams impacted by acid mine drainage chlorophytes, Chlorella sp. (Brock, 1978) and euglenoids, Euglena mutabilis (Hutchinson et al., 1978) can be found in waters as low as pH 2.0. In Yellowstone where acidic streams have pH values of 1.0 or less, monospecific populations of the chlorophyte Cyanidium caldarium can be found (Smith and Brock, 1973). Such ecosystems tend to be very simple in community makeup with very low species diversity (Lind and Campbell, 1970). More common forms of plankton and higher plants are lacking (Johnson et al., 1970; Hargreaves et al., 1975; Parsons, 1977).

Brownwater lakes and streams are typically associated with peatlands and cypress swamps (Janzen, 1974; Moore and Bellamy, 1974). Brownwater habitats contain an increased diversity of organisms than that found in inorganic acidotrophic waters, due largely to a lack of temperature, pH extremes, and higher nutrient levels (Smith, 1961). They are characterized by reduced light penetration, high concentrations of dissolved organic matter, low ion concentrations, and low dissolved oxygen. Their acidity is derived from organic acids leached from decayed gymnosperm plant materials. A major producer of these ecosystems, the moss Sphagnum, further exacerbates the low pH by releasing hydrogen ions in exchange for nutrient ions such as Ca^{2+} (Clymo, 1963, 1967). Many brownwater lakes become dystrophic supporting no aquatic plant communities at all (McLachlan and McLachlan, 1975) although the aquatic fern Isoetes and several genera of vascular plants, notably Alternanthera, Ceratophyllum, Juncus, Limnobium, Nuphar, Potamogeton, and Utricularia, have been reported by others (Griffiths, 1973; Stoneburner and Smock, 1980). The shoreline plant community is often extensive and has been well defined for northern bogs (Heinselman, 1970; Vitt and Slack, 1975). Species representing most phytoplankton phyla have been reported albeit in low densities (Birge and Juday, 1927; Bricker and Gannon, 1976; Stoneburner and Smock, 1980). Diatoms and desmids are part-icularly characteristic (Woelkerling and Gough, 1976; Stoneburner and Smock, 1980). Often species diversity is relatively high in brownwater environments because the increased metal solubility upwards to toxic concentrations, normally associated with acidi-fication, does not occur due to chelation by humic acids (Patrick et al., 1981).

Clearwater, Low Alkalinity Waters

Low alkalinity softwater environments will be reviewed in greater detail. Their chemical vulnerability is more directly associated with acidic deposition (Beamish, 1976; Galloway et. al., 1976). A number of observations provided here have been discussed in other recent reviews (Nilssen, 1980; Haines, 1981; Conway and Hendrey, 1982).

Most of the pristine ultraoligotrophic lakes and streams of the world occur in the northern latitudes. Those located in the northeastern United States, the Canadian Shield region, and Scandinavia are often found in regions of igneous rock where glaciation has removed younger calcareous layers exposing the more weather-resistant granitic and siliceous bedrock (Ryder, 1964; Galloway and Cowling, 1978). Such ecosystems are small, characterized by low nutrient concentrations, low dissolved ion concentrations, and low primary and secondary productivity. The absence of carbonate rocks results in water of low carbonate-bicarbonate buffering capacity. The buffering ability of the water is strained with even low acidic additions into the drainage basin. On a global scale, these clearwater, low alkalinity, acid-susceptible environments are the most vulnerable of all bodies of water (Drabløs and Tollan, 1980; National Research Council of Canada, 1981). From numerous studies it is clear that these lakes have become recently acidified (Cogbill, 1975). Although much of the published literature is widely scattered and casually documented, it would appear that historically these lakes contained, prior to acidification, a great diversity of taxa at all levels even though overall productivity was usually low (Hutchinson, 1967).

Phytoplankton

Diversity. The phytoplankton of low alkalinity oligotrophic lakes represent a diversity of taxa. In temperate lakes of circum-neutral pH, the community may consist of several hundred species (Kalff and Knoechel, 1978). Crysophytes, diatoms, desmids, and other green algae are characteristic (Hutchinson, 1967; Schindler and Holmgren, 1971; Schindler, 1972). Bluegreen algae and dino-flagellates are typically not significant (Schindler and Holmgren, 1971; Duthie and Ostrofsky, 1974; Ostrofsky and Duthie, 1975a, 1975b). As hydrogen ion concentration increases in these ecosystems, species diversity decreases. In specific studies, the number of desmid species declined (Coesel et al., 1978). In Dutch moor-land acidified pools, diatom species diversity has declined between the 1920's and 1978 but has not in nonacidified pools (Van Dam and Kooyman-Van Blokland, 1978; Van Dam et al., 1980). Diatom samples, collected from Norwegian acidified lakes between 1949 and the 1970's, have shown a definite increase in acidophilic species (Leivestad et al., 1976; Davis and Berge, 1980).

Not only is there a change in phytoplankton species diversity as pH declines, but also a change occurs in the composition of dominant species. Almer et al. (1974, 1978) showed in Swedish lakes greater than pH 6 that all planktonic groups are present in approximately equal numbers. At lower pH, the proportion of cyanophytes and chlorophytes become reduced while the proportion of chrysophytes and dinoflagellates increases (Almer et al., 1974, 1978). A similar pattern has been observed in Adirondack lakes

(Hendrey et al., 1981) and in Canadian Shield lakes in Ontario to the extent that some of the same species were dominant in both countries (Yan and Stokes, 1978; Yan, 1979). In contrast, Schindler et al. (1980) and Schindler and Turner (1982) found that after experimental acidification, no changes were seen in phytoplankton composition. In this study, however, the pH was only reduced to 5.6. The studies of Conroy et al. (1976) and Kwiatkowski and Roff (1976) are in even further contrast. In seven metal contaminated lakes in the Sudbury area of Ontario with pH values as low as 4.4, cyanophytes comprised as high as 70% of the community species composition.

Many of the data supporting the change to acidophilic species within the plankton community in the recent past have come from an assessment of sediment diatom stratographic analysis. Such studies from Scandinavia (Davis and Berge, 1980; Huttunen and Merilainen, 1983), Europe (Van Dam et al., 1980), and the United States (Del Prete and Schofield, 1981; Brugam, 1983; Del Prete, 1983) are all consistent in their conclusions. Over the last 100 years, many currently acidified oligotrophic clearwater lakes worldwide now support an abundance of acidophilic chrysophyte species which once contained predominately circumneutral species.

With the phytoplankton, several factors contribute to community changes. Many species are intolerant to low pH and cease to play dominate roles while acid tolerant species formerly numerous, such as the chrysophytes, continue to dominate. On the other hand, the dinoflagellates, which are not numerous in circumneutral oligotrophic waters, now become significant (Almer et al., 1978; Yan, 1979). At present, there are insufficient data to conclude that species changes are strictly due to either low pH tolerance or intolerance (Foy and Gerloff, 1972; Cassin, 1974; Baker et al., 1983).

Altered grazing pressures certainly occur because of changes in planktonic zooplankton and benthic invertebrates (Porter, 1973; Almer et al., 1974, 1978; Hendrey and Wright, 1976; McCauley and Briand, 1979). There is experimental evidence to support the reported effects of acidification on algae, zooplankton, and invertebrates are the result of changing predator-prey relationships brought about by the disappearance of fish from these lakes rather than from toxic effects of acidity or metal ions. Effects such as species reduction and altered species dominance can be experimentally produced by the removal of fish from circumneutral lakes (Eriksson et al., 1979; Henrikson et al., 1980). When diversity, productivity, and biomass of the phytoplankton all decrease, the base of the aquatic food chain becomes both qualitatively and quantitatively impoverished. Little is known to what extent low zooplankton and invertebrate diversity is a result of a decrease in phytoplankton diversity and biomass. Species reduced in acidified streams are often herbiovores and scrapers (Sutcliffe and Carrick, 1973; Friberg et al., 1980) suggesting a reduction in a more taxonomically varied periphyton community. Its replacement by a less palatable community dominated by cyanophytes and bryophytes may be responsible for reductions in these organisms (Haines, 1981).

Productivity. Relatively little research has been done on the productivity of phytoplankton in acidified waters, and the data that do exist are contradictory. Conclusive data are not yet available which would show that increased acidity actually reduces primary productivity. In studies in the Canadian Shield region and the Adirondacks, primary productivity values were two to three times lower than those in comparable circumneutral lakes (Kwiatkowski and Roff, 1975; Conroy et al., 1976; Conway and Hendrey, 1982). Several studies showed a concomitant decrease in algal biomass in the water column (Stokes et al., 1973; Wright et al., 1976; Conroy et al., 1976). However, Schindler (1980) found that phytoplankton biomass and productivity did not decline in a lake experimentally acidified from pH 6.6 to 5.6.

It has been shown that in many of the studied lakes where species diversity is clearly reduced, planktonic productivity and biomass appear to be more closely aligned to nutrient levels, especially phosphorus, than pH. Phosphorus limitation becomes significant under such conditions because aluminum, found at elevated levels in acidified waters, removes dissolved phosphorus by flocculation and precipitation (Dickson, 1978). When acidic and nonacidic lakes with similar phosphorus levels are compared, biomass and productivity are similar (Hendrey et al., 1976; Almer et al., 1978; Yan and Stokes, 1978; Dillon et al., 1979; Yan, 1979; Raddum et al., 1980; DeCosta et al., 1983). When phosphorus was added to acidic lakes in Ontario, phytoplankton biomass increased (Dillon et al., 1979; Wilcox and DeCosta, 1982). However, in a lake acidified by mine drainage, phosphorus additions did not increase phytoplankton biomass levels unless the water was simultaneously buffered (DeCosta and Preston, 1980).

If water column chlorophyll concentration is considered as a measure of algal biomass, such measurements have shown a decrease with lower pH in Ontario and Adirondack lakes of similar phosphorus levels (Kwiatkowski and Roff, 1976; Boylen et al., 1982; Conway and Hendrey, 1982; Singer et al., 1983a). When pH and biomass or pH and productivity relationships are as contradictory as they sometimes appear, the conclusion that pH does not control phytoplankton biomass directly must be considered (Conway and Hendrey, 1982).

Periphyton

As with the phytoplankton -- with an increase in acidity, the number of species in the periphyton community decreases (Almer et al., 1974; Muller, 1980). However, as changes occur in species composition within both the epiphytic and epilithic communities, successful members often thrive as the pH drops below 5.0 (Hendrey et al., 1976; Leivestad et al., 1976). A striking feature, observed in lakes of pH below 5.0 in both Scandinavia and North America, is the decline in the typically diatom-dominated periphyton community and the development of a filiform algal community. In such communities, the chlorophytes Mougeotia and Spirogyra often dominate; however, Batrachospermum and Dinobryon also can be found densely covering stems and leaves of any object positioned in the water column, e.g., macrophytes, twigs, branches, etc. (Almer et al., 1978; Nilssen, 1980; Stokes, 1981; Singer et al., 1983b). Similar results can be produced in streams and lake enclosures undergoing artificial acidification (Hendrey, 1976; Hall and Likens, 1980, Hall et al., 1980; Muller, 1980). Schindler (1980) and Stokes (1981) noted that Mougeotia first became abundant at approximately pH 5.5. Although an increase in standing crop biomass is observed with decreasing pH, when experimentally measured, no concomitant increase in productivity occurs (Hendrey, 1976; Muller, 1980). It has been suggested that the increase in algal standing crops at low pH are related to a reduction in heterotrophic activity (Haines, 1981; Conway and Hendrey, 1982).

Early descriptions of acidic Swedish lakes by Grahn et al. (1974) and Hultberg and Grahn (1976) documented the extensive growth of fungi on the sediment surface. Similar observations were made by Hendrey et al. (1976) and Hendrey and Barvenik (1978). Hendrey and Vertucci (1980) and Stokes (1981) reported that such observations of the so-called fungal mat development were incorrect and that the mats are actually composed primarily of filamentous algae. As more and more acidic lakes are studied, this type of mat appears to be extraordinarily common. The cyanophytan mat differs from that produced by the chlorophytes in several characteristics: a) it is more tightly cohesive being only 2-5 mm thick and grows on the bottom as opposed to the cloudlike chlorophyte mass that forms on and around material suspended within the water column; b) it is found throughout the littoral zone and in particular can be quite extensive at substantial depths as opposed to the chlorophyte clouds which tend to be restrictive to the more shallow portions less than three m deep; and c) this mat may play a significant role in the sequestering of reactive phosphorus within the aquatic ecosystem making it unavailable to the phytoplankton (Gorham, 1976; Cole and Stewart, 1983).

In certain acidified lakes in Sweden, these felt-like mats have been shown to consist of bluegreen algae of the genera Lyngbya, Oscillatoria, and Pseudoanabaena (Lazarek, 1980). In others, the cyanophytan mat is composed mainly of Hapalosiphon (Lazarek 1982a, 1982b). In the Adirondacks, the mat is predominantly Phormidium tenue (Hendrey and Vertucci, 1980) or Plectonema (Singer et al., 1983b) while in Canada, it is reported to be Scytonema and Phormidium spp. (cited in Stokes, 1981).

Because of their extensive growth in acidified lakes, it is suspected that they serve to maintain high total lake productivity even as the phytoplankton and macrophyte communities have been extensively reduced. Unfortunately there are no data to support that hypothesis. At present, the significance of the benthic algal production or various aspects of the acidic aquatic environment is highly conjectural. Nonetheless, mats existing on sediment cores taken from an acid lake in the Adirondacks and incubated under various conditions in the laboratory were highly active in their removal of phosphate and nitrate (Singer et al., 1983c).

The predominance of the cyanophytan mat in acidified lakes is somewhat an enigma. While their position within the planktonic community is much reduced as water pH is reduced, they dominate the benthic community. Basing his observations on acidified waters associated with geothermal activity worldwide, Brock (1973) suggested that pH 4 was the physiological acid limit for cyano- phytes and that they would be absent in waters of less than pH 5.0. However, as can be seen, acidified softwater oligotrophic waters are new ecosystems to which these organisms have readily adapted.

Macrophytes

Three rather general, but by no means universal, observations can be made of macrophyte communities in acidified water: a) species diversity declines, b) the littoral zone increases due to enhanced water clarity, and c) increased growth of mosses such as Sphagnum competes effectively to reduce the presence of higher plants. In terms of whole lake macrophyte biomass and produc- tivity, reported changes attributed to acidification are not as clearly shown.

The flora of oligotrophic temperate softwater lakes is highly diverse (Fassett, 1930; Moyle, 1945; Ogden et al., 1976; Sheldon and Boylen, 1977; Pip, 1979). Contributions to species diversity include floating-leaved, filiform, and rosette species. As a group, the Potamogetonaceae are very numerous in clearwater oligo- trophic lakes (Hellquist, 1980). However, on a species basis, they rarely become a major component of the macrophyte community. The dominant flora of such temperate lakes in both North America and Europe are the rosette species (Hutchinson, 1975) often repre- senting genera of single or a very few species such as Eriocaulon septangulare, Isoetes sp., Lobelia dortmanna, and Saggitaria sp.

The alteration of macrophyte community structure due to acidification was first noted in Scandinavia (Grahn et al., 1974; Hultberg and Grahn, 1976; Grahn, 1977; Halvorsen, 1977). Macrophyte communities originally dominated by Lobelia and Littorella were later dominated by Sphagnum sp. or Juncus bulbosus, a species common to Scandinavia but one not reported in North America (Nilssen, 1980). Hendrey and Vertucci (1980) were the first to report a similar dominance pattern by Sphagnum in Lake Colden in the Adirondacks. The conclusions of these earlier studies are probably an oversimplification. Invasions by Sphagnum into North American lakes have not been a widespread observation (Gorham, 1976; Wile, 1981; Wile and Miller, 1983). In the study of 20 clearwater Adirondack lakes, of those with pH £ 5.5, less than half had a dominant Sphagnum community (Singer et al., 1983a). In some acidic lakes in southern Canada, other mosses such as Drepanocladus fluitans and Leptodictyum riparium are seen to cover the bottom (Gorham and Gordon, 1963).

In studies encompassing whole lake macrophyte community mapping, Roberts et al. (1985) observed that the dominant species in low alkalinity clearwater lakes in the Adirondacks are the same regardless of lake pH. These species include Eriocaulon septangulare, Lobelia dortmanna, Myriophyllum tenellum, Utricularia spp., and several genera of water lillies. The major trend seen in this relatively small data set (nine lakes) was a reduction in the total number of species in the acidified lakes and a resultant reduction in diversity. The species eliminated from some lakes were, for the most part, minor components of the plant community while certain distinctly acid-tolerant species such as Potamogeton confervoides (a species common to the northeast United States but not reported in Scandinavia), U. geminiscapa and Sphagnum spp. appeared. A similar community profile was reported for acidic Lake Colden which is also in the Adirondacks (Hendrey and Vertucci, 1980). For larger data sets, results were not as clearly associated strictly with pH differences (Wile, 1981; Wile and Miller, 1983).

The recent use of SCUBA as a sampling technique showed that the depth extension of the littoral zone is much greater than earlier thought (Sheldon and Boylen, 1977, 1978). In Scandinavian lakes where acidification has reduced plankton biomass, water clarity has increased significantly, thereby extending the depth at which rooted plants are found. In Sweden (Grahn, 1977) and Norway (Nilssen, 1980), inland lakes showed an increased transparency by more than 2-fold over the last 2-3 decades coupled with an increase in the depth extension of the littoral zone. Singer et al. (1983b) observed in Silver Lake in the Adirondacks the occurrence of Utricularia geminiscapa at 18 m.

An historic comparison of changes within the same body of water is largely lacking. Grahn (1977) compared the current dominance of Sphagnum in Swedish lakes with its abundance 30-50 years earlier. Hendrey and Vertucci (1980) made similar comparisons on Lake Colden with reference to 1932 data. Such community changes may well be indicative of pH alterations of the water when all other factors (nutrients, land use patterns, etc.) can be assumed relatively constant. In a recent study of Deer Pond, a circumneutral lake in the Huntington Forest in the Adirondacks, Roberts et al. (1985) found that the presence of macrophyte species, their abundance and relative depth preferences were basically unaltered since an earlier investigation in 1940 (Heady, 1942).

The conditions found in acidic lakes favor rosette, reduced and fine-leaved plant forms generally successful in other oligotrophic environments Morphology appears to play a significant role in their success in such waters. Isoetid plants such as Lobelia dortmanna and Littorella uniflora absorb CO_2 directly from the sediment (Wium-Anderson, 1971; Sondergaard and Sand-Jenzen, 1979), an adaptation to low inorganic carbon concentrations in the water column. Best and Peverly (1981) analyzed the elemental composition of numerous macrophyte species growing in acidic lakes and found the phosphorus levels to be close to limiting levels. Therefore, other sediment-enriched nutrients are also taken up through extensive root systems (Carignan and Kalff, 1979, 1980; Barko and Smart, 1981).

Direct laboratory effects of acidity on aquatic plant productivity have not been extensively conducted. Laake (1976) showed that the growth of L. dortmanna was reduced by 75 percent at pH 4 as compared to controls grown at higher pH values (4.3 to 5.5). Lobelia appears to thrive in acidic waters at pH 4.4 to 4.8 (Roberts et al., 1985). Roberts et al. (1982) found reduced growth of Potamogeton robbinsii, Myriophyllum sp., and Vallisneria americana below pH 5. These species, although common in circumneutral softwaters of northeastern United States, are not found normally in acidic lakes.

CONCLUSIONS

Interpreting the aquatic acidification literature is difficult because complex physical, chemical, and biological relationships are being altered. The effects on all components of the aquatic ecosystem are varied. Many show dramatic alterations with dominant species disappearing or becoming dominant while other taxa undergo more subtle changes. Short-lived organisms with high reproductive potential respond quickly and markedly to change whereas long-lived organisms respond more slowly. Survey data comparing different bodies of water are often difficult to interpret because the changes in biota occur at different rates due to different metabolic responses to acidity. Water acidification occurs at different rates. Many circumneutral, low alkalinity waters have been undergoing acidification worldwide over the past few decades. Originally these had different biota from the beginning, so changes are often manifested in each lake in entirely different ways.

The consensus of recent reviewers (Haines, 1981; Conway and Hendrey, 1982) is that the literature is highly varied in experimental rigor and scientific merit. Many of the early studies of the past decade lack quantification. Although such synoptic lake surveys concentrating on changes in the kinds and numbers of species are important, the lack of knowledge of alterations to productivity and how such rates are influenced by pH make it difficult to assess the effects on energy processing phenomena. When studies are qualitative and based on limited sample observations, overgeneralizations are easy to make and extremely misleading. Statements such as "the most striking change [with acidification] is the disappearance of diatoms and bluegreen algae" (Almer et al., 1978) and "in acidified lakes, large areas of the sediment surface are made up of a dense felt of fungus hyphae" (Hultberg and Grahn, 1976) are simply incorrect. With more intensive observations, the so-called fungal mat was actually composed of bluegreen algae (Hendrey and Vertucci, 1980), a characteristic community now found in many acidified lakes.

In international literature of interest to scientists of varied backgrounds and disciplines, conclusions as to the effects of acidification on producer communities have been drawn from observations and experiments involving many different approaches. These include field observations, _in situ_ manipulations, and laboratory studies of communities as well as single isolated species. Acidification of freshwaters is a complex process of which an increase in acidity is only a single variable. Often only a few variables are measured in any given study so that discrepancies in data between studies make cause and effect relationships difficult to ascertain. Organism tolerance and adaptation must be evaluated in terms of other well-documented changes which include increased water clarity, loss of top predators and other changes in trophic interactions, increased metal solubility, and increased detritus accumulation and its consequent effects on nutrient cycling and availability.

Very little experimentation has been done to understand the effects of increased metal solubilization on growth of primary producers. Macrophytes and benthic algae are known to accumulate aluminum, toxic to invertebrates and fish, but the levels found in plant and algal tissue are not known to affect productivity (Stokes et al., 1973; Best and Peverly, 1981). Stokes et al.

(1983) reported filamentous algae to be active heavy metal accumulators in acid-stressed lakes.

Presently an incomplete picture has emerged concerning the community dynamics of producer organisms in acidic environments. It is hoped in the near future that definitive studies will be done to sort out the most influential parameters on community structure: limits of tolerance to acidity and extent of niche breadth, effects of grazing pressures and nutrient availability, and the degree of metal toxicity exhibited in acidified waters.

REFERENCES

Almer, B., Dickson, L., Ekstrom, C., and Hornstrom, E., 1974, Effects of acidfication on Swedish lakes. Ambio 3:30-36.

Almer, B., Dickson, L., Ekstrom, C., and Hornstrom, E., 1978, Sulfur pollution and the aquatic ecosystem, pp. 271 to 311. In: Sulfur in the Environment, Part II. J. Nriagu, ed., John Wiley and Sons, New York, NY.

Baker, M.D., Mayfield, C.I., Inniss, W.E., and Wong, P.T.S., 1983, Toxicity of pH, heavy metals and bisulfite to a freshwater green algae. Chemosphere 12:35-44.

Barko, J.W. and Smart, R.M., 1981, Sediment-based nutrition of submersed macrophytes. Aquatic Bot. 10:339-352.

Beamish, R.J., 1976, Acidification of lakes of Canada by acid precipitation and the resulting effects on fish. Water, Air, Soil Pollu. 6:541-514.

Best, M.D. and Peverly, J.H., 1981, Water and sediment chemistry and elemental composition of macrophytes in thirteen Adirondack lakes (Abstract). In: Proc. Intl. Symp. Acidic Precipitation Fishery Impacts in N.E. North America, N.E. Div. Am. Fisheries Soc., Cornell Univ., Ithaca, NY.

Birge, E.A. and Juday, C., 1927, The organic content of the water of small lakes. Bull. U.S.A. Bureau Fish. 42:185-205.

Boylen, C.W., Singer R., Roberts, D.A., and Shick, M.O., 1982, Chemical and biological surveys of Adirondack lakes, pp. 81 to 82. In: New York State Symposium on Atmospheric Deposition Proceedings, J.S. Jacobson, ed., Center Environmental Research, Cornell Univ., Ithaca, NY.

Bricker, F.J. and Gannon, J.E., 1976, Limnological investigation of Hoop Lake - a northern Michigan Bog. The Michigan Academician 9:25-41.

Brock, T.D., 1973, Lower pH limit for the existence of blue-green algae: Evolutionary and ecological implications. Science 179:480-483.

Brock, T.D., 1978, Thermophile Organisms and Life at High Temperatures. Springer-Verlag, New York, NY. 465 pp.

Brugam, R.B., 1983, The relationship between fossil diatom assemblages and limnological conditions. Hydrobiol. 98:223-235.

Carignan, R., and Kalff, J., 1979, Quantification of the sediment
    phosphorus available to aquatic macrophytes. J. Fish. Res.
    Board Can. 36:1002-1005.
Carignan, R., and Kalff, J., 1980, Phosphorus sources for aquatic
    weeds: Water or sediments? Science 207:987-988.
Cassin, P.E., 1974, Isolation, growth, and physiology of acido-
    philic Chlamydomonads. J. Phycol. 10:439-447.
Clymo, R.S., 1963, Ion exchange in Sphagnum and its relation to
    bog ecology. Ann. Bot. 27:309-324.
Clymo, R.S., 1967, Control of cation concentrations and in partic-
    ular of pH in Sphagnum dominated communities, pp.273 to 284.
    In: Chemical Environment in the Aquatic Habitat, H.L.
    Golterman and R.S. Clymo, eds., Noord-Hollandche, Uitgevers
    Maatschappij, Amsterdam, The Netherlands.
Coesel, P.F.M., Kwakkestein, R., and Verschoor, A., 1978, Oligo-
    trophication and eutrophication tendencies in some Dutch
    moorland pools as reflected in their desmid flora.
    Hydrobiol.   61:21-31.
Cogbill, C.V., 1975, History and character of acid precipitation
    in eastern North America, pp. 363 to 370. In: Proc. First
    Symp. Acid Precipitation and the Forest Ecosystem, C.S.
    Dochinger and T.A. Seliga, eds., Techn. Rep. NE-23, USDA
    Forest Service, Ohio State Univ., Columbus, OH.
Cole, C.V. and Stewart, J.W.B., 1983, Impact of acid deposition in
    P cycling. Envir. Exp. Bot. 23:235-241.
Conroy, N., Hawley, K., Keller, W., and LaFrance, C., 1976, Influ-
    ences of the atmosphere on lakes in the Sudbury area.
    J. Great Lakes Res. 2:146-165.
Conway, H.L. and Hendrey, G.R., 1982, Ecological effects of acid
    precipitation on primary producers, pp. 277 to 295. In:
    Acid Precipitation: Effects on Ecological Systems, F.M.
    D'Itri, ed., Ann Arbor Science, Ann Arbor, MI.
Davis, R.B. and Berge, F., 1980, Atmospheric deposition in Norway
    during the last 300 years as recorded in SNSF lake sediments.
    II. Diatom stratigraphy and inferred pH, pp. 270 to 271.
    In: Ecological Impact of Acid Precipitation, D. Drabløs and
    A. Tollan, eds., Proceedings International Conference,
    Sandefjord, Norway. SNSF Project, Ås-NLH, Norway.
DeCosta, J. and Preston, C., 1980, The phytoplankton productivity
    of an acid lake. Hydrobiol. 70:39-49.
DeCosta, J., Janicki, A., Shellito, G., and Wilcox, G., 1983, The
    effect of phosphorus additions in enclosures on the phyto-
    plankton and zooplankton of an acid lake. Oikos 40:283-294.
Del Prete, A., 1983, A comparison of seven Adirondack lakes by
    multivariate analysis of their diatom assemblages, pp. 65 to
    70. In: The Lake George Ecosystem, Vol. III, C.D. Collins,
    ed., The Lake George Association, Lake George, NY.
Del Prete, A. and Schofield, C.L., 1981, The utility of diatom
    analyses of lake sediments for evaluating acid precipitation
    effects on dilute lakes. Arch. Hydrobiol. 91:332-340.

Dickson, W.T., 1978, Some effects of the acidification of Swedish
     lakes. Verh. Internat. Verein. Limnol. 20:851-856.
Dillon, P.S., Yan, N D., Scheider, W.A., and Conroy, N., 1979,
     Acidic lakes in Ontario, Canada: characterization, extent,
     and responses to base and nutrient additions, Arch.
     Hydrobiol. Beih. Ergebn. Limnol. 13:317-336.
Drabløs, D. and Tollan, A., 1980, Ecological Impacts of Acid
     Precipitation, Proceedings International Conference, Sande-
     fjord, Norway. SNSF Project, Ås-NLH, Norway. 383 pp.
Duthie, H.C. and Ostrofsky, M.L., 1974, Plankton, chemistry and
     physics of lakes in the Churchill Falls region of Labrador.
     J. Fish. Board Canada 31:1105-1117.
Eriksson, M., Henrikson, L.Nilssen, B., Oscarson, H., and Stenson,
     A., 1979, Predator-prey relations important for the biotic
     changes in acidfied lakes. Ambio 9:248-249.
Fassett, N.C., 1930, The plants of some northeastern Wisconsin
     lakes. Trans. Wis. Acad. Arts, Science Letters 25:157-168.
Foy, C. and Gerloff, G., 1972, Response of Chlorella pyrenoidosa
     to aluminum and low pH. J. Phycol. 8:268-271.
Friberg, F., Otto, C , and Svensson, B., 1980, Effects of acidifi-
     cation on the dynamics of allocthomous leaf material and
     benthic invertebrate communities in running waters, pp. 304
     to 305. In: Ecological Impact of Acid Precipitation,
     D. Drabløs and A. Tollan, eds., Proceedings International
     Conference, Sandefjord, Norway. SNSF Project, Ås-NLH,
     Norway.
Galloway, J.N. and Cowling, E.B., 1978, The effects of precipita-
     tion on aquatic and terrestrial ecosystems: A proposed
     precipitation chemistry network. J. Air Pollu. Cont. Assoc.
     28:229-235.
Galloway, J.N., Likens, G.E., and Edgerton, E.S., 1976, Acid
     precipitation in the Northeastern United States pH and
     acidity. Science 194:722-724.
Gorham, E., 1976, Acid precipitation and its influence upon
     aquatic ecosystems--An overview. Water, Air, Soil Pollu.
     6:457-481.
Gorham, E. and Gordon, A.G., 1963, Some effects of smelter pollu-
     tion upon aquatic vegetation near Sudbury, Ontario. Can. J.
     Bot. 41:371-378.
Grahn, O., 1977, Macrophyte succession in Swedish lakes caused by
     deposition of airborne acid substances. Water, Air, Soil
     Pollu. 7:295-305.
Grahn, O., Hultberg, H., and Landner, L., 1974, Oligotrophication
     - a self-accelerating process in lakes subjected to excessive
     supply of acid substances. Ambio 3:93-94.
Griffiths, D., 1973, The structure of an acid moorland pond commu-
     nity. J. Animal Ecol. 42:263-283.
Haines, T.A., 1981, Acidic precipitation and its consequences for
     aquatic ecosystems: A review. Trans. Am. Fish. Soc.
     110:669-707.

Hall, R.J. and Likens, G.E., 1980, Ecological effects of experi-
    mental acidification on a stream ecosystem,, pp. 375 to 376,
    In: Ecological Impact of Acid Precipitation, D. Drabløs and
    A. Tollan, eds., Proceedings International Conference,
    Sandefjord, Norway. SNSF Project, Ås-NLH, Norway.
Hall, R.J., Likens, G.E., Fiance, S.B., and Hendrey, G.R., 1980,
    Experimental acidification of a stream in the Hubbard Brook
    Experimental Forest, New Hampshire. Ecology 61:967-989.
Halvorsen, K., 1977, Makrofyttvegetajonen i endel vann pa Agder,
    SNSF Project, TN32/77, Oslo, Norway. 154 pp.
Hargreaves, J., Lloyd, E., and Whitton, B., 1975, Chemistry and
    vegetation of highly acidic streams. Freshwater Biol.
    5:563-576.
Heady, H.F., 1942, Littoral vegetation of the lakes on the
    Huntington Forest. Roosevelt Wildlife Bull. 8:1-37.
Heinselman, M.L., 1970, Landscape evolution, peatland types, and
    the environment in the Lake Agassiz Peatlands Nature Area,
    Minnesota. Ecol. Monogr. 40:235-261.
Hellquist, C.B., 1980, Correlation of alkalinity and the distribu-
    tion of Potamogeton in New England. Rhodora 82:331-344.
Hendrey, G.R., 1976, Effects of low pH on the growth of periphytic
    algae in artificial stream channels. SNSF Project, IR 25/76,
    Oslo, Norway. 50 pp.
Hendrey, G.R. and Barvenik, F.W., 1978, Impacts of acid precipita-
    tion on decomposition and plant communities in lakes, pp. 92
    to 103. In: Scientific Papers from the Public Meeting on
    Acid Precipitation, H.H. Igard and J.S. Jacobson, eds.,
    Center for Environmental Research, Cornell University,
    Ithaca, NY.
Hendrey, G.R. and Vertucci, F.A., 1980, Benthic plant communities
    in acidic Lakes Colden, New York: Sphagnum and the algal
    mat, pp. 314 to 315. In: Ecological Impact of Acid
    Precipitation, D. Drabløs and A. Tollan, eds., Proceedings
    International Conference, Sandefjord, Norway. SNSF Project,
    Ås-NLH, Norway.
Hendrey, G.R. and Wright, R.F., 1976, Acid precipitation in
    Norway: Effects on aquatic fauna. J. Great Lakes Res.
    2:192-207.
Hendrey, G.R., Yan, N.D., and Baumgartner, L.J., 1981, Responses
    of freshwater plants and invertebrates to acidification,
    pp. 89 to 101. In: Proceedings International Symposium for
    Inland Waters and Lake Restoration, Portland, ME.
Hendrey, G.R., Baalsrud, K., Traaen, T.S., Laake, M., and Raddum,
    G., 1976, Acid precipitation: Some hydrobiological changes.
    Ambio 5:224-227.
Henrikson, L., Oscarson, H., and Stenson, J., 1980, Does the
    change of predator system contribute to the biotic
    development in acidified lakes? p. 316. In: Ecological

Impact of Acid Precipitation, D. Drabløs and A. Tollan, eds.,
     Proceedings International Conference, Sandefjord, Norway.
     SNSF Project, Ås-NLH, Norway.
Hultberg, H. and Grahn, O., 1976, Effects of acid precipitation on
     macrophytes in oligotrophic Swedish lakes. J. Great Lakes
     Res. 2:208-217.
Hutchinson, G.E., 1957, A Treatise on Limnology, Vol. I.
     Geography, Physics, and Chemistry. John Wiley and Sons, New
     York, NY. 1015 pp.
Hutchinson, G.E., 1967, A Treatise on Limnology, Vol. II. Intro-
     duction to Lake Biology and the Limnoplankton. John Wiley
     and Sons, New York, NY. 1115 pp.
Hutchinson, G.E., 1975, A Treatise on Limnology, Vol. III.
     Limnological botany. John Wiley and Sons, New York, NY.
     660 pp.
Hutchinson, T.C., Gizyn, W., Havas, M., and Zobens, V., 1978,
     Effect of long-term lignite burns on arctic ecosystems at the
     Smoking Hills N W.T., pp. 317 to 332. In: Trace Substances
     in Environmental Health-XII, D.D. Hemphill, ed., University
     of Missouri, Columbia, MO.
Huttunen, P. and Merilainen, J., 1983, Interpretation of lake
     quality from contemporary diatom assemblages. Hydrobiol.
     103:91-97.
Janzen, D.H., 1974, Tropical blackwater rivers, animals, and mast
     fruiting by the diterocarpacea. Biotropica 6:69-103.
Johnson, M.G., Michalski, M.F.P., and Christie, A.E., 1970,
     Effects of acid mine wastes on phytoplankton communities of
     two northern Ontario lakes. J. Fish. Res. Board Can.
     27:426-444.
Kalff, J. and Knoechel, R., 1978, Phytoplankton and their dynamics
     in oligotrophic and eutrophic lakes. Ann. Rev. Ecol. System
     9:475-495.
Kwiatkowski, R.E. and Roff, J.C., 1976, Effects of acidity on the
     phytoplankton and primary productivity of selected northern
     Ontario lakes. Can. J. Bot. 54:2546-2561.
Laake, M., 1976, Effekter av lav pH pa produksjon, nedbrytning og
     stoffkretslop i littoralsonnen. SNSF Project, IR 29/76,
     Oslo, Norway. 75 pp.
Lazarek, S., 1980, Cyanophycean mat communities in acidified
     lakes. Naturwissenschaften 67:97-87.
Lazarek, S., 1982a, Structure and function of a cyanophytan mat
     community in an acidified lake. Can. J. Bot. 60:2235-2240.
Lazarek, S., 1982b, Structure and productivity of epiphytic algal
     communities on Lobella dortmanna L.in acidified and limed
     lakes. Water, Air, Soil Pollu. 18:333-342.
Leivestad, H., Hendrey, G.R., Muniz, I.P., and Snelvik, E., 1976,
     Effect of acid precipitation on freshwater organisms, pp. 87
     to 111. In: Impact of Acid Precipitation on Forest and
     Freshwater Ecosystems in Norway, F.H. Braekke, ed., FR 6/76,
     SNSF Project, Oslo, Norway.

Lind, O. and Campbell, R.S., 1970, Community metabolism in acid
    and alkaline strip-mine lakes. Trans. Am. Fish. Soc.
    99:577-582.
McCauley, E. and Briand, F., 1979, Zooplankton grazing and phyto-
    plankton species richness: Field tests of the predation
    hypothesis. Limnol. Oceanogr. 24:243-252.
McLachlan, A.J. and McLachlan, S.M., 1975, The physical environ-
    ment and bottom fauna of a bog lake. Arch. für. Hydrobiol.
    76:198-217.
Moore, P.D. and Bellamy, D.J., 1974, Peatlands. Elek Science,
    London, U.K. 221 pp.
Moyle, J.B., 1945, Some chemical factors influencing the distribu-
    tion of aquatic plants in Minnesota. Am. Midland Nat.
    34:402-420.
Muller, P., 1980, Effects of artificial acidification on the
    growth of periphyton. Can. J. Fish. Aquatic Science
    37:355-363.
National Research Council of Canada, 1981, Acidification in the
    Canadian Aquatic Environment: Scientific Criteria for
    Assessing the Effects of Acidic Deposition on Aquatic
    Ecosystems. NRCC 18475, Environmental Secretariat, Ottawa,
    Canada. 369 pp.
Nilssen, J.P., 1980, Acidification of a small watershed in
    southern Norway and some characteristics of acidic aquatic
    environments. Int. Rev. Ges. Hydrobiol. 65:177-207.
Ogden, E.C., Dean, J.K., Boylen, C.W., and Sheldon, R.B., 1976,
    Field guide to the aquatic plants of Lake George, New York.
    State Education Department, Albany, NY. 65 pp.
Ostrofsky, M.L. and Duthie, H.C., 1975a, Primary productivity and
    phytoplankton of lakes on the Eastern Canadian Shield. Verh.
    Int. Verein. Limnol. 19:732-738.
Ostrofsky, M.L. and Duthie, H.C., 1975b, Primary productivity,
    phytoplankton, and limiting nutrient factors in Labrador
    lakes. Int. Rev. Ges. Hydrobiol. 60:145-158.
Parsons, J.D., 1977, Effects of acid mine wastes on aquatic
    ecosystems. Water, Air, Soil Pollu. 7:333-354.
Patrick, R., Binetti, V.P., and Halterman, S.G., 1981, Acid lakes
    from natural and anthropogenic causes. Science 211:446-448.
Pip, E., 1979, Survey of the ecology of submerged aquatic macro-
    phytes in central Canada. Aquatic Bot. 7:339-357.
Porter, K.G., 1973, Selective grazing and differential digestion
    of algae by zooplankton. Nature 244:179-180.
Raddum, G.G., Hobaek, A., Lomsland, E., and Johnsen, T., 1980,
    Phytoplankton and zooplankton in acidified lakes in southern
    Norway, pp. 332 to 333. In: Ecological Impact of Acid
    Precipitation, D. Drabløs and A. Tollan, eds., Proceedings
    International Conference, Sandefjord, Norway. SNSF Project,
    Ås-NLH, Norway.
Roberts, D.A., Boylen, C.W., and Singer, R., 1982, Acid precipita-
    tion and the Adirondacks: Acid effects on the biota indigen-

ous to Lake George, pp. 106 to 113. In: Proceedings of the
    Second Lake George Research Symposium, M.H. Schadler, ed.,
    The Lake George Association, Lake George, NY.
Roberts, D.A., Singer, R., and Boylen, C.W., 1985, The submersed
    macrophyte communities of Adirondack (New York, USA) lakes of
    varying degrees of acidity. Aquatic Bot. In Press.
Ryder, R.A., 1964, Chemical characteristics of Ontario lakes as
    related to glacial history. Trans. Am. Fish. Soc.
    93:260-268.
Schindler, D.W., 1972, Production of phytoplankton and zooplankton
    in Canadian shield lakes, pp. 311 to 331. In: Proceedings
    Symp. Productivity Problems of Freshwater, Z. Kajak and H.
    HillbrichtIlkowska, eds., IBP-UNESCO, Warsaw-Krakow, Poland.
Schindler, D.W., 1980, Experimental acidification of a whole lake:
    A test of the oligotrophication hypothesis, pp. 370 to 374.
    In: Ecological Impact of Acid Precipitation, D. Drabløs and
    A. Tollan, eds., Proceedings International Conference,
    Sandefjord, Norway. SNSF Project, Ås-NLH, Norway.
Schindler, D.W. and Holmgren, S.K., 1971, Primary production and
    phytoplankton in the experimental lakes area, Northeastern
    Ontario, and other low carbonate waters and a liquid scintil-
    lation method for determining C activity in photosynthesis.
    J. Fish. Res. Board Canada 28:189-201.
Schindler, D.W. and Turner, M.A., 1982, Biological, chemical, and
    physical responses of lakes to experimental acidification.
    Water, Air, Soil Pollu. 18:259-271.
Schindler, D.W., Wageman, R., Cook, R.B., Ruszcynski, R., and
    Prokopowich, J., 1980, Experimental acidification of
    Lake 223, Experimental Lakes Area: Background data and the
    first three years of acidification. Can. J. Fish. Aquatic
    Science 37:342-354.
Sheath, R.G., Havas, M., Hellebust, J.A., and Hutchinson, T.C.,
    1982, Effects of long-term natural acidification on the algal
    communities of tundra pools at the Smoking Hills, N.W.T.,
    Canada. Can. J. Bot. 60:58-72.
Sheldon, R.B. and Boylen, C.W., 1977, Maximum depth inhabited by
    aquatic vascular plants. Am. Midland Nat. 97:248-254.
Sheldon, R.B. and Boylen, C.W., 1978, An underwater survey method
    for estimating submerged macrophyte population density and
    biomass.Aquatic Bot. 4:65-72.
Singer, R., Roberts, D.A., and Boylen, C.W., 1983a, Biological
    effects of acidification on small oligotrophic Adirondack
    lakes, pp. 189 to 197. In: Second New York State Symposium
    on Atmospheric Deposition, Proceedings, L.S. Raymond, ed.,
    Cornell University, Ithaca, NY.
Singer, R., Roberts, D.A., and Boylen, C.W., 1983b, The macro-
    phytic community of an acidic Adirondack (New York, U.S.A.)
    lake: A new depth record for aquatic angiosperms. Aquatic
    Bot. 16:49-57.

Singer, R., Roberts, D.A., and Boylen, C.W., 1983c, The role of
    the benthic algal mat in annual nutrient cycling between the
    sediment and water of an acidic Adirondack lake, pp. 79 to
    84. In: The Lake George Ecosystem, Vol. III, C.D. Collins,
    ed., The Lake George Association, Lake George, NY. 178 pp.
Smith, D.W. and Brock, T.D., 1973, Water status and the distribu-
    tion of Cyanidium caldarium in soil. J. Phycol. 9:330-332.
Smith, M.W., 1961, A limnological reconnaissance of a Nova Scotian
    brown-water lake. J. Fish. Res. Board Can. 18:463-478.
Sondergaard, M. and Sand-Jenzen, K., 1979, Carbon uptake by leaves
    and roots of Littorella uniflora (L.) Aschers. Aquatic Bot.
    6:1-12.
Stokes, P.M., 1981, Benthic algal communities in acidic lakes, pp.
    119 to 138. In: Effects of Acidic Precipitation on Benthos,
    R. Singer, ed., North American Benthological Society,
    Hamilton, NY.
Stokes, P.M., Hutchinson, T.C., and Krauter, K., 1973, Heavy metal
    tolerance in algae isolated from polluted lakes near the
    Sudbury, Ontario smelters. Water Pollu. Res. Can. 8:177-202.
Stokes, P.M., Drier, S.I., Farkas, M.O., and McLean, R.A.N., 1983,
    Mercury accumulation by filamentous algae: A promising
    biological monitoring system for methyl mercury in acid-
    stressed lakes. Envir. Pollu. 5:255-271.
Stoneburner, D.L. and Smock, L.A., 1980, Plankton communities of
    an acidic brown-water lake. Hydrobiol. 69:131-137.
Sutcliffe, D.W. and Carrick, T.R., 1973, Studies on mountain
    streams in the English Lake District. Freshwater Biol.
    3:437-462.
Van Dam, H. and Kooyman-Van Blokland, H., 1978, Man-made changes
    in some Dutch Moorland Pools, as reflected by historical and
    recent data about diatoms and macrophytes. Int. Revue Ges.
    Hydrobiol. 63:587-607.
Van Dam, H., Suurmond, G., and ter Braak, C., 1980, Impact of acid
    precipitation on diatoms and chemistry of Dutch Moorland
    Pools, pp. 298 to 299. In: Ecological Impact of Acid Pre-
    cipitation, D. Dabløs and A. Tollen, eds., Proceedings Inter-
    national Conference, Sandefjord, Norway. SNSF Project,
    Ås-NLH, Norway.
Vitt, D.H. and Slack, N.G., 1975, An analysis of the vegetation of
    Sphagnum dominated kettle-hole bogs in relation to environ-
    mental gradients. Can. J. Bot. 53:332-359.
Wilcox, G. and DeCosta, J., 1982, The effect of phosphorus and
    nitrogen addition on the algal biomass and species
    composition of an acidic lake. Arch. Hydrobiol. 94:393-424.
Wile, I., 1981, Macrophytes, pp. 216 to 223. In: Acidification
    in the Canadian Aquatics Environment. Report No.18475,
    National Res. Council, Environmental Secretariat, Ottawa,
    Canada.

Wile, I. and Miller, G.E., 1983, The macrophyte flora of the 46
    acidified and acid-sensitive soft water lakes in Ontario.
    Ministry of the Environment, Rexdale, Ontario, Canada.
    35 pp.

Wium-Andersen, S., 1971, Photosynthetic uptake of free CO_2 by the
    roots of Lobelia dortmanna.   Physiol. Plant 25:245-248.

Woelkerling, W.J. and Gough, S.B., 1976, Wisconsin Desmids III.
    Desmid community composition and distribution in relation to
    lake type and water chemistry. Hydrobiol. 51:3-32.

Wright, R.F., Dale, L., Gjessing, T., Hendrey, G.R., Henriksen,
    A., Johannessen, M., and Muniz, I.P., 1976, Impact of acid
    precipitation on freshwater ecosystems in Norway. Water,
    Air, Soil Pollu. 6:483-499.

Yan, N.D., 1979, Phytoplankton community of an acidified, heavy
    metal-contaminated lake near Sudbury, Ontario: 1973-1977.
    Water, Air, Soil Pollu. 11:43-55.

Yan, N.D. and Stokes, P.M., 1978, Phytoplankton of an acid lake,
    and its responses to experimental alterations of pH.
    Environ. Conserv. 5:93-100.

ACIDIFICATION IMPACTS ON FISH POPULATIONS: A REVIEW

Joan P. Baker[a] and Carl L. Schofield[b]

[a]Acid Deposition Program, North Carolina State
University, Raleigh, North Carolina and [b]Department of
Natural Resources, Cornell University, Ithaca, New York

ABSTRACT

The clearest evidence for impacts of acidic deposition is the
documentation of adverse effects on fish populations. Loss of fish
populations associated with acidification of surface waters has
been documented for five areas--the Adirondack region of New York
State, the LaCloche Mountain region of Ontario, Nova Scotia,
southern Norway, and southern Sweden.   In other regions of the
world with low alkalinity waters receiving acidic deposition,
acidification of surface waters does not appear to have progressed
to levels clearly detrimental to fish. Three major mechanisms for
the disappearance of fish populations with acidification have been
proposed: (1) decreased food availability and/or quality, (2) fish
kills during episodic acidification, and (3) recruitment failure.
Each probably plays some role, although recruitment failure has
been hypothesized as the most common cause of population loss.

INTRODUCTION

The clearest evidence for impacts of acidic deposition is the
documentation of adverse effects on fish populations.   The
literature is extensive and varied.   Available data on effects of
acidification on fish are of at least seven types:

1)   historic records of declining fish populations in lakes
and rivers, coupled with historic records of increasing
acidity;

2) historic records of declining fish populations in lakes and rivers currently acidic but with no historic records on levels of acidity;

3) regional lake survey data and correlations of present-day fish status with present-day acidity levels in lakes and rivers;

4) data on success/failure of fish stocking efforts related to acidity of the surface water;

5) experimental acidification of aquatic ecosystems and observations of biological responses;

6) results of *in situ* exposures of fish to acidic waters; and

7) laboratory bioassay data on survival, growth, behavior, and physiological responses of fish to low pH, elevated aluminum concentrations, and other water quality conditions associated with acidification.

Each of these data sets has been reviewed. Combined, they provide strong evidence that acidification of surface waters has adverse effects on fish, and in some cases results in the gradual extinction of fish populations from acidified lakes and rivers.

Loss of fish populations from acidified surface waters is not, however, a simple process and cannot be accurately summarized as: "X" pH results in the disappearance of "Y" species of fish. At the very least, biological and chemical variation within and between aquatic ecosystems must be taken into account. For example, tolerance of fish to acidic conditions varies markedly, not only between different species but also between different strains or populations of the same species and among individuals within the same population. In addition, the water chemistry within an acidified aquatic system typically undergoes substantial temporal and spatial fluctuations. The survival of a population of fish may be more closely keyed to the timing and duration of acid episodes in relation to the presence of particularly sensitive life history stages, or to the availability of "refuge areas" during acid episodes, or to the availability of spawning areas with suitable water quality than to any expression of the annual average water quality. Because of these complexities, summary of the effects of acidification on fish in one or a few simple concluding tables can be misleading. In addition, our understanding of functional relationships between acidification and fish responses is still incomplete.

FIELD OBSERVATIONS

By themselves, field observations often fail to establish cause-and-effect responses definitively. Most extensive field observations are simply correlations between acidity of surface waters and absence of various fish species. Unfortunately, only in a few instances are historic records available that provide concurrent documentation of the decline of a fish population and the gradual increase in water acidity. Clear demonstration that the absence of fish resulted from high acidity requires supporting evidence from experiments conducted in the field or laboratory. A review of observed fish population changes apparently related to acidification does, however, serve to establish the nature and extent of the potential impact of acidification on fish.

Loss of Populations

Adirondack Region of New York State. The Adirondack region of New York State is the largest sensitive (low alkalinity) lake district in the eastern United States where extensive acidification has been reported (Galloway et al., 1983). The region encompasses approximately 2877 individual lakes and ponds (114,000 surface ha) (Pfeiffer and Festa, 1980), and an estimated 9350 km (6700 ha) of significant fishing streams (Colquhoun et al., 1981). Twenty-two fish species are native to the region, including brook trout (Salvelinus fontinalis), lake trout (Salvelinus namaycush), brown bullhead (Ictalurus nebulosus), white sucker (Catostomus commersoni), creek chub (Semotilus atromaculatus), lake chub (Couesious plumbeus), and common shiner (Notropis cornutus) (Greeley and Bishop, 1932). In addition, a variety of other species (e.g., smallmouth bass, Micropterus dolomieui; yellow perch, Perca flavescens) have been introduced into Adirondack waters, especially into the larger, more accessible lakes. Brook trout are frequently the only game fish species resident in the many small headwater ponds that are located at high elevations and are particularly susceptible to acidification (Pfeiffer and Festa, 1980). Although native to the Adirondacks, in some waters brook trout populations were introduced and must be maintained by stocking due to a lack of suitable spawning area.

Information relevant to effects of acidification on Adirondack fish populations evolves primarily from three sources: (1) a comprehensive survey of water quality and fish populations in many Adirondack surface waters conducted by the New York State Conservation Department in the 1920's and 1930's (Greeley and Bishop, 1932) followed by sporadic sampling of lakes and rivers up until the 1970's (data maintained on file by NY State DEC); (2) in 1975, a complete survey of all lakes (214) located above an elevation of 610 m (Schofield, 1976a); and (3) from 1978 to the present, accelerated sampling by the New York State Department of Environmental Conservation (DEC) of low alkalinity lakes or lakes that contain particularly valuable fisheries resources (Pfeiffer and Festa, 1980). In addition, a preliminary survey of fish populations and water quality for 42 Adirondack streams was completed by the DEC in 1980 (Colquhoun et al., 1981). None of these efforts has involved intensive studies of individual aquatic systems.

Evaluations of Adirondack data to date are limited to correlations of present-day fish status with present-day pH levels and, for a limited number of lakes, a comparison of current data with historic data or pH and fish population status. Each of the studies concluded that the geographic distribution of fish is strongly correlated with pH level, and that the disappearance of fish populations appears to have been associated with declines in pH. Indices of fish populations in Adirondack streams were statistically ($p < 0.05$) correlated with pH measurements (taken in the spring of 1980; Colquhoun et al., 1981). Schofield (1976b, 1981, 1982) noted fewer fish species in lakes with pH levels below 5.0 (Figure 1). Schofield and Trojnar (1980) also observed that poor stocking success for brook trout stocked into 53 Adirondack lakes was significantly ($p < 0.01$) correlated with low pH and elevated aluminum levels.

In many of the acid waters surveyed in the 1970's, no fish species were found. In high elevation lakes, about 50 percent of the lakes had pH less than 5.0 and 82 percent of these acidic lakes were devoid of fish. Thus, of the total lakes surveyed, 48 percent had no fish. High elevation lakes, however, constitute a particularly sensitive subset of Adirondack lakes, and these percentages do not apply to the entire Adirondack region. Unfortunately, neither a complete survey nor a random subsampling of all Adirondack lakes and streams has yet been attempted.

All lakes now devoid of fish need not, however, have lost their fish populations as a result of acidification or acidic deposition. A portion of these lakes never sustained fish populations. In addition, if earlier fish populations have disappeared, it must be demonstrated that acidification was the cause.

For 40 of the 214 high elevation lakes, historic records are available for the 1930's (Schofield, 1976b). In 1975, 19 of these 40 lakes had pH levels below 5.0 and had no fish. An additional two lakes with pH 5.0 to 5.5 also had no fish. Thus, 52 percent had no fish. In the 1930's, only three lakes had pH levels below 5.0 and, again, none of these had fish at that time. One additional lake with a pH 6.0 to 6.5 also had no fish. Thus, in the 1930's, only 10 percent of the 40 lakes were devoid of fish.

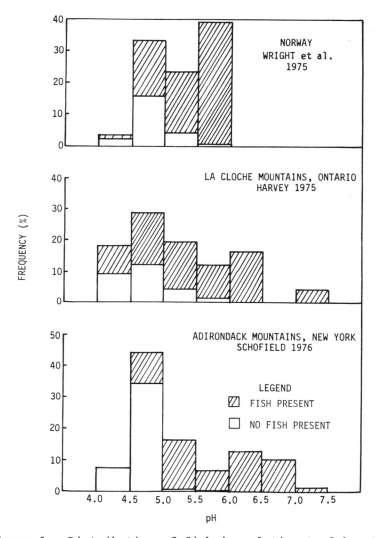

Figure 1:   Distribution of fish in relation to lake pH.

This implies that 17 lakes (or 42 percent) have lost their fish populations over the 40-year period. If this holds true for high elevation lakes in general, then 39 percent (83 lakes) of the high elevation Adirondack lakes may have actually lost fish populations. However, this assumes that the subset of 40 lakes represents an unbiased subsample of the 214 high elevation Adirondack lakes.

For Adirondack lakes in general, the DEC reported that about 180 lakes (6 percent of the total), representing some 2900 ha

(3 percent of the total), have lost their fish populations (Pfeiffer and Festa, 1980; Schofield, 1981). The basis for this estimation has not, however, been clearly delineated. Presumably, there are 180 lakes for which recent (1970's) fish sampling efforts have yielded no fish and for which historic records of fish surveys (1930's to 1960's) are available that indicate the presence of fish in earlier years. All are listed as former brook trout ponds (Pfeiffer and Festa, 1980). Because the names of these lakes have not been published and the data are available only in DEC files, this important conclusion cannot be validated.

It is also necessary to demonstrate that the loss of fish from Adirondack lakes has occurred as a result of acidic deposition and/or acidification of surface waters. Retzsch et al. (1982) argued that "although precipitation acidity cannot be excluded as a possible cause, it represents only one of a number of factors that may alter fish populations in the Adirondacks." They consider that loss of fish populations in the Adirondacks may also be a result of (1) natural acidification with the development of naturally acidic wetlands adjacent to lakes; (2) declines in the number of fish stocked into Adirondack lakes and changes in management practices; (3) introductions of non-native fish species; (4) increased recreational use and fishing pressure; and (5) construction of dams (manmade or beaver) along with manipulations of lake levels and stream flow.

All of these reasons sound feasible, yet the DEC argued that loss of fish has occurred in the absence of alternative explanations other than acidification of surface waters (N.Y. DEC, 1982). For example, inadvertent introductions of non-native fish species occur primarily in accessible low elevation waters that are generally not, at present, impacted critically by acidification. Nongame fish species, not subject to stocking, management, or fishing pressure, have also been reduced or eliminated. In addition, numerous waters located in the immediate proximity of high-use public campgrounds in the Adirondacks have maintained excellent trout populations throughout the years despite heavy fishing pressure (N.Y. DEC, 1982). Dean et al. (1979) evaluated the impact of black fly larvacide on 42 Adirondack stream fish populations and found no significant differences in occurrence and density of fish in treated versus untreated streams. By default, acidification has been implicated as a factor causing the loss of fish in a number of lakes and streams.

A detailed analysis of field data for individual lakes has not, however, been published that examines evidence for loss of fish populations and potential explanations for these losses, including acidification. Still, the data in total are sufficient to conclude that loss of fish in the Adirondacks, at least for some surface waters, was associated with acidification. The number of fish populations adversely impacted, and the significance of these losses relative to the total resource available in the Adirondacks is, however, inadequately quantified at the present time.

Other regions of the eastern United States. Schofield (1982) summarized available data relating water acidity and fish population status for areas in the eastern United States with waters potentially acidified by acidic deposition. Very few of these studies, with the exception of studies in the Adirondack region, included comprehensive inventories of fish populations or historic changes in fish population status with time. Davis et al. (1978) noted that biological effects have not yet been detected in Maine lakes. Haines (1981a) discussed the potential for adverse effects of acidification on Atlantic salmon (Salmo salar) rivers of the eastern United States. Although the rivers were defined as "vulnerable," no discernable effect on salmon returns was reported. Crisman et al. (1980) sampled gamefish populations in the two most acidic lakes (pH 4.7 and 4.9) in the Trail Ridge area of northern Florida. Populations of largemouth bass (Micropterus salmoides) and bluegull sunfish (Lepomis macrochirus) exhibited no clear evidence of stress directly related to low pH values or elevated aluminum concentrations. In Pennsylvania, some fish species have disappeared from a few headwater stream systems (Arnold et al., 1980), but no consistent trends in the data conclusively demonstrated acidification impacts (Schofield, 1982). Jones et al. (1983) investigated fish kills in fish-rearing facilities in the Raven Fork watershed at Cherokee, North Carolina. Episodic pulses of low pH and elevated aluminum levels were identified as the cause of death, but the forest-soils complex, rather than acidic deposition, appeared to be the primary factor controlling hydrogen and aluminum ions in the stream following storm events.

No adverse effects of acidic deposition and/or acidification on fish have been definitely identified in regions of the United States other than the Adirondack Mountain area of New York State. Discussions generally referred only to "potential impact."

LaCloche Mountain Region of Ontario. Information collected on fish populations in the LaCloche Mountain region of Ontario provides some of the best evidence of adverse effects of acidification on fish. The principal source of acid entering the LaCloche area is sulfur dioxide emitted from the Sudbury smelters located about 65 km northeast (Beamish, 1976). Large additions have resulted in relatively rapid acidification of many of the region's lakes--acidification rapid enough that fish population declines, and in some cases extinctions, have occurred over the course of the 15 years that the lakes have been monitored by researchers from the University of Toronto (H. Harvey, R. Beamish, and associates; see references below).

The LaCloche Mountains cover 1300 km² along the north shore of Lake Huron. Contained within this area are 212 lakes, approximately 150 of which have been surveyed for chemical characteristics; 68 for fish populations. Fish populations in several of the lakes have been studied in detail since the late 1960's and early 1970's (Beamish and Harvey, 1972; Beamish, 1974a,b; Beamish et al., 1975; Harvey 1975). Major sport fishes common in these lakes include lake trout, smallmouth bass, and walleye (Stizostedion vitreum. Other fish occurring very frequently are yellow perch, pumpkinseed sunfish (Lepomis gibbosus), rock bass (Ambloplites rupestris), brown bullhead, lake herring (Coregorus artedii), and white sucker. LaCloche Mountain lakes in general have waters with low ionic content and are quite clear, indicative of low organic acid content (Harvey, 1975). Of 150 lakes surveyed in 1971, 22 percent had pH levels below 4.5 and 25 percent were in the pH range of 4.5 to 5.5 (Beamish and Harvey, 1972).

Harvey (1975) noted that the number of species of fish in 68 LaCloche Mountain lakes was significantly (p < 0.005) correlated with lake pH (Figure 1). In addition, however, the number of species of fish in each lake was also significantly correlated with lake area and other physical features. Because small lakes tend to have low pH values, the effects of these two independent variables on fish may be confounded. A covariate analysis based on data presented by Harvey (1975) indicated, however, that the correlation with lake pH was still significant (p < 0.005) even after adjustment for differences in lake area. Of the 31 lakes with pH less than 5.0, 14 had no fish. Fourteen lakes had pH values of 6.0 or greater; all of these had at least one species of fish with usually seven or more species being present.

For the 68 LaCloche Mountain lakes surveyed during 1972-73, 38 lakes are known or are suspected to have had reductions in fish species composition (Harvey, 1975). Based on historic fisheries information, some 54 fish populations are known to have been lost, including lake trout populations from 17 lakes, smallmouth bass from 12 lakes, largemouth bass from four lakes, walleye from four lakes, and yellow perch and rock bass from two lakes each. Assuming that lakes with current pH < 6.0 originally contained the same number of species as lakes with an equal surface area and pH greater than 6.0, an estimated 388 fish populations have been lost from the 50 lakes surveyed with pH < 6.0 (Harvey and Lee, 1982).

The gradual disappearance of fish populations with time and with increased acidity has been described in detail for Lumsden Lake, George Lake, and O.S.A. Lake (Beamish and Harvey, 1972; Beamish, 1974b; Beamish et al., 1975). Lake pH levels measured in 1961 by Hellige color comparator were 6.8, 6.5, and 5.5 in Lumsden, George, and O.S.A. lakes, respectively. In 1971-73, pH levels measured in the three lakes with a portable pH meter were

4.4, 4.8 to 5.3, and 4.4 to 4.9, respectively. In the 1950's, eight species of fish were reported in Lumsden Lake. Over the period of 1961-71, a drastic decline in the abundance of both game and nongame fish occurred. In George Lake, during the interval 1961-73, lake trout, walleye, burbot, and smallmouth bass disappeared from the lake, and from 1967 to 1972 the white sucker population decreased in number by 75 percent and in biomass by 90 percent. For O.S.A. Lake in 1961, local residents reported good catches of lake trout and smallmouth bass. In 1972, intensive fish sampling yielded only four yellow perch, two rock bass, and eight lake herring. By 1980, no fish were present (Harvey and Lee, 1980). Beamish (1976) concluded that increased acidity was the principal factor resulting in the loss of fish populations.

Other areas of Ontario. Harvey (1980) estimated that approximately 200 lakes in Ontario have lost their fish populations. For the most part, however, these lakes are in the vicinity of Sudbury, Ontario. Studies that suggest fish loss in response to acidification for other areas of Ontario are very limited. Although the Muskoka-Haliburton region of Ontario receives large additions of acidic deposition and decreases in alkalinity have been suggested for some lakes (Dillon et al., 1978), no adverse effects on fish populations have been documented; pH values apparently have not decreased to levels harmful to fish.

Nova Scotia. In Nova Scotia, rivers with pH < 5.4 occur only in areas underlain by granitic and metamorphic rock; all flow in a southerly direction to the Atlantic Coast (Watt et al., 1983). Thirty-seven rivers within this region have historic records indicating that they sustained anadromous runs of Atlantic salmon. For 27 of these rivers, angling catch records are available from annual reports of Federal Fishery Offices for the period 1936 to 1980. Of these 27, five rivers have undergone major alterations since 1936 that could have potentially impacted salmon stocks. For the 22 remaining rivers, 12 presently have pH > 5. Statistical analysis of angling catch from 1936 to 1980 indicated that only one of these 12 rivers had experienced a significant (p < 0.01) decline in salmon catch since 1936, one river a significant (p < 0.05) increase, and 10 no significant trend in angling catch with time. In contrast, of the 10 rivers with current pH < 5.0, nine have had significant (p < 0.02) declines in success since 1936, while one experienced no significant trend.

Salmon angling records for rivers with pH < 5.0 vs pH > 5.0 are compared in Figure 2. From 1936 through the early 1950's, angling catch in the two groups of rivers was similar. After the 1950's, angling catch in rivers with pH < 5.0 declined, while salmon catch in rivers with pH > 5.0 continued to show no significant trend with time.

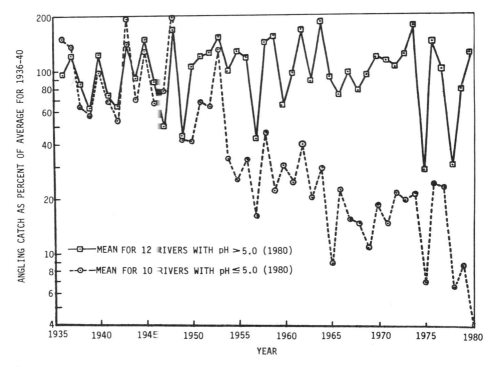

Figure 2:   Average angling success for Atlantic salmon in 22 Nova
            Scotia rivers since 1936. Data were collected from
            reports of federal fishery offices and normalized by
            expressing each river's angling catch as a percentage of
            the average catch in that river during the first
            five years of record (1936-40; Watt et al., 1983).

        Year-to-year variations in salmon catch are considerable
reflecting the many factors affecting angling success and reporting
accuracy.   Between the two groups of rivers (pH < 5.0 and pH
> 5.0), however, occurrence of high and low success years generally
corresponds.   Both groups of rivers are well distributed along the
500 km Atlantic coastline of Nova Scotia.   Tag return data
suggested that salmon stocks in this area all share a common marine
migratory pattern.   Biological and physical factors leading to
greater or lesser angler success (e.g., sea survival, river
discharge rates, or juvenile year-class survival) probably act
uniformly over the entire area (Watt et al., 1983).   Decreases in
salmon catch over time are, on the other hand, clearly correlated
with present-day pH values of 5.0 and below.

Watt et al. (1983) concluded that seven former salmon rivers in Nova Scotia, with mean annual pH < 4.7, no longer support salmon runs. An electrofishing survey in the summer of 1980 failed to find any signs of Atlantic salmon reproduction in any of these seven rivers. Farmer et al. (1980), however, observed that for the most part these rivers are also naturally somewhat acidic (highly colored waters, indicating the presence of organic acids) and historically had relatively low fish production. Peat deposits and bogs are common to much of this area. These materials probably contribute to the low pH levels and have some impact on salmon production. Historical records of pH for a few rivers within this area (Thompson et al., 1980) indicated that acidity has increased from the mid-1950's to early 1970's. Acidic conditions and acidification, therefore, probably contribute to the loss of Atlantic salmon populations in Nova Scotia.

The estimated lost (rivers with pH < 4.7) or threatened (rivers with pH 4.7 to 5.0) Atlantic salmon production potential represents 30 percent of the Nova Scotia resource and two percent of the total Canadian potential. Atlantic salmon rivers in New Brunswick, Prince Edward Island, and other areas of Nova Scotia generally have pH levels above 5.4 and are not under any immediate acid threat (Watt, 1981).

Norway. Extensive information on acidification and loss of fish populations in Norwegian waters has been collected under the auspices of the joint research project SNSF--"Acid Precipitation - Effects on Forest and Fish," 1972-1980. Documentation of the effects of acidification on fish is derived principally from (1) yearly records of catch of Atlantic salmon in 75 Norwegian rivers from 1876 to the present; (2) a survey of water chemistry and fish population status in 700 small lakes in southern Norway in 1974-75; (3) collation of information on fish population status (current and historic) for some 5000 lakes in southern Norway, validated with test fishing in 93 lakes during 1976-79; and (4) detailed analyses of historic changes in fish population status related to land use changes with time in selected watersheds. Together these data provide strong evidence that acidification has had profound impacts on fish.

Statistical data for the yearly salmon catch from major salmon rivers in Norway have been recorded since 1876 (Jensen and Snekvik, 1972; Leivestad et al., 1976; Muniz, 1981). While catch in all rivers declined slightly from 1900 until the 1940's, in 68 northern rivers the decline was followed by a marked increase, and catch in the 1970's equalled or exceeded that around 1900. In contrast, in seven southern rivers, annual catch dropped sharply over the years 1910-17, has declined steadily since then, and is now near zero. This decrease is reflected in all seven rivers and cannot be explained by known changes in exploitation practices. Massive fish

kills of Atlantic salmon were reported in these rivers as early as
1911. Efforts over the last 50 years to restock with hatchery-
reared fry and fingerlings have been unsuccessful. In the
seven southern rivers pH levels averaged 5.12 in 1975, as compared
to an average pH of 6.57 for 20 of the 68 northern rivers.
Leivestad et al. (1976) reported that acidity in southern rivers
has been steadily increasing. From 1966 to 1976 hydrogen ion
concentration increased by 99 percent.

In 1974-75, the SNSF project completed a synoptic (nonrandom)
survey of water chemistry and fish population status in 700 small
to medium-sized lakes in Sørlandet (the four southernmost counties
of Norway; Wright and Snekvik, 1978). Based on interviews with
local residents, fish populations in lakes were classified as
barren, sparse population, good population, and overpopulated. The
principal species of fish was brown trout (<u>Salmo</u> <u>trutta</u>). Other
important species were perch (<u>Perca</u> <u>fluviatilis</u>), char (<u>Salvelinus</u>
<u>alpinus</u>), pike (<u>Esox</u> <u>lucius</u>), rainbow trout (<u>Salmo</u> <u>gairdneri</u>), and
brook trout. About 20 percent of the 700 lakes were reported as
barren of fish, and an additional 40 percent had sparse
populations. Fish status was clearly related to water chemistry
(Figure 1); most low pH, low conductivity lakes were either barren
or had only sparse populations.

The original data base on fish populations in Sørlandet
collected by Jensen and Snekvik (1972) and Wright and Snekvik
(1978) has gradually been extended to the whole country. By 1980,
data on fish in more than 5000 lakes in the southern half of Norway
had been collected by interviewing fisheries authorities, local
landowners, local fishermen's associations, and other local experts
(Sevaldrud et al., 1980; Overrein et al., 1980; Muniz and
Leivestad, 1980a). Interview data were validated for 93 lakes by
comparison with results from a standardized testfishing program.
Interview data provided an accurate assessment of actual fish
stocks for over 90 percent of the lakes (Rosseland et al., 1980).

At present, fish population damage has apparently occurred in
an area of 33,000 km^2 in southern Norway. Twenty-two percent of
the lakes at low elevations below 200 m have lost their brown trout
populations; 68 percent of the trout populations in high altitude
lakes above 800 m are now extinct. In about 40 percent of this
area, fish populations in all lakes are extinct or near extinction.

Besides information on the current status of fish populations
in these 5000 lakes, the SNSF project has also compiled available
historic information on changes in fish populations with time. For
almost 3000 lakes in Sørlandet, the population status of brown
trout has been recorded by local fishermen since about 1940. The
time trend for loss of populations is diagrammed in Figure 3. The
rate of disappearance of brown trout from lakes in Sørlandet has

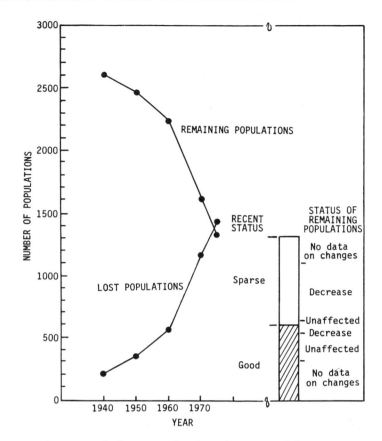

Figure 3:   Time trend for population losses of brown trout in the
            affected areas in the four southernmost counties
            (Rogaland, Vest-Agder, Aust-Agder, and Telemark) in
            Norway (Sevaldrud et al., 1980).

been particularly rapid since 1960. Today, more than 50 percent of
the original populations have been lost, and approximately
60 percent of the remaining are in rapid decline (Sevaldrud et
al., 1980). Attempts at restocking acidified lakes containing
reduced populations have largely failed (Overrein et al., 1980).

    A relationship between water acidity and fish population
status or even water acidification and concurrent loss of fish
populations does not necessarily implicate acidic deposition as the
primary cause for adverse effects on fish. However, several
studies have been completed in Norway that examine alternate
explanations for acidification, e.g., changes in land use,
specifically as they relate to historic changes in fish populations
(Drabløs and Sevaldrud, 1980; Drabløs et al., 1980). In each of

three study areas, correlations were not found between shifts in land use and human activities and changes in fish status. Areas that have experienced changes in land use (e.g., abandonment of pasture farms or discontinuance of lichen harvests) do not have any higher proportion of lakes with declines in fish population than do areas without such land use changes. In contrast, fish population declines are correlated with acidic deposition.

Sweden. Sweden has about 90,000 lakes, many of which have low alkalinity and are potentially sensitive to acidic deposition. Extensive surveys of acidification and fish population status have not, however, been completed. In southern Sweden, 100 lakes with pH 4.3 to 7.5 were sampled in the 1970's (Almer et al., 1978). Apparently as a result of acidification (i.e., disappearance of fish was associated with current low pH levels in lakes), 43 percent of the minnow (Phoxinus phoxinus) populations, 32 percent of the roach (Rutilus rutilus), 19 percent of the artic char, and 14 percent of the brown trout populations had been lost. In a study of six lakes in southern Sweden, Grahn et al. (1974) cited historic pH data suggesting a pH decline of 1.4 to 1.7 units since the 1930's-40's and the simultaneous elimination of minnows, roach, pike, and brown trout from two or more of these six lakes. Disappearance of populations of roach in lakes in southwestern Sweden was recorded as early as the 1920's and 1930's (although not definitely correlated with acidification; Dickson, 1975). In eastern Sweden, loss of roach from Lake Ärsjön near Stockholm occurred in association with a decrease in pH readings: pH 5.1 to 5.3 in 1974 as compared to pH 6.0 measured colorimetrically in the 1940's (Milbrink and Johansson, 1975).

Population Structure

The well-being of a population can be judged in part by examination of its age composition (National Research Council Canada, 1981). Theoretically, age one fish should be more numerous than age two fish; age two fish more numerous than age three fish; age three fish more numerous than age four fish, etc. Two factors commonly alter this theoretical distribution: gear selectivity and large natural variations in year class strength. Almost all procedures for sampling fish populations are size selective. Small, young fish are often poorly sampled. In addition, relative numbers of fish in each age group may fluctuate greatly from year to year as a consequence of natural environmental and biological factors (e.g., year-to-year temperature variations and competition between age groups). The frequent absence of one or several age groups within a population may, however, be indicative of a population under stress or undergoing change. Studies of fish populations in acidic water frequently reveal reduced or missing age groups.

Deviations from the expected age class distribution in acidic lakes result in some cases from the absence of young fish; in others from the absence of older fish. A population with only fairly large, fairly old individuals, suggests that recruitment and/or reproduction have failed. A population with only young fish may imply the occurrence of a mortality factor acting only on fish after a certain age (e.g., after sexual maturity). Both types of distributions have been observed in acidic waters, although the absence of young fish has been cited as a primary factor leading to the gradual extinction of fish populations in acidic waters (Schofield, 1976b; Overrein et al., 1980; Haines, 1981b).

Studies of lakes in the LaCloche Mountain region of Ontario by Beamish, Harvey, and others provide detailed observations of the structure of fish populations in acidic and acidifying lakes. White suckers were last reported in Lumsden Lake in 1969 at a pH of 5.0 to 5.2 (Beamish and Harvey, 1972). Intensive sampling in 1967 yielded no young-of-the-year and very few age one fish, suggesting poor recruitment of white suckers in both 1966 and 1967. In 1972, O.S.A. Lake had a pH of about 4.5. Intensive sampling yielded only a small number of very old fish--eight lake herring aged six to eight years, four yellow perch aged eight years, and two rock bass aged 13 years (Beamish, 1974b). By 1980, there were no fish in O.S.A. Lake.

In addition to these intensive studies of individual lakes in the LaCloche Mountain region, Ryan and Harvey (1977, 1980) surveyed (through rotenone applications) the age distribution of populations of yellow perch and rock bass in 32 and 20 LaCloche Mountain lakes, respectively. For both species, lakes with lower pH levels had a higher frequency of populations missing young-of-the-year fish. The most acidic lake yielding young-of-the-year yellow perch was characterized by a pH of 4.4, for rock bass by a pH of 4.8.

Absence of young age groups in fish populations from acidic and acidifying lakes has also been documented for a few lakes in the Adirondack region and in Scandinavia. In South Lake in the Adirondacks, white suckers netted in 1957-68 (pH 5.3 in 1968) ranged in length from 15 to 51 cm, suggesting a wide range of age classes. However, by 1973-75 (pH 4.9 in 1975) recruitment of young fish appeared to have ceased. White suckers collected ranged from 30 to 49 cm in length. Five suckers captured in 1975 were aged six to eight years (Schofield, 1976a; Baker, 1981). In Lake Skarsjön in Sweden, prior to lake liming (pH 4.5-5.5) only very large, old perch remained in the lake. One year after liming (pH 6.0), reproduction was reestablished and two size classes of perch were present, both very large, old fish and new group of small, one year old perch (Muniz and Leivestad, 1980a).

Recruitment failure may result either from acid-induced mortality of fish eggs and/or larvae or because of a reduction in numbers of eggs spawned. Beamish and Harvey (1972) attributed the lack of reproduction in fish populations in LaCloche Mountain Lakes to a failure of adult fish to spawn. Failure of female fish to spawn has not, however, been reported elsewhere. From a survey of 88 lakes in Norway, Rosseland et al. (1980) noted that female fish remaining in acidic lakes had normal gonads, and indications of unshed or residual eggs were rare. Studies conducted in Scandinavia and the United States (Schofield, 1976b; Muniz and Leivestad, 1980a) suggested that increased mortality of eggs and larvae in acidic waters is the primary cause of recruitment failures.

In addition to the lack of young fish in a population, associated with recruitment failure as described above, in some cases loss of older fish has been observed in acidic waters. Three lakes in the Tovdal River, Norway, were test fished from 1976 to 1979 (Rosseland et al., 1980). Before 1975, brown trout populations in these lakes were stunted and grew to 8-10 years of age. In 1975, the Tovdal River had a severe fish kill. Since 1976, no post-spawning brown trout (age five and up) have been found, and the population is dominated by young fish. Test fishing in autumn indicated the presence of maturing recruit-spawners. By each subsequent year, however, this age group had disappeared while their offspring survived. Researchers speculated that stress associated with spawning activities, coupled with acid-induced stress, resulted in significant post-spawning mortality (Muniz and Leivestad, 1980a).

Growth

Observations on fish growth in acidic waters and changes in growth rate over time with acidification suggested that indirect effects of acidification resulting from changes in food availability are generally insignificant for adult fish. In very few cases have reduced growth rates been reported. For the most part, fish in acidic and/or acidified waters grow at the same rate or faster than fish in circumneutral waters in the same region.

Decreases in fish growth rate associated with acidification have been documented only for acidic lakes in the LaCloche Mountain region, Ontario. Beamish et al. (1975) reported that growth rates for white suckers in acidic George Lake (pH 4.8 to 5.3, 1972-73) had declined over the period 1967 to 1973, and this was apparently associated with lake acidification. In more recent surveys, however, this trend appears to have reversed. Fish collected in 1978 and 1979 were larger (at a given age) than fish in 1972 and similar in size to fish collected in 1967 to 1968 (Harvey and Lee,

1980). Therefore, even in this instance consistent decreases in growth over time with increased water acidity have not occurred.

On the other hand, several studies suggested increased fish growth in acidic waters and/or with acidification. For two acidic lakes in the Adirondacks sampled in the 1950's and 1970's, numbers of brook trout caught decreased over the 20 year period, and significant increases in fish growth were observed (Schofield, 1981). Roach in acidic lakes (pH 4.6 to 5.5) in Sweden grew at substantially faster rates than roach in circumneutral lakes (pH 6.3 to 6.8; Almer et al., 1974, 1978). Growth of rock bass in 25 LaCloche Mountain lakes was also significantly (p < 0.05) faster in lakes with greater acidity, even after adjustment for effects of lake depth on fish growth (Ryan and Harvey, 1977, 1981).

Rosseland et al. (1980), on the other hand, in a survey of 88 lakes in southern Norway, found no obvious tendency for increase in growth in sparse populations in acidic lakes despite the fact that fish from acidic lakes had higher proportions of full stomachs and were in better condition (i.e., weighed more for a given length). Ryan and Harvey (1980, 1981) observed that yellow perch in 39 LaCloche Mountain lakes grew more quickly in more acidic waters up to age three years, but thereafter grew more slowly.

Fish growth response to acidification may be a complex function of two factors: acid-induced metabolic stress and food availability. Reduced growth in acidic waters as a result of physiological stress has been noted frequently in laboratory experiments. Presumably, similar responses occur in acidic lakes and streams. Observations of increased or unchanged growth in acidified surface waters, however, suggest that adverse effects of acidity on fish metabolism and physiology are counterbalanced, in part or totally, by changes in food availability.

Acidification is associated with substantial changes in the structure and, in some cases, the function of lower trophic levels (Magnuson et al., 1983). Despite the fact that some important prey organisms are sensitive to acidic conditions and, as a result fish may be required to shift their predation patterns, in most acidic lakes food does not seem to be a significant limiting factor for adult fish (Beamish et al., 1975; Hendrey and Wright, 1976). With decreased fish density resulting from recruitment failures or fish kills, decreased interspecific and/or intraspecific competition for food supplies may possibly lead to increased food availability for the remaining fish. Increased food availability may balance any negative effects of acid-induced metabolic stress. Detailed studies of effects of food availability on fish, at all life history stages, in acidic waters are not available. Therefore, the conclusion that shifts in food availability with acidification have no adverse effects on fish survival or production is preliminary.

Episodic Fish Kills

Observations of dead and dying fish in acidifying waters are not common. Mechanisms of population extinction (e.g., recruitment failure) are often too subtle to be easily detected. However, instances of massive acute mortalities of adult and young fish have occurred, typically associated with rapid decreases in pH resulting from large influxes of acid into the system during spring snowmelt or heavy autumn rains. In general, organisms are less tolerant of rapid increases in toxic substances than they are of chronic exposure and gradual changes in concentration. As a result, the rapid fluctuations in acidity associated with short-term acidification may be particularly lethal to fish and may play an important role in the disappearance of fish from acidified lakes and streams.

Fish kills apparently associated with acidic episodes have been reported numerous times in the streams and rivers of southern Norway (Jensen and Snekvik, 1972; Muniz, 1981). The first records of mass mortality of Atlantic salmon date from 1911 and 1914 and coincide closely with the sharp drop in salmon catch recorded for rivers in southern Norway over the years 1910-17. Additional observations of mass mortality were reported in 1920, 1922, 1925, 1948, and 1969 following either heavy autumn rains or rapid snowmelt, particularly in May to June.

A similar episode occurred in the Tovdal River (Norway) in the spring of 1975 (Leivestad et al., 1976). Dead fish were first observed at the end of March. During the first weeks of April thousands of dead trout covered a 30 km stretch of the river. The Tovdal River valley is sparsely populated and has no industry. Veterinary tests failed to find signs of any known fish diseases. The pH of the river was about 5.0 in March. At two stations downstream a drop in water pH was recorded apparently associated with a period of snowmelt at altitudes below 400 m. At higher altitudes, dead fish were not found and temperatures probably never rose above freezing.

Fish kills attributed to short-term acidification have been reported for only one location outside of Norway. During each spring (1978-1981) coincident with spring run-off, dead and dying fish, especially pumpkinseed sunfish, were observed in Plastic Lake in the Muskoka-Haliburton region of Ontario (Harvey, 1979; Harvey and Lee, 1982). Measured pH levels were 5.5 at the lake surface and 3.8 in the major inlet. Field experiments to verify these toxic conditions in Plastic Lake were completed in 1981.

In addition to these observations of mass mortalities of fish attributed to acidic episodes under natural field conditions, several instances of unusually heavy fish mortality have been reported within fish hatcheries receiving water directly from lakes or rivers. In Norway, poor survival of eggs and newly-hatched larvae of Atlantic salmon attributed to water acidity were reported as early as 1926 in hatcheries on rivers in Sørlandet (Muniz, 1981). In Nova Scotia, 19 to 38 percent mortality of Atlantic salmon fry occurred in 1975 to 1978 at the Mersey River hatchery (Farmer et al., 1981). In Norway and Nova Scotia, neutralization of the water by passage through limestone alleviated the problem. In the Adirondacks, adult, yearling, and larval brook trout, which had been maintained without incident over the winter of 1976-77 in water from Little Moose Lake, experienced distress and mortality during the first major winter thaw in early March (Schofield and Trojnar, 1980). The minimum pH measured was 5.9 on March 13 (with 0.39 mg Al l^{-1}). Mortalities occurred over a 5-day period from March 13 to 17. Deaths included three adult brook trout, 25 yearlings (132 to 167 mm), and an undetermined number of recently hatched fry. Eyed brook trout eggs exposed to the same water did not experience significant mortality.

All of the above observations of fish kills were associated with episodic increases in acidity. Grahn (1980), however, recorded fish kills in two lakes in Sweden associated with decreases in acidity. In June 1978 in Lake Ransjön and in June 1979 in Lake Ämten, large numbers of dead ciscoe (Coregonus albula) were discovered. A weather pattern of heavy rainfall, decreasing pH levels, and increasing aluminum concentrations in the lakes, followed by a long period of dry, sunny weather preceded fish kills in both lakes. pH levels in the lake epilimnion increased from approximately 4.9 and 5.4 to 5.4 and 6.0, respectively. Grahn (1980) hypothesized that the increase in pH level precipitated aluminum hydroxide, and that ciscoe, migrating into the epilimnion to feed, were exposed to these lethal conditions. Laboratory experiments by Baker and Schofield (1982) also noted that aluminum is particularly toxic to fish as it precipitates. Dickson (1978) reported that acidic lake waters immediately after liming (pH values increased to 5.5 and above) were toxic to trout. Concentrations of aluminum were still high and, presumably, aluminum would be actively precipitating.

FIELD EXPERIMENTS

Correlations between fish population status and acidity of surface waters, and field observations of declines in fish populations concurrent with acidification of a lake, river, or stream strongly imply that acidification has serious detrimental effects on fish. Such observations, however, rarely prove cause-and-effect relationships. In experiments, one variable is changed and the response to that change is recorded. Thus, the cause and its effect are clearly delineated.

Whole-ecosystem acidification experiments have been carried out at two locations: Lake 223 in the Experimental Lakes Area, Ontario and Norris Brook in the Hubbard Brook Experimental Forest, New Hampshire. In both cases, acid was added directly to the water and pH levels were held fairly constant. Despite these deviations from the process of acidification in nature, results from these two experiments demonstrate important biological changes associated with increased water acidity.

Experimental acidification of Lake 223, Ontario

Lake 223 is a small, oligotrophic lake on the Precambrian Shield of western Ontario. Prior to acidification, surface waters had an average alkalinity of about 80 μeq l^{-1} and pH of 6.5 to 6.9. Five species of fish were present: lake trout, white sucker, fathead minnow (Pimephales promelas), pearl dace (Semotolus margarita), and slimy sculpin (Cottus cognatus). Beginning in 1976, additions of sulfuric acid to the lake epilimnion gradually reduced lake pH. Early in each ice-free season, lake pH was decreased to a predetermined value and then maintained at that value through the following spring, at which time pH was again reduced. Mean pH values were 6.8 in 1976, 6.1 in 1977, 5.8 in 1978, 5.6 in 1979, 5.4 in 1980, and 5.1 in 1981. Biological responses to the acidification have been described by Schindler et al. (1980), Schindler (1980), Malley et al. (1982), Schindler and Turner (1982), NRCC (1981), U.S./Canada MOI (1982), and Mills (1984), and are summarized in Table 1.

A number of important biological changes occurred at pH values of 5.8 to 6.0, notably the disappearance of the opossum shrimp (Mysis relicta), a benthic/planktonic crustacean, and the collapse of the fathead minnow population. Although both these species were important prey for lake trout in the lake, no effects on trout populations were detected. Lake trout density and population structure remained stable, and year-class recruitment failures were not detected until 1981 at a pH of 5.1. At the onset of acidification (1976), fathead minnows were abundant while pearl dace were rare. With the collapse and eventual extinction of the fathead minnow population as the pH declined to 5.5, pearl dace abundance increased dramatically (perhaps in response to the loss of its closest competitor). The increased abundance of pearl dace and a succession of strong year classes of white suckers in 1978 to 1980 apparently provided adequate food alternatives for the lake trout.

Despite many changes in lower trophic levels, lake trout and white sucker populations showed no definite indications of stress until 1981 when the pH was 5.1 and reproductive failures occurred. During the early years of acidification population numbers of both species increased and growth rates were relatively unchanged. The

Table 1. Biological changes in Lake 223 in response to
          experimental acidification (Schindler and Turner, 1982;
          Mills, 1984).

pH	Recorded change
Below 6.5	Increased bacterial sulfate reduction partially neutralize acid additions Increased abundance of Chlorophyta (green algae) Decreased abundance of Chrysophyceans (golden brown algae Increased abundance of rotifers Increased dipteran emergence
5.8-6.0	Disappearance of the opossum shrimp (Mysis relicta) Reproductive impairment of the fathead minnow (Pimephales promelas) Possible increased embryonic mortality of lake trout (Salvelinus namaycush) Inhibition of calcification of exoskeleton of crayfish (Orconectes virilis) Disappearance of the copepod Diaptomus sicilis
5.3-5.8	Increased hypolimnetic primary production Development of Mougeotea algal mats along shoreline Increased infestation of crayfish with a parasite Thelohania sp. Collapse of the fathead minnow population Increased abundance of the pearl dace minnow (Semotilus margarita) Decreased abundance of the slimy sculpin (Cottus cognatus) Decreased abundance of crayfish Increased abundance of white sucker (Catostomus commersoni) Increased abundance of lake trout Disappearance of copepod Epischura lacustris First appearance of the cladoceran Daphnia catawba schoedleri
Below 5.3	Recruitment failure of lake trout Recruitment failure of white sucker

primary food source for white suckers, benthic dipterans, increased in abundance. Although types of prey available to lake trout changed dramatically, suitable food remained abundant. Both species spawned successfully during the years of study prior to 1981, and there were no indications of egg resorption or skeletal malformations.

The population of bottom-dwelling slimy sculpin gradually declined throughout the acidification period from 1976 to 1981. Potential reasons for the decline include direct adverse effects of increased acidity and/or increased trout predation associated with an increase in water clarity.

Among the fish, fathead minnow seemed to be most sensitive to acidification. Fathead minnows are ubiquitous in lakes in northern North America and form an important part of aquatic food chains. The population in Lake 223 disappeared extremely quickly, probably as a result of two factors: its particular sensitivity to acidity and its short life span. Recruitment failure occurred initially at pH 5.8 in 1978. Prior to acidification, fathead minnow in Lake 223 typically lived only three years. Natural mortality rates during their second and third years of life were extremely high, over 50 percent per year, presumably as a result of heavy trout predation. Few individuals remained after the second year of life. Year-class failure in 1978, therefore, left few spawning adults (ages two and three the following year. Successive year-class failures in 1978 and 1979 assured the rapid disappearance of this species from Lake 223.

In summary, experimental acidification of Lake 223 resulted in severe changes in fish populations at pH values as high as 5.8 to 6.0. Adverse effects on fish and loss of populations occurred primarily as a result of recruitment failures rather than as a result of increased mortality of adult fish or reductions in food supplies.

Experimental Acidification of Norris Brook, New Hampshire

Norris Brook, a third order stream in the Hubbard Brook Experimental Forest, New Hampshire, was experimentally acidified to pH 4.0 from April to September 1977 (Hall et al., 1980; Hall and Likens, 1980a,b). Brook trout were observed in the study section before and after acid addition. Small numbers of trout confined in the study section during low water in June, July, and August were exposed continuously to water at pH 4.0 to 5.0 and total aluminum levels up to about 0 23 mg l^{-1}. Trout captured at pH 4.0, 5.0, and 6.4 in August showed no evidence of pathological changes in gill structure. Most of the trout, however, moved downstream to areas of higher pH at the onset of acid addition in the spring. Fish mortality was not observed, only a general avoidance reaction. Potential effects on young-of-the-year trout and reproductive success were not included in this study.

Exposure of Fish to Acidic Surface Waters

    In addition to the above field experiments involving acidification of an entire ecosystem, smaller scale field experiments have been conducted involving the transfer of fish into acidic lakes and streams. It is important to distinguish these small-scale field experiments from similar exposures of fish to acid waters in laboratory experiments for two reasons: (1) water quality conditions in field experiments may undergo substantial natural fluctuations while conditions are usually held rather constant in laboratory experiments, and (2) many laboratory experiments create acidic water by diluting strong acids (H_2SO_4, HNO_3, HCl) into nonacidic background water. These artificially acidic waters may not precisely mimic acidified surface waters, and, as a result, fish responses recorded in laboratory bioassays may not always accurately represent what would occur in the field.

    Excessive mortality of adult fish has been observed in a number of in situ experiments with fish held in cages in acidic waters. Following observation of fish kills in Plastic Lake (LaCloche Mountain region) in 1979 and 1980, rainbow trout (Salmo gairdneri) were held in cages at four locations in Plastic Lake and at four locations in a control nonacidic lake during the spring of 1981 (Harvey et al., 1982). Mortality did not occur at any of the cage sites in the control lake (pH 6.09 to 7.34). In Plastic Lake, however, mortality ranged from 12 percent at the lake outlet (pH 5.0 to 5.85) to 100 percent at the inlet (pH 4.03 to 4.09).

    During the winter (December to April) 1971-72, Hultberg (1977) placed seatrout and minnows (Phoxinus phoxinus), both with a mean length of 6.5 cm, at ten test stations ranging in pH from 4.3 to 6.0 within the watershed of Lake Ålevatten, Sweden. At all but three of the test stations native minnow populations had disappeared within the ten years preceding the experiment. Fifty-three percent of the seatrout and 91 percent of the minnows died during the four-month test. Most of the mortalities (68 percent of the seatrout total mortality; 59 percent of the minnows) coincided with periodic drops in pH level.

    A number of studies have also examined survival of fish eggs incubated in waters from acidic lakes and streams (Table 2). Hatching success and egg survival of brook trout ova decreased sharply between pH levels 5.0 and 4.6. For brown trout, hatching was near 100 percent at pH levels 6.2 and 6.5, but none hatched at pH 4.8 and 5.1. The critical pH for hatching of Atlantic salmon eggs appears to be 5.0 to 5.6; for walleye about pH 5.4 and for roach something above pH 5.7.

In three studies, results from in situ incubation experiments were compared with concurrent surveys of occurrence of fish species within the same waters. Leivestad et al. (1976) reported that no brown trout eggs hatched and few trout fry were found (by electrofishing) in an acidic tributary (pH 4.8), formerly an important spawning ground. By contrast, in a second tributary with inferior spawning conditions but pH 6.2, numerous trout fry were collected. Harriman and Morrison (1982) reported no survival of Atlantic salmon eggs incubated in acidic streams (pH 4.2 to 4.4) draining forested catchments in Scotland and the absence of fish from the same streams in an electrofishing survey. Finally, Milbrink and Johansson (1975) incubated perch (Perca fluviatilis) and roach eggs in situ in Lakes Mälaren (pH 7.5), Stensjön (pH 5.7), and Trehörningen (pH 4.7) in Sweden. While some perch eggs hatched in all three lakes (89, 50, and 28 percent, respectively), very few or no roach eggs hatched in the two acidic lakes (14 percent in Lake Stensjön, none in Trehörningen). Likewise, perch populations occurred in all three lakes, although extremely few perch were collected in the most acidic lake, Trehörningen. Roach, on the other hand, have apparently disappeared from Lake Trehörningen. Roach are still prevalent in both Lake Stensjön and Mälaren.

LABORATORY EXPERIMENTS

One of the best ways to prove cause and effect is to conduct experiments in a carefully controlled environment, i.e., the laboratory. Experimental conditions and fish response can be clearly quantified and dose-response relationships developed with a minimum of time and effort. Unfortunately, laboratory experiments have several drawbacks. For one, the simplified, controlled environment of the laboratory may differ from the natural environment in essential attributes. Factors that cannot be easily incorporated into laboratory experiments include: (1) the temporal and spatial variability in the field environment; and (2) the potential for compensatory mortality, i.e., shifts in the efficacy of natural mortality factors (e.g., predation, starvation) resulting from the addition of acid-induced mortality and/or stress. Consequently, results from laboratory experiments cannot be translated automatically into an expected response in the field.

The more closely the laboratory environment simulates the field experience, the more realistic the observed response. Laboratory bioassays conducted to date vary substantially in their use of conditions appropriate to the problem of acidification of surface waters. Most laboratory experiments concerned with acidification have focused on the effects of low pH on fish. With acidification, however, other factors also change in association with decreasing pH. Increased aluminum concentrations in acidic waters, in particular, have been shown to affect fish adversely (Baker, 1982). Unfortunately, most of the bioassay results to date have failed to include aluminum. Thus, these results must be interpreted with caution. In addition to aluminum concentration, other factors change with acidification, e.g., increased manganese and zinc concentrations and perhaps a decrease in dissolved organic carbon (Galloway et al., 1983). The importance of these other changes to fish populations in acidified waters has yet to be delineated in either laboratory or field experiments.

Unfortunately, because of the large number and diversity of laboratory experiments that have been conducted, it is not possible to thoroughly summarize these results within the constraints of this paper. A complete review is given in Baker (1983). Data for ten species are noted in Table 2. Interpretation of laboratory results must consider that fish response in a bioassay is a function of testing conditions (e.g., temperature, flow-through or static water supply), background water quality (e.g., water hardness, concentrations of dissolved gases), and characteristics of the fish tested (e.g., prior exposures and stress, size, age, and condition).

SUMMARY

Extent of Impact

    Loss of fish populations associated with acidification of surface waters has been documented for five areas--the Adirondack region of New York State, the LaCloche Mountain region of Ontario, Nova Scotia, southern Norway, and southern Sweden. The following summarizes major points from the available literature:

°   The best evidence that loss of fish has occurred in response to acidification is derived from observations of lakes in the LaCloche Moutain region, Ontario. Twenty-four percent of 68 lakes surveyed had no fish. Fifty-six percent of the 68 lakes are known or suspected to have had reductions in fish species composition (Harvey, 1975). Based on historic fisheries information, 54 fish populations are known to have been lost, including lake trout populations from 17 lakes, smallmouth bass from 12 lakes, largemouth bass from four lakes, walleye from four lakes, and yellow perch and rock bass from two lakes each (Harvey and Lee, 1982). The principal source of acidic deposition to the LaCloche area is sulfur dioxide emitted from the Sudbury smelters located about 65 km to the northeast. Large additions have resulted

Table 2. Summary of field observations, field experiments, and laboratory experiments relating water pH to fish response.

	Brook Trout	Lake Trout	Brown Trout	Rainbow Trout	Atlantic Salmon	Smallmouth Bass	Largemouth Bass	Walleye	White Sucker	Fathead Minnow
FIELD OBSERVATIONS										
Recruitment failure	5.0-5.5[a]	5.2-5.5[b] 6.9[e] 5.7[g]	<5.0[c] 4.7-5.1[f]		4.7-5.0[d]	5.5-6.0[d] 5.0[e]	5.1[e] 5.0[g]	5.5-6.0[b] 5.4[e] 6.5[g]	5.2[a] 4.7-5.2[b] 5.0[e]	5.8-6.0[o] 5.3-5.8[o]
Population loss	5.0[a] 4.5-4.8[h]	5.0-5.5[a] 4.4[e]	4.5-4.8[h] 4.5-5.0[g] 4.7[j]	5.5-6.0[h]	5.1[i]	4.4[e] 6.0[g]	4.4[c]	5.2[e] 5.5[g]	5.1[a] 4.3[e]	
Fish kill	5.9[l]		4.9-5.1[k] 4.6[k] 5.0[i]		3.9-4.2[m] 5.0[n]					
FIELD EXPERIMENTS										
Recruitment failure		5.1-5.3[o]							5.1-5.3[o]	
Population extinction										
Adult mortality	4.8-5.2[l] 4.7-5.1[q] 4.4-4.6[r]			4.0-5.0[p]						
Embryo mortality	4.5-4.6[q]		4.5[m] 4.8[i] 5.1[v]		4.5-5.0[s] 5.0-5.5[m] 4.9[i]			5.4[t]		
LABORATORY EXPERIMENTS										
Adult mortality-acute (2 day)	3.8[w] 3.5[aa] 3.6[cc] 4.4[cc] 4.5[gg]		3.8[x]	4.0[y] 3.8-4.8[bb] 4.0-4.2[dd]					3.9[z]	
Adult mortality-chronic (>20 day)	*4.1[hh]		<4.8[ee]	<5.0[ee]						<4.6[ff]
Embryo mortality	6.1[gg] 4.4[cc] <4.6[mm] <5.0[ii] 4.0-5.0[mm]		4.1[ii] 4.0-4.5[jj]		4.1[ii] 5.5[kk] 5.5[jj] 4.5[nn]				*>5.6[hh] 5.2[ll]	5.9[ff]
Fry mortality	*4.4-4.9[l] *4.4-4.8[hh] 6.1[gg] 4.2[cc] <5.4[ii] <5.0[mm]		4.4[ii]		3.7-4.0[oo] 5.0[ii]				*5.4-5.6[hh] 5.0-5.4[ll]	<5.9[ff]
Reduced production of viable eggs	5.1[gg]									
Reduced growth	6.5[gg]		4.8[ee]	4.8[ee]						
Avoidance	4.5[rr]		<5.0[pp] 4.3-4.8[qq]		5.3[ss]					6.6[ff]

REFERENCES

a   Schofield, 1976c
b   Beamish, 1976
c   Almer et al., 1978
d   Watt et al., 1983
e   Harvey, 1979
f   Hultberg, 1977
g   Beamish et al., 1975
h   Grande et al., 1978
i   Leivestad et al., 1976
j   Grahn et al., 1978
k   Overrein et al., 1980
l   Schofield and Trojnar, 1980
m   Jensen and Snevik, 1972

n   Farmer et al., 1981
o   Mills, 1982
p   Harvey et al., 1982
q   Schofield, 1965
r   Dunson and Martin, 1973
s   Harriman and Morrison, 1982
t   Hulsman and Powles 1981 as
    reported in MOI 1982
u   Milbrink and Johnasson, 1975
v   Muniz and Leivestad, 1980b
w   Johnson, 1975
x   Brown, 1981
y   Kwain, 1975
z   Beamish, 1972

aa   Robinson et al., 1976
bb   McDonald et al., 1980
cc   Swarts et al., 1978
dd   Lloyd and Jordan, 1964
ee   Edwards and Hjeldnes, 1977
ff   Mount, 1973
gg   Menendez, 1976
hh   Baker and Schofield, 1982
ii   Johansson et al., 1977
jj   Carrick, 1979
kk   Peterson et al., 1980a
ll   Trojnar, 1977b
mm   Trojnar, 1977a
nn   Peterson et al., 1980b
oo   Daye and Garside, 1976
pp   Jacobsen, 1977
qq   Nelson, 1982
rr   Johnson and Webster, 1977
ss   Hoglund, 1961

*Refers to laboratory experiments taking into account both low pH and inorganic aluminum (at the expected concentration for that pH; based on Driscoll, 1980).

in relatively rapid acidification of many of the region's lakes.

° In Norway, sharp decreases in the catch of Atlantic salmon in southern rivers began in the early 1900's and are associated with current low pH levels and a recorded doubling of the hydrogen ion concentration in one of these rivers from 1966 to 1976 (Jensen and Snekvik, 1972, Leivestad et al., 1976). For almost 3000 lakes in Sørlandet (southernmost Norway) data on the status of brown trout has been recorded since about 1940 (Sevaldrud et al., 1980). Today, more than 50 percent of the original populations have been lost, and approximately 60 percent of the remaining are in rapid decline (Sevaldrud et al., 1980). Fish population declines have been correlated with acidity, acidification, and/or acidic deposition (Wright and Snekvik, 1978).

° In Nova Scotia, records of angling catch of Atlantic salmon in rivers date back, in some cases, to the early 1900's. Of ten rivers with current pH < 5.0 and historic catch records, nine have had significant declines in angling success over the time period 1936 to 1980. For 12 rivers with pH > 5.0, only one experienced a significant decrease in salmon catch. Decrease in salmon catch over time is correlated with present-day pH values 5.0 and below (Watt et al., 1983), and apparent acidification of rivers in the area between 1954 and 1974 (Thompson et al., 1980).

° Finally, fish populations in Adirondack lakes and streams have also declined over the last 40 to 50 years. The New York State Department of Environmental Conservation reported that about 180 lakes (2900 ha) out of a total of 2877 lakes (114,000 ha) in the Adirondacks have lost their fish populations (especially brook trout; Pfeiffer and Festa, 1980). The absence of fish in Adirondack lakes and streams is clearly correlated with low pH levels (Schofield, 1976b), but records have not been published to substantiate that loss of fish in these 180 lakes resulted from acidification. Historical data are available for very few individual lakes that suggest both lake acidification and simultaneous loss of fish. Acidification probably contributed to the disappearance of fish for at least some surface waters, but exactly how many lakes and streams (perhaps substantially less than or more than 180) have been impacted cannot be satisfactorily evaluated at this time.

° In other regions of the world with low alkalinity waters receiving acidic deposition (e.g., Muskoka-Haliburton area of Ontario and Maine), acidification of surface waters does not appear to have progressed to levels clearly detrimental to fish (Schofield, 1982). Damage to fish populations has not been reported.

Mechanism of Effect

Three major mechanisms for the disappearance of fish populations with acidification have been proposed: (1) decreased food availability and/or quality; (2) fish kills during episodic acidification; and (3) recruitment failure. Each probably plays some role, although recruitment failure has been hypothesized as the most common cause of population loss (Schofield, 1976b; Harvey, 1980; Overrein et al., 1980; Haines, 1981b; NRCC, 1981). The following summarizes major points from the available literature:

° The influence of food chain effects on decreases in fish populations in acidified waters has received little attention to date, but available information suggests it plays a relatively minor role (Beamish, 1974b; Hendrey and Wright, 1976; Muniz and Leivestad, 1980a; Rosseland et al., 1980). With acidification, or in comparisons between acidic and circumneutral lakes, fish growth is often unaffected or increased with increasing acidity.

° Fish kills have been observed during episodic acidification of surface waters and in certain instances may play an important role in the disappearance of fish from acidified surface waters. For example, in the Tovdal River, Norway, thousands of dead adult trout were observed in association with the first major snow melt in the spring of 1975 (Leivestad et al., 1976). Dead and dying fish are, however, seldom reported in acid-stressed waters relative to the large number of lakes, streams, and rivers with fish populations apparently impacted by acidification. In contrast, a substantial portion of fish populations examined in acidified lakes lack young fish and apparently have experienced recruitment failure.

° Recruitment failure may result either from acid-induced mortality of eggs and/or larvae or because of a reduction in numbers of eggs spawned. The number of eggs spawned could be reduced as a result of disruption of reproductive physiology and ovarian maturation or inhibition of spawning behavior. Evidence exists that supports each one of these proposed mechanisms. It is likely that each one of these factors plays some role in recruitment failure but the importance of each factor probably varies substantially among aquatic systems, depending on the particular circumstances.

° More research is necessary to define clearly the specific mechanism for population decline in a given water. However, many studies in the United States and Scandinavia (Schofield, 1976b; Muniz and Leivestad, 1980a) emphasize increased mortality of eggs and larvae in acidic waters as the primary cause of recruitment failures, and recruitment failure as a common cause for the loss of fish populations with acidification of surface waters.

Relationship Between Water Acidity and Fish Population Response

To quantitatively assess the impact of acidification on fish resources, the functional relationship between acidification and fish population response must be understood. Unfortunately, loss of fish populations from acidified surface waters is not a simple process. The mechanism by which fish are lost seems to vary between aquatic systems and probably within a given system from year-to-year. The water chemistry within a given aquatic system is likewise extremely variable both spatially and temporally, and these variations are very important to the survival or decline of fish populations. Fish species differ not only in their ability to tolerate acidic conditions but also in their ability to exploit these chemical variations in their environment (e.g., spawning time and location). Serious gaps exist in the understanding of how to use laboratory results in the quantitative prediction of fish response in the field. It is therefore not surprising that the development of an accurate functional relationship between acidification and fish response is impossible at this time.

First steps, however, in developing such a relationship are to: examine in a semi-quantitative manner all of the available information connecting acidity with fish populations (summarized in Table 2), produce an initial approximation of the dose-response relationships (Figure 4), and then assess patterns and reasons for deviations from this initial approximation. In large part, the analysis of deviations and variations must be done on a lake-by-lake, population-by-population basis, and is the subject for further research. Several points are, however, obvious. Acidification adversely affects fish populations. Sensitivity of fish to acidity is species-dependent, and determined by aluminum and calcium concentrations, in addition to pH values. Loss of fish populations need not be associated with large declines in annual average pH, but could result from indirect effects on aluminum chemistry or episodic acidification.

Acknowledgements - The research described in this article has been funded in part by the NCSU Acid Deposition Program (a cooperative agreement between the U.S. Environmental Protection Agency and North Carolina State University). It has not been subjected to EPA's required peer and policy review, and therefore does not

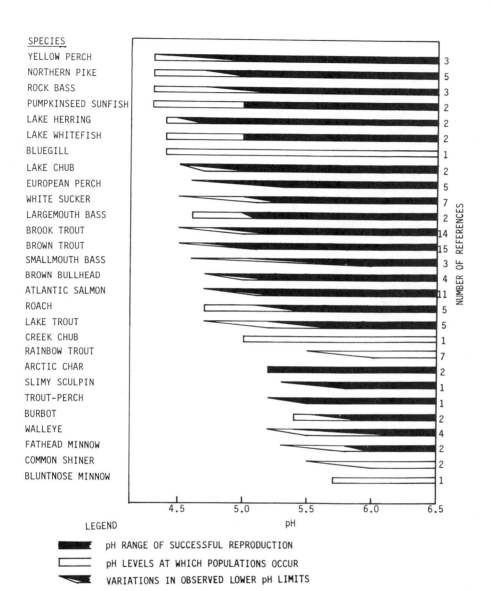

Figure 4:   Initial estimates of relationship between acidity and
            fish response, based on references in Table 2 (from
            Baker, 1983).

necessarily reflect the views of the Agency and no official
endorsement should be inferred.

REFERENCES

Almer, D., Dickson, L., Ekstrom, C., and Hornstrom, E., 1974,
    Effects of acidification on Swedish lakes. Ambio 3:30-36.
Almer, B., Dickson, L., Ekstrom, C., and Hornstrom, E., 1978,
    Sulfur pollution and the aquatic ecosystem, pp. 273 to 311.
    In: Sulfur in the Environment, Part II. J. Nriagu, ed.,
    John Willey and Sons, New York, NY.
Arnold, D.E., Light, R.W., and Dymond, V.J., 1980, Probable effects
    of acid precipitation on Pennsylvania waters.
    EPA-600/3-80-012, U.S.E.P.A. Environmental Laboratory,
    Corvallis, OR. 24 pp.
Baker, J.P., 1981, Aluminum toxicity to fish as related to acid
    precipitation and Adirondack surface water quality. Ph.D.
    thesis, Cornell University, Ithaca, NY. 441 pp.
Baker, J.P., 1982, Effects on fish of metals associated with
    acidification, pp. 165 to 176. In: Acid Rain/Fisheries,
    R. Johnson, ed. American Fisheries Society, Bethesda, MD.
Baker, J.P, 1983, Fishes, pp. 5-76 to 5-136. In: The Acidic
    Deposition Phenomenon and Its Effects: Critical Assessment
    Review Papers, I. Linthurst and A. Altshuller, eds.,
    EPA-600/8-83-016B, U.S.E.P.A., Washington, DC.
Baker, J.P. and Schofield, C.L., 1982, Aluminum toxicity to fish in
    acidic waters. Water, Air, Soil Pollu. 18:289-309.
Beamish, R.J., 1972, Lethal pH for the white sucker Catostomus
    commersoni (Lacepede). Trans. Am. Fish. Soc. 101:355-358.
Beamish, R.J., 1974a Growth and survival of white suckers
    (Catostomus commersoni) in an acidified lake. J. Fish. Res.
    Board Can. 31:49-54.
Beamish, R.J., 1974b Loss of fish populations from unexploited
    remote lakes in Ontario, Canada as a consequence of
    atmospheric fallout of acid. Water Res. 8:85-95.
Beamish, R.J., 1976, Acidification of lakes in Canada by acid
    precipitation and the resulting effects on fishes. Water,
    Air, Soil Pollu 6:501-514.
Beamish, R.J. and Harvey, H.H., 1972, Acidification of the LaCloche
    mountain lakes, Ontario, and resulting fish mortalities. J.
    Fish. Res. Board Can. 29:1131-1143.
Beamish, R.J., Lockart, W.L., Loon Van, J.C., and Harvey, H.H.,
    1975, Long-term acidification of a lake and resulting effects
    on fishes. Ambio 4:98-102.
Brown, D.J.A., 1981, The effects of various cations on the survival
    of brown trout, Salmo trutta, at low pH's. J. Fish Biol.
    18:31-40.

Carrick, T.R., 1979, The effect of acid water on the hatching of
    salmonid eggs. J. Fish Biol. 14:165-172.
Colquhoun, J.R., Symula, J., Pfeiffer, M., and Feurer, J., 1981,
    Preliminary report of stream sampling for acidification
    studies 1980. Technical Report 81-2, NY State Dept. Envir.
    Conservation, Albany, NY. 116 pp.
Crisman, T.L., Schulze, R.L., Brezonik, P.L., and Bloom, S., 1980,
    Acid Precipitation: The biotic response in Florida lakes,
    pp. 296 to 297. In: Ecological Impact of Acid Precipitation,
    D. Drabløs and A. Tollan, eds., Proceedings International
    Conference, Sandefjord, Norway. SNSF Project, Ås-NLH, Norway.
Davis, R.B., Smith, M.O., Bailey, J.H., and Norton, S.A., 1978,
    Acidification of Maine (USA) lakes by acidic precipitation.
    Verh. Int. Verein. Limnol. 20:532-537.
Daye, P.G. and Garside, E.T., 1976, Histopathologic changes in
    surficial tissues of brook trout, Salvelinus fontinalis
    (Mitchill), exposed to acute and chronic levels of pH. Can.
    J. Zool. 54:2140-2155.
Dean, H.J., Coloquhoun, J.R., and Symula, J., 1979, Preliminary
    survey of some Adirondack stream fish populations in relation
    to the blackfly control program and acid precipitation.
    Unpublished report, NY State Dept. Envir. Conservation,
    Albany, NY. 15 pp.
Dickson, W., 1975, The acidification of Swedish Lakes. Rep. Inst.
    Freshw. Res. Drottningholm 54:8-20.
Dickson, W., 1978, Some effects of the acidification of Swedish
    lakes. Verh. Int. Verein. Limnol. 20:851-856.
Dillon, P.J., Jeffries, D.S., Snyder, W., Reid, R., Yan, N.D.,
    Evans, D., Moss, J., and Scheider, W.A., 1978, Acid
    precipitation in south-central Ontario: recent observations.
    J. Fish. Res. Board. Can. 35:809-815.
Drabløs, D. and Sevaldrud, I., 1980, Lake acidification, fish
    damage and utilization of outfields. A comparitive survey of
    six highland areas in Hedmark County, eastern Norway,
    pp. 354 to 355. In: Ecological Impact of Acid Precipitation,
    D. Drabløs and A. Tollan, eds., Proceedings International
    Conference, Sandefjord, Norway. SNSF Project, Ås-NLH, Norway.
Drabløs, D., Sevaldrud, I., and Timberlid, J.A., 1980, Historical
    land-use changes related to fish status development in
    different areas in southern Norway, pp. 367 to 369. In:
    Ecological Impact of Acid Precipitation, D. Drabløs and
    A. Tollan, eds., Proceedings International Conference,
    Sandefjord, Norway. SNSF Project, Ås-NLH, Norway.
Dunson, W.A. and Martin, R.R., 1973, Survival of brook trout in a
    bog-derived acidity gradient. Ecology 54:1370-1376.
Edwards, D.J. and Hjeldness, S., 1977, Growth and survival of
    salmonids in water of different pH. Report FR 10/77, SNSF-
    project, Ås-NLH, Norway. 12 pp.
Farmer, G.J., Goff, T.R., Ashfield, D., and Samant, H.S., 1980,
    Some effects of the acidification of Atlantic salmon rivers in

Nova Scotia. Can. Tech. Rep. of Fish. Aquat. Sci. No. 972, 13 pp.

Farmer, G.J., Goff, L.R., Ashfield, D., and Samant, H.S., 1981, Some effects of the acidification of Atlantic salmon rivers in Nova Scotia, pp. 73 to 91. In: Acid Rain and the Atlantic Salmon. Spec. Publ. 10, International Atlantic Salmon Foundation.

Galloway, J.N., Anderson, D.S., Church, M.R., Cronan, C.S., Davis, R.B., Dillon, P.J., Driscoll, C.T., Norton, S.A., and Schafran, G.C., 1983, Effects on aquatic chemistry, pp. 4-1 to 4-162. In: The Acidic Deposition Phenomenon and its Effects: Critical Assessment Review Papers, R. Linthurst and A. Altshuller, eds., EPA-600/8-83-016B, U.S.E.P.A., Washington, DC.

Grahn, O., 1980, Fish kills in two moderately acid lakes due to high aluminum concentration, pp. 310 to 311. In: Ecological Impact of Acid Precipitation, D. Drabløs and A. Tollan, eds., Proceedings International Conference, Sandefjord, Norway. SNSF Project, Ås-NLH, Norway.

Grahn, O., Hultberg, H., and Landner, L., 1974, Oligotrophication - A self-accelerating process subjected to excessive supply of acid substances. Ambio 3:93-94.

Grande, M., Muniz, I.P., and Andersen, S., 1978, The relative tolerance of some salmonids to acid waters. Verh. Int. Verein. Limnol. 20:2076-2084.

Greeley, J.R. and Bishop, S.C., 1932, Fishes of the area, pp. 54 to 76. In: A Biological Survey of the Oswegatchee and Black River Systems. Supplement to 21st Annual Report, NY State Dept. Conservation, Albany, NY.

Haines, T.A., 1981a, Effects of acid rain on Atlantic salmon rivers and restoration efforts in the United States, pp. 57 to 63. In: Acid Rain and the Atlantic Salmon. Spec. Publ. 10, International Atlantic Salmon Foundation.

Haines, T.A., 1981b, Acidic precipitation and its consequences for aquatic ecosystems: A Review. Trans. Am. Fish. Soc. 110:669-707.

Hall, R.J. and Likens, G.E., 1980a, Ecological effects of experimental acidification on a stream ecosystem, pp. 375 to 376. In: Ecological Impact of Acid Precipitation, D. Drabløs and A. Tollan, eds., Proceedings International Conference, Sandefjord, Norway. SNSF Project, Ås-NLH, Norway.

Hall, R.J. and Likens, G.E., 1980b, Ecological effects of whole-stream acidification, pp. 443 to 451. In: Atmospheric Sulfur Deposition, Environmental Impact and Health Effects, D.S. Shriner, C.R. Richmond, and S.E. Lindberg, eds., Ann Arbor Science Publishers, Ann Arbor, MI.

Hall, R.J., Likens, G.G., Fiance, S.B., and Hendrey, G.R., 1980, Experimental acidification of a stream in the Hubbard Brook Experimental Forest, New Hampshire. Ecology 61:976-989.

Harriman, R. and Morrison, B.R., 1982, Ecology of streams draining
    forested and non-forested catchments in an area of central
    Scotland subject to acid precipitation. Hydrobiol.
    88:251-263.
Harvey, H.H., 1975, Fish populations in a large group of acid-
    stressed lakes. Verh. Int. Verein. Limnol. 19:2406-2417.
Harvey, H.H., 1979, The acid deposition problem and emerging
    research needs in the toxicology of fishes, pp. 115 to 128.
    In: Proc. Fifth Ann. Aquatic Toxicity Workship, Hamilton,
    Ontario. Tech. Rep. 862, Fish. Mar. Serv.
Harvey, H.H., 1980, Widespread and diverse changes in the biota of
    North American lakes and rivers coincident with acidification,
    pp. 93 to 98. In: Ecological Impact of Acid Precipitation,
    D. Drabløs and A. Tollan, eds., Proceedings International
    Conference, Sandefjord, Norway. SNSF Project, Ås-NLH, Norway.
Harvey, H.H. and Lee, C., 1980, Fishes of the LaCloche Mountain
    lakes of Ontario, 1965-1980. Report to the Ontario Ministry
    of Natural Resources (cited in NRCC, 1981).
Harvey, H.H. and Lee, C., 1982, Historical fisheries changes
    related to surface water pH changes in Canada, pp. 45 to 56.
    In: Acid Rain/Fisheries, R. Johnson, ed., Amer. Fisheries
    Soc., Bethesda, MD.
Harvey, H.H., Dillon, P.J., Fraser, G.A., Somers, K.M., Fraser,
    P.G., and Lee, C., 1982, Elevated metals and enhanced metal
    uptake in fishes in acid-stressed waters. Abstract, 185th
    National Meeting, American Chemical Society 22:438-441.
Hendrey, G.R. and Wright, R.F., 1976, Acid precipitation in Norway:
    Effects on aquatic fauna. J. Great Lakes Res. 2(suppl. 1):
    192-207.
Hoglund, L.B., 1961, The reactions of fish in concentration
    gradients. Chapter 3, Reaction to pH and CO_2. Rep. Inst.
    Freshw. Res. Drottningholm 43:95-147.
Hulsman, P.F., Powles, P.M., and Gunn, J.M., 1983, Mortality of
    walleye eggs and rainbow trout yolk-sac larvae in low-pH
    waters of the LaCloche Mountain area, Ontario. Trans. Amer.
    Fish. Soc. 112:680-688.
Hultberg, H., 1977, Thermally stratified acid water in late winter
    - a key factor inducing self oligotrophication processes which
    increase acidification. Water, Air, Soil Pollu. 7:279-294.
Jacobsen, O.J., 1977, Brown trout (Salmo trutta L.) growth at
    reduced pH. Aquaculture 11:81-84.
Jensen, K.W. and Snekvik, E., 1972, Low pH levels wipe out salmon
    and trout populations in southernmost Norway. Ambio
    1:223-225.
Johansson, N., Runn, P., and Milbrink, G., 1977, Early development
    of three salmonid species in acidified water. Zoon 5:127-132.
Johnson, D.W., 1975, Spawning behavior and strain tolerance of
    brook trout (Salvelinus fontinalis) in acidified water. M.S.
    thesis, Cornell University, Ithaca, NY. 108 pp.

Johnson, D.W. and Webster, D.A., 1977, Avoidance of low pH in
     selection of spawning sites by brook trout (Salvelinus
     fontinalis). J. Fish. Res. Board Can. 34:2215-2218.
Jones, H.C., Noggle, J.C., Young, R.C., Kelly, J.M., Olem, H.,
     Ruane, R.J., Pasch, R.W., Hyfantis, G.J., and Parkhurst, W.J.,
     1983, Investigators of the cause of fish kills in fish-
     rearing facilities in Raven Fork watershed. TVA/ONR/WR-83/9,
     Division of Air and Water Resources, Tennessee Valley
     Authority, Oak Ridge, TN.
Kwain, W., 1975, Effects of temperature on development and survival
     of rainbow trout Salmo gairdneri, in acid water. J. Fish.
     Res. Board Can. 32:493-497.
Leivestad, H., Hendrey, G., Muniz, I.P., and Snekvik, E., 1976,
     Effects of acid precipitation of freshwater organisms,
     pp. 87 to 111. In: Impact of Acid Precipitation on Forest and
     Freshwater Ecosystems in Norway, F.H. Braekke, ed., SNSF
     Project, FR 6/76, Ås-NLH, Norway.
Lloyd, R. and Jordan, D.H.M., 1964, Some factors affecting the
     resistance of rainbow trout (Salmo gairdneri, Richardson) to
     acid waters. Int. J. Air Water Pollu. 8:393-403.
Magnuson, J.J., Baker, J.P., Daye, P.G., Driscoll, C.T., Fischer,
     K., Guthrie, C.A., Peverly, J.H., Rahel, F.J., Schafran, G.C.,
     and Singer, R., 1983, Effects on aquatic biota, pp. 5-1 to
     5-203. In: The Acidic Deposition Phenonmenon and its
     Effects: Critical Assessment Review Papers, R. Linthurst and
     A. Altshuller, eds., EPA-600/8-83-016B, U.S.E.P.A.,
     Washington, DC.
Malley, D.F., Findlay, D.L., and Chang, P.J., 1982, Ecological
     effects of acid precipitation on zooplankton, pp. 297 to 327.
     In: Acid Precipitation Effects on Ecological Systems, F.M.
     D'Itri, ed., Ann Arbor Science, Ann Arbor, MI.
McDonald, D.G., Hobe, H., and Wood, C.M., 1980, The influence of
     calcium on the physiological responses of the rainbow trout,
     Salmo gairdneri, to low environmental pH. J. Exp. Biol.
     88:109-131.
Menendez, R., 1976, Chronic effects of reduced pH on brook trout
     (Salvelinusi fortinalis). J. Fish. Res. Board Can.
     33:118-123.
Milbrink, G. and Johansson, N., 1975, Some effects of acidification
     on roe of roach, Rutilus rutilus L., and perch, Perca
     fluiatilis L. - with special reference to the Avaå Lake system
     in eastern Sweden. Rep. Inst. Freshwater Res., Drottningholm
     54:52-62.
Mills, K.H., 1984, Fish population responses to experimental
     acidification of a small Ontario lake, pp. 117 to 132. In:
     Early Biotic Responses to Advancing Lake Acidification, G.R.
     Hendrey, ed., Butterworth Publishers, Boston, MA.
Mount, D.I., 1973, Chronic effect of low pH on fathead minnow
     survival, growth and reproduction. Water Res. 7:987-993.

Muniz, I.P., 1981, Acidification and the Norwegian salmon,
     pp. 65 to 72. In: Acid Rain and The Atlantic Salmon. Spec.
     Publ. 10, International Atlantic Salmon Foundation.
Muniz, I.P. and Leivestad, H., 1980a, Acidification effects on
     freshwater fish, pp. 84 to 92. In: Ecological Impact of Acid
     Precipitation, D. Drabløs and A. Tollan, eds., Proceedings
     International Conference, Sandefjord, Norway. SNSF Project,
     Ås-NLH, Norway.
Muniz, I.P. and Leivestad, H., 1980b, Toxic effects of aluminum on
     the brown trout, Salmo trutta L, pp. 320 to 321. In:
     Ecological Impact of Acid Precipitation, D. Drabløs and
     A. Tollan, eds., Proceedings International Conference,
     Sandefjord, Norway. SNSF Project, Ås-NLH, Norway.
Nelson, J., 1982, Physiological observations on developing rainbow
     trout, Salmo gairdneri (Richardson), exposed to low pH and
     varied calcium ion concentrations. J. Fish. Biol. 20:359-372.
New York State Department of Environmental Conservation, 1982,
     Clarification of issues raised concerning the effects of acid
     precipitation on fish resources in the Adirondack region of
     New York State. Testimony presented before the Senate
     Committee on Environment and Public Works, Washington, DC.
     (May 27, 1982).
NRCC, 1981, Acidification in the Canadian aquatic environment:
     Scientific criteria for assessing the effects of acidic
     deposition on aquatic ecosystems. National Research Council
     Canada, Report No. 18475, Environmental Secretariat, Ottawa,
     Ontario, Canada.
Overrein, L.N., Seip, H.M., and Tollan, A., 1980, Acid
     precipitation - effects on forest and fish. Final Report of
     the SNSF Project 1972-1980. Ås-NLH, Norway.
Peterson, R.H., Daye, P.G., and Metcalfe, J.L., 1980a, Inhibition
     of Atlantic salmon hatching at low pH. Can. J. Fish. Aquat.
     Sci. 37:770-774.
Peterson, R.H., Daye, P.G., and Metcalfe, J.L., 1980b, The effects
     of low pH on hatching of Atlantic salmon eggs, p. 328. In:
     Ecological Impact of Acid Precipitation, D. Drabløs and
     A. Tollan, eds., Proceedings International Conference,
     Sandefjord, Norway. SNSF Project, Ås-NLH, Norway.
Pfeiffer, M. and Festa, P., 1980, Acidity status of lakes in the
     Adirondack region of New York in relation to fish resources.
     FW-P168 (10/80), New York State Department Environmental
     Conservation, Albany, NY. 72 pp.
Retzsch, W., Everett, A., Duhaime, P., and Nothwanger, R., 1982,
     Alternative explanations for aquatic ecosystems effects
     attributed to acidic deposition. Utility Air Regulatory
     Group, Everett and Associates, Rockville, MD. 77 pp.
Robinson, G.D., Dunson, W.A., Wright, J.E., and Mamolito, G.E.,
     1976, Differences in low pH tolerance among strains of brook
     trout (Salvelinus fontinalis). J. Fish. Biol. 8:5-17.

Rosseland, B.O., Seva_durd, I., Svalastog, D., and Muniz, I.P.,
    1980, Studies on freshwater fish populations-effects of
    acidification on reproduction, population structure, growth
    and food selection, pp. 336 to 337. In: Ecological Impact of
    Acid Precipitation, D. Drabløs and A. Tollan, eds.,
    Proceedings International Conference, Sandefjord, Norway.
    SNSF Project, Ås-NLH, Norway.
Ryan, P. and Harvey, H., 1977, Growth of rock bass, Ambioplites
    rupestris, in relation to the morphoedaphic index as an
    indicator of an environment stress. J. Fish. Res. Board Can.
    34:2079-2088.
Ryan, P.M. and Harvey, H.H., 1980, Growth responses of yellow
    perch, Perca flavescens (Mitchill), to lake acidification in
    the LaCloche Mountain Lakes of Ontario. Env. Biol. Fish.
    5:97-108.
Ryan, P.M. and Harvey, H.H., 1981, Factors accouting for variation
    in the growth of rock bass (Ambloplites rupestris) and yellow
    perch (Perca flavescens) in the acidifying LaCloche Mountain
    lakes of Ontario, Canada. Verh. Int. Verein. Limnol.
    21:1231-1237.
Schindler, D.W., 1980, Experimental acidification of a whole lake:
    A test of the oligotrophication hypothesis, pp. 370 to 374.
    In: Ecological Impact of Acid Precipitation, D. Drabløs and
    A. Tollan, eds., Proceedings International Conference,
    Sandefjord, Norway. SNSF Project, Ås-NLH, Norway.
Schindler, D.W. and Turner, M.A., 1982, Biological, chemical, and
    physical responses of lakes to experimental acidification.
    Water, Air, Soil Pollu. 18:259-271.
Schindler, D.W., Wagemann, R., Cook, R.B., Ruszczynski, T., and
    Prokopowich, J. 1980, Experimental acidification of Lake 223,
    Experimental Lakes Area: Background data and the first three
    years of acidification. Can. J. Fish. Aquat. Sci. 37:342-354.
Schofield, C.L., 1965, Water quality in relation to survival of
    brook trout, Salvelilnus fontinalis (Mitchill). Trans. Am.
    Fish. Soc. 94:217-235.
Schofield, C.L., 1976a, Dynamics and management of Adirondack fish
    populations. Project Report April 1, 1975 - March 31, 1976,
    No. F-28-R-4, Dept. Environmental Conservation, Albany, NY.
    28 pp.
Schofield, C.L., 1976b, Acid precipitation: Effects on fish.
    Ambio 5:228-230
Schofield, C.L., 1976c, Lake acidification in the Adirondack
    Mountains of New York: causes and consequences, p. 477. In:
    Proc. First Int Symp. Acid Precipitation and the Forest
    Ecosystem, L.S. Dochinger and T.A. Seliga, eds., Tech. Rep.
    NE-23, USDA Forest Serv., Upper Darby, PA.
Schofield, C.L., 198_, Acid rain and the Adirondack trout,
    pp. 93 to 96. In: Acid Rain and the Atlantic Salmon. Spec.
    Publ. 10, International Atlantic Salmon Foundation.

Schofield, C.L., 1982, Historical fisheries changes as related to
    surface water pH changes in the United States, pp. 57 to 68.
    In: Acid Rain/Fisheries. R. Johnson, ed., American Fisheries
    Society, Bethesda, MD.

Schofield, C.L. and Trojnar, J.R., 1980, Aluminum toxicity to brook
    trout (Salvelinus fontinalis) in acidified waters,
    pp. 341 to 362. In: Polluted Rain, T.Y. Toribara, M.W.
    Miller, and P.E. Morrow, eds., Plenum Press, New York, NY.

Sevaldrud, I., Muniz, I.P., and Kalvenes, S., 1980, Loss of fish
    populations in southern Norway. Dynamics and magnitude of the
    problem, pp. 350 to 351. In: Ecological Impact of Acid
    Precipitation, D. Drabløs and A. Tollan, eds., Proceedings
    International Conference, Sandefjord, Norway. SNSF Project,
    Ås-NLH, Norway.

Swarts, F.A., Dunson, W.A., and Wright, J.E., 1978, Genetic and
    environmental factors involved in increased resistance of
    brook trout to sulfuric acid solutions and acid mine polluted
    waters. Trans. Am. Fish. Soc. 107:651-677.

Thompson, M.E., Elder, F.C., Davis, A.R., and Whitlow, S., 1980,
    Evidence of acidification of rivers of eastern Canada,
    pp. 244 to 245. In: Ecological Impact of Acid Precipitation,
    D. Drabløs and A. Tollan, eds., Proceedings International
    Conference, Sandefjord, Norway. SNSF Project, Ås-NLH, Norway.

Trojnar, J.R., 1977a, Egg and larval survival of white suckers
    (Catostomus commersoni) at low pH. J. Fish. Res. Board Can.
    34:262-266.

Trojnar, J.R., 1977b, Egg hatchability and tolerance of brook trout
    (Salvelinus fontinalis) fry at low pH. J. Fish. Res. Board
    Can. 34:574-579.

U.S.-Canada Memorandum of Intent on Transboundary Air Pollution,
    1982, Impact Assessment Group 1, Phase III Final Report,
    Vol. II. January 1983. 519 pp.

Watt, W.D., 1981, Present and potential effects of acid
    precipitation on the Atlantic salmon in eastern Canada.
    pp. 39 to 45. In: Acid Rain and the Atlantic Salmon. Spec.
    Publ. 10, International Atlantic Salmon Foundation.

Watt, W.D., Scott, C.D., and White, W.J., 1983, Evidence of
    acidification of some Nova Scotian rivers and its impact on
    Atlantic salmon, Salmo salar. Can. J. Fish. Aquat. Sci.
    40:462-473.

Wright, R. F. and Snekvik, E., 1978, Acid precipitation: Chemistry
    and fish populations in 700 lakes in southernmost Norway.
    Verh. Intern. Verein. Limnol. 20:765-775.

Wright, R.F., Dale, T., Gjessing, E.T., Hendrey, G.R.,
    Henriksen, A., Johannessen, M., and Muniz, I.P., 1975, Impact
    of acid precipitation on freshwater ecosystems in Norway.
    Research Report FR 3/75, Ås-NLH, Norway. 16 pp.

ACID DEPOSITION AND FOREST SOILS: POTENTIAL IMPACTS

AND SENSITIVITY

Ivan J. Fernandez

Department of Plant and Soil Sciences
University of Maine at Orono
1 Deering Hall
Orono, ME 04469

ABSTRACT

Acid deposition can alter soil properties and thus affect forest productivity and the quality of the water in associated aquatic ecosystems. Impacts on soil properties most likely to occur include nutrient additions to forest soils mainly as nitrogen and sulfur, increased rates of cation leaching, changes in the rate of soil acidification, increased mobilization of metals such as aluminum, changes in mineral weathering rates, altered soil biology, and the closely associated process of heavy metal deposition and accumulation. Assessing the sensitivity of forest soils to acid deposition requires a clear definition of the specific effect of concern, along with the realization that the mechanism of impact on forest productivity is not yet fully understood in the scientific community.

INTRODUCTION

Concern regarding the impacts of acid deposition on natural resources first focused on the threat to aquatic ecosystems and more recently has expanded to include forests. Agricultural systems are also being studied by the scientific community but present concern for potential acid deposition impacts has been tempered by what is believed to be the overriding effects of fertilizer and lime on soil chemistry. Generally, it is thought that these soil amendments control soil acidity in farmlands and outweigh the effects of the increment of acidity from the atmosphere. Nevertheless, direct acid deposition impacts on crops, or atmospheric additions of heavy metals, may also affect agricultural production and require further investigation. Whether aquatic, agricultural or forested ecosystems are examined with respect to the issue of acid deposition, it is clear that the soil resource is a key component in assessing potential mechanisms of impact.

The following discussion deals with the consequences of acid deposition, or more appropriately atmospheric deposition, for forest soils. While the same considerations may apply to both forest productivity and water quality concerns, this discussion is primarily directed towards the potential consequences of atmospheric deposition on tree growth.

A PERSPECTIVE ON THE FOREST DECLINE ISSUE

Before dealing with the impacts of acid deposition on forest soils in greater detail, it is important to provide some perspective on the role soil effects may play in the widespread and poorly understood modern phenomena of forest decline currently being experienced in forests worldwide, including southern Scandanavia, Central Europe and the eastern United States (Johnson and Siccama 1983; Tomlinson, 1983). Trees rarely, if ever, exist without some level of acute or chronic stress due to natural or anthropogenic environmental factors. It is likely that at any one location a combination of environmental factors affect forest productivity, including a unique suite of modern air pollution stress factors.

Evidence suggests natural stress factors such as drought or frost may play an important role in the timing of the development of forest decline (Johnson and Siccama, op. cit.). Smith (1981) pointed out that ozone (O_3) is the most widespread gaseous air contaminant influencing United States forests today. Organic atmospheric materials such as photochemical oxidants or hydrocarbons can be important components in air pollution induced stress on trees. The harvesting of forests also contributes to soil acidification where intensive utilization of biomass is practiced (Ulrich, 1983). Physiological and climatic consequences of elevated atmospheric carbon dioxide (CO_2) may contribute to stresses on future forest growth, and could even be an incipient stress to presently consider (Lemon, 1984). These factors and others must be evaluated in assessing the mechanisms of forest decline along with the indirect effects of acid deposition on tree growth through soil processes often emphasized in current hypotheses.

What emerges from the current scientific literature on forest decline is a series of studies which suggest forest decline may progress in three phases. First, forests are subjected to chronic stress due to acid deposition along with other pollutants (e.g., O_3, heavy metals). Biomass harvesting intensities should also be considered. It is in this phase that slow, cumulative alterations in the soil probably would be most important in reducing the vigor of individual trees. The second phase of the decline may result from acute stress factors which can be natural phenomena such as drought or frost. Finally, the third phase of decline occurs when secondary pathogens and insects further weaken the already stressed trees.

Primary concern in this discussion will be centered on the potential impacts of atmospheric and acidic deposition on forest soils. The volume of literature in these areas is both large and growing exponentially; however, several references are particularly useful for summarizing current knowledge on forest soil effects (Drabløs and Tollan, 1980; Hutchinson and Havas, 1980; Smith, 1981; U.S.-Canada M.O.I., 1981; Borghi and Adler, 1982; Fernandez, 1983a; Johnson and Siccama, 1983; Linthurst, 1983; Nat. Swed. Env. Board, 1983; Tomlinson, 1983; Ulrich and Pankrath, 1983).

POTENTIAL IMPACTS ON FOREST SOILS

While there is little evidence of direct effects on foliage due to acid deposition at doses similar to those experienced by forests on a regional basis, a great deal of attention has been given to possible alterations in soil chemistry and biology. From our current understanding, it is evident that several specific phenomena seem most likely to result from acid deposition.

Nutrient Additions

The primary acid forming components of acid deposition are sulfur (S) and nitrogen (N), both of which are essential plant nutrients. It can be expected that for terrestrial ecosystems deficient in either of these nutrients, acid deposition would have a fertilizing effect and actually increase growth. Indeed, fertilization experiments on forests have shown N to be the most commonly limiting nutrient throughout the world. The effects of additional nitrogen from acid deposition probably have a favorable impact on forest productivity, and the initial effect of acid deposition may be to increase the rate of tree growth. It is also possible that additions of nitrogen could alter the physiological status of trees making them more susceptible to environmental factors such as frost or drought. On the other hand, sulfur is rarely limiting to growth in forests with exceptions being documented in the northwestern United States and eastern Australia (D. Johnson et al., 1982). It is possible that more cases of S

deficiency in forests have not been found due to the modern
atmospheric source of this nutrient. The characteristic low
concentration and continual supply of nutrients from atmospheric
deposition is particularly advantageous in meeting plant
requirements throughout the growing season as compared to the more
inefficient one-time fertilizer applications used in forest
management.

Other nutrients are also supplied for forests through
atmospheric deposition. Significant amounts of several nutrients
will arise from atmospheric deposition as shown in Table 1, where
the contribution from precipitation is expressed as a percentage of
the annual nutrient requirement for each site. These values are
considered minimal since dry fallout and increased deposition due
to canopy filtering are not included and could more than double the
atmospheric contributions of nutrients shown here (Fowler, 1980;
Johnson and Siccama, 1983).

Increased Rates of Base Cation Leaching

Genesis of soils in humid, temperate climates results in the
loss of basic cations from the solum (i.e. the upper, biogeo-
chemically altered soil layers) as primary minerals weather. The
weathering of aluminosilicate and calcareous minerals consumes
protons and releases bases which are then either taken up by the
biota, adsorbed by soil particles or leached from the soil in
solution. Therefore, the loss of basic cations is a natural
process, but it is the rate of this process which is of concern
relative to potential acid deposition effects.

Acid deposition contains both protons and the anionic forms of
S and N as SO_4^{2-} and NO_3^-. Interaction between soil exchange
surfaces and these strong acid components includes (a) an exchange
of H^+ for cations such as Ca^{2+}, Mg^{2+} or K^+ from soil colloids, and
(b) the movement of these basic cations in soil solution with SO_4^{2-}
and NO_3^- anions. In soil solutions a charge balance is required
which causes the leaching of anions from the soil to also demand
the leaching of cations on a chemically equivalent basis. Thus, as
SO_4^{2-} and NO_3^- move through the soil in solution, they may be
accompanied by basic cations which are displaced into solutions as
a result of ion exchange with H^+. The selectivity of soil exchange
sites for specific cations is a function of a number of factors,
one of which is the relative abundance of a specific cation. In
general, the less abundant a cation is in the soil, the less
susceptible it will be to leaching losses with SO_4^{2-} or NO_3^- (Johnson
and Richter, 1984).

Since a majority of the soil colloids have a net negative
charge, anions are not held in the soil by ion exchange mechanisms.
If there were no other options for anions in soil solution besides

Table 1. Contributions of nutrients from atmospheric deposition to
          forested ecosystems[a]

Site	N	K	Ca	P	Mg
Hubbard Brook, New Hampshire (Northern Hardwood Forest)					
Requirement (kg ha^{-1}yr^{-1})	89.7	37.8	37.7	7.4	6.6
Precipitation (kg ha^{-1}yr^{-1})	6.5	0.9	2.2	0.04	0.6
Minimum Atmospheric Contribution (%)	7.3	2.4	5.8	0.5	9.1
Solling, West Germany (Norway Spruce Plantation)					
Requirement (kg ha^{-1}yr^{-1})	49.2	25.9	20.2	4.5	1.6
Precipitation (kg ha^{-1}yr^{-1})	21.8	3.7	12.6	0.5	2.6
Minimum Atmospheric Contribution (%)	44.3	14.3	62.4	11.1	162.5
Thompson Research Center, Washington (Douglas-Fir Plantation)					
Requirement (kg ha^{-1}yr^{-1})	14.8	31.0	13.3	5.9	6.8
Precipitation (kg ha^{-1}yr^{-1})	1.7	2.2	2.2	2.3	0.5
Minimum Atmospheric Contribution (%)	4.1	7.1	16.5	38.9	7.4

[a]Adapted from Cole and Rapp (1980).

adsorption on cation exchange sites or leaching, the threat of basic cation losses would be much more severe. This is not the case. Due to tree requirements for nitrogen, a nutrient which is usually limiting in forests, much of the NO_3^- resulting from acid deposition is taken up by plant roots. This significantly limits the loss of nitrogen from the system. Abrahamsen (1981) estimated that an average of 70% of the nitrogen in precipitation was retained by forest ecosystems. On the other hand, sulfur is usually not in limiting supply and, therefore, poses the major threat to accelerating the loss of basic cationic nutrients from forest soils by leaching (Cronan et al., 1977). Nitrogen can pose a more significant threat to accelerated leaching of bases in soils where nitrogen-fixing species are abundant or prolonged, high levels of N deposition have enriched the nitrogen status of the soil to a point where much greater leaching losses of NO_3^- can occur.

Soils high in iron (Fe) and aluminum (Al) oxides, but relatively low in organic matter, can adsorb significant amounts of SO_4^{2-} and thus reduce the leaching rate of both sulfate and all potential accompanying cations (e.g., Al^{3+}, H^+, Ca^{2+}, Mg^{2+}, K^+).

To a lesser extent the clay mineral kaolinite can also adsorb a limited amount of SO_4^{2-} in some soils. Adsorption of sulfate by iron and aluminum oxides may also increase cation exchange capacity (CEC) which would increase the buffering capacity of soils. This would aid in preventing further alterations in base status from acid deposition.

Sulfate adsorption in soils is dependent on the concentration of SO_4^{2-} in soil solution. As additions of H_2SO_4 increase, more SO_4^{2-} will be adsorbed until a steady-state is reached (Johnson and Cole, 1980). The steady-state condition reflects the ability of soils to adsorb sulfate at equilibrium. When such additions decrease, sulfate may desorb from the soil leaching both sulfate and accompanying cations until a new equilibrium is established governed by the lower sulfate concentrations in soil solution. During the desorption period, SO_4^{2-} additions would be less than losses until the new equilibrium is attained, at which time steady-state conditions would be reestablished. This desorption phenomena would only occur with reversibly adsorbed sulfate. Irreversibly adsorbed SO_4^{2-} would presumably not readily undergo desorption upon decreased loadings of sulfate.

Soil Acidification and Metal Mobilization

Soil acidification is well recognized by scientists as being a natural process of soil development in humid climates. Soils supporting forest growth are typically further acidified by the significant uptake of cationic bases by trees, the acidifying nature of forest litter and the lack of any type of soil amendment such as lime so widely employed in agriculture. This natural soil acidification process is accompanied by both the leaching loss of basic cations and an enrichment of more mobile phases of aluminum in soils. In the northeastern U.S., typical podzolic soil profiles can exhibit highly acid pH values with surface organic horizons often in the range of pH between 3.0 to 4.0. It is important to note that many of the most prevalent tree species such as the spruces or balsam fir found in these regions are shallow rooted with most of their root biomass typically found in the highly acid surface organic and mineral soil horizons (U.S.D.A., 1965).

With this in mind, then the major concern regarding acid deposition is that it may increase the rate of soil acidification which could potentially affect tree growth. In areas where soil pH values are already much below average ambient precipitation pH, further acidification due to acid deposition seems unlikely, since the lower the pH, the more buffered to further acidification is the soil (i.e. pH is a logarithmic scale) and equilibrium with higher pH precipitation could suggest soil pH should increase. Many soils, while considered acid, are not as acid as the ambient precipitation. The risk of accelerated soil acidification at these sites is then a function of the buffer capacity of the soil (i.e., the ability of the soil to resist changes in pH).

These relationships are somewhat more complex, however, since not only pH (i.e. hydrogen ions) but also the quantity of other materials in precipitation (e.g., calcium ions) must also be considered in predicting precipitation impacts on soil chemistry. Ulrich (1984) stated that a decline in soil pH has been demonstrated in all regions of West Germany where pH trends in forest soils have been examined over the last few decades. Through examinations of ion flux balances, he believed acid deposition has played the major role in this soil acidification trend. Similarly, Hanson et al. (1982) examined forest floors along a geographical precipitation pH gradient from southern New England to Canada. They found that decreasing concentrations of calcium, manganese, and pH were correlated with increased precipitation acidity and concluded these two trends were directly related.   Linzon and Temple (1980), on the other hand, investigated the pH of six Ontario soils after an 18 year period and found no change in pH for samples from the major soil horizons.

As soil pH decreases, many of the metals in soils become more mobile. For soils rich in potentially toxic heavy metals (e.g., Hg, Pb, Cd, Cu) this could mean that soil acidification is accompanied by a greater risk of biological effects from excessive heavy metal levels in soil solutions. Many of these metals are immobilized by particulate phases of organic matter in soils which would also have to be considered in toxicity assessments. A better understanding of the magnitude of heavy metal transport in soil solutions and solubilization/precipitation mechanisms for forested regions impacted by acid deposition is clearly needed.

A significant amount of attention has been given to the concern for increased Al mobility in soil solutions as a result of acid deposition.   Ulrich et al. (1980) studied soil solution chemistry in a beech forest of West Germany and found aluminum concentrations increased from less than 10% of the equivalent cation sum in 1966 to 30% in 1979 for soil layers below the 20 cm depth. They believed that the high Al concentrations (2 mg l^{-1}) continuously found in these soil solutions were a result of acid deposition and are primarily responsible for the forest decline being experienced in that region. Cronan (1980) also reported evidence of elevated Al concentrations in soil solutions from eastern North America.

It is difficult to interpret the existing evidence relative to the biological consequences of elevated soil solution Al concentrations. While Al concentration is important, the concentration of dissolved organic carbon or other ions such as calcium may play a major role in determining the toxicity of Al in soil solutions. Also, some research suggested many tree species are able to tolerate Al concentrations much higher than those found in the West German studies (McCormick and Steiner, 1978). Johnson and Siccama (1983) reported evidence for Al concentrations in fine roots of healthy and declining red spruce which suggested Al toxicity is not a causative factor in the tree mortality experienced in Vermont and New Hampshire. While these studies and others have resulted in a significant amount of research being developed to examine the soil aluminum question, the results of these efforts are not yet available to help clarify a currently poorly defined issue.

Lastly, D. Johnson et al. (1982) made an important distinction between soil acidification and cation leaching. Although both processes may occur simultaneously as a result of acid deposition, each process can occur independently. Soils can adsorb both SO_4^{2-} and H^+ which would increase CEC, decrease pH and base saturation, and yet have no effect on net cationic nutrient losses. Similarly, soil organic matter humification increases the soil CEC but does not contribute to soil cation content. Thus, it is possible to acidify a soil without increased leaching losses of base cations. On the other hand, acid deposition could accelerate the release of bases from mineral weathering which might offset any declines in pH or base saturation. The result would be accelerated cation leaching without simultaneous acidification of the soil.

Mineral Weathering

One of the potential effects of acid deposition on soil chemical processes is the alteration of the rate of mineral weathering. Weathering is essentially an acid consuming process for many of the common soil minerals. Typical weathering reactions which occur in temperate climate forest soils are shown in reactions 1 and 2 for a common aluminosilicate and carbonate mineral, respectively:

$$2KAlSi_3O_8 \text{ (microcline)} + 2H^+ + 9H_2O \quad <\!\!-\!\!-\!\!> \tag{1}$$
$$Al_2Si_2O_5(OH)_4 \text{ (kaolinite)} + 4H_4SiO_4 + 2K^+$$

$$CaCO_3 \text{ (calcite)} + 2H^+ \quad <\!\!-\!\!-\!\!> \quad Ca^{2+} + CO_2(g) + H_2O \tag{2}$$

During these reactions H^+ is consumed and basic cations are released into soil solution; however, there is still a poor quantitative understanding of the present rate of mineral soil weathering in forests. It is possible that increased mineral weathering rates, as a result of acid deposition, could offset some of the cation leaching losses due to sulfate. However, mineral weathering is an extremely slow process, and changes in the weathering rate may prove a minor factor in overall nutrient balances.

Bache (1982) recently discussed the critical factors controlling the rate of mineral decomposition, and thus the rate of acid neutralization by this mechanism. The two general factors are (a) rock composition or content of the weatherable minerals, and (b) the access of percolating waters to rock surfaces. The key to understanding rock weathering capacity to neutralize acids is the pattern of water flow and the residence time for solutions. While the capacity for rocks and minerals to neutralize acids is infinite, without adequate contact time soil solutions would retain their electrolytes and could actually become further acidified by soil components.

Soil Biology

Organisms inhabiting the soil play an intimate role in the processes of organic matter decomposition, nutrient cycling, and other factors such as soil structure and pathological relationships. Examples of important groups of soil organisms include the invertebrate fauna, fungi and bacteria. Of lesser concern regarding forest productivity are the algae, protozoa, rotifers and nematodes. While little or no research has described acid deposition impacts on earthworms, these organisms are known to be intolerant to acid soils (i.e., not often found in acid forest soils) and would be expected to be detrimentally affected by declines in pH.

Particularly important for forest soils, with regard to the supply of nutrients to tree roots, is the rate of organic matter decomposition. It is this process which provides much of the nutrients to forest species and which, if decreased by acid deposition, could limit nutrient supply. Likewise, the supply of nitrogen in soils is dependent upon a series of microbially mediated transformation processes (e.g., ammonification, nitrification) which, if altered by acid deposition, could have an impact on N supply and tree growth. A specialized group of fungi, called mycorrhiza, form symbiotic relationships with tree roots which has been shown to be advantageous to growth. If these organisms were negatively affected by acid deposition, clearly tree growth could suffer due to problems in nutrient supply and other factors.

The literature on the subject of acid deposition impacts on soil organisms is not uniform. Results from studies on acid deposition impacts to organic matter decomposition and micro-biological response show a range from positive to negative or no detectable impacts (Fernandez, 1983a; Linthurst, 1983). At present, it appears that the role of heavy metals in forest soils may be more important to examine, both alone and in combination with acid deposition, to understand potential impacts on soil microflora processes.

Heavy Metal Deposition

In discussing the potential impacts of acid deposition on forested ecosystems, it is clear that the role of heavy metals (e.g., Pb, Zn, Cu, Cd, Fe) must also be considered in order to develop meaningful hypotheses for potential impacts on forest productivity. Recent studies in the northeastern United States showed high levels of heavy metals in forest floors with strong evidence that the major portion of these metals are atmospherically derived (Reiners et al., 1975; Andresen et al., 1980; Hanson et al., 1982; A. Johnson et al., 1982; Johnson and Siccama, 1983; Friedland et al., 1984). These studies often reported that metals such as copper, zinc, or nickel may be accumulating in forest floors, while each reported the accumulation of lead. Heavy metals also appear in greater concentrations in the forest floor at higher elevations suggesting a correlation between metal additions and increased precipitation typically found along elevational gradients. A. Johnson et al. (1982) pointed out that the level of lead in the forest floor is a function of: (a) amount of precipitation, (b) proximity to lead emitting industries, (c) distance from urban corridors, (d) elevation, (e) forest floor age, and (f) nature of the underlying soil. Research also showed that almost all of the lead deposited over the last century has been retained in the soil (Andresen et al., 1980; Hanson et al., 1982). Thus, even as lead deposition rates decline, accumulation and potential biological impacts still exist.

One of the processes that may be first affected by the accumulation of heavy metals in the forest floor is organic matter decomposition. Studies in Sweden near point sources of heavy metal emissions have shown even moderate amounts of metals can depress organic matter decomposition, nitrogen and phosphorus mineralization, and soil urease and acid phosphatase activity (Ruhling and Tyler, 1973; Tyler, 1975). Therefore, it is critical to include heavy metal deposition and accumulation factors in the development of any hypothesis on atmospheric deposition impacts to forest soils or the forest decline phenomena.

FOREST SOIL SENSITIVITY

In soil sensitivity assessments regarding acid deposition the specific impact being considered must be clearly defined to permit the development of meaningful sensitivity ratings (i.e., sensitivity to what?). From current understanding it seems that

many of the same soil parameters would be utilized to assess soil
sensitivity for the major effects of concern, yet the interpretation of these data could differ significantly for each of the
different effects. For example, an acid soil with a low base
saturation might be considered insensitive to accelerated losses of
bases and, therefore, not of concern regarding declines in forest
productivity due to nutrient losses. This same soil, however,
could permit aluminum mobilization and leaching and be considered
very sensitive for potential effects on nearby aquatic ecosystems.
Therefore, soil sensitivity ratings should refer to: (a) forest
productivity alterations, (b) aquatic ecosystem acidification, (c)
soil acidification, (d) base cation leaching losses, or some other
concern such as an ability to accumulate heavy metals in
biologically active soil layers. Furthermore, the role of acid
deposition and soil processes as related to the forest decline
phenomena are still poorly understood. Indeed, more than one
mechanism of acid deposition impact on forests seems likely.
Therefore, only the probable sensitivity of forest soils to acid
deposition induced alterations in forest productivity can be
determined at this time based on currently favored hypotheses
regarding forest decline phenomena.

Various sensitivity rating schemes have been developed using
lake chemistry, bedrock geology, soils or vegetation information;
only those dealing with soils will be discussed. McFee (1980)
developed one of the first sensitivity criteria based on soil
properties utilizing CEC, flooding occurrence and the presence or
absence of carbonates (Table 2). While this approach is attractive
due to the availability of information needed on soil properties,
it is too general for site-specific assessments and it fails to
include base saturation.

Table 2. Soil sensitivity classification after McFee (1980).

Sensitivity Rating	Description
Non-sensitive areas	Soils which are calcareous, subject to frequent flooding, or have a CEC > 15.4 meq/100 g in the top 25 cm of soil
Slightly sensitive areas	Soils with a CEC between 15.4 to 6.2 meq/100 g in the top 25 cm of soil
Sensitive	Soils which have a CEC < 6.2 meq/100 g in the top 25 cm of soil

A more detailed description of soil chemical response to acid deposition has been developed by Ulrich (1980) and describes sequential buffering ranges whereby a soil could be categorized in reference to sensitivity (Table 3). The dominant acid neutralizing mechanism is considered to remain in a steady-state as long as the rate of H^+ deposition and internal production equals the rate of neutralization. Once H^+ addition and production exceeds the rate of neutralization, the dominant mechanism of soil buffering will shift to the next lower range.

Work has been carried out in Canada to assess soil sensitivity by Wang and Coote (1981) and Coote et al. (1981) for agricultural soils. These workers used a similar approach as McFee (1980) but included soil pH, base saturation and soil texture in arriving at their sensitivity classes. Cowell et al. (1981) also developed one of the first soil sensitivity rating schemes specifically aimed at forest productivity affects.

What most of these schemes do not address is the issue of how well a soil can adsorb SO_4^{2-} or NO_3^- anions. D. Johnson et al. (1982) examined forest soil sensitivity based on the anion mobility concept. This considers the major factor controlling the impact of

Table 3. Classification of soil buffering ranges to acid deposition as defined by Ulrich (1980).

Buffering Range	pH Range	Description
Carbonate Buffering Range	6.5 - 8.3	Soil buffering of incoming H^+ primarily due to $CaCO_3$ dissolution
Silicate Neutralization Range	5.0 - 6.5	Soil buffering of incoming H^+ due to silicate mineral dissolution
Cation Exchange Range	4.2 - 5.0	Soil buffering of incoming H^+ due to displacement of bases
Aluminum Buffering Range	3.0 - 4.2	Soil buffering of incoming H^+ due to Al^{3+} release from soil polymeric hydroxy Al compounds
Iron Buffering Range	below 3.0	Soil buffering of incoming H^+ due to the release of Fe ions from Fe-oxides

acid deposition on soil cation leaching losses to be the mobility of the associated anions (i.e., SO_4^{2-} and NO_3^-) in soil solutions.

Fernandez (1983b) developed a program for the forest products industry, drawing on the state-of-knowledge at that time, which identified information necessary to adequately assess the sensitivity of soils to acid deposition impacts on forest productivity. The major drawbacks for this program were (a) information is not available on some of the key parameters, and (b) without a better understanding of the mechanism of forest decline, it is difficult to classify sensitivity. The program did appear to identify the soil measurements needed both initially and, if made periodically, over time to provide a better understanding of the effects of acid deposition on forest soils. The physical properties of soils considered important include effective soil depth, texture, bulk density, porosity, drainage and percent coarse fragments. Most of these parameters are commonly measured on soils and are useful in a sensitivity assessment by defining the degree of interaction which could be expected between soil solutions and soil particles. Topographic properties such as slope and aspect should also be included. One particularly critical factor in these assessments would be elevation. Factors thought to be involved in the forest decline phenomena, such as total precipitation levels, incidence of cloud cover, and heavy metal depositions, are all strongly influenced by elevation.

Unlike the physical properties desirable for soil sensitivity assessments, not all of the necessary soil chemical parameters are typically measured in conventional soil characterizations. The most important soil chemical properties for assessing sensitivity to acid inputs include:

○ Sulfate adsorption capacity (SAC) and adsorbed sulfate - an index of soil susceptibility to cation leaching losses as a result of the mobile SO_4^{2-} anion;

○ Cation exchange capacity (CEC) and base saturation - the larger the CEC the more buffered (i.e., resistant) a soil is to changes in exchangeable cation composition, while base saturation is a measure of the exchangeable base cation status of the soil (which may decline due to accelerated leaching) and is used to determine whether basic or acidic cations are more likely to accompany mobile anions in the leaching process;

○ pH - a valuable measurement for monitoring soil acidity over time and determining, as is the closely correlated base saturation, whether basic or acidic cations are more likely to accompany mobile anions in the leaching process;

° Organic matter content and quality – an important indicator of
  microbiological processes in the soil and also directly
  related to the soil SAC, CEC, and pH;

° Heavy metal content – necessary to determine the potential for
  direct effects of heavy metals on soils and plants along with
  the potential for interactive effects with acid deposition or
  other stress factors;

° Nitrogen – used to determine soil susceptibility to
  accelerated base cation leaching losses due to the mobile NO_3^-
  anion at N-rich sites (e.g., alder stands) and possible
  physiological stress on trees due to excessive levels of
  nitrogen accumulation in the soils (i.e., N loading); and

° Uptake by vegetation – this is an essential flux to quantify
  since only when the uptake of elements from the soil by plants
  is known can a meaningful assessment of potential alterations
  to the soil cation status as a result of acid deposition be
  carried out.

A great deal of the research currently underway on acid
deposition and forest decline will provide valuable information to
improve the precision by which the sensitivity of soils can be
defined.   It is important that present knowledge is used to
estimate the resources at risk for research and policy development,
without appearing to draw final conclusions on current or potential
impacts of acid deposition as related to changes in soil properties
or forest productivity.

CONCLUSIONS

The acid deposition issue has served to clearly identify our
current shortcomings regarding basic knowledge on forest ecosystem
biogeochemical cycling.   What appears to be needed is the expansion
of current efforts to carefully monitor, on a long-term basis, the
growth and elemental fluxes which take place in various soil-stand
type combinations.   This research would provide basic knowledge of
forest ecosystem functioning which not only would help define the
effects of atmospheric deposition, but also will be a prerequisite
for forest managers to make appropriate decisions on future
intensive management practices (e.g. whole tree harvests) and other
related forestry issues.

REFERENCES

Abrahamsen, G., 1981, Effects of air pollution on forests,
     pp. 433 to 446. In: Beyond the Energy Crisis - Opportunity

and Challenge, R.A. Fazzolare and C.B. Smith, eds., Pergamon
    Press, New York, NY.

Andresen, A.M., Johnson, A.H., and Siccama, T.G., 1980, Levels of
    lead, copper, and zinc in the forest floor in the northeastern
    United States. J. Environ. Qual. 9:293-296.

Bache, B.W., 1982, The implications of rock weathering for acid
    neutralization, pp. 175 to 187. In: Ecological Effects of
    Acid Deposition. Report and Background Papers for 1982
    Stockholm Conference on Acidification of the Environment -
    Expert Meeting I. National Swedish Envir. Protection Board,
    Stockholm, Sweden.

Borghi, L. and Adler, D., 1982, The effects of air pollution and
    acid rain on fish, wildlife, and their habitats: Forests.
    Air Pollution and Acid Rain Report No. 6, Fish Wildlife
    Service, U.S. Dept. Interior, Washington, DC. 83 pp.

Cole, D.W. and Rapp, M., 1980, Elemental cycling in forest
    ecosystems, pp. 341 to 409. In: Dynamic Properties of Forest
    Ecosystems, D.E. Reichle, ed., Cambridge University Press,
    Boston, MA.

Coote, D.R., Siminovitch, D., Singh, S.S., and Wang, C., 1981, The
    significance of acid rain to agriculture in eastern Canada.
    Contribution No. 19, Land Resource Res. Inst., Agriculture
    Canada, Ottawa, Canada. 26 pp.

Cowell, D.W., Lucas, A.E., and Rubec, C.D.A., 1981, The development
    of an ecological sensitivity rating for acid precipitation
    impact assessment. Lands Directorate Working Paper No. 10,
    Environment Canada, Ottawa, Canada. 42 pp.

Cronan, C.S., 1980, Consequences of sulfuric acid inputs to a
    forest soil, pp. 335 to 343. In: Atmospheric Sulfur
    Deposition, D.S. Shriner, C.R. Richmond and S.E. Lindberg,
    eds., Ann Arbor Science, Ann Arbor, MI.

Cronan, C.S., Reiners, W.A., Reynolds, R.C., Jr., and Lang, G.E.,
    1977, Forest floor leaching: Contributions from mineral,
    organic, and carbonic acids in New Hampshire subalpine
    forests. Science 200:309-311.

Drabløs, D. and Tollan, A., eds., 1980, Ecological Impact of Acid
    Deposition, Proceedings International Conference, Sandefjord,
    Norway. SNSF Project, Ås-NLH, Norway. 383 pp.

Fernandez, I.J., 1983a, Acidic deposition and its effects on forest
    productivity - A review of the present state of knowledge,
    research activities, and information needs. Second Progress
    Report. Tech. Bull. No. 392, National Council Paper Industry
    for Air and Stream Improvement, New York, NY. 104 pp.

Fernandez, I.J., 1983b, Field study program elements to assess the
    sensitivity of soils to acidic deposition induced alterations
    in forest productivity. Tech. Bull. No. 404, National Council
    Paper Industry for Air and Stream Improvement, New York, NY.
    176 pp.

Fowler, D., 1980, Removal of sulphur and nitrogen compounds from
    the atmosphere in rain and by dry depositions, pp. 22 to 32.

In: Ecological Impact of Acid Precipitation, D. Drabløs and
   A. Tollan, eds., Proceedings International Conference,
   Sandefjord, Norway. SNSF Project, Ås-NLH, Norway.
Friedland, A.J., Johnson, A.H., Siccama, T.G., and Mader, D.L.,
   1984, Trace metal profiles in the forest floor of New England.
   Soil Sci. Soc. Am. J. 48:422-425.
Hanson, D.W., Norton, S.A., and Williams, J.S., 1982, Modern and
   paleolimnological evidence for accelerated leaching and metal
   accumulation in soils in New England, caused by atmospheric
   deposition. Water, Air, Soil Pollu. 18:227-238.
Hutchinson, T.C. and Havas, M., eds., 1980, Effects of Acid
   Precipitation on Terrestrial Ecosystems. Plenum Press, New
   York, NY. 654 pp.
Johnson, A.H. and Siccama, T.G., 1983, Acid deposition and forest
   decline. Environ. Sci. Tech. 17:294A-305A.
Johnson, A.H., Siccama, T.G., and Friedland, A.J., 1982, Spatial
   and temporal patterns of lead accumulation in the forest floor
   in the northeastern United States. J. Environ. Qual.
   11:577-580.
Johnson, D.W. and Cole, D.W., 1980, Anion mobility in soils:
   Relevance to nutrient transport from forest ecosystems.
   Environ. Internat. 3:79-80.
Johnson, D.W. and Richter, D.D., 1984, The combined effects of
   atmospheric deposition, internal acid production, and
   harvesting on nutrient gains and losses from forest
   ecosystems, pp. 36 to 46. In: U.S.-Canadian Conference on
   Forest Responses to Acidic Deposition, L. Breece and
   S. Hasbrouck, eds., Land and Water Resources Center,
   University of Maine, Orono, ME.
Johnson, D.W., Turner, J., and Kelly, J.M., 1982, The effects of
   acid rain on forest nutrient status. Water Resour. Res.
   18:449-461.
Lemon, E.R., 1984, CO_2 and Plants - the Response of Plants to
   Rising Levels of Atmospheric Carbon Dioxide. Westview Press,
   Boulder, CO. 280 pp.
Linthurst, R.A., ed., 1983, The acidic deposition phenomenon and
   its effects. Critical assessment review papers (draft), Vol.
   II. Effects Sciences. Pub. No. EPA-600/9-83-016B.
   U.S.E.P.A., Washington, DC. 687 pp.
Linzon, S.N. and Temple, P.J., 1980, Soil resampling and pH
   measurements after an 18-year period in Ontario, pp. 176 to
   177. In: Ecological Impact of Acid Precipitation, D. Drabløs
   and A. Tollan, eds., Proceedings International Conference,
   Sandefjord, Norway. SNSF Project, Ås-NLH, Norway.
McCormick, L.H. and Steiner, K.C., 1978, Variation in aluminum
   tolerance among six genera of trees. For. Sci. 24:565-568.
McFee, W.W., 1980, Sensitivity of soil regions to acid
   precipitation. EPA-60013-80-013, U.S.E.P.A., Washington, DC.
   179 pp.

National Swedish Environment Board, 1983, Ecological effects of
    acid deposition. Report and background papers. 1982
    Stockholm Conference on Acidification of the Environment.
    Expert Meeting I. Report SNV pm 1636, Stockholm, Sweden.
    340 pp.
Reiners, W.A., Marks, R.H., and Vitousek, P.M., 1975, Heavy metals
    in subalpine and alpine soils of New Hampshire. Oikos
    26:264-275.
Ruhling, Å. and Tyler, G., 1973, Heavy metal pollution and
    decomposition of spruce needle litter. Oikos 24:402-416.
Smith, W.H., 1981, Air Pollution and Forests. Springer-Verlag, New
    York, NY. 377 pp.
Tomlinson, G.H., 1983, Air pollutants and forest decline. Environ.
    Sci. Tech. 17:246A-256A.
Tyler, G., 1975, Effect of heavy metal pollution on decomposition
    and mineralization rates in forest soils, pp. 217 to 220. In:
    Symposium Proc. on Int. Conf. on Heavy Metals in the Environ.,
    T.C. Hutchinson, ed., Toronto, Canada.
Ulrich, B., 1980, The production and consumption of hydrogen ions
    in the ecosphere, pp. 255 to 282. In: Effects of Acid
    Precipitation on Terrestrial Ecosystems, T.C. Hutchinson and
    M. Havas, eds., Plenum Press, New York, NY.
Ulrich, B., 1983, A concept of forest ecosystem stability and of
    acid deposition as driving force for destabilization, pp. 1 to
    29. In: Effects of Accumulation of Air Pollutants in Forest
    Ecosystems, B. Ulrich and J. Pankrath, eds., D. Reidel Publ.,
    Boston, MA.
Ulrich, B., 1984, Effects of accumulation of air pollutants in
    forest ecosystems. Institut für Bodenkunde und Waldernährung
    der Universität Göttingen, Göttingen, Federal Republic of
    Germany. 20 pp.
Ulrich, B. and Pankrath, J., eds., 1983, Effects of Accumulation of
    Air Pollutants in Forest Ecosystems. D. Reidel Publ., Boston,
    MA. 389 pp.
Ulrich, B., Mayer, R., and Khanna, P.K., 1980, Chemical changes due
    to acid precipitation in a loess-derived soil in central
    Europe. Soil Sci. 130:193-199.
U.S.-Canada Memorandum of Intent on Transboundary Air Pollution,
    1981, Impact assessment working group I. Phase II summary
    report. U.S.D.A., Washington, DC. 12 pp.
U.S. Department of Agriculture, 1965, Silvics of forest trees of
    the United States. Agricultural Handbook No. 271, U.S.D.A.,
    Washington, DC. 762 pp.
Wang, C. and Coote, D.R., 1981, Sensitivity classification of
    agricultural land to long-term acid precipitation in eastern
    Canada. Contribution No. 98, Land Resource Res. Inst.,
    Agriculture Canada, Ottawa, Canada. 9 pp.

ACID RAIN INTERACTIONS WITH LEAF SURFACES: A REVIEW

David S. Shriner and J. William Johnston, Jr.

Environmental Sciences Division
Oak Ridge National Laboratory
Oak Ridge, TN   37831

ABSTRACT

The terrestrial regions of North America and Europe that experience elevated levels of acidity in rain are typically covered with at least one layer of vegetation. Those vegetation surfaces are the primary receptors of rain-deposited pollutants. As primary receptors, vegetation surfaces may be affected by and affect acids in rain. Alterations of leaf surface structure and function may be manifested by changes in whole plant function. However, interactions of vegetation surfaces with acid rain involve more than potential plant effects. Since the chemistry of incident precipitation is altered as it passes through vegetation canopies, the response of secondary receptors, such as soil and aquatic systems, may be affected by previous interactions between acid rain and vegetation surfaces. This chapter provides a review and discussion of literature pertinent to the interactions between vegetation surfaces and acidic deposition.

INTRODUCTION

Emissions of sulfur dioxide and nitrogen oxides from fossil fuel combustion and metal smelting have led to acidification of precipitation by sulfuric and nitric acid over large areas of North America and Europe. The potential for adverse effects on vegetation, soil, and aquatic systems has been recognized, and research is in progress to identify problem areas and quantify effects.

The frequent and abundant rainfall, characterizing much of the areas currently impacted by industrial pollution in eastern North America and western Europe, is one of the factors that created conditions favorable to the establishment of abundant vegetation. In fact, most of the areas that experience elevated levels of acidity in rainfall are covered with at least one, and often several, layers of vegetation. Therefore, vegetation surfaces are the primary receptors for most of the incident precipitation that can potentially affect not only internal plant processes, but also soil and aquatic systems in subsequent reactions. Recognition of that fact raises two important issues regarding the interactions of water droplets (rain) and plant surfaces:

○   effects of incident precipitation chemistry on the structure and function of the receptor surface, and

○   effects of receptor surface on the chemistry of the incident precipitation.

DISCUSSION

Most of the research directly concerning interactions between acid deposition and vegetation has been conducted during the last decade. Although this is a relatively short time duration for agricultural and ecological research, a great deal of progress toward and understanding of the acid deposition phenomenon has been made. In addition, there is a large body of information that is pertinent to acid deposition-vegetation interactions that was developed independent of acid deposition research objectives. These two bodies of information are brought together in the discussion that follows so that a better understanding of the issues mentioned above can be achieved.

Effects of Acidity on Leaf Structure and Functional Modifications

Based on experimental evidence with simulated rain, a wide range of plant species is believed to be sensitive to direct injury from some elevated level of wet acidic deposition (Evans et al., 1981; Shriner, 1984). Other species have been noted to be tolerant of equally elevated levels (to pH 2.5 for up to ten hours total exposure) without visible injury (Haines et al., 1980). These results suggested that generalizations about sensitivity to injury may be difficult, and some understanding of the mechanisms by which injury may occur is necessary. The sensitivity of an individual species of vegetation appears to be influenced by structural features of the vegetation, which (1) influence the foliage wettability; (2) make the foliage more vulnerable to injury (e.g., through differential permeability of the cuticle); or (3) retain rainwater due to leaf size, shape, or attachment angle. In those instances where one or more of the above conditions renders a plant potentially sensitive to acidic deposition, effects may be manifested in alterations of leaf structure or function.

Injury to foliage by simulated acidic precipitation largely depends on the effective dose to which sensitive tissues are exposed. The effective dose - that concentration and amount of hydrogen ion and time period responsible for necrosis of an epidermal cell - for example, is influenced by the contact time of an individual water droplet or film on the foliage surface (Evans et al., 1981; Shriner, 1984). Contact time, in turn, can be regulated by the wettability of the leaf or by leaf morphological features that affect runoff of water from the surface. Physical characteristics of the leaf surface (e.g., roughness, pubescence, waxiness) or the chemical composition of the cutin and epicuticular waxes determine the wettability of most leaves (Martin and Juniper, 1970).

For injury to occur at the cellular level, the ions responsible must penetrate these protective physical and chemical barriers or enter through stomata (Evans et al, 1981). Crafts (1961) postulated that cuticle penetration occurs through micropores. Evidence indicates that these micropores are most frequent in areas such as at the bases of trichomes and other specialized epidermal cells (Schnepf, 1965). However, the occurrence of such micropores is not well documented for all plant cuticles (Martin and Juniper, 1970). Hull (1974) demonstrated that basal portions of trichomes are more permeable than adjacent areas; cuticles of guard and subsidiary cells around stomata are preferred absorption sites (Dybing and Currier, 1961; Sargent and Blackman, 1962). In addition, Linskens (1950) and Leonard (1958) found that the cuticle near veins is apparently a preferential site for absorption of water-soluble materials.

Perhaps as important as the greater density of micropores associated with these specialized cells is Rentschler's (1973) evidence that, at least in certain species, epicuticular wax is less frequently present on certain of these specialized epidermal cells. Such an absence of wax, in combination with increased cuticular penetration at those sites, would tend to maximize the sensitivity of those sites. From field observations, Evans et al. (1977a,b; 1978) determined that approximately 95% of the foliar lesions on those plant species occurred near the bases of such specialized epidermal cells as trichomes, stomatal guard and subsidiary cells, and along veins. Stomatal penetration by precipitation, on the other hand, is thought to be infrequent (Adam, 1948; Gustafson, 1956, 1957; Sargent and Blackman, 1962) and is considered a relatively insignificant route of entry of leaf surface solutions (Evans et al., 1981).

Solution pH has also been shown to influence the rate of cuticular penetration in studies with isolated cuticles (Orgell and Weintraub, 1957; McFarlane and Berry, 1974). The rate of penetration of acidic substances increased with decreasing pH, while the rate of penetration of basic substances increased with increasing pH (Evans et al., 1981). Preliminary work by Shriner (1974) suggested that, in addition to the physical abrasion of superficial wax structure by raindrops, leaves exposed to simulated sulfuric acid rainfall of pH 3.2 over the course of a full growing season appeared to weather more rapidly during treatments than did control plant leaves exposed to pH 5.6. However, it was impossible to determine from those experiments whether chemical processes at the wax surface were responsible for the differences or whether the acidic rain induced physiological changes that retarded regeneration of the waxes and subsequent recovery from mechanical damage. The latter explanation may be the most tenable since the waxes would be expected to resist chemical reaction with dilute strong acids (Evans et al., 1981) and because of the numerous reports of physiological imbalance resulting from acidic precipitation exposure (as reviewed by Shriner, 1984). However, Hoffman et al. (1980) proposed a mechanism by which precipitation acidity can act as a chemical factor in weathering epicuticular waxes. They pointed out that the wax composition, as polymeric structures of condensed long-chain hydroxy-carboxylic acids, may result in an imperfect wax matrix in which the uncondensed sites containing hydroxy functional groups are more readily weathered. Strong acid additions to such a system would oxidize and release a wide range of organic acids from the basic waxy matrix, conceivably yielding the type of change in weathering rate Shriner (1974) observed. This hypothesis has not been verified by experimental evidence.

Rentschler (1973) and, more recently, Fowler et al. (1980) showed relationships between the superficial wax layer of plants and plant response to gaseous air pollution. The work of Fowler et al. (op. cit.) compared the rate of epicuticular wax degradation of Scots pine needles from field studies of polluted and unpolluted sites. The polluted sites included exposure to both dry deposition of gaseous pollutants and wet deposition as acid rain, making it impossible to distinguish between relative effects of the two forms of deposition. Needles at the polluted site showed greater epicuticular wax structure degradation during the first eight months of needle expansion. Determining the quantity of wax, per unit leaf area, showed very small differences between polluted and clean air sites. By scanning electron microscopy, Fowler et al. (op. cit.) concluded that observed differences were "due more to changes in form that gross loss of wax." Since the fine structure of the wax layer is controlled largely by the chemical composition of the wax (Jeffree et al., 1975), the observed changes may also reflect stress-induced changes in wax synthesis. Fowler et al. (op. cit.) estimated that increased water loss due to accelerated breakdown of cuticular resistance would only influence trees if water were a limiting factor. They concluded that "the extra water loss may reduce the period (or degree) of stomatal opening" and that the magnitude of the effect on dry matter productivity would not be greater than five percent at their polluted site.

Histological studies of foliar injury caused by acidic precipitation revealed evidence of modification of leaf structure associated with plant exposure to acidic precipitation (Evans and Curry, 1979). Quercus palustris (pin oak), Tradescantia sp. (spiderwort), and Populus sp. (yellow poplar) exposed to simulated acidic precipitation exhibited abnormal cell proliferation and cell enlargement. In Quercus (oak) and Populus (poplar) leaves, prolonged exposure to treatment at pH 2.5 produced hypertrophic[1] and hyperplastic[2] responses of mesophyll cells. Lesions developed, followed by enlargement and proliferation of adjacent cells, resulting in formation of a gall on adaxial leaf surfaces. In poplar test plants, this response involved both palisade and spongy mesophyll parenchyma cells, while in oak test plants only spongy mesophyll cells were affected (Evans and Curry, 1979). Because other similar histological studies have not been reported, it is impossible to evaluate how frequent or widespread such structural modifications may be. Because species that have been reported to show hyperplastic and hypertrophic response of leaf tissues were consistently injured less than species that did not show these responses, gall formation may be linked to characteristics common to species tolerant to acidic precipitation exposure.

Several studies have reported modification of various physiological functions of the leaf as a result of exposure to simulated acidic precipitation. Sheridan and Rosenstreter (1973), Ferenbaugh (1976), Hindawi et al. (1980), and Jaakkola et al. (1980) reported reduced chlorophyll content as a result of tissue exposure to acidic solutions. Ferenbaugh (op. cit.), however, observed that significant reduction in chlorophyll content did not occur at pH 2.0 and that chlorophyll content slightly increased at pH 3.0. Irving (1979) also reported higher chlorophyll content of leaves exposed to simulated precipitation at pH 3.1. Hindawi et al. (op.cit.) observed a steady reduction of chlorophyll content in the range between pH 3.0 to 2.0 and found no change in the ratio of chlorophyll a to chlorophyll b.

Ferenbaugh (1976) determined photosynthesis and respiration rates of bean plants exposed to simulated acidic precipitation. Respiration and photosynthesis were significantly increased at pH 2.0. Ferenbaugh (op. cit.) concluded that because growth of the

---

[1] Abnormal cell enlargement
[2] Abnormal cell proliferation

plants was significantly reduced, photophosphorylation was uncoupled by these acidic treatments. Irving (1979) reported increased photosynthetic rates in some soybean treatments, attributing them to increased nutrition from sulfur and nitrogen components of the rain simulant, which overcame any negative effect of the pH 3.1 treatment. Jacobson et al. (1980) reported a shift in photosynthate allocation from vegetative to reproductive organs as a result of acidic rain treatments of pH 2.8 and 3.4, also suggesting that the primary effect was not on the photosynthetic process itself.

Effect of Vegetation Surfaces on Foliar Leaching and the Chemistry of Canopy Throughfall

Rain, fog, dew, and other forms of wet deposition play important roles as sources of nutrients for vegetation and as mechanisms of removal from vegetative surfaces of inorganic nutrients and a variety of organic compounds: carbohydrates, amino acids, and growth regulators (Kozel and Tukey, 1968; Lee and Tukey, 1971; Hemphill and Tukey, 1973; Tukey, 1975). Tukey (1970, 1975, 1980) and Tukey and Morgan (1963) extensively reviewed the leaching of substances from plants as the result of water films on plant surfaces.

During those periods between episodic precipitation events, the vegetation canopy serves as a sink, or collection surface, upon which dry particulate matter, aerosols, and gaseous pollutants accumulate by gravitational fallout, impaction, and absorption. Throughfall can be defined as that portion of the gross, or incident, precipitation that reaches the forest floor through openings in the forest canopy and by dripping off leaves, branches, and stems (Patterson, 1975). Throughfall generally accounts for 70 to 90 percent of gross rainfall with the balance divided between stemflow and interception loss to the canopy. At high elevations where vegetation intercepts cloud water, throughfall may exceed gross deposition measured as rain (Olson et al., 1981).

Chemical enrichment of throughfall has been well documented for a broad variety of forest species (Tamm, 1951; Madgwick and Ovington, 1959; Nihlgard, 1970; Patterson, 1975; Lindberg and Harriss, 1981). This enrichment has three potential sources: (1) exchange reactions on the leaf surface in which cations on exchange sites of the cuticle are exchanged by hydrogen from rainfall; (2) movement of cations directly from the translocation stream within the leaf into the surface film of rainwater, dew, or fog by diffusion and mass flow through areas devoid of cuticle (Tukey, 1980); and/or (3) washoff of atmospheric particulate matter that has been deposited on the plant surfaces (Patterson, 1975; Parker et al., 1980; Lindberg and Harriss, 1981).

The exchange of hydrogen ions in precipitation for cations on the cuticle exchange matrix can result in significant scavenging of hydrogen ions by a plant canopy. Eaton et al. (1973), for example, found the forest canopy to retain 90% of the incident hydrogen ions from pH 4.0 rain (growing season average), resulting in less acidic (~pH 5.0) solutions reaching the forest floor.

Separation of the relative contribution of internal (leached) and external (washoff) fractions of throughfall enrichment is difficult and has been attempted infrequently. Parker et al. (1980) reviewed these attempts to estimate the importance of dry deposition of sulfur to throughfall enrichment of sulfate-sulfur (Table 1). For those studies that have attempted such an analysis, the estimated percentage contribution of dry deposition to through-fall enrichment ranged from 13 to 100 percent, or from 0.3 to 14.4 kg ha^{-1} yr^{-1}. Parker et al. (op. cit.) concluded that for temperate hardwood forests in industrialized regions, 40 to 60 per-cent of annual net throughfall (throughfall enrichment) for sulfate is due to washoff of dry deposition with 30 to 50 percent being typical for conifers of the same regions. For hardwoods and conifers in regions typified by low background levels of dry sulfur deposition, washoff may range from 0 to 20 percent of throughfall enrichment. Similar data have been developed for several trace elements (Lindberg and Harriss, 1981).

Through the application of simulated rainfall in controlled experiments, precipitation acidity has been studied as a variable influencing the rate of leaching of various cations and organic carbon from foliage (Wood and Bormann, 1974; Fairfax and Lepp, 1975; Abrahamsen et al., 1976). Foliar losses of potassium, magnesium, and calcium from bean and maple seedlings were found to increase as the acidity of simulated rain increased. Tissue injury occurred at pH levels equal to or below 4.0 (Wood and Bormann, 1974). Abrahamsen and Dollard (1979) observed that Norway spruce [Picea abies (L.) Karst] lost greater quantities of nutrients under their most acidic treatments, but no related change in foliar cation content occurred. Wood and Bormann (1977) noted similar results for eastern white pine (Pinus strobus L.).

SUMMARY

Interactions of vegetation surfaces and acidic rain must be viewed from two perspectives. One concern is the effect of acid rain on the structure and function of the plant surface. Morpho-logical features of the leaf surface such as pubescence or leaf shape that enhance shedding of water, reduce the amount of solution that the leaf retains, thereby enhancing tolerance to acidic rain. Chemical properties of the leaf cuticle and the presence or absence of epicuticular waxes affect the wettability, and therefore the

Table 1. Reported values for sulfate-sulfur deposition rates for canopy throughfall and incident precipitation in world forests

Forest System	S deposition (kg ha⁻¹ yr⁻¹)		Precipitation amount (cm)	Reference
	Incident	Throughfall		
Mixed oak, Tennessee	8.7[a]	15.0	154	Kelly, 1979
Mixed oak, Tennessee	11.3[a,b]	14.0	75	Kelly, 1979
Hemlock, British Columbia	11.0[a]	40.0	245[c]	Feller, 1977
Subalpine balsam fir, New Hampshire	24.4	46.4	203[d]	Cronan, 1978
Beech, central Germany	24.1[d]	47.6	106	Heinrichs and Mayer, 1977
Chestnut oak, Tennessee	13.2[b,e]	32.0	143	Lindberg et al., 1979
Spruce, central Germany	24.1	80.0	106	Heinrichs and Mayer, 1977
Hemlock-spruce, southeastern Alaska	0.0	16.4	270	Johnson, 1975
Tropical rain forest, Costa Rica	12.5	23.3	390	Johnson, 1975
Douglas fir, Washington	4.0	5.2	165	Johnson, 1975
Subalpine silver fir, Washington	16.8[f]	5.3	300	Johnson, 1975
Hardwoods, Amazonian Venezuela	44.5-46.6	16.7-19.6	391-412	Jordan et al., 1980
Hard beech, New Zealand	8.4	10.4	135	Miller, 1963
Beech, southern Sweden	7.9[d]	18.5	95	Nihlgard, 1970
Spruce, southern Sweden	7.9[d]	54.2	95	Nihlgard, 1970
Oak, southern France	16.4	22.6	NA	Rapp, 1973
Loblolly pine, North Carolina	7.9[a]	9.9	NA	Wells et al., 1975

[a] Scaled up from a subannual estimate
[b] In vicinity of factory or power plant
[c] Mean of extreme estimates
[d] Includes stem flow
[e] Several years data
[f] Little throughfall

sensitivity of leaf surfaces to acidic precipitation. Effects of acidic precipitation on leaves that may be manifested in whole plant responses include necrosis of leaf tissue, leaching of nutrient ions from leaves, chlorophyll degradation, or alteration of leaf structure through abnormal cell growth.

The second concern regarding interactions of acidic rain and vegetation surfaces involves the modifying influence of vegetation on the chemistry of rain that penetrates a plant canopy prior to contact with a secondary receptor such as a soil surface. The washoff of dry-deposited materials from leaf surfaces may further acidify or neutralize the acidity of incident rain, depending on the chemical characteristics of the particulates on the leaf and in the rain. The contribution of washoff to canopy throughfall chemistry may be substantial. Foliar leaching may also affect the chemistry of throughfall by exchange of basic cations in leaves with hydrogen ions in the rain. Leaching of organic acids from plant tissues may contribute to the acidity of throughfall. Vegetation may significantly alter the chemistry of rain solutions and may, therefore, affect the response of subsequent receptors such as soil or aquatic systems.

In summary, the first surface that comes into contact with acidic rain is probably a leaf. The function of the contacted leaf may be affected by the chemistry of the rain. It is also likely that the chemistry of the rain will be changed by contact with the leaf. Therefore, plant surfaces may play a primary role in land-scape response to atmospheric deposition, which may or may not be correlated with direct changes at the individual plant level.

Acknowledgments – This research was sponsored by the Office of Fossil Energy Planning and Environment, U.S. Department of Energy, under Contract No. DE-AC05-84OR21400 with Martin Marietta Energy Systems, Inc.; Publication No. 2410, Environmental Sciences Division, Oak Ridge National Laboratory. By acceptance of this article, the publisher or recipient acknowledges the U.S. Government's right to retain a nonexclusive, royalty-free license in and to any copyright covering the article.

REFERENCES

Abrahamsen, G. and Dollard, G.J., 1979, Effects of acid precip-
    itation on forest vegetation and soil, pp. 1 to 17. In:
    Ecological Effects of Acid Precipitation, G. Howells, ed.,
    Report No. EPRI SOA77-403, Electric Power Research Institute,
    LaJolla, CA.

Abrahamsen, G., Bjor, K., Horntvedt, R., and Tveite, B., 1976, Effects of acid precipitation on coniferous forest, pp. 37 to 63. In: Impact of Acid Precipitation on Forest and Fresh Water Ecosystems in Norway, F.H. Braekke, ed., Research Report 6, SNSF-Project, Oslo, Ås-NLH, Norway.

Adam, N.K., 1948, Principles of penetration of liquids into solids. Discuss. Faraday Soc. 3:5-11.

Crafts, A.S., 1961, The Chemistry and Mode of Action of Herbicides. Interscience Publ., New York, NY. 269 pp.

Cronan, C.S., 1978, Solution chemistry of a New Hampshire subalpine ecosystem: Biogeochemical patterns and processes. Ph.D. Dissertation, Dartmouth College, Hanover, NH. 248 pp.

Dybing, C.D. and Currier, H.B., 1961, Foliar penetration by chemicals. Plant Physiol. 36:169-174.

Eaton, J.S., Likens, G.E., and Bormann, F.H., 1973, Throughfall and stemflow chemistry in a northern hardwood forest. J. Ecology 61:495-508.

Evans, L.S. and Curry, T.M., 1979, Differential responses of plant foliage to simulated acid rain. Am. J. Bot. 66:953-962.

Evans, L.S., Hendrey, G.R., Stensland, G.J., Johnson, D.W., and Francis, A.J., 1981, Acid precipitation: Considerations for an air quality standard. Water, Air, Soil Pollu. 16:469-509.

Evans, L.S., Gmur, N.F., and DaCosta, F., 1977a, Leaf surface and histological perturbations of leaves of Phaseolus vulgaris and Helianthus annuus after exposure to simulated acid rain. Am. J. Bot. 64:903-913.

Evans, L.S., Gmur, N.F., and Kelsch, J.J., 1977b, Perturbations of upper leaf surface structures by acid rain. Envir. Exp. Bot. 17:145-149.

Evans, L.S., Gmur, N.F., and DaCosta, F., 1978, Foliar response of six clones of hybrid poplar to simulated acid rain. Phyto-pathology 68:847-856.

Fairfax, J.A.W. and Lepp, N.W., 1975, Effect of simulated "acid rain" on cation loss from leaves. Nature 255:324-325.

Feller, M.C., 1977, Nutrient movement through western hemlock-western red cedar ecosystems in southwestern British Columbia. Ecology 58:1269-1283.

Ferenbaugh, R.W., 1976, Effects of simulated acid rain on Phaseolus vulgaris L. (Fabaceae). Am. J. Bot. 63:283-288.

Fowler, D., Cape, J.N., Nicholson, I.A., Kinnaird, J.W., and Paterson, I.S., 1980, The influence of a polluted atmosphere on cuticle degradation in Scots pine (Pinus sylvestris), p. 146. In: Ecological Impact of Acid Precipitation, D. Drabløs and A. Tollan, eds., Proceedings, International Conference, Sandefjord, Norway. SNSF Project, Ås-NLH, Norway.

Gustafson, F.G., 1956, Absorption of Co^{60} by leaves of young plants and its translocation through the plant. Am. J. Bot. 43:157-160.

Gustafson, F.G., 1957, Comparative absorption of cobalt-60 by upper and lower epidermis of leaves. Plant Physiol. 32:141-142.

Haines, B., Stefani, M., and Hendricks, F., 1980, Acid rain:
    Threshold of leaf damage in eight plant species from a
    southern Appalachian forest succession. Water, Air, Soil
    Pollu. 14:403-407.
Heinrichs, H. and Mayer, R., 1977, Distribution and cycling of
    major and trace elements in two central European forest
    ecosystems. J. Envir. Qual. 6:402-407.
Hemphill, D.D. and Tukey, H.B., Jr., 1973, The effect of inter-
    mittent mist on absisic acid content of Euonymus alatus Sieb,
    'Compactus'. J. Am. Soc. Hort. Sci. 98:416-420.
Hindawi, I.J., Rea, J.A., and Griffis, W.L., 1980, Response of bush
    bean exposed to acid mist. Am. J. Bot. 67:168-172.
Hoffman, W.A., Jr., Lindberg, S.E., and Turner, R.R., 1980, Some
    observations of organic constituents in rain above and below
    the forest canopy. Envir. Sci. Tech. 14:999-1002.
Hull, H.M., 1974, Leaf structure as related to penetration of
    organic substances, pp. 45 to 93. In: Absorption and Trans-
    location of Organic Substances in Plants, J. Hacskaylo, ed.,
    7th Ann. Symposium, Southern Sec. Am. Soc. Plant Physio-
    logists, Emory Univ., Atlanta, GA.
Irving, P.M., 1979, Response of field-grown soybeans to acid
    precipitation alone and in combination with sulfur dioxide.
    Ph.D. Dissertation, Univ. Wisc., Milwaukee, WI. 169 pp.
Jaakkola, S., Katainen, H., Kellomäki, S., and Saukkola, P., 1980,
    The effect of artificial acid rain on the spectral reflectance
    and photosynthesis of Scots pine seedlings, pp. 172 to 173.
    In: Ecological Impact of Acid Precipitation, D. Drabløs and
    A. Tollan, eds., Proceedings, International Conference,
    Sandefjord, Norway. SNSF Project, Ås-NLH, Norway.
Jacobson, J.S., Troiano, J., Colavito, L.J., Heller, L.I., and
    McCune, D.C., 1980, Polluted rain and plant growth, pp. 291 to
    299. In: Polluted Rain, T.Y. Toribara, M.W. Miller, and
    P.E. Morrow, eds., Plenum Publ., New York, NY.
Jeffree, C.E., Baker, E.A., and Halloway, P.J., 1975, Ultrastruc-
    ture and recrystallization of plant epicuticular waxes. New
    Phytol. 75:539-541.
Johnson, D.W., 1975, Processes of elemental transfer in some
    tropical, temperate, alpine and northern forest soils:
    Factors influencing the availability and mobility of major
    leaching agents. Ph.D. Dissertation, University of Wash-
    ington, Seattle, WA. 169 pp.
Jordan, C.F., Golley, F., Hall, J., and Hall, J., 1980, Nutrient
    scavenging of rainfall by the canopy of an Amazonian rain
    forest. Biotropica 12:61-66.
Kelly, J.M., 1979, Camp Branch and Cross Creek experimental water-
    shed projects: Objectives, facilities, and ecological charac-
    teristics. Report No. EPA-600/7-79-053, U.S.E.P.A.,
    Washington, DC. 162 pp.

Kozel, P.C. and Tukey, H.B., Jr., 1968, Loss of gibberellins by
    leaching from stems and foliage of Chrysanthemum morifolium
    'Princess Ann'. Am. J. Bot. 55:1184-1189.
Lee, C.I. and Tukey, H.B., Jr., 1971, Effect of intermittent mist
    on development of fall color in foliage of Euonymus alatus
    Lieb 'Compactus'. J. Am. Soc. Hort. Sci. 97:97-101.
Leonard, O.A., 1958, Studies on the absorption and translocation of
    2,4-D in bean plants. Hilgardia 28:115-160.
Lindberg, S.E. and Harriss, R.C., 1981, The role of atmospheric
    deposition in elemental cycling in an eastern United States
    deciduous forest. Water, Air, Soil Pollu. 16:13-31.
Lindberg, S.E., Harriss, R.C., Turner, R.R., Shriner, D.S., and
    Huff, D.D., 1979, Mechanisms and rates of atmospheric deposi-
    tion of selected trace elements and sulfate to a deciduous
    forest watershed. Report No. ORNL/TM-6674, Oak Ridge National
    Laboratory, Oak Ridge, TN. 514 pp.
Linskens, H.F., 1950, Quantitative Bestimmung der Benetzbarkeit von
    Blattoberflächen. Planta 38:591-600.
Madgwick, H.A.I. and Ovington, J.J., 1959, The chemical composition
    of precipitation in adjacent forest and open plots. Forestry
    32:14-22.
Martin, J.T. and Juniper, B.E., 1970, The Cuticles of Plants.
    St. Martins Press, New York, NY. 247 pp.
McFarlane, J.C. and Berry, W.L., 1974, Cation penetration through
    isolated leaf cuticles (Apricot). Plant Physiol. 53:723-727.
Miller, R.B., 1963, Plant nutrients in hard beech III. The cycle
    of nutrients. New Zealand J. Sci. 6:388-413.
Nihlgard, B., 1970, Precipitation, its chemical composition and
    effect on soil water in a beech and a spruce forest in South
    Sweden. Oikos 21:208-217.
Olson, R.K., Reiners, W.A., Cronan, C.S., and Lang, G.E., 1981,
    The chemistry and flux of throughfall and stemflow in sub-
    alpine balsam fir forests. Holarct. Ecol. 4:291-300.
Orgell, W.H. and Weintraub, R.L., 1957, Influence of some ions on
    foliar absorption of 2,4-D. Bot. Gaz. 119:88-93.
Parker, G.G., Lindberg, S.E., and Kelly, J.M., 1980, Atmosphere-
    canopy interations of sulfur in the southeastern United
    States, pp. 477 to 493. In: Atmospheric Sulfur Deposition:
    Environmental Impact and Health Effects, D.S. Shriner,
    C.R. Richmond, and S.E. Lindberg, eds., Ann Arbor Science, Ann
    Arbor, MI.
Patterson, D.T., 1975, Nutrient return in the stemflow and through-
    fall in individual trees in the piedmont deciduous forest,
    pp. 800 to 812. In: Mineral Cycling in Southeastern Ecosys-
    tems, F.G. Howell, J.B. Gentry, and M.H. Smith, eds., Report
    No. CONF-740513, ERDA Symposium Series. Available through
    NTIS, Springfield, VA.
Rapp, M., 1973, Le cycle biogeochimique du soufre dans une foret de
    Quercus ilex L. du sud de la France. Oecol. Plant. 8:325-334.

Rentschler, I., 1973, Significance of the wax structure in leaves
    for the sensitivity of plants to air pollutants, pp. A139 to
    A142. In: Proceedings of the Third International Clean Air
    Congress, H. Schackmann, ed. International Union of Air
    Pollution Prevention Associations, Dusseldorf, Germany.
    (ORNL-TR-4120, Oak Ridge National Laboratory Translation, Oak
    Ridge, TN).
Sargent, J.A. and Blackman, G.E., 1962, Studies on foliar penetra-
    tion. I. Factor controlling the entry of 2,4-dichlorophenoxy-
    acetic acid. J. Exp. Bot. 13:348-368.
Schnepf, E., 1965, Licht and elektronenmikroskopische Beobachtungen
    an den Trichom-Hydathodes von Cicer arietinum. Z. Pflanzen-
    physiol. 53:245-254.
Sheridan, R.P. and Rosenstreter, R., 1973, The effect of hydrogen
    ion concentrations in simulated rain on the moss Tortula
    ruralis (Hedw.) Sm. Bryologist 76:168-173.
Shriner, D.S., 1974, Effects of simulated rain acidified with
    sulfuric acid on host-parasite interactions. Ph.D. Disserta-
    tion, North Carolina State Univ., Raleigh, NC. 79 pp.
Shriner, D.S., 1984, Terrestrial vegetation air pollutant inter-
    actions: Nongaseous pollutants, wet deposition. In: Air
    Pollutants and Their Effects on Terrestrial Ecosystems,
    A. Legge and S.V. Krupa, eds., Wiley Interscience, New York,
    NY. (In press).
Tamm, C.O., 1951, Removal of plant nutrients from tree crowns by
    rain. Physiol. Plant. 4:184-188.
Tukey, H.B., Jr., 1970, The leaching of substances from plants.
    Ann. Rev. Plant Physiol. 21:305-324.
Tukey, H.B., Jr., 1975, Regulation of plant growth by rain and
    mist. Int. Plant Prop. Soc., Proc. 25:403-406.
Tukey, H.B., Jr., 1980, Some effects of rain and mist on plants,
    with implications for acid precipitation, pp. 141 to 150. In:
    Effects of Acid Precipitation on Terrestrial Ecosystems,
    T.C. Hutchinson and M. Havas, eds., Plenum Press, New York,
    NY.
Tukey, H.B., Jr., and Morgan, J.V., 1963, Injury to foliage and its
    effect upon the leaching of nutrients from above-ground plant
    parts. Physiol. Plant. 16:557-563.
Wells, C.G., Nicholas, A.K., and Buol, S.W., 1975, Some effects of
    fertilization on mineral cycling in loblolly pine, pp. 754 to
    764. In: Mineral Cycling in Southeastern Ecosystems,
    F.G. Howell, J.B. Gentry, and M.H. Smith, eds., Report
    No. CONF-740513, ERDA Symposium Series. Available through
    NTIS, Springfield, VA.
Wood, T. and Bormann, F.H., 1974, The effects of an artificial acid
    mist upon the growth of Betula alleghaniensis Britt. Envir.
    Pollu. 7:259-268.
Wood, T. and Bormann, F.H., 1977, Short-term effects of simulated
    acid rain upon the growth and nutrient relations of Pinus
    strobus L. Water, Air, Soil Pollu. 7:479-488.

FOREST VULNERABILITY AND THE CUMULATIVE EFFECTS OF ACID DEPOSITION

George H. Tomlinson

Domtar Research Centre
P.O. Box 300
Senneville, Quebec, Canada   H9X 3L7

ABSTRACT

Wide spread die-back and death of trees in the Federal
Republic of Germany have been attributed to the deleterious
effects of acid deposition. Similar damage is now occurring in
eastern North America and possible mechanisms that could explain
this serious development are discussed. Periodic warm dry years
can augment the cumulative effect of many years of acid deposition
on the leaching of calcium and magnesium from the soil. The
symptoms of tree decline and death are consistent with an inadequate uptake of these essential nutrients and their chemical and
physiological role in the growth and well-being of the tree.

INTRODUCTION

During the last few years, it has become apparent that many
forests, both in Europe and in eastern North America, are in a
serious condition of decline and die-back; at least in a few areas
complete collapse has occurred. The rate at which forest damage is
now increasing is a cause for some alarm, and it is important to
develop a concensus as to the cause so that appropriate action can
be taken. Recent reports indicated the extent of this damage as
described in the discussion.

DISCUSSION

The discussion will address the degree of forest damage
observed in Europe and the northeastern United States and the

serious nature of the problem. The probable causes of tree die-
back and the chemical reactions which occur within the soils
surrounding rootlets will be examined in detail.

Tree Die-Back and Regeneration Problems in the Federal Republic of Germany

The Federal Minister of Food, Agriculture and Forests (1983)
of West Germany, Herr Ignaz Kiechle, issued a statement on November 9, 1983 which described the extent of forest damage as determined in the ministry's most recent official survey. It was found
that the area of damage, as a percentage of total forested area in
West Germany, had increased from eight percent as measured in the
1982 survey (Federal Minister of Food, Agriculture and Forests,
1982) to 34 percent in 1983. Damage was particularly serious in
the south with 49 percent and 46 percent of the total forest area
of Baden-Württemberg and Bavaria, respectively, being affected.
The damage, seen initially only in mountainous areas, is now
appearing at lower elevations. Virtually all species are affected.
The ministry's statement points out that the experts are unanim-
ously of the opinion that many factors are involved. These include
air pollutants, particularly SO_2, but also heavy metals, nitrogen
oxides and photo-oxidants which may be associated with factors such
as frost, dryness, pests and forest structure. "However, in
relation to a single factor, the clues indicate that air pollutants
and their conversion products are the fundamental cause. All clues
indicate that without air pollutants, the forest damage would not
have occurred" (Federal Minister of Food, Agriculture and Forests,
1983).

A matter of particular concern is the fact that in many areas
of Central Europe the natural regeneration of the forest is no
longer possible. In the upper elevations of the Krkonose Park in
Czechoslovakia, where the spruce forest has collapsed, natural
regeneration ceased in about 1977 (Tomlinson and Silversides, 1982)
although a few seedlings can be found growing in rotted stumps
which contain nutrients from an earlier period. Hüttermann and
Ulrich (1984) and Ulrich and Matzner (1983) noted that, in Germany,
although apparently satisfactory germination of beech occurred
following the last heavy seed year in 1976, the seedlings eventually died off in many areas. Examination of root systems has
shown an inability of the tap root to penetrate and maintain
viability in the mineral soil. In order to establish the cause,
beech seedlings were simultaneously planted in test plots in
problem areas -- directly in the soil without pretreatment, in the
soil after it had been mixed with calcium carbonate, and in a
greenhouse soil mixture which replaced the original soil. As
before, the beech seedlings could not establish a normal tap root
system in the untreated soil whereas they experienced no problems

in the other two cases. These and other experiments give strong evidence that the problem of lack of regeneration in areas where, in the past, magnificent beech forests have developed is related not to direct effects of air pollutants, such as ozone, on the foliage but to a change in the soil since earlier times. This provides important clues that cumulative effects on soils are involved with long-term deposition of acidic substances.

Tree Die-Back in North America

In North America, forest damage surveys have been largely confined to studies by university groups who had carried out tree inventories in the mid 1960's at various elevations on mountains in New York State and Vermont, thus making it possible by returning to the same sites to estimate the changes which have occurred (Johnson and Siccama, 1983; Scott et al., in press). Tree damage is now very heavy on these mountains. For instance, between 1965 and 1979, the change in forest composition in the spruce-fir forest in the "transition zone" at mid elevation on Jay Peak, Mt. Abraham and Bolton Mt. in Vermont (Johnson and Siccama, 1983) was as follows:

	Tree Diameter Changes (1965-1979)	
	>10 cm	2-10 cm
Red spruce	-87%	-90%
Balsam fir	-24	-90
White birch	-73	+33

On some of the mountains, small and apparently healthy fir appear to be in good supply; but growth during the first ten years, at least in one survey on Whiteface Mountain, as measured from stem diameter at the 10th annual ring, was only about 20% of that found at the tenth year on 60 year old fir. The percentage loss in red spruce was abnormally high on all the mountains examined, and Johnson and Siccama (1983) reported that of the remaining spruce, most of which are in various stages of die-back, growth slowed down in the mid sixties. These reports indicated that the trees have been subject to stress since that time. As in Europe, the fine root systems of spruce were found to be badly damaged with necrotic tips (Siccama et al., 1982; Klein, 1985).

Less severe tree damage is now apparent with several species at lower elevations and over wide areas. Although no comprehensive tree damage surveys in these areas have been reported, state environment and forestry officials in New York, Vermont, and New Hampshire are concerned about the rapid development of damage that is now visually apparent. As a result, surveys dealing with the decline and mortality of red spruce and balsam fir are being presently conducted by these states in cooperation with the U.S. Forest Service (Anon, 1984).

The Scientific Debate with Regard to Cause

Whereas in the Federal Republic of Germany there is a general concensus among scientists and government officials alike that deleterious effects from air pollution are involved and that a major reduction in air pollution is required, there is no such agreement in North America. However, at least in the mountain areas of New York State and New England, it is clear that such previously known causes of tree damage as insects and fungi are not involved (Klein, 1985). Since tree die-backs of hardwoods such as yellow birch have occurred in the past during relatively warm dry years, it is possible that such periods in the 1950's and mid-1960's resulted in a drought condition that could be responsible. Johnson and Siccama (1983) indicated, in fact, that drought could be the triggering mechanism in relation to the current problem. However, they noted that the levels of acid deposition on the mountains are very high and stated in their conclusion that "conceivably acid deposition could enhance drought stress, or vice-versa".

Klein (1983) noted that spruce of all ages are dying on Camels Hump Mountain in Vermont, including young trees not in existence in the mid-1960's when drought conditions could have prevailed. Following snow melt on this mountain in the spring, the release of accumulated acid from the snow resulted in an increase in soil-water acidity and in a relatively high Al^{3+} content, while at the same time the Ca^{2+} concentration was reduced (Klein, 1985). When spruce seedlings were grown hydroponically in nutrient solution having concentrations of Ca^{2+} and H^+ similar to that of the mountain soil-water but lacking Al^{3+}, normal growth was obtained. However, if Al^{3+} at 5.0 mg l^{-1}, the value found in mountain soils, was added in an otherwise duplicate experiment, the ability of the seedlings to absorb water as indicated by transpiration rates was diminished, and the seedling subsequently died. Klein (1985) concluded that these and other observations strongly support the theory that atmospheric additions of acids to the soil, leaching of calcium, magnesium, and other nutrients required by the tree, and solubilization of aluminum all interfere with fine root function and development. This creates stress in the tree and ultimately death. This mechanism lends support to an hypothesis previously developed by Ulrich (1980a).

Regardless of the findings of Klein (op. cit.) and others, the fact that the forest has recovered following previous tree die-back episodes, which occurred at a time when it was unlikely that air pollution was a factor, makes the drought theory attractive. It can be argued that since the forest recovered before, it will again, regardless of air pollution levels. In order to test this assumption, it is important to review research findings from previous die-back episodes and consider these in relation to present conditions.

Soil Temperature and Tree Die-Back

During the 1950's, a serious die-back of yellow birch took place in Nova Scotia. According to Redmond (1955), the first visible symptoms of decline occurred in the periphery of the crown with curled or chlorotic leaves at the tips. Later, buds died or failed to open. Still later, branches and whole portions of the tree died, with final death, if it occurred, taking place from three to five years after first visible symptoms. Because this damage progressed without direct correlations between periods of climatic drought and the reduction of moisture in the tree, it was thought that the problem started with the roots. Investigation by Greenridge (1953) showed, at initial stages, that the fine roots died and the number of dead rootlets increased as the deterioration of the crown progressed. There had been a summer warming period of 1.0-1.4°C above long-term averages in preceding years, coupled with slightly below average summer precipitation (Redmond, 1955,1957). Since soil temperature also increases during periods of reduced rain-fall, it occurred to Redmond (1955) that an increase in soil temperature could result in the loss of rootlets, and thus their inability to extract moisture from the soil. To test this hypothesis, a site in New Brunswick with moist soil, in which none of the yellow birch had shown die-back symptoms, was selected. An electrically heated cable was placed in the soil and the temperature measured at four-inch intervals from the cable. Heating was applied for three summer months.

The results of Redmond's (1955) experiments are shown in Figure 1. The upper curve provides temperature as a function of distance from the cable while the lower curve, on the same basis, shows root mortality. At a distance of 12 inches, there was a 2°C temperature rise which resulted in a 60% mortality of fine roots. Measurements during and after the experiment indicated that soil moisture was well above a value which could cause wilting. This proved Redmond's (op. cit.) hypothesis that soil temperature, rather than inadequate soil moisture-content, could cause rootlet mortality; and it was this lack of fine roots that resulted in the drying of the tree as the die-back progressed.

In reviewing the Nova Scotia yellow birch die-back and subsequent hardwood die-back episodes, Manion (1981), noted that "climatic or site factors are almost always major predisposing or inciting factors in the decline syndrome. Feeder roots and mycor-

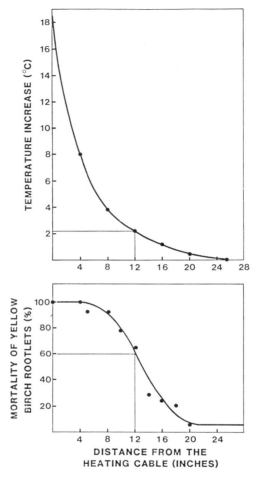

Figure 1.   Increase in soil temperature
            (top) and rootlet mortality
            (bottom) in relation to dis-
            tance from heated cable (adapted
            from Redmond, 1955)

rhiza degenerate prior to onset of symptoms in the above-ground
portion of the trees."   He referred to ash and maple die-back
episodes and to birch die-back associated with stand openings by
logging, where climatic drought and presumably soil temperatures
were involved.   Manion (op. cit.) observed that stress on the tree,
resulting from an inadequate root system, makes it vulnerable to
such problems as frost and insect and fungal attack, which he
regarded as secondary or tertiary damaging factors.

Whereas Redmond (1955) and Manion (1981) both found that degeneration of the fine roots induced by climatic change initiates tree die-back, they could not explain the reason for root sensitivity. Ulrich and his colleagues (Ulrich and Matzner, 1983; Hüttermann and Ulrich, 1984) also observed rootlet mortality with increased soil temperature and established a chemical explanation. They found that an abnormal elevation in soil temperature, occurring as a consequence of an unusually warm summer, resulted in the production of a high concentration of nitric acid in the soil; this had a damaging effect on the fine roots. As part of the normal cycle, nitrogen in the form of protein in forest litter and humus and in rotting wood and roots of acidic forest soils, decomposes during the summer months to form nitric acid. The nitric acid is normally reabsorbed through the fine roots of the tree to form the protein necessary for its growth. The nitric acid formation reaction is biotic in nature, and its rate is controlled by the number of microorganisms. This number, in turn, is markedly increased by soil temperature. In northern latitudes with relatively low soil temperatures, the normal decomposition rate is relatively low and large stores of protein nitrogen have built up in the soil. The average summer soil temperature at any given location controls the microorganism population and hence the quantity of stored nitrogen. In a healthy forest, the ability of the tree to utilize nitric acid is normally in excess of its formation rate. Thus, nitric acid and nitrogen oxides reaching the forest ecosystem as part of acidic atmospheric deposits have been largely utilized as fertilizer by healthy forests in the past and, at least in some well-buffered soils, acid rain has been found to have a growth-stimulating effect. Ulrich and Matzner (1983) and Hüttermann and Ulrich (1984) observed that in periods of increased soil temperature, the large quantities of nitric acid formed were substantially in excess of that which could be consumed by the tree. Under these conditions, atmospheric deposition of nitric acid has a deleterious effect on the soil, and the situation is exacerbated by the concurrent death of fine roots which further reduces the tree's ability to consume the nitric acid that is formed.

The effect of nitric acid in the soil, when in excess of that which can be utilized by the tree, is similar and additive to that of sulfuric acid; aluminum is dissolved from solid phases, and calcium and magnesium are leached from the soil. When soil temperatures return to normal or below normal, the biotic reaction resulting in nitric acid formation slows down; when this occurs in a buffered soil, viable trees can establish new roots and recovery sets in. This was the pattern of die-back episodes in the past. However, the continuing year-after-year accumulative effects of acid deposition have resulted in the leaching of nutrients from the soil and reduction of its buffering capacity. Forest ecosystems subject to continued heavy acid deposition thus become increasingly vulnerable to damage, particularly during and following warming periods.

The year 1982 was particularly warm in Germany. The following
year tree damage had increased to 34% of the total forest area.
Ulrich and his associates measured soil temperature and observed
changes in soil chemistry and rootlet biomass (Hüttermann and
Ulrich, 1984). The average soil temperature at 10 cm depth in the
Solling spruce forest during the June to October period of 1982 was
11.5°C, 1.4°C higher than for the same months in 1981. The effects
of this increase in soil temperature are shown in Figure 2. Note
that in June of both 1981 and 1982, the levels of nitric acid in
the soil solution were essentially identical. Following the month
of June 1981, the concentration of nitric acid dropped, and the pH
increased from 3.7 to 4 as more nitrogen (as nitric acid) was
consumed by the roots than was generated. Following June 1982,

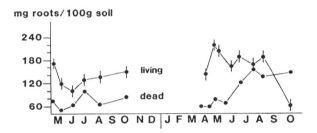

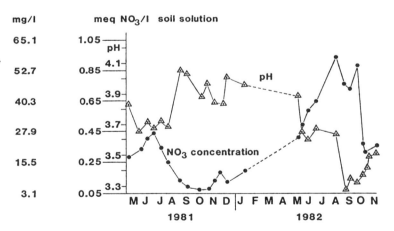

Figure 2.   Living and dead spruce fine rootlet mass
            (top) and soil solution nitrate concentra-
            tions and pH values (bottom) for Solling,
            Germany. Measurements were collected from
            a 10-cm soil depth (from Hüttermann and
            Ulrich, 1984)

with the higher soil temperature, nitric acid concentration increased instead of dropping and reached a value of approximately 60 mg l^{-1} while the pH dropped to 3.3. As seen in the upper curves of Figure 2, the biomass of living fine roots dropped rapidly between August and October to about one third the quantity observed during the April to August period. The year 1983 was also warm and dry, and increased tree damage has been reported in some areas in 1984.

Changes in Biological and Chemical Transfers and Reactions which Occur During Periods of Reduced Rainfall

The increase in soil acidity resulting from elevations in soil temperature, whether from increased air temperature or from a lack of rain, appears to be a natural phenomenon that has occurred in the past and will occur in the future, regardless of sulphur dioxide emissions. However, as discussed above, its effects on the forest become substantially greater as the soil loses its buffering capacity, thus reducing the chance of soil recovery. In addition to the effect of soil warming, examination of available data indicates that during periods of low rainfall, i.e., climatic drought, air pollution can initiate a chain of events uniquely deleterious to the ecosystem. The transfers and chemical changes that result are as follows:

° An increase in the proportion of dry versus wet deposition. Sulfur dioxide is dry-deposited on tree foliage, and subsequently chemically converted to sulfuric acid which is buffered by the calcium salts of organic acids present in the leaves. The sulfate-containing aerosols are also dry-deposited and, under certain conditions, acidic fog and cloud droplets are intercepted by the foliage. Sulfate ions, which accumulate on the leaves, are periodically washed by rain or snow to the forest floor as throughfall. The quantity of sulfate reaching the forest floor is substantial. For instance, at Wood's Lake in the Adirondack Mountains, the total sulfate deposited from throughfall under stands of spruce was 118 kg ha^{-1} yr^{-1} (Vasudevan and Clesceri, 1983) versus 37 kg ha^{-1} yr^{-1} as measured by wet deposition in an adjacent open area (Johannes et al., 1981). Monthly values are shown in Figure 3. Because of the large surface area of forest foliage, it acts as a major sink for SO_2 and its conversion products. Spruce forests with their jagged canopy and retention of foliage in the winter are subject to higher annual sulfate deposition rates than beech.

During periods of climatic drought, a reduced proportion of sulphur dioxide is taken up by cloud water to be later deposited as wet fallout; thus, dry deposition contributes a greater share. As a result of dry deposition, sulfuric acid

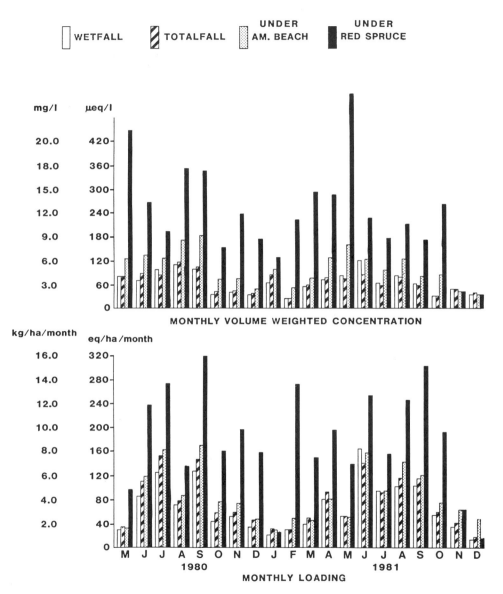

Figure 3. Seasonal variation of atmospheric sulfate loading to the
Woods Lake watershed, Adirondack Park, New York (from
Vasudevan and Clesceri, 1983)

is formed directly at very high concentrations on and within the leaf surface, whereas much of the acid in rainfall is shed by the waxy surficial coating. Thus, acid resulting from dry deposition is more damaging than a like amount in rainfall.

° The increased demand for maintenance and replenishment of buffering capacity of the leaves. The cell walls of needles and leaves contain calcium pectates, a hemicellulosic material containing glucuronic acid groups in the form of calcium salts (the normal calcium concentration of needles is approximately 0.4%). Sulfuric acid formed by dry deposition is chiefly buffered by the calcium salts, as shown in the following reaction:

$$H_2SO_4 + (RCOO)_2Ca \rightarrow CaSO_4 + 2\ RCOOH \qquad (1)$$

In order to maintain the buffering capacity of the leaves, particularly important during prolonged periods of drought, the calcium content is replenished by calcium bicarbonate carried in transpiration water from the roots (Ulrich, 1983a). On reaching the leaves, bicarbonate reacts with glucuronic acid regenerating its calcium salt. A mechanism for introduction of calcium to the root may be as follows. Carbon dioxide resulting from respiration in the root is discharged at the root surface where, with water and under suitable pH conditions, carbonic acid is formed:

$$CO_2 + H_2O \rightarrow H_2CO_3 \qquad (2)$$

Carbonic acid formed in equation (2) will ionize, with its hydrogen ions exchanging with calcium adsorbed on clay particles or combined with humic acids in the soil:

$$\boxed{\begin{array}{c} \text{Exchange} \\ \text{site} \end{array}} Ca^{2+} + H_2CO_3 \rightarrow \boxed{\begin{array}{c} \text{Exchange} \\ \text{site} \end{array}} \begin{array}{l} H^+ \\ H^+ \end{array} + Ca(HCO_3)_2 \quad (3)$$

The calcium released from exchange sites in the soil moves into the fine roots and this is then carried as bicarbonate to the leaves. Hydrogen ions neutralized in the foliage are balanced by an equal amount deposited in the soil in the root microenvironment. Hüttermann and Ulrich (1984) estimated that with soils in the aluminum buffer range, i.e., pH less than 4.2, as in the Solling, West Germany, 20% of the total hydrogen ions deposited on the spruce foliage are indirectly transferred to the soil. With beech, the corresponding figure was 30%. With soils in the exchange buffer range of pH 4.2-5.0, as in Hamburg, 50% of the total hydrogen ions deposited on beech foliage were indirectly transferred to the soil. It is suggested that even higher percentages of hydrogen ions would be transferred when beech is growing in calcareous soils.

Because of the instability of calcium bicarbonate at low pH, some mechanism other than that described in equation (3) is required in strongly acidic soils such as those in the aluminum buffer range. Nitrate ions, required for tree growth, presumably bring calcium into the tree roots as $Ca(NO_3)_2$ during periods of growth. If the calcium or other cationic requirement of the leaf is in excess of the nitrate required for leaf growth, surplus nitrate brought to the leaf must be converted to some other form, say protein, in order to release calcium to react with the glucuronic acids. This could explain the high nitrogen content reported in the foliage of stressed trees which White (1974) believed was responsible for subsequent insect attacks.

The change in ion concentration in the root microenvironment. Ulrich (1980b) observed that the concentration of ions in the root sap of healthy roots was substantially different from that in the soil solution surrounding them. The cortex (bark), through which calcium and other ions enter the fine roots, has ion-specific transfer properties. Prenzel (1979) calculated the mass flow coefficient, M, for beech (Table 1). The mass flow coefficient is related to concentration by the equation:

$$C_1 \times M = C_2 \tag{4}$$

where C_1 is the concentration of an ion in soil water as it enters the root membrane and C_2 is the concentration in the sap. Calcium, which the tree requires, has a coefficient of 2.9 while Al^{3+}, which the tree wishes to reject, has a value of 0.076. Thus, as water is drawn into the roots by transpiration, the ratio of Ca^{2+} to Al^{3+} in the soil solution will diminish in the root microenvironment as calcium is accepted and aluminum rejected. Although Prenzel (op. cit.) did not give a coefficient for H^+, it is obvious that it would also be rejected since a low pH in the roots would liberate CO_2 gas from soluble bicarbonate thus forming a gas embolism in the liquid column bringing inorganic nutrient solutions to the leaves. Because of exclusion of acid by the root membrane, the H^+ concentration would also increase in the microenvironment of the fine root. Ulrich (1983b) observed that the pH of soil under a beech tree and near the roots is highly variable, ranging in one case from 5.6 to 3.7.

Even though this undesirable high concentration of H^+ and

Table 1. Mass flow coefficients of ions present in soil-water and
          in fine root sap of beech as a result of transpiration in
          the tree

Ions Subject to Selective Discrimination	Coefficient*
Aluminum	0.076
Chloride	0.086
Sodium	0.34
Sulfur	0.77
Ions Subject to Selective Uptake	
Iron	1.5
Magnesium	1.7
Manganese	2.2
Calcium	2.9
Potassium	8.3
Nitrogen	11.0
Phosphorus	120.0

*Concentration in soil water x coefficient = concentration in the
root sap of the tree

Al^{3+} near the root cortex would occur regardless of weather
conditions, during a period of frequent rains the concentrated
soil solution surrounding the roots would be periodically
flushed. However, with lower soil water volumes prevailing in
periods of infrequent rains and given high rates of transpiration due to rapid evaporation from leaves during periods of
low atmospheric humidity, Al^{3+} and H^+ ions would be concentrated near the roots to very high levels and for prolonged
periods of time. Where rootlet die-back occurs, the increased
demand for transpiration, with fewer available roots, will
further increase concentrations of undesired ions in the soil
environment surrounding the remaining fine roots.

o An increase in nitric acid as the forest canopy opens. As
   discussed above, the concentration of nitric acid in the soil
   was found to increase dramatically with elevated soil temperatures, thus exacerbating the problem of chemicals concentrating in the root environment. If forest die-back leads to

an opening of the canopy, the soil can undergo a second rise
in temperature due to the effect of direct sunlight, with
additional nitric acid formation. Likens et al. (1970, 1978)
observed the effects of clear-cutting on the chemistry of
stream water which drained two sites in New Hampshire. The
effect of increased nitric acid formation resulting from
higher soil temperature, coupled with the loss of most of the
living vegetation which could have utilized this nitrogen
source, resulted in a major export of nitric acid. As a
result of leaching, large amounts of calcium were also lost
from the watershed. A normal commercial harvest was carried
out at one site near Gale River while in an experimental
clear-cut at Hubbard Brook, a herbicide was used to inhibit
growth of successional vegetation. The losses were as
follows:

	Two Years Export in Stream Water	
	Nitric Acid $(kg\ N\ ha^{-1})$	Calcium $(kg\ Ca\ ha^{-1})$
Gale River, commercial cut	95	89
Hubbard Brook, experimental cut with herbicide	236	169

Similar conditions will develop as the canopy opens up in a
die-back episode where increased soil temperatures, combined
with the inability of damaged trees to utilize excess nitric
acid, result in a major pulse of lost calcium.

The question of drought versus air pollution as a possible
cause of tree decline was also raised in Germany. Ekstein et al.
(1983) and Bauch (1983) measured growth patterns on the basis of
the width of annual rings of Norway spruce and white fir from
different areas of Germany. They found damaged trees exhibited
reduced growth which started many years previous to the time when
visual damage became apparent. They observed that the year of
initiation of growth reduction which, in some cases was thirty
years ago, was variable and not related to specific years associated with climatic drought. Nevertheless, there is evidence that,
at least in certain cases, a dry warm year such as 1975 resulted in
an accelerated decrease in growth. This is consistent with the
analysis of the combined effects of acid deposition and soil
warming discussed previously. Decrease in growth of red spruce on
the mountains of the northeastern United States started during a
period between 1950 and 1978 (Johnson and Siccama, 1983). In many
cases, initiation started in the mid-1960's characterized by warm
dry summers. This pattern of reduced growth rate, starting over a

period of decades and continuing for several years prior to tree death, is similar to that observed in Germany.

It is apparent that the deleterious chain of transfers and reactions which occur in the soil as a result of acid additions from dry and wet deposition will occur, regardless of whether or not it is initiated by or accompanied by an increase in soil temperature and/or a decrease in rainfall. The continuing effect of acid deposition, which has occurred over many decades, is resulting in a loss of buffering capacity and of nutrients, with the result that the trees become increasingly vulnerable to future warming periods. The large stores of protein that have accumulated in the soil in past years can be rapidly converted to nitric acid as found by Hüttermann and Ulrich (1984) and illustrated in Figure 2. The assumption that there is no need for concern in relation to acid deposition, since forest growth has previously recovered following early die-back episodes, does not stand up to scientific evidence resulting from the insights and experimental findings of Redmond in the 1950's and those of Ulrich and others in subsequent years.

Accumulation of Sulfate and Aluminum Ions in, and Their Release from Soils

The rates at which sulfate enters the soil and leaves the rooting zone do not necessarily coincide. Thus, in certain soils such as at Coweeta, NC, sulfate is accumulating in the soil, whereas at other locations as at Hubbard Brook, NH there is a greater rate of loss than entry (Johnson et al., 1980). Ulrich and coworkers (Ulrich and Matzner, 1983; Hüttermann and Ulrich, 1984) found that sulfate may be temporarily stored in the soil in the form of $AlOHSO_4$, but that continued acid additions can trigger its release as soil acidity is increased. $AlOHSO_4$ reacts with sulfuric acid at pH values lower than about 4.0-4.2 as follows:

$$2 \ AlOHSO_4 + 2H^+ + SO_4^{2-} \quad \rightarrow \quad 2 \ Al^{3+} + 3SO_4^{2-} + 2H_2O \qquad (5)$$

Aluminum is leached along with sulfate from the soil. However, because of the adsorption properties of trivalent Al^{3+}, as compared with divalent Ca^{2+} and Mg^{2+}, these latter ions which might other-wise have been adsorbed on soil or organic matter are displaced and removed from the system by the large concentration of aluminum. This reduces the Ca^{2+}/Al^{3+} ratio in the soil solution, which in turn has a deleterious effect on root growth.

The effect of this reaction can be seen from Figure 4, adapted from bar charts of Matzner and Ulrich (1984). The annual deposi-tion of sulfate in an open area and in canopy throughfall in adjacent spruce and beech forests is shown in Figure 4a, while the change in soil storage of various elements, i.e., the difference

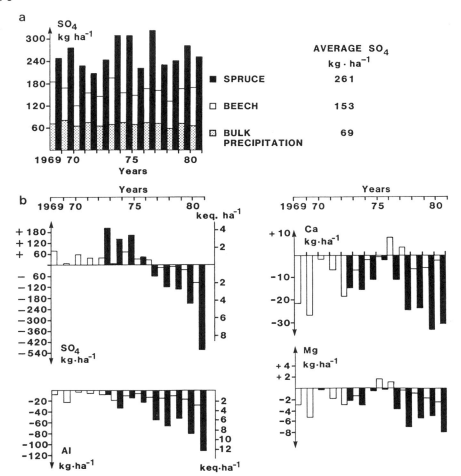

Figure 4.   Top (a) – Sulfate deposition rates under spruce and
            beech canopies and in bulk precipitation collectors
            for Solling, Germany
            Bottom (b) – Change in the storage of ions within the
            soil of Solling, Germany (from Matzner and Ulrich, 1984)

between atmospheric additions and losses from leaching, is provided
in Figure 4b.   The data are for the Solling district of Germany.
Under spruce, the average sulfate addition to the forest floor at
this location is 261 kg ha^{-1} yr^{-1} (expressed as SO_4).   During the
period 1973 to 1976, a portion of this sulfate was retained in the
soil; but starting in 1977, as a consequence of the solubilization
of $AlOHSO_4$, substantially more sulfate left the soil each year than
entered (equation 5).   The loss rate accelerated, with the net loss

from the soil being about 450 kg SO_4 ha^{-1} in 1981, nearly twice the input (Figure 4b). The increased net loss in sulfate was accompanied by increased net losses of Al^{3+}, Ca^{2+} and Mg^{2+}, the values being approximately 110, 30, and 8 kg ha^{-1}, respectively, in 1981. In addition to these problems in the forest watersheds, the large quantities of Al^{3+} solubilized and released from solid phases in the soil will eventually reach streams and lakes with harmful effect on fish.

Although retention of sulfate in the soil can, for a period, result in a decreased leaching rate of nutrient cations, a greatly accelerated loss can result during subsequent phases. Thus, stores of sulfur and nitrogen that have accumulated in the soil in past years can be rapidly converted to sulfuric and nitric acid as observed by Hüttermann and Ulrich (1983) and Matzner and Ulrich (1984). This is illustrated in Figures 4 and 2, respectively.

The Effect of Sulfur Dioxide Concentration Levels

Forest damage due to the effects of sulfur dioxide and its conversion products has been recognized for decades although, until recently, it was assumed that such problems could be expected only at relatively high SO_2 concentrations. In Germany, serious damage to forest trees had occurred in 1966 in the industrial Ruhr area where atmospheric SO_2 concentrations had averaged 240 µg m^{-3} (Knabe, 1972, 1976). In 1970, signs of tree damage outside the original "emission impact zone" were observed in areas where SO_2 concentrations were substantially less. In these areas, annual average concentrations had exceeded 80 µg m^{-3} for three of the previous four years; but after one or even two years, damage had not been observed. Symptoms were clearly absent in all forms of vegetation in areas having an average concentration of 60 µg m^{-3}, and it was assumed that this concentration was below a threshold value at which damage could occur (Knabe, 1976). On the basis of this, it was considered appropriate to discharge the pollutant gases from high stacks so that sulfur dioxide would be diluted as a result of turbulent mixing before reaching ground level. The German Environment Office set up a rural monitoring network for measuring atmospheric concentrations to assure that they did not reach the previously-assumed critical level (Anon, 1982).

Following the 1982 survey of forest damage (Federal Minister for Food, Agriculture and Forests, 1982), Stienen and Rademacher (1983) developed a map showing SO_2 concentrations prevailing in areas within, or adjacent to, forests showing substantial damage. This map, which also provided pH concentrations in the upper soil horizons in the area of damage, was reproduced by Tomlinson (1983). The average annual atmospheric SO_2 values for the period 1973-1981, obtained by the federal monitoring studies, were in the range of only 7-21 µg m^{-3}. This suggested a serious error in the earlier

decision with regard to safe emission levels. Policy at that time
had been based on the assumption that sulfur dioxide-related damage
was confined to direct effects on the foliage. It had ignored the
cumulative indirect effects of dry and wet deposition carried by
throughfall to the soil and the effect that this could have on tree
roots and vitality (Ulrich, 1983c). The sites of major forest
damage shown in this map have very acidic soils, from which
virtually all available calcium and magnesium reserves have been
leached. Moreover, further quantities of these ions formed from
the breakdown of litter and humus or by weathering of rock have
been transported by sulfate from the potential rooting zone. Since
these soils are in the aluminum buffering range, aluminum is
released as acidification proceeds. Such soils are the most
vulnerable to present and future acidic additions, where acidification proceeds as follows:

$$2Al(OH)_3 + 3H_2SO_4 \quad \rightarrow \quad 2Al^{3+} + 3SO_4^{2-} + 6H_2O \qquad (6)$$

Thus, when trees were dying or dead in areas of heavy acid deposition 15-20 years ago, unseen damage was also occurring in areas of
lower rates of input. In these soils, similar symptoms are now
being observed (Bauch, 1983). Thus, as degradation of the soil
proceeds, they become increasingly vulnerable to low annual inputs
of acids.

Importance of Calcium and Magnesium to Tree Vitality and Uptake
Inhibition by Aluminum

Calcium is an essential element for the vitality and growth of
a tree. It is particularly involved in the development and growth
of the apical meristem (Wallace, 1961). The apical meristem
consists of the primary specialized cells that make up the growing
tips of the root and rootlets, lateral branches, and the main stem
(Wallace, 1961; Panshin and DeZeeuw, 1970). Calcium pectate acts
as a bonding agent for both immature fibers as they are formed and
also for the fibres in the phloem, or inner bark, through which
sugars and other organic building blocks are transmitted from the
leaves to the growing points including root tips (Panshin and
DeZeeuw, 1970). Because of the importance of calcium to the cells
of the growing root tip, its deficiency prevents normal cell
division and growth. Calcium is also involved in the synthesis of
lignin which bonds the fibres in the xylem, or wood, through which
water and nutrients are transmitted from the roots to the leaves
(Westermark, 1982). Recent studies indicated that calcium in the
tips of the roots is responsible for their geotropism, i.e., the
signal that guides normal roots in a downward direction (Lee et
al., 1982). In addition, its presence in leaves is involved in the
neutralization of sulfuric acid formed from the dry deposition of
sulfur dioxide.

Magnesium is a vital constituent of the chlorophyll molecule which is essential for photosynthesis (Wallace, 1961). It is also involved as a carrier for phosphorus, and as an element it is equally vital in the production of DNA. It is an important constituent of seeds, not only for the production of the first leaves but also in activating the germination. A deficiency of magnesium results in chlorosis in the leaves (Wallace, op. cit.).

At pH values above about 5, aluminum is either insoluble or complexed with organic molecules in such a manner that it has no adverse effects on plant growth. However, at a lower pH and particularly below 4.0, the aluminum ion, Al^{3+}, is formed in soil solutions; under certain conditions this can inhibit the pick-up of calcium by the roots. Rost-Siebert (1983) carried out hydroponic studies on the effect of aluminum at different concentrations on the length and appearance of Norway spruce rootlets. At pH 3.8 the following results were obtained:

| | Concentration of Ca^{2+} (mg l^{-1}) | |
Concentration of Al^{3+} (mg l^{-1})	Resulting in Reduced Rootlet Growth (2-7 mm)	Resulting in Normal Rootlet Growth (12-15 mm)
0	0.5	1
2	2.0	8
8	4.0	10

Very low calcium concentrations inhibited rootlet growth even without aluminum. However, at calcium concentrations where normal growth was obtained, addition of Al^{3+} resulted in stunted growth. At high concentrations of calcium, the deleterious effect of Al^{3+} disappeared. At low Ca^{2+}/Al^{3+} ratios, the rootlet lost control of apical growth and a number of nodular buds formed on the main root; sometimes a bottle-brush-like root-form developed. The rootlet cortex (bark) partially sloughed off and necrosis of the main and side roots occurred. When similar experiments were carried out at pH 5.0 instead of pH 3.8, negative effects on rootlet growth or appearance were not observed since aluminum was present as insoluble hydroxides. This is a matter of great importance in relation to future forest management.

Hüttermann (1983) and Hüttermann and Ulrich (1984) explained the lack of normal apical rootlet growth where calcium utilization is inhibited by aluminum. This was demonstrated in hydroponic experiments with spruce seedlings using radio-active ^{45}Ca as a tracer. Addition of aluminum prevented the normal pick-up of calcium observed in the control. Cells in the rootlet meristem and those between the cortex and the central cylinder showed necrosis, a phenomenon seen in both controlled laboratory experiments and in the roots of damaged trees in the field. It is in these cells, normally bonded with calcium pectate (Panshin and DeZeeuw, 1970), that lysis or cell disintegration occurred in the absence of calcium.

In relating the interactions between aluminum and calcium, Bauch (1983) found that when a splint of fresh wood was placed in an acidic aluminum solution, the Al^{3+} ion displaced calcium from the cell structure, further illustrating its antagonistic effect. Roots of spruce and fir from several areas of Germany were also examined. Trees growing in acidic soils and showing visible damage had seriously retarded root systems. Calcium and magnesium normally present in the cortex of healthy roots were completely missing from the rootlets of these damaged trees, indicating that conditions in the root environment had inhibited their incorporation within the rootlets. These studies in Germany and those of Klein (1985) in the U.S. indicated the serious effects which the loss of calcium, combined with the mobilization of aluminum, can have on the vitality of trees.

The combination of low calcium concentrations and the presence of Al^{3+} in the root environment coupled with the lack of an adequate and properly functioning rootlet system can seriously reduce the nutrient reserves in the tree. Depending on the soil, either calcium or magnesium might be depleted first, resulting in the following symptoms:

° A lack of calcium to supply the pectate essential to the growth of rootlets and terminal buds on the main and lateral stems of the tree could explain the die-back syndrome. As seen in damaged spruce in eastern North America and in parts of Germany such as Lower Saxony, this involves a die-back from the top and ends of branches.

° A lack of magnesium to replace that leached from needles could explain the loss of green chlorophyll and thus a premature shedding of needles, a phenomenon seen in spruce growing in Bavaria.

Calcium Stores and Losses in Eastern North American Soils

Rost-Siebert's (1983) studies showed that acid conditions do not present a problem for root development unless calcium levels are very low; with adequate calcium, moderate amounts of aluminum can probably be tolerated. Thus, the question of calcium stores and the rate of loss is important in assessing the status of a forest soil. Unfortunately, relatively little information on this subject is available.

Weetman and Webber (1972) presented data on the calcium needs of red spruce and fir forests and the quantity of total and available calcium present in the soil. In studies carried out near St.-Jovite, Quebec, their values were:

	Calcium in Soil		Calcium in Biomass
	Total (kg ha^{-1})	Available (kg ha^{-1})	Full Tree (kg ha^{-1})
Humus	253	111	---
Top 26 cm of mineral soil	766	6	---
Above ground biomass			413

These investigators noted with concern the small quantity of available calcium in relation to the requirements for biomass following full tree harvesting. The nonavailable calcium in the mineral soil, consisting of rocky till lying above boulders, was present in granite and gneiss which were dissolved for analysis by hydrochloric and perchloric acids. The authors questioned whether any substantial amount of this calcium would become available during the next cycle of tree growth. However, they assumed that the nonavailable calcium in the humus layer would be largely present in woody matter which would be usable as needed. Thus, the total store of calcium in the humus plus the available calcium in the mineral soil was 259 kg ha^{-1}, 154 kg ha^{-1} less than that required following a full-tree harvest for the next forest cycle. However, the authors believed, but without experimental evidence, that there was adequate available calcium to start the next cycle. They suggested that the incoming calcium in dust and rain would make up this deficit, together with that from leaching losses as the trees grew. At the time of this study in 1972, little information was available concerning sulfate and nitrate leaching of soils by acid deposition. Ulrich (1983) noted that additions of airborne calcium had been a factor in supplying the necessary forest nutrient at earlier times in Germany. However, it is clearly evident from his current data, as reproduced in Figure 4b, that this is no longer the case, and that the acidic nature of precipitation results in a net loss of calcium from the soil. Similarly, studies at Hubbard Brook, NH, showed that the atmospheric contribution of calcium to the watershed was 2.2 kg ha^{-1}, while stream losses were 13.7 kg ha^{-1}, a net loss of 11.5 kg ha^{-1}. Values were expressed as an annual average over the period 1967-1974 (Likens et al., 1977).

A high net loss of calcium from Adirondack soil has also been observed, as described later. The assumption that airborne calcium could supply nutrient requirements for forests with an inadequate calcium supply for the next biomass rotation seems no longer justified. In fact, subsequent die-back and death of red spruce in higher elevation tills could be related to low calcium availability. Even though observations are lacking for available and nonavailable calcium in the soil, Raynal et al. (1980) reported values on available calcium in various soil horizons for the Huntington Forest in the Adirondacks. From this information, Tomlinson (1983) calculated the available calcium inventory, which amounted to only 102.5 kg ha^{-1} in the top 60 cm of humus and soil. Even though the loss of calcium from the A horizon through leaching was less than input from throughfall, it amounted to 5.9 kg ha^{-1} over a 105-day period. This is a considerable quantity compared to the small inventory and forest needs. Aluminum content was approximately twice that of calcium, on an equivalent basis, in the soil water extracted from lysimeters under the A horizon (Mollitor and Raynal, 1983). The growth rate of red spruce at this site had decreased since 1965 (Raynal et al., 1980) and die-back symptoms are now apparent. The relatively high rate of leaching presents a serious threat to the nutrient status of Adirondack soils (Raynal et al., op. cit.).

Soil Solution Equilibria and Ion Loss

In addition to mineral acids, organic acids are formed in the humus and soil from the decomposition of litter and roots. Referring to these in New Hampshire soils, Cronan (1980) stated: "These soluble organics tend to peak in concentration in the forest floor and A2 horizon solution, are mostly immobilized or metabolized in the B horizon, and generally are absent from spring water." In other words, the humic acids are immobilized by precipitation within the soil while the simpler organic acids are metabolized to CO_2 and H_2O. Humic acids have played an important role in promoting an efficient recycling of calcium and magnesium needed for tree growth. If release of these cations from decomposing litter does not occur at a time when they are required for active tree growth, they may form solid-phase salts with humic acid and be retained in the soil until exchanged by H^+ or required by the root. These insoluble humic acids thus act in a manner similar to the solid ion-exchange resins used in water softeners which will remove Ca^{2+} and Mg^{2+} when water containing these ions is introduced in one cycle, and release them when a relatively concentrated salt or mineral acid solution is used for the recharging cycle. The soil acts as a giant ion-exchange column, where over several decades additions of H^+ from the atmosphere have been gradually increasing the acidity, layer by layer, from the surface downwards. Ca^{2+} and Mg^{2+} are gradually displaced from exchange sites on clays and humic acids and replaced by H^+ and by Al^{3+} formed at low pH. Thus, when soils reach a pH less than 4, relatively little exchange capacity for Ca^{2+} and Mg^{2+} is left and these ions are carried from the rooting zone along with the mobile sulfate ion.

The soil, at any horizon, will contain a soil solution in which anions establish an equilibrium with available cations. Thus, negatively charged anions $RCOO^-$, SO_4^{2-}, NO_3^-, and Cl^- are balanced by an equivalent amount of positively charged cations Al^{3+}, Ca^{2+}, Mg^{2+}, K^+, NH_4^+, H^+, etc. Cronan (1980) reported that the export of negative anions from the soil, as measured in spring water, was dominated by SO_4^{2-} with very little organic acids; of the positive cations exported, the concentration of Al^{3+} was approximately twice that of Ca^{2+}, followed in decreasing amounts by H^+, Mg^{2+}, and K^+. Loss of these cations is a result of mineral acid additions and not from organic acids produced in the soil.

Export of sulfuric and nitric acids, in excess of that entering from canopy throughfall, will produce an increased loss of calcium. This can be seen by comparing percolating water with canopy throughfall values for June to October at Lake Laflamme, Quebec, during 1982 (Robitaille, 1983):

	Bulk Precipitate ($mg \ l^{-1}$)	Throughfall ($mg \ l^{-1}$)	Percolate ($mg \ l^{-1}$)
Sulfate	2.5	3.5	5.3
Nitrate	0.45	0.01	10.6
Calcium	0.14	0.5	2.5

Thus, the long-term effect of atmospheric acidic deposition has been to decrease the ability of soils to retain nutrients required by the tree, regardless of whether they are released from litter, from mineral weathering or from incoming dust and rain. The continuing additions of mineral acids will further deplete the nutrients essential for tree growth.

CONCLUSIONS

Studies in Germany indicate that growth rates of trees, which are presently exhibiting visible symptoms of stress, have decreased substantially in recent years; the loss of an adequate rootlet system is a predisposing phase. Experiments carried out by German scientists determined that soils are losing Ca^{2+} and Mg^{2+} and gaining Al^{3+} as a result of acid deposition. This is presently occurring at an accelerating rate.

Laboratory investigations showed that Al^{3+} inhibits the incorporation of Ca^{2+} into the roots. This leads to a loss of the growing rootlet tip and a sloughing of the outer rootlet layer. It is speculated that this problem results from an inability of the root to produce calcium pectate, a substance essential for bonding the phloem cells of the inner bark of rootlets and stems. Calcium pectate is also essential in the rootlet tip cells where growth should occur. Damage to the rootlets interferes with transmission of water and mineral nutrients to the above-ground portions of the tree.

Examination of the outer layer of rootlets from damaged trees in the field indicates that calcium and magnesium, normal constituents of the cell wall, are absent. This indicates that conditions in the root environment have hindered root development, a situation similar to that seen in the laboratory where similar effects have been obtained with low Ca^{2+}/Al^{3+} ratios. Since calcium and magnesium are unique elements for the synthesis of pectate and chlorophyll, respectively, and no other element can substitute for them, the inability of the tree to take up these nutrients can lead to its death.

All of these factors give strong support to the concept that acid deposition, which results in deleterious changes in the soil, is a primary cause of tree death. Experiments in North America and Germany illustrated that a warming of the soil can result in rootlet mortality. Convincing evidence has been obtained that a relatively small increase in soil temperature can result in the production of large amounts of nitric acid; this acid can have an adverse effect which is additive to that of sulfuric acid entering the soil from acid deposition. The increase in dry versus wet deposition occurring during periods of climatic drought and the reduction of soil-flushing by rain can be particularly damaging to the root environment.

Ten years ago when forest damage and collapse occurred at SO_2 concentrations of about 90 µg m^{-3} near sources of emission, the linkage between cause and effect was perceived by scientists and administrators even though the reasons were not apparent. Similar damage is now appearing in sensitive soils at much greater distances from emission sources and in areas where sulfur dioxide concentrations are substantially less. It is important that changes which may have occurred in the soil be considered in explaining the mechanisms involved.

Loss of nutrient cations and buffering capacity as a result of the cumulative and continuing effects of acid deposition makes soils increasingly sensitive to present and future conditions of increased temperature, for whatever reason. An elevation in soil temperature can result from such causes as increased sun exposure following harvesting, insect defoliation, or for many other reasons. Rapid recovery from harvesting has been observed in the past; however, it is possible that this will no longer be the case for soils in sensitive areas. For these reasons, it is believed that scientists, foresters, and administrators should give serious consideration to the problem of forest vulnerability so that ameliorative programs can be developed and implemented.

REFERENCES

Anon, 1982, Grossräumige Luftverunreinigung in der Bundesrepublik Deutschland. Texte 33, Umweltbundesamt, Bonn, Federal Republic of Germany. 87 pp.

Anon, 1984, New forest survey. Announcement of cooperative survey of red spruce and balsam fir declines and mortality in New Hampshire, New York, and Vermont. Acid Precip. Digest 2:49.

Bauch, J., 1983, Biologische Veränderungen in Stamm und Wurzeln umweltbelasteter Waldbäume; SO$_2$ und die Folgen. Gesellschaft für Strahlen und Umweltforschung, MBH, München, Federal Republic of Germany. GSF-Bericht, München-Neuherberg A3/83:49-57.

Cronan, C. S., 1980, Solution chemistry of a New Hampshire subalpine ecosystem. Oikos 34:271-281.

Eckstein, D., Aniol, R. W., and Bauch, J., 1983, Dendroklimotologische Untersuchungen zum Tannensterben. Europ. J. Forest Pathology 13:279-288.

Federal Minister for Food, Agriculture, and Forests. Schriftenreihe des Bundesministers für Ernährung, Landwirtschaft und Forsten, 1982, Forest damage due to air pollution, Waldschäden durch Luftverunreinigung. Landwirtschaftsverlag, GmBH, Münster-Hiltrup, Federal Republic of Germany. 65 pp.

Federal Minister for Food, Agriculture and Forests, 1983, Bericht des Bundesministers für Ernährung, Landwirtschaft und Forsten anlässlich der Waldschädenserhebung, 1983, Report of the Federal Minister for Food, Agriculture and Forest on the Forest Damage Level, Bonn, Republic of Germany. 17 pp.

Greenridge, K. N. H., 1953, Further studies on birch die-back in Nova Scotia. Can. J. Bot. 31:548-559.

Hüttermann, A., 1983, Auswirkungen "sauer Deposition" auf die Physiologie des Wurzelraumes von Waldökosystemen. Allgem. Forst. Zeit. 26/27:663-664.

Hüttermann, A. and Ulrich, B., 1984, Solid phase/solution/root interactions in soil subjected to acid deposition. Phil. Trans. Royal Soc., London B305:353-368.

Johannes, A. H., Altwicker, E. R., and Clesceri, N. L., 1981, Characteristics of acidic precipitation in the Adirondack region, pp. 6 to 34. In: EA-1826 Research Project 1155-1, Electric Power Research Institute, Palo Alto, CA.

Johnson, A. H. and Siccama, T. G., 1983, Acid deposition and forest decline. Envir. Science Tech. 19:294A-305A.

Johnson, D. W., Hornbeck, J. W., Kelly, J. M., Swank, W. T., and Todd, D. E., 1980, Regional patterns of soil sulfate accumulation, pp. 507 to 520. In: Atmospheric Sulfur Deposition,

D. S. Shriner, C. R. Richmond, and S. E. Lindberg, eds., Ann Arbor Science, Ann Arbor, MI.

Klein, R. M., 1983, Deposition of Richard M. Klein, Univ. Vermont, Burlington, VT. Hearings before the Committee on Evironment and Public Works, U.S. Senate, 98th Congress, First Session, U.S. Government Printing Office, Washington, DC. 658 pp.

Klein, R. M., 1985, Effect of acidity and metal ions on water movement through red spruce, pp. 303 to 322. In: Acid Deposition: Environmental, Economic, and Policy Issues, D. D. Adams, ed., Plenum Publishing Co., New York, NY.

Knabe, W., 1972, Immissionbelastung und Immissiongefährdung der Wälder im Ruhrgebeit: Mitteilungen der Forstlichen Bundes-Versuchstanstalt. Wien 97:53-87.

Knabe, W., 1976, Effects of sulfur dioxide on terrestrial vegetation. Ambio 5:213-218.

Lee, J. S., Mulkey, T. J., and Evans, M. L., 1982, Reversible loss of gravitropic sensitivity in maize roots after tip application of calcium chelators. Science 220:1375-1376.

Likens, G. E., Bormann, F. H., Pierce, R. S., and Rieners, W. A., 1978, Recovery of a deforested ecosystem. Science 199:492-496.

Likens, G. E., Bormann, F. H., Johnson, N. M., Fisher, D. W., and Pierce, R. S., 1970, Effect of forest cutting and herbicide treatment on nutrient budgets. Ecological Monographs 40:23-47.

Likens, G. E., Bormann, F. H., Pierce, R. S., Eaton, J. S., and Johnson, N. M., eds., 1977, Biogeochemistry of a Forested Ecosystem. Springer-Verlag, New York, NY. 146 pp.

Manion, P. D., 1981, Decline diseases of complex biotic and abiotic origin, pp. 324 to 339. In: Tree Disease Concepts, P. D. Manion, ed., Prentice-Hall, Englewood Cliffs, NJ.

Matzner, E. and Ulrich, B., 1984, Raten der Deposition, der internen Production und des Umsatzes von Protonen in zwei Waldökosystemen. Z. Pflanzenernähr. Bodenkd. 147:290-308.

Mollitor, A. V. and Raynal, D., 1983, Acid precipitation and ionic movement in Adirondack forest soils. Soil Science Soc. Am. J. 46:137-141.

Panshin, A. J. and DeZeeuw, C., 1970, Tree growth, pp. 24 to 37. In: Text Book of Wood Technology, Vol. 1. McGraw-Hill Book Co., New York, NY.

Prenzel, J., 1979, Mass to the root system and mineral uptake of a beech stand calculated from 3-year field data. Plant Soil 51:39-49.

Raynal, D. J., Leaf, A. l., Manion, A. D., and Wang, G. J. K., 1980, Actual and potential effects of acid precipitation in the Adirondack Mountains. Report No. 56/ES-/HS/79, New York State Energy Res. Development Authority, Albany, NY. 266 pp.

Redmond, D. R., 1955, Studies in forest pathology, XV. Rootlets, mycorrhiza, and soil temperature in relation to birch die-back. Can. J. Bot. 33:595-627.

Redmond, D. R., 1957, The future of birch from the viewpoint of disease and insects. For. Chron. 33:25-30.

Robitaille, G., 1983, Excursion Guide, Lake Laflamme Catchment Basin. Acid Rain and Forest Resource Conference, Quebec City. Can. Forestry Service, Ste. Foy, Quebec, Canada. 20 pp.

Rost-Siebert, K., 1983, Aluminumtoxizität und Toleranz an Keimpflanzen von Fichte und Buche. Allgem. Forst. Zeit. 26/27:686-689.

Scott, J. T., Siccama, T. G., Johnston, A. H., and Breisch, A. R., 1984, Decline of red spruce in the Adirondacks. Bul. Torrey Bot. Club (in press).

Sicamma, T. G., Bliss, M., and Vogelmann, H. W., 1982, Decline of red spruce in Green Mountains of Vermont. Bul. Torrey Bot. Club 109:162-168.

Stienen, H. and Rademacher, P., 1983, Abiotische und biologische Aspekte des Waldsterbens. Natur und Mus. 113:157-166.

Tomlinson, G. H. 1983, Air pollutants and forest decline. Envir. Science Tech. 17:246A-258A.

Tomlinson, G. H. and Silversides, C. R., 1982, Acid deposition and forest damage - The European linkage. Domtar Research Centre, Senneville, Quebec, Canada. 35 pp.

Ulrich, B. 1980a, Die Wälder in Mitteleuropa: Messergebnisse ihrer Umweltbelastung, Theorie ihrer Gefährdung, Prognose ihrer Entwicklung. Algemeine Forstzeitschrift 45:1198-1202.

Ulrich, B., 1980b, Production and consumption of hydrogen ions, pp. 255 to 282. In: Effects of Acid Precipitation on Terrestrial Ecosystems, Hutchinson and M. Havas, eds., Plenum Press, New York, NY.

Ulrich, B., 1983a, Forest ecosystem stability and acid deposition, pp. 25 to 26. In: Effects of Accumulation of Air Pollutants in Forest Ecosystems, B. Ulrich and J. Pankrath, eds., D. Reidel Pub. Co., Boston, MA.

Ulrich, B., 1983b, Soil acidity and its relation to acid deposition, pp. 182 to 183. In: Effects of Accumulation of Air Pollutants in Forest Ecosystems, B. Ulrich and J. Pankrath, eds., D. Reidel Pub. Co., Boston, MA.

Ulrich, B., 1983c, An ecosystem oriented hypothesis on the effect of air pollution on forest ecosystems, pp. 221 to 223. In: Ecological Effects of Acid Deposition. 1982 Stockholm Meeting, Report PM 1636, National Swedish Environment Protection Board, Solna, Sweden.

Ulrich, B. and Matzner, E., 1983, Abiotische Folgewirkungen der weiträumigen Ausbreitung von Luftverunreinigungen, pp. 146 to 157. In: Luftreinhaltung Forschungsbericht. Report No. 104 02 615, Umweltbundesamt des Bundesministers des Innern, Bonn, Federal Republic of Germany.

Vasudevan, C. and Clesceri, N. L., 1983, Interrelationship
    between atmospheric deposition and a forested ecosystem.
    Paper presented at Annual Meeting of Amer. Assoc. Adv.
    Science, Detroit, MI.
Wallace, T., 1961, Essential points in the nutrition of plants,
    pp. 11 to 12. In: The diagnosis of mineral deficiencies in
    plants. Her Majesty's Stationery Office, London, U.K.
Weetman, G. F. and Webber, B., 1972, The influence of wood
    harvesting on the nutrient status of two spruce stands:
    Can. J. Forst. Res. 2:351-369.
Westermark, U., 1982, Calcium promoted phenolic coupling by
    superoxide radical - a possible lignification reaction in
    wood. Wood Sci. Technol. 16:71-78.
White, T. C. R., 1974, A hypothesis to explain outbreaks of
    looper caterpillars. Oecologia 16:279-301.

ATMOSPHERIC DEPOSITION AND IONIC MOVEMENT IN ADIRONDACK FORESTS

Dudley J. Raynal, Frances S. Raleigh,
and Alfred V. Mollitor[a]

State University of New York
College of Environmental Science and Forestry
Syracuse, New York   13210

ABSTRACT

The nature of atmospheric deposition and ionic movement in hardwood and conifer dominated forests of the central Adirondack Mountains is characterized in this paper. Weighted mean precipitation pH measured 4.15 during the period 1979-1982. Concentration of major ions in precipitation showed the following patterns: for cations, $H^+ > NH_4^+ > Ca^{2+} > Na^+ > Mg^{2+} > K^+$ and for anions, $SO_4^{2-} > NO_3^- > Cl^- > PO_4^{3-}$. Canopy throughfall and soil solution chemistry revealed that patterns of ion movement were similar in hardwood and conifer forests, but ionic concentrations were greater in the conifer system. The extent to which any nutrient leaching from soils in these forests represents chronic losses must be evaluated in the context of forest type, soil characteristics, stand age and development, and other factors influencing nutrient dynamics.

INTRODUCTION

The Adirondack Mountain region of New York State is a receptor region for acidic deposition. Prevailing weather systems that track from the west, southwest, northwest, and southeast over heavily populated and industrialized pollutant-producing areas of the United States and Canada transport air pollutants which are largely responsible for precipitation acidification and the depo-

---

[a]Present Address: School of Forest Resources and Conservation, University of Florida, Gainesville, FL.

sition of acidifying substances, including both wet and dry fallout. The Adirondack Mountain area is thought to be susceptible to
the effects of acidic deposition because of its meteorological
conditions, geographical location, and geological, edaphic, and
hydrological characteristics. The actual and potential effects of
acidic deposition on aquatic and terrestrial ecosystems of the
region are of great concern to natural scientists, private landowners, recreationists, governmental bodies, as well as to businesses and industries whose welfare depends upon the natural
resources of the region. Recognition of acidic deposition as a
very serious environmental problem has fostered major research
efforts devoted to assessment of the influences of acid precipitation on natural ecosystems in North America, Europe, and Scandinavia. In the United States, perhaps the greatest intensity of
research activity focuses on the Adirondack Mountain region.

Assessment of the influences of atmospheric deposition on
natural ecosystems is complex because of the dynamic equilibrium
that exists between the components of the ecosystem: atmosphere,
water, soil, and biota. Despite considerable research effort, an
understanding of these interactions is incomplete. In terrestrial
ecosystems, however, it is generally recognized that the transfer
of the various chemical constituents from the atmosphere to the
earth's surface and vegetation canopies is a complex and important
linkage in the relationships between ecosystem components. Therefore, any study that seeks to characterize the influences of acidic
deposition on forest ecosystems must reasonably include an accurate
documentation of the chemical constituents of atmospheric deposition, a characterization of solution chemistry as precipitation is
intercepted by the forest canopy and passes through it, and a
description of ionic movement as canopy throughfall penetrates the
forest floor and percolates through mineral soil horizons. The
purpose of this paper is to characterize atmospheric deposition and
ionic movements in hardwood and conifer dominated forests at
Huntington Wildlife Forest in the central Adirondack Mountains.

METHODS

Studies were conducted at Huntington Wildlife Forest, a
6,066 hectare forest property held in trust for the State University of New York College of Environmental Science and Forestry.
The forest is located near the village of Newcomb, western Essex
County, New York (latitude 44°00", longitude 74°13", lying entirely
with the Adirondack State Park; Figure 1).

Atmospheric Deposition Monitoring

In accordance with the guidelines and procedures prescribed by
the National Atmospheric Deposition Program (NADP; Bigelow, 1982),

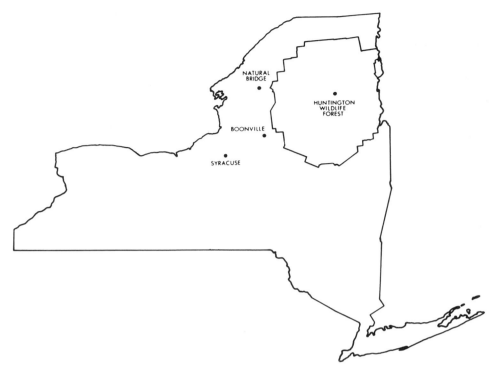

Figure 1.   Outline map of New York State showing the boundary of
the Adirondack State Park and, within, the location of
Huntington Wildlife Forest.

an atmospheric deposition monitoring station was established at the
Huntington Forest meteorological station in October 1978.   An
Aerochem Metrics model 301 (Aerochem Metrics Corp., Miami, FL)
automatic sensing wet/dry precipitation collector, designed to
collect rain and snow in a polyethylene bucket which is open only
during precipitation events, was utilized to collect wet deposition
samples.   A paired bucket, open between precipitation episodes,
collected dry-deposited substances.

Wet deposition samples were collected weekly, dry deposition
samples bimonthly.   Wet deposition samples of 70 g or more were
analyzed on site for pH and conductivity by removing a 20-ml
aliquot of sample with a clean calibrated syringe.   The remainder
of the sample was shipped by United Parcel Service to the NADP
Central Analytical Laboratory (CAL) at the Illinois State Water
Survey, Champaign, IL for chemical analysis.

At CAL, wet deposition samples were first weighed to determine the total precipitation amount. Specific conductance and pH were determined prior to filtration. Samples were then filtered through a 0.45 μm membrane filter to separate the soluble fraction from the insoluble material. The filtrate (50 ml) was transferred to polypropylene bottles and stored at room temperature for subsequent chemical analysis. Potassium, sodium, calcium, and magnesium were determined using flame atomic absorption methodology. Sulfate, nitrate, chloride, phosphate, and ammonium concentrations were determined using a Technicon Auto-analyzer system (Stensland et al., 1980).

Upon receipt of sample chemical analyses from CAL, field and laboratory measured data for each wet and dry deposition sample were computer-archived. In describing monthly, yearly, and total sampling period patterns in the ionic composition of wet and dry deposition, volume-weighted concentration means were calculated using the following equation similar to that described by Barrie and Sirois (1982):

$$C = \frac{\sum\limits_{i=1}^{N} C_i P_i}{\sum\limits_{i=1}^{N} P_i}$$

where C is the volume-weighted concentration mean, C_i is the ionic concentration for the time interval (one week for wet samples, two months for dry samples), P_i is the precipitation amount for the time interval, and N is the number of samples for which data were available. The dry deposition volume was 250 ml (the amount of distilled water used by CAL to collect the dry deposition sample) plus any amount of precipitation that was collected in the dry bucket. Ion deposition was calculated by multiplying concentrations by the precipitation volume and summing the values for the desired period.

Ionic Movement in Hardwood and Conifer Forests

Forest canopies are not simply passive recipients of atmospheric deposition, but rather the nature of the forest canopy strongly influences the nature and magnitude of atmospheric input to the forest floor (Gosz et al., 1983; Graustein and Armstrong, 1983). To characterize ionic additions and movement in the Adirondack forests, two adjacent contrasting forest stands were chosen for study at Huntington Forest: a northern hardwood stand dominated by Acer saccharum Marsh. (sugar maple), Fagus grandifolia Ehrh. (American beech), and Betula alleghaniensis Britton (yellow birch) and a lake margin conifer stand composed primarily of Tsuga canadensis L. Carr (eastern hemlock), Betula alleghaniensis Britton, and Picea rubens Sarg. (red spruce). A 20- by 20-m study plot was selected within each of two stands. The soil supporting these stands is a coarse-loamy, mixed, frigid Typic Fragiorthod. Morphological and chemical characteristics of study site soils are reported elsewhere (David et al., 1982; Mollitor and Raynal, 1982). Vegetation composition has also been described (Raynal et al., 1980).

Precipitation and canopy throughfall were sampled with continuously-open funnel collectors from 2 May 1979 to 7 May 1980, a period of 371 days. Sample collection and analysis have been described by Mollitor and Raynal (1983).

Ionic movement in the first floor and mineral soil horizons of the hardwood and conifer stands were studied using lysimetry. Zero-tension collectors were used to collect samples beneath the forest floor and tension lysimeters were employed beneath the mineral soil horizons. Specific methods of sample collection and solution analyses are found in Mollitor and Raynal (1982).

RESULTS AND DISCUSSION

Ionic concentrations in atmospheric precipitation as well as total deposition of individual ions (concentration x volume) will be described. Their concentrations within various strata of the forest will also be evaluated. Lastly, the flux of ions through various portions of the forest ecosystem will be presented.

Concentration of Ions in Precipitation

The volume-weighted mean annual pH of precipitation at Huntington Forest measured at the field station from November 1978 through September 1982 was 4.15. For the years 1979, 1980, and 1981, pH means were 4.13, 4.05, and 4.17, respectively (Table 1). Huntington Forest precipitation samples measured at the Central Analytical Laboratory (CAL) generally had a higher pH than those measured in the field (overall weighted mean of 4.33; Table 1), presumably due to chemical changes associated with the time lapse between sample collection and analysis. The pH of precipitation at Huntington Forest is similar to that recorded at Hubbard Brook Experimental Forest, New Hampshire, during 1964-1974 when the mean annual volume-weighted precipitation pH ranged from 4.03 to 4.21 (Likens et al., 1977). The Adirondack Mountains are outside the area in the U.S. where precipitation is most acidic, a region extending from eastern Illinois through Indiana, Ohio, Pennsylvania, and southern New York (Semonin, 1981). The Huntington Forest pH values are similar to those found throughout New England,

Table 1.   Volume-weighted mean annual wet deposition pH [Central Analytical Laboratory
(CAL)-measured and Huntington Wildlife Forest (HF)-measured values], percentage of samples less than the annual and overall mean values (CAL-measured),
and specific conductance (μmho cm^{-1}) at Huntington Forest

YEAR	n	pH (CAL-measured)	pH (HF-measured)	% < annual mean	% < overall mean	Specific Conductance (HF-measured)	Specific Conductance (CAL-measured)
1978	7	4.31	4.49	57.1	71.4	15.6	17.3
1979	52	4.37	4.13	51.9	48.1	22.7	21.7
1980	53	4.25	4.05	41.5	50.9	29.3	29.2
1981	52	4.37	4.17	51.9	46.2	22.5	25.1
1982	36	4.35	4.30	44.4	38.9	24.6	24.9
Overall (1978–1982)	200	4.33	4.15	--	--	24.3	24.7

Quebec and eastern Ontario, and the lower Michigan peninsula (Semonin, op.cit.).

At Huntington Forest, between 41.5% and 51.9% (n=200) of the precipitation samples, during years for which complete sample analyses are available (1979-1981), had a pH of less than the annual mean pH (Table 1). The lowest field-measured pH value for a weekly precipitation sample was 3.2, the highest was 7.2. Hydrogen ions accounted for 61.5% of the total cationic strength, nearly 10% less than that recorded at Hubbard Brook during 1963-1974 (Likens et al., 1977). Hydrogen ion concentration was 3.2 times greater than the next most abundant cation, ammonium.

Hydrogen ion concentration exhibited a pronounced seasonal pattern with spring/summer maxima and winter minima (Figure 2). This trend reflected changes in the ionic strength of precipitation seen in the measurements of specific conductance which showed a similar seasonal pattern (Figure 3). Field-measured sample conductivity ranged from 7.2 to 162.7 μmho cm^{-1} with a volume-weighted mean for the total sample period of 24.3 μmho cm^{-1} (24.7 μmho cm^{-1}, CAL-measured; Table 1).

Ions which greatly influence the precipitation pH and serve as a forcing function for free hydrogen ion concentration are sulfate and nitrate. Sulfate concentrations exhibited a strong seasonal pattern with summer maxima and winter minima (Figure 4). On an equivalent basis, sulfate concentrations over the sampling period were about two times greater than nitrate, although during the period June through August mean sulfate to nitrate ratios ranged from 2.5 to 3.7 (Table 2). Unlike sulfate, nitrate showed little seasonal concentration pattern (Figure 5). Taken together, sulfate and nitrate formed about 95.1% of the anionic strength of precipitation. Chloride contributed 4.3% of the total anions and phosphate 0.6%. The yearly volume-weighted concentrations of ions in wet deposition at Huntington Forest during the sampling period is given in Table 3.

Together, the acidic cations, H^+ and NH_4^+, contributed about 80% of the total cationic strength of precipitation at Huntington Forest. The ammonium ion, comprising 18.7% of the total cationic strength during the sample period, showed a variable but seasonal pattern of concentration with warm season maxima and cold season minima. Of the remaining cations, calcium comprised 8.9% of the total, sodium 6.8%, magnesium 3.2%, and potassium 0.9% (Table 3).

Characterization of Ionic Deposition

Precipitation in the Adirondack Mountains is rather evenly distributed over the year. During the sampling period of this study, precipitation ranged from a minimum of 89.7 cm in 1980 to a

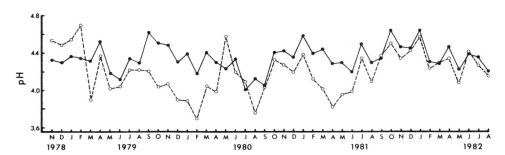

Figure 2.   Volume-weighted mean monthly precipitation pH at
            Huntington Forest.   Open circles are measurements
            at the monitoring site immediately following collec-
            tion; filled circles indicate measurements by the
            NADP Central Analytical Laboratory

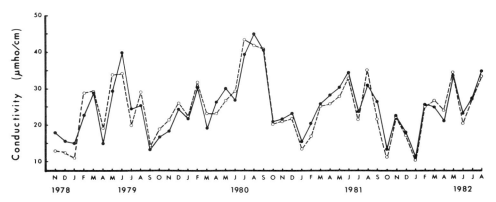

Figure 3.   Volume-weighted mean monthly specific conductance
            (μmho cm^{-1}) of precipitation at Huntington Wildlife
            Forest.   Open and closed circles are defined in
            Figure 2

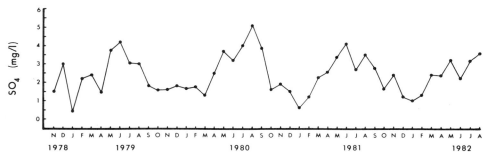

Figure 4.   Volume-weighted mean monthly sulfate ion concentration
            (mg l^{-1}) of precipitation at Huntington Wildlife Forest

Table 2. Ratio of sulfate to nitrate ions (on an equivalent basis) in precipitation for each month, 1979-1982, at Huntington Forest

Month	1979	1980	1981	1982	MEAN
January	0.6	1.0	0.7	1.1	0.9
February	1.7	0.5	1.2	0.5	1.0
March	1.2	1.0	0.9	1.2	1.1
April	1.3	1.3	1.5	1.9	1.6
May	1.6	2.1	1.9	1.8	1.9
June	3.3	1.8	2.5	2.3	2.5
July	3.1	1.9	2.4	2.7	2.5
August	2.1	7.8	2.3	2.6	3.7
September	2.9	2.0	2.1	2.2	2.4
October	1.7	3.2	1.7	1.5	2.0
November	1.4	0.8	1.2	1.1	1.1
December	0.8	0.7	0.7	1.0	0.8

maximum of 109.5 cm in 1979. Seasonal trends in ionic deposition are influenced to a greater extent by seasonal ionic concentration trends than by seasonal precipitation patterns. Deposition of sulfate in precipitation measured 23-26 kg ha^{-1} yr^{-1} and total wet and dry sulfate deposition amounted to about 27-30 kg ha^{-1} yr^{-1}. Nitrate deposition was 14-18 kg ha^{-1} yr^{-1} in precipitation with total wet and dry nitrate deposition about 17-20 kg ha^{-1} yr^{-1}. Comparatively smaller amounts of other ions were deposited in wet and dry fallout (Table 4).

For the most abundant ions, dry deposition accounted for a relatively small portion of the total amount deposited (2-9% of H^+, 13-18% of SO_4^{2-}, and 11-19% of NO_3^- for the years 1979-1981; Table 4). During 1979-1981, chloride and sodium dry deposition ranged from 11% to 26% of total deposition of these ions, while this was 23% to 41% for ammonium, calcium, and magnesium and 56% to 85% for potassium and phosphate. The contribution from dry fallout must be viewed with caution, however, since sampling methodology

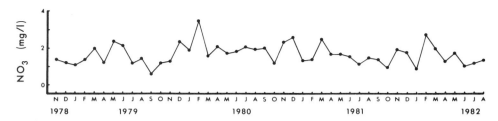

Figure 5.   Volume-weighted mean monthly nitrate ion concentration
(mg l^{-1}) of precipitation at Huntington Forest

using the open bucket method represents at best a crude estimate.
The accuracy of these dry deposition samples is questionable
because of poor efficiency for atmospheric scavenging associated
with the collection method and the likelihood of localized contamination due to the long period of sample bucket exposure.

Ionic Concentration in Forest Strata

     The pattern of ion movement from bulk precipitation through
the hardwood and conifer forest ecosystem strata is depicted in
Figures 6 and 7 (Mollitor and Raynal, 1982; 1983). Because forest
vegetation is probably more efficient at capturing airborne particles and aerosols than most precipitation samplers, precipitation
sampling systems underestimate atmospheric additions to forest
ecosystems (Swank and Henderson, 1976). Ions reaching the forest
floor in throughfall may be derived from a variety of sources
(Parker, 1983). These include dissolved materials from distant and
local sources, dust and aerosols deposited on vegetation, particulates scavenged by rainfall, and ions leached from tree foliage.
Concentrations of sulfate, potassium, calcium, and magnesium were
generally increased by contact with the hardwood canopy and were
greatly increased by contact with the conifer canopy (Table 5;
Mollitor and Raynal, 1983).

     The hardwood canopy altered nitrate concentration very little,
but the conifer canopy significantly increased its concentration.
Hydrogen ion concentration was reduced in hardwood throughfall but
showed little changes with conifers. Mean volume-weighted pH of
bulk precipitation, hardwood throughfall, and conifer throughfall
was 4.2, 4.5, and 4.2, respectively, from early May to December
1979 and 4.1, 4.2, and 4.0 over the annual cycle, May 1979 to May
1980. Sodium was essentially unaffected by either forest canopy
type.

Table 3. Volume-weighted yearly concentration of ions in wet deposition at Huntington Forest, 1979–1981, and the overall sample period mean, November 1981 through September 1982

Ion	1979		1980		1981		OVERALL MEAN	
	mg l^{-1}	µeq l^{-1}	mg l^{-1}	µeq l^{-1}	mg l^{-1}	µeq l^{-1}	mg l^{-1}	µeq l^{-1}
Ca^{2+}	0.11	5.28	0.15	7.51	0.17	8.44	0.13	6.75
Mg^{2+}	0.02	1.83	0.03	2.26	0.04	3.64	0.03	2.42
K^+	0.02	0.48	0.04	1.13	0.02	0.57	0.03	0.66
Na^+	0.16	7.12	0.15	6.32	0.07	2.97	0.12	5.15
NH_4^+	0.15	8.35	0.40	22.14	0.26	14.38	0.25	14.12
H^+	0.04	42.82	0.06	56.61	0.04	42.36	0.05	46.49
NO_3^-	1.45	23.41	2.00	32.20	1.47	23.66	1.57	25.31
Cl^-	0.12	3.32	0.18	5.12	0.11	2.99	0.12	3.42
SO_4^{2-}	2.26	47.17	2.77	57.80	2.42	50.43	2.43	50.63
PO_4^{3-}	0.002	0.07	0.05	1.69	0.004	0.12	0.02	0.47
Σ cations		65.88		95.97		72.36		75.59
Σ anions		73.97		96.81		77.20		79.83

Table 4. Wet and dry deposition of ions (kg ha^{-1} yr^{-1}) and total yearly deposition (Σ) at Huntington Forest, 1970–1981

Ion	1979			1980			1981		
	Wet	Dry	Σ	Wet	Dry	Σ	Wet	Dry	Σ
Ca^{2+}	1.20	0.60	1.80	1.35	0.79	2.14	1.62	0.48	2.10
Mg^{2+}	0.25	0.13	0.38	0.25	1.17	0.42	0.42	0.14	0.56
K^+	0.22	0.45	0.67	0.40	0.50	0.90	0.21	0.27	0.48
Na^+	1.86	0.22	2.08	1.31	0.22	1.53	0.65	0.17	0.82
NH_4^+	1.71	0.79	2.50	3.59	2.15	5.74	2.48	1.18	3.66
H^+	0.50	0.03	0.53	0.51	0.01	0.52	0.43	0.04	0.47
NO_3^-	16.50	3.00	19.50	17.95	2.15	20.10	14.02	3.21	17.23
Cl^-	1.34	0.17	1.51	1.63	0.37	2.00	1.01	0.36	1.37
SO_4^{2-}	25.75	4.77	30.52	24.94	5.46	30.40	23.14	3.46	26.60
PO_4^{3-}	0.03	0.15	0.18	0.48	0.64	1.12	0.04	0.24	0.28

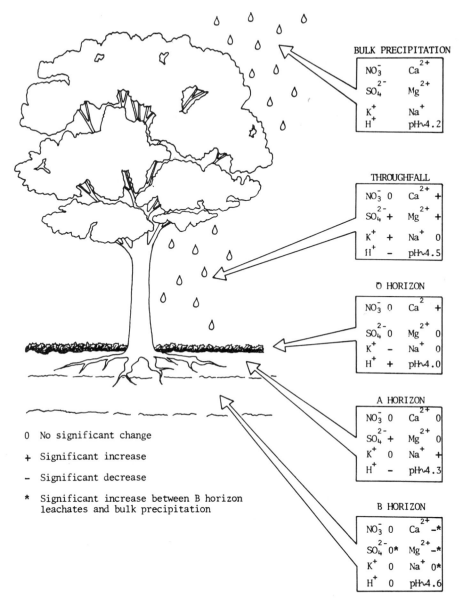

Figure 6. Pattern of ionic movement from the atmosphere (bulk precipitation), through the canopy (throughfall), forest floor (O horizon), and upper (A horizon) and lower (B horizon) mineral soil horizons in a hardwood forest at Huntington Forest (after Mollitor and Raynal, 1982, 1983). Legend refers to statistical analysis of original data (P ≤ 0.05)

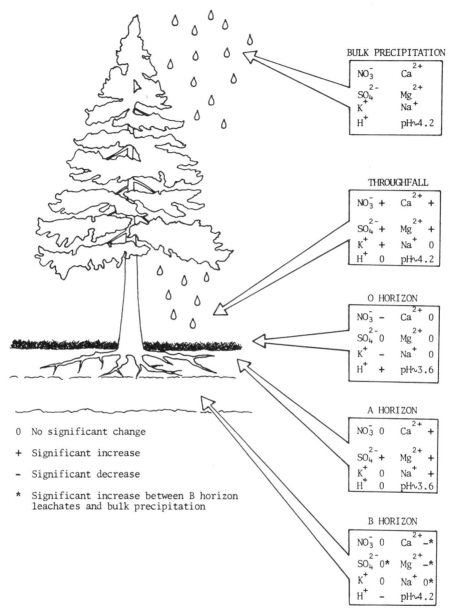

Figure 7.   Pattern of ionic movement from the atmosphere (bulk
             precipitation), through the canopy (throughfall),
             forest floor (O horizon), and upper (A horizon) and
             lower (B horizon) mineral soil horizons in a conifer
             forest at Huntington Forest (after Mollitor and Raynal,
             1982, 1983). Legend refers to statistical analysis of
             original data (P ≤ 0.05)

Table 5. Mean, standard deviation (μeq l^{-1}) and number of samples (in parentheses) for ionic concentrations in bulk precipitation, hardwood throughfall, and conifer throughfall for the 371-day study period (from Mollitor and Raynal, 1983)

Ion	Bulk Precipitation		Hardwood Throughfall		Conifer Throughfall	
NO_3^-	31 ± 22	(67)	34 ± 29	(87)	44 ± 41	(80)
SO_4^{2-}	66 ± 37	(66)	96 ± 54	(88)	157 ± 100	(76)
H^+	88 ± 56	(64)	66 ± 74	(82)	96 ± 82	(79)
K^+	6 ± 15	(70)	52 ± 57	(94)	80 ± 65	(91)
Ca^{2+}	13 ± 19	(70)	39 ± 37	(94)	75 ± 66	(90)
Mg^{2+}	5 ± 9	(70)	25 ± 34	(94)	40 ± 47	(90)
Na^+	4 ± 8	(70)	6 ± 10	(94)	7 ± 12	(90)

The two forest canopy types affected fluxes (Table 6) in much the same way that concentrations were affected. During a one-year period, about 74 cm and 65 cm of water were intercepted by the hardwood and conifer canopies, respectively; but fluxes of sulfate, potassium, calcium, and magnesium were much greater in throughfall than in precipitation. Fluxes of nitrate and sodium were little affected by the tree crowns. The hardwood canopy retained hydrogen, while hydrogen flux was decreased only slightly by the conifer canopy. Fluxes of sulfate and calcium were greater in conifer than in hardwood throughfall.

Total deposition as determined by analysis of bulk precipitation samples agreed quite well with deposition collected by the NADP collector (Table 6). The discrepancies in hydrogen ion deposition measurements were most likely due to differences in observed pH values associated with measurement time. NADP pH determinations were made both a few minutes and a few days after sample collection. Bulk precipitation pH was determined within 24 hours after sample collection. Apparently, pH increased during the elapsed time between sample collection and pH determination. Observed potassium deposition was higher in bulk precipitation than in the NADP collector, although potassium concentration levels were highly variable. Sodium deposition collected by the NADP samples was higher than observed in bulk precipitation.

Table 6. Deposition of water (cm) and ions (kg ha^{-1}) from May 2, 1979 to May 7, 1980 in bulk precipitation, hardwood throughfall, and conifer throughfall at Huntington Forest. NADP collector results are also included

Substance	Bulk Precipitation[a]	Hardwood Throughfall[a]	Conifer Throughfall[a]	NADP Collector Wet	Dry
Water	96.9 ± 1.7	73.9 ± 2.4	65.7 ± 5.0	100.7	–
Nitrate	15.28 ± 0.84	12.71 ± 1.06	18.91 ± 2.08	16.92	2.66
Sulfate	26.10 ± 1.98	30.21 ± 3.45	49.84 ± 4.65	23.81	4.41
Hydrogen	0.70 ± 0.06	0.39 ± 0.03	0.61 ± 0.03	0.95[b]	0.01
Potassium	1.65 ± 1.08	13.72 ± 2.12	16.56 ± 3.63	0.21	0.35
Calcium	1.55 ± 0.23	4.30 ± 0.53	7.74 ± 0.65	1.24	0.65
Magnesium	0.40 ± 0.11	1.64 ± 0.26	2.12 ± 0.31	0.25	0.13
Sodium	0.68 ± 0.11	0.71 ± 0.12	0.77 ± 0.11	1.68[c]	0.24

[a] Mean and 95% confidence intervals (p = 0.05)

[b] Hydrogen deposition as determined in the field was 0.95 kg ha^{-1}; as determined in the lab was 0.46 kg ha^{-1}

[c] About 1.1 kg ha^{-1} Na was collected in a single weekly sample with 156 μeq l^{-1} Na and 3.1 cm water

As water passed through the forest floor (0 horizon) in the hardwood forest, the concentration of potassium decreased, calcium increased substantially, and magnesium and sodium were unchanged (Figure 6). Water passing through the A horizon showed no further significant change in potassium or calcium, but magnesium and sodium concentrations increased. In the B horizon, calcium and magnesium decreased while sodium and potassium remained unchanged. Concentrations of calcium, magnesium, and sodium were significantly greater in B horizon leachates than in bulk precipitation (Figure 6).

The conifer canopy influenced cation concentration much like the hardwood canopy (Figure 7), but the magnitudes of change were generally greater (Mollitor and Raynal, 1982). Calcium, magnesium, and sodium concentrations increased significantly as water passed through the A horizon. Water moving through the B horizon showed decreases in calcium and magnesium concentration but no further change in sodium. Concentrations of calcium, magnesium, and sodium were significantly higher in B horizon leachates than in bulk precipitation (Figure 7).

Canopy throughfall measurements indicate that hydrogen ion acidity was effectively neutralized in these forests much as Cronan and Reiners (1983) found for hardwood and conifer forests in the White Mountains of New Hampshire. Nitrate and potassium were conserved in the Adirondack forests. Unlike Cronan and Reiners (1983), but like some other studies (Parker, 1983), the ionic concentrations determined beneath a red spruce-eastern hemlock-yellow birch canopy exceeded values found beneath a beech-sugar maple-yellow birch canopy. This may have been due to greater dry deposition collection efficiency of the conifer-dominated canopy.

It is unknown whether or not leaching of calcium, magnesium, and sodium from the soil represents chronic losses of these elements from the pool of exchangeable cations. It is likely that these ions will be largely replaced by ions released from mineral weathering and mineralization of organic matter. Cation-anion charge balances for the two forest types indicate anion deficits in water leaving the soils for both hardwood and conifer sites - these amount to about 200 μeq l^{-1} and 400 μeq l^{-1}, respectively (Mollitor and Raynal, 1982). These deficits are presumably largely accounted for by organic anions which, along with sulfate, influence cation mobility. Whether or not long term additions of anthropogenically produced sulfate to these forests accelerate the leaching of natural sulfate and organic anions will require further investigation in the context of forest type, stand age and development, soil characteristics, and other factors influencing nutrient cycling.

SUMMARY AND CONCLUSIONS

Atmospheric deposition from wet and dry fallout was measured at Huntington Forest in cooperation with the National Atmospheric Deposition Program monitoring network during 1978-1982. Field measured volume-weighted mean pH was 4.15; lab measured pH was 4.33. Precipitation pH at Huntington Forest is comparable to that recorded throughout New England, eastern Canada, and the Michigan peninsula. Hydrogen ion concentration contributed 61.5% of the total cationic strength, 3.2 times more than the next most abundant cation, ammonium. Of the anions, sulfate and nitrate contributed 95% of the total anionic strength. Concentration of major cations in wet deposition showed the following pattern:

$$H^+ > NH_4^+ > Ca^{2+} > Na^+ > Mg^{2+} > K^+,$$

while anion concentrations showed the following:

$$SO_4^{2-} > NO_3^- > Cl^- > PO_4^{3-}.$$

Precipitation was rather evenly distributed over the year; but deposition of sulfate, which measured about 30 kg ha^{-1} yr^{-1}, peaked during the warm months. Nitrate deposition, which measured about 20 kg ha^{-1} yr^{-1}, showed no pronounced seasonal trend.

A one-year investigation of bulk precipitation, canopy throughfall, and soil solution chemistry in adjacent Adirondack hardwood and conifer forests identified the nature of ionic movement through ecosystem strata. Patterns of movement for most ions were similar for both study sites, but concentrations were generally greater in the conifer system. Cation leaching from soil horizons was apparently influenced by sulfate and organic anion leaching. The extent to which nutrient leaching from the soils in these forests represents chronic losses due to the mobility of externally-generated sulfate must be evaluated in the context of depositional additions, forest type, soil characteristics, stand age and development, and other factors influencing nutrient dynamics.

Acknowledgements – Gratitude is expressed to the National Atmospheric Deposition Program (NADP) and the staff at the Central Analytical Laboratory for expert assistance and analysis. Mr. Raymond D. Masters, Site Operator of the NADP monitoring station at Huntington Forest, is thanked for faithful sample collections, analysis, and data reporting. Mr. Cary D. Dustin is gratefully acknowledged for preparation of the drawings for Figures 6 and 7. Research was supported by New York State Energy Research and Development Authority and New York State Electric and Gas Corporation.

REFERENCES

Barrie, L.A. and Sirois, A., 1982, An Analysis and Assessment of
    Precipitation Chemistry Measurements Made by CANSAP (The
    Canadian Network for Sampling Precipitation): 1977-1980.
    Atmospheric Environment Service Report AQRB-82-003-T, Downs-
    view, Ontario, Canada. 163 pp.

Bigelow, D.S., 1982, National Atmospheric Deposition Program
    Manual: Site Selection and Installation. Natural Resource
    Ecology Laboratory, Colorado State University, Fort Collins,
    CO. 21 pp.

Cronan, C.S. and Reiners, W.A., 1983, Canopy processing of acidic
    precipitation by coniferous and hardwood forests in New
    England. Oecologia 59:216-223.

David, M.B., Mitchell, M.J., and Nakas, J.P., 1982, Organic and
    inorganic sulfur constituents of a forest soil and their
    relationship to microbial activity. Soil Sci. Soc. Am. J.
    46:847-852.

Gosz, J.R., Brookins, D.G., and Moore, D.I., 1983, Using stron-
    tium isotope ratios to estimate inputs to ecosystems.
    Bioscience 33:23-30.

Graustein, W.C. and Armstrong, R.L., 1983, The use of strontium-
    87/strontium-86 ratios to measure atmospheric transport into
    forested watersheds. Science 219:289-292.

Likens, G.E., Bormann, F.H., Pierce, R.S., Eaton, J.S., and John-
    son, N.M., 1977, Biogeochemistry of a Forested Ecosystem.
    Springer-Verlag, New York, NY. 146 pp.

Mollitor, A.V. and Raynal, D.J., 1982, Acid precipitation and
    ionic movements in Adirondack forest soils. Soil Sci. Soc.
    Am. J. 26:137-141.

Mollitor, A.V. and Raynal, D.J., 1983, Atmospheric deposition and
    ionic input in Adirondack forests. J. Air Poll. Contr.
    Assoc. 33:1032-1036.

Parker, G.G., 1983, Throughfall and stemflow in the forest
    nutrient cycle, pp. 57 to 173. In: Advances in Ecological
    Research, Vol. 13, A. Macfadyen and E. D. Ford, eds.,
    Academic Press, New York, NY.

Raynal, D.J., Leaf, A.L., Manion, P.D., and Wang, C.J.K., 1980,
    Actual and Potential Effects of Acid Precipitation on a
    Forest Ecosystem in the Adirondack Mountains. Report 80-28,
    New York State Energy Research and Development Authority,
    Albany, NY. 242 pp.

Semonin, R.G., 1981, Seasonal precipitation concentrations and
    depositions for North America from the CANSAP/NADP networks,
    pp. 56 to 119. In: Study of Atmospheric Pollution Scaven-
    ging, R.G. Semonin, V.C. Bowersox, D.F. Gatz, M.E. Peden,
    and G.J. Stensland, eds., Report No. 252, Illinois Institute
    Natural Resources, State Water Survey Division, Champaign,
    IL.

Stensland, G.J., Semonin, R.G., Peden, M.E., Bowersox, V.C.,
     McGurk, F.F., Skowron, L.M., Slater, M.J., and Stahlhut,
     R.K., 1980, National Atmospheric Deposition Program Quality
     Assurance Report, Central Analytical Laboratory.ˈ Natural
     Resource Ecology Laboratory, Colorado State University, Fort
     Collins, CO. 26 pp.
Swank, W.T. and Henderson, G.S., 1976, Atmospheric input of some
     cations and anions to forest ecosystems in North Carolina
     and Tennessee. Water Resources Res. 12:541-546.

EFFECT OF ACIDITY AND METAL IONS ON WATER MOVEMENT THROUGH RED

SPRUCE

Richard M. Klein

Botany Department
University of Vermont
Burlington, VT   05405

ABSTRACT

   Field observations suggested that death of red spruce of all
age classes in the northern coniferous forest of Camels Hump
Mountain, Vermont could be due in part to inadequate movement of
water through the trees. Short term laboratory experiments were
conducted with young red spruce to evaluate the possibility that
soil solutions and acidic metal ion solutions interfered with water
movement. Highly acidic (pH 3.1) spring-collected soil solutions,
but not the less acidic (pH 4.5) soil solutions collected in fall,
reduced water movement. Aluminum, cadmium and lead ions, at
concentrations close to those in spring-collected soil solutions,
reduced transpirational water flow. The apparent absence of
mycorrhizal associations of red spruce would further limit water
uptake and it was found that acidity and metal ions repressed the
growth of several species of mycorrhizal fungi.

INTRODUCTION

   Evaluation of the impact of acidic, metal-containing
depositions on forested ecosystems has proved to be a difficult
task. Research on freshwater aquatic systems has been more
productive; aquatic biota have relatively short half lives
measurable in a few years or even a matter of months. Although
assessing the long-term responses of aquatic ecosystems is far from
simple, the difficulties pale in comparison with terrestrial
systems where trees grow for more than a century, where turnover

303

and cycling of litter is measured in decades and where perenniating forbs persist for years.

As a research base, there are three sets of facts. Wet and dry depositions are acidic and contain biologically significant quantities of metals including cadmium, copper, lead and zinc (Galloway et al., 1980; Lindberg and Harris, 1981). The snow in northern and mountainous regions, although nominally the same acidity as rain, forms highly acidic snowmelt which may drop the pH of the soil solution a full pH unit lower than the bulk snow (Johannessen and Henriksen, 1978). The cloud water that sweeps across upper elevations and condenses on leaves and needles is five times more acidic than rain and contains higher concentrations of metals (Scherbatskoy and Bliss, 1984). Since cloud water is an important source of water for high elevation ecosystems (Lovett et al., 1982), they receive much higher acid and metal loadings.

Information on soils forms a second set of facts. Several reports have documented the accumulation of metals in soils exposed to acidic, metal-containing depositions (Reiners et al., 1975; Siccama et al., 1980). Indigenous and deposited metals are being solubilized and moved through the affected ecosystems (Cronan and Schofield, 1979). The evidence is less conclusive on changes in soil texture and structure (Strayer and Alexander, 1981; Volk et al., 1982), on responses of soil biota (Firestone and McColl, 1982; Coleman, 1983) and on nitrogen transformations (Alexander, 1982), but is sufficiently impressive to allow the tentative conclusion that there are alterations of consequence.

Documentation of alterations in forest biota form a third set of facts. As reported by Ulrich (1981) and groups in North America, there are significant alterations in the health of forest ecosystems. Siccama et al. (1982) found that red spruce populations on Camels Hump and other Vermont mountains, and possibly on other elevated areas throughout its growing range, has been drastically reduced. Similar findings have been reported by others (Raynal et al., 1980; Johnson and Siccama, 1984; Scott, unpublished studies on Whiteface Mountain, New York). Sugar maple, beech and other hardwood species have declined significantly and other tree species in the northern coniferous forest have also been adversely affected. Litter is accumulating (Moloney et al., 1983), soil organic matter has been altered, and coverage of the forest floor by mosses has decreased (Klein and Bliss, 1984). These and other changes in the forests have implications in nutrient cycling, in soil composition, and in many other aspects of the flow of energy and materials through ecosystem compartments (Klein, 1983). There is no doubt that in the past 25 years a set of disturbing floristic and ecological facts have accumulated.

A key question in research on the possible effects of acid

rain on terrestrial ecosystems is whether there are cause and effect relationships between and among these three sets of facts. There is not, at present, unequivocal proof that the meteorological and the soils facts show cause and effect relationships; some reports have stated that there may not be any relationship between them (Krug and Frink, 1983), although the topic is still unresolved (Nilsson et al., 1982). But if there is to be any modulation of the interplay between human activity and ecosystem stability, the scientific community must demonstrate a strong probability that cause and effect relationships have sufficient coherence to call for legal, political and economic adjustments. The concept of "a clear and present danger" must be established or negated.

Research findings are now forming a pattern that will lead to an effective working hypothesis. Although no firm consensus has yet developed - and this in spite of 15 years of work in North America and over 20 years in Europe - there is a developing consensus that acidic, metal-containing depositions could be implicated in the observed and measured alterations in soils, soil life and the health and vigor of plants. Puckett (1982) has recently suggested a cause and effect relationship between acidic depositions and growth decline in hardwood forests. If cause and effect relationships do in fact exist, the causal complex of acids and metals must be directly and indirectly affecting many physiological and chemical processes rather than impacting a single process or component in the complex, interdependent and integrated "Web of Life" that comprises an ecosystem (Klein, 1983).

To evaluate the working hypothesis that acidic, metal-containing depositions are implicated in changes in flora and soils, it is essential that research combine field studies with parallel and comparable laboratory simulation experiments. It is obvious that the laboratory is not reality, and at the same time there is altogether too much noise in nature to allow definitive conclusions to be drawn. Laboratory studies can, however, demonstrate - as field studies cannot - that one or more of the factors in a presumed causal complex are capable of producing responses that may be homologous to those seen in the field. Field-derived information forms the base for laboratory simulations and, where possible, the information obtained in the laboratory must be tested or evaluated in the field.

TRANSPIRATION EXPERIMENTS

Present studies indicate that a spruce tree progresses from visibly healthy to morbid in approximately 3-5 years although reversible or irreversible injury may have existed prior to this time. The sequence of visible changes begins with needles drying from the tips, browning and falling off. This is followed by twig death. These symptoms usually begin at the top of the tree and at branch extremities and proceed basipetally down the stem. Root systems become discolored, the young root tips that take up water and nutrients die and as death is followed by decay, the tree is destabilized and easily blown over (Fig. 1). This syndrome suggests that one proximate cause of spruce decline could be a stress on the water/mineral absorbing and conducting processes, collectively called total transpiration.

One-year-old red spruce (_Picea_ _rubrum_ Sarg.), grown from seed in greenhouse soil in plastic pots, were gently freed of adhering soil and inserted into plastic vials. They were supplied with a balanced nutrient solution (Ingestad, 1959) with the concentrations of macroelements adjusted to reflect the composition of lysimeter-collected soil water in the conifer zone of Camels Hump Mountain (Table 1). The solution pH was adjusted to either 3.0 (that of the soil solution in the spring) or to 4.5 (that of the soil water in late summer). Transpiration was measured by the gravimetric (lysometric) method (Witham et al., 1971). Solutions were changed

Figure 1.   The northern coniferous forest of Camels Hump Mountain, Vermont in 1982. Photograph courtesty of David Like.

Table 1. Comparison of Ingestad's balanced nutrient solution for
          spruce with the composition of soil solutions obtained as
          lysimeter collections from northern coniferous forest
          soils. Concentrations given as mg l^{-1}.

Element	Ingestad 1X	Ingestad 0.1X	Camels Hump June 1982	Mt. Moosilauke, NH (average)[a]
NO_3-N	3.6	0.36	0.9	0.13
P	0.64	0.06	0.5	–
K	0.96	0.10	0.06	0.16
Na	0.63	0.06	0.10	0.70
Ca	1.0	0.1	0.08	0.25
Mg	0.6	0.06	0.04	0.15
SO_4	0.6	0.06	–	0.16
pH	5.0	–	3.1	4.3

[a]Data from Cronan (1980)

daily to simulate moving soil water or were not changed for the
duration of an experiment. The test unit of vial, solution and
plant was weighed daily. Four to six plants were used for each
variable. Total transpiration is here considered to be the average
weight loss per plant per day with subtraction of appropriate
blanks. Metal ions were presented as their chlorides, with metal
concentration calculated as the free cationic form. Most
experiments ran for four days although some were extended to
two weeks.

The environmental conditions were carefully controlled. Vials
were placed in a sealed chamber containing dishes of saturated
$Ca(NO_3)_2$ to maintain the relative humidity within the chamber at
53% (Klein and Klein, 1970). The chamber was equilibrated to 20°C.
Continuous radiation was supplied from a cool-white fluorescent
luminaire providing 300 microEinsteins m^{-2} sec^{-1} (PAR) at plant
level.

Variation among replicated units was calculated as the
statistical variance which averaged 10-15% about the mean. There
were differences in total transpiration from experiment to
experiment because of changing conditions over the duration of the
study. For this reason, the data are calculated as the alterations
in transpiration within an experiment expressed as a percentage of
the appropriate control.

As determined in preliminary experiments, acidity per se did not affect transpiration of young spruce trees. Lysimeter-collected soil solutions from the June sampling period reduced transpiration while soil water collected in October was innocuous (Table 2). It is important to note that conifer trees in mountainous ecosystems show their major growth flush at the time that they are exposed to the transpiration-inhibiting soil water collected in June. Aluminum ion at 5 mg l^{-1}, a concentration found in spring-collected soil water, reduced water movement when the nutrient solution was at either pH 3.0 or 4.5, but was without effect at 0.5 mg l^{-1}, a concentration found in soil water in October (Table 3). Repressions of transpiration occurred whether or not the solutions were changed daily to simulate rapid or slow soil water movement. Solutions containing 10% of the balanced nutrients, but enriched with cadmium and lead ions, also reduced transpiration (Table 4). Zinc ions did not affect transpiration. In contrast to aluminum, changing solutions daily was more effective. These results are compatable with the hypothesis that the solubilization of metals by acidic depositions and particularly by highly acidic snowmelt water results in a soil solution that causes reduction in water and mineral flow through the tree. Since test solutions at pH 3.0 and 4.5 without metal additions were without effect, it is reasonable to assume that acidity facilitated the uptake of toxic concentrations of metal ions. It is not yet known whether these reductions in water movement through spruce are due to alteration in water uptake, water movement or leaf transpiration.

Since it is known that litter has doubled in depth since 1965 at Camels Hump Mountain (Moloney et al., 1983), the effects of litter extracts on transpiration of red spruce were examined to evaluate the possibility that acidic litter extracts contained one or more toxic substances. Air-dried, mixed balsam fir:red spruce needles (2:1 by weight) were extracted at pH 4.0 with a simulated acid rain composed of H_2SO_4:HNO_3 in a 60:40 hydrogen ion ratio or with an HCl solution at pH 4.0 as a control. The simulated acid rain solution extracted substances that repressed transpiration of red spruce trees (Table 5).

It has been postulated that trees stressed by drought are more susceptable to decline. This was tested in simulation experiments where elevated soil moisture tension was mimicked with an osmoticum of mannitol. Water stress and aluminum ion on plant development appeared to be interactive in that the level of deterioration was visibly more than the sum of the two stresses (Fig. 2). This may be provisional evidence that both water and aluminum stresses may be acting at the same cellular or tissue site.

Another approach to the question of the role of metals in transpiration is by examination of plant growth. Research using

Table 2.   Effect of lysimeter-collected soil solution on total
            transpiration of red spruce trees. Soil solution
            collected in northern coniferous zone of Camels Hump
            Mountain, VT in June 1982 (spring collection, pH 3.1) or
            in October 1982 (fall collection, pH 4.2) and stored at
            -20°C until used. Four replicates measured daily for
            three days. Data are averages of two experiments.

SPRING SOIL SOLUTION, pH 3.1

	24 hr.	48 hr.	72 hr.
Control	1.20 ± .10	0.80 ± .03	1.03 ± .06
Soil solution	0.73 ± .07	0.60 ± .05	0.74 ± .07
% control	72*	75*	71*

FALL SOIL SOLUTION, pH 4.2

	24 hr.	48 hr.	72 hr.
Control	1.00 ± .07	1.15 ± .09	1.04 ± .14
Soil solution	1.08 ± .05	1.27 ± .11	1.25 ± .08
% control	108	110	96

*p = 0.05

crop plants has shown that metals, particularly aluminum, prevent
the morphogenesis of normal tissue systems in roots resulting in
depressed water metabolism (Clarkson, 1965; Fleming and Foy, 1968).
Spruce trees were grown under hydroponic conditions (Table 6).
Nutrients were provided either at optimum levels (Ingestad, 1959)
or at 10% of this concentration (Table 1) which is a more adequate
simulation of soil solutions. Solutions were adjusted to either
pH 3.0 or pH 4.5, aerated, and were changed three times weekly.
The experiment was maintained in the greenhouse with supplementary
illumination with cool-white fluorescent lamps providing
1500 microEinsteins $m^{-2} sec^{-1}$ (PAR) on a 17 hour photoperiod.

The trees were harvested after eight weeks. Injury caused by
aluminum was pH dependent, with more rapid and more severe injury
at pH 3.0 than at pH 4.5 (Table 6). Root deterioration, seen as
browning of root tips and failure of lateral root development,
preceeded needle browning and abscission (cf. Ligon and Pierre,
1932). As reported by Rorison (1980), more severe damage was seen
at pH 3.0 when the nutrient solution contained optimum levels of

Table 3. Effect of aluminum ion on total transpiration of red
          spruce trees. Ingestad's nutrient solution, 0.1 X
          adjusted to pH 3.0 or 4.5 with HCl. Five plants per
          replicate measured daily for four days. Data are
          averages of two experiments. Test solutions changed
          daily or not changed for the four days of the experiment.

ALUMINUM ION AT 5 mg l^{-1} (Concentration in spring soil solutions)

		pH 3.0	pH 4.5
Solutions	Control	0.91 ± .09	0.82 ± .08
changed	+ Al	0.40 ± .03	0.36 ± .04
daily	% control	36*	43*
Solutions	Control	0.84 ± .07	0.84 ± .06
not	+ Al	0.31 ± .05	0.44 ± .06
changed	% control	36*	64*

ALUMINUM ION AT 0.5 l^{-1} (Concentration in fall soil solutions)

		pH 3.0	pH 4.5
Solutions	Control	1.05 ± .08	0.89 ± .09
changed	+ Al	1.01 ± .12	0.86 ± .08
daily	% control	96	96
Solutions	Control	0.84 ± .10	0.83 ± .10
not	+ Al	0.84 ± .06	0.75 ± .11
changed	% control	99	91

* $p = 0.05$

nitrogen that when this element was less available. Evaluation of
the possible protective role of calcium in ameliorating damage
caused by aluminum (Ulrich, 1981) was confounded by pH and will
require additional study. Cadmium at 1 mg l^{-1} had little effect.

    Most forest trees in the temperate zone show optimum
development only when mycorrhized (Mikola and Laiho, 1962) and
water uptake is greatly facilitated by these associations (Reeves
et al., 1979; Dixon et al., 1980). The presence of mycorrhizae on
red spruce roots taken from the coniferous zone of Camels Hump has
not been seen either directly or through microscopic examination,
although spruce roots from the lower, hardwood zone showed

Table 4.   Effect of cadmium, lead and zinc ion on total
           transpiration of red spruce trees. Ingestad's nutrient
           solution, 0.1 X, adjusted to pH 3.0 or 4.5 with HCl.
           Five plants per replicate measured daily for three or
           four days. Solutions changed daily or not changed during
           an experiment.

CADMIUM ION AT 1 mg l^{-1}

		pH 3.0	pH 4.5
Solutions	Control	1.12 ± .14	1.15 ± .13
changed	+ Cd	0.84 ± .11	0.85 ± .07
daily	% control	75*	74*
Solutions	Control	1.36 ± .17	1.08 ± .12
not	+ Cd	1.10 ± .16	1.02 ± .14
changed	% control	81	94

LEAD ION AT 1 mg l^{-1}

		pH 3.0	pH 4.5
Solutions	Control	1.27 ± .17	1.15 ± .15
changed	+ Pb	0.93 ± .14	1.23 ± .25
daily	% control	73*	107
Solutions	Control	1.06 ± .17	0.96 ± .21
not	+ Pb	1.02 ± .17	0.96 ± .20
changed	% control	96	100

ZINC ION AT 1 mg l^{-1}

		pH 3.0	pH 4.5
Solutions	Control	1.19 ± .17	1.24 ± .21
changed	+ Zn	1.02 ± .09	1.25 ± .23
daily	% control	86	101
Solutions	Control	0.90 ± .09	1.03 ± .21
not	+ Zn	1.02 ± .24	1.25 ± .31
changed	% control	113	121

* $p = 0.05$

Table 5.   Effect of litter extract on total transpiration of red
           spruce trees. Balsam fir: red spruce needles (2:1 w/w,
           air dried) extracted for 24 hours with simulated acidic
           precipitation constituted with HC1 or with H_2SO_4:HNO_3
           (60:40 hydrogen ion ratio). Extracts filtered and used
           full strength after pH 4.0 adjustment. Solutions changed
           daily for four days. Data are average of three trees.

Extracting solution	Control	Extract	% control
HC1	0.91 ± .10	0.91 ± .13	100
H_2SO_4:HNO_3	1.28 ± .22	0.69 ± .02	54*

* $p = 0.05$

mycorrhizal Hartig-net formation. As part of our evaluation of
spruce decline, a study on mycorrhizae was started. Axenic
cultures of Cenococcum graniforme, ATCC 12746, and Polyporus
circinatus, ATCC 9389, both occurring on spruce (Trappe, 1962),
were grown at 21°C as shake cultures in liquid media (Mikola, 1948;
Schenck, 1982). Growth was not greatly affected by pH (cf.
Aaltonen, 1948). There was, however, a strong interaction between
initial medium pH and concentrations of four metal ions each
presented in several concentrations (Figs. 3 and 4). The tested
aluminum concentrations were higher than have found in soil
solutions from Camels Hump, although the lowest tested
concentration was within the range reported by others (cf. Ulrich,
1981). The concentrations of the other tested metals are within
the range that occur on Camels Hump. Work in progress will
determine the effects of acidity and metal ions on the
establishment of mycorrhizae, since it is known that repression of
the growth of the fungal partner reduces or prevents mycorrhizal
associations (Scott, 1969; Sands and Theodorou, 1978). As a first
approximation, it is reasonable to postulate that acidity and
metals are involved in the repression of mycorrhizae and that this
can affect the water relations of impacted red spruce.

DISCUSSION

     This same pattern of acid-metal interaction was found with two
nitrogen-fixing blue green algae and the free-living, N-fixing
bacterium Azotobacter, with a common mountain moss (Klein and
Bliss, 1984), and with the activity of litter degrading microorganisms (Moloney et al., 1983). There seems little doubt that

Figure 2.   Interactive response of red spruce to aluminum ion and
            water stress. Young trees grown in Ingestad's full
            strength nutrient solution for three weeks with
            solutions changed twice weekly. Al^{3+} at 5 mg l^{-1};
            mannitol as osmoticum to maintain a water potential of
            -4 bars.

acid-metal   interactions   can   impose   growth,   biochemical   and
physiological stresses on plants (cf. Hutchinson, 1980) in the
impacted ecosystems of Camels Hump.

    Water movement through seedling spruce trees was repressed by
acidic soil solutions and by comparable nutrient solutions
containing realistic concentrations of H$^+$ and metal ions.   These
results are consistant with those of other studies (Baker, 1974;
Carlson et al., 1975; Jarvis et al., 1976; Lamoreaux and Chaney,
1977; Jastrow and Koeppe, 1980).   Field observations on the absence

Table 6. Viability responses of red spruce trees to acidity and
          aluminum ion under hydroponic conditions. Nutrient
          solutions changed three times weekly in aerated cultures.
          Experiment terminated after eight weeks.

Nutrient Solution	Ca (mM)	Al (mM)	Ca/Al	Visible growth responses[a] pH 3	pH 4.5
1 X Ingestad nutrient	1.0	0.0	–	4/4	4/4
+ 5 mg l^{-1} Al^{3+}	1.0	0.185	5.4	0/0	1/1
+ 0.5 mg l^{-1} Al^{3+}	1.0	0.018	0.54	1/0	2/2
0.1 Ingestad nutrient	0.1	0.0	–	4/4	4/4
+ 5 mg l^{-1} Al^{3+}	0.1	0.185	0.54	0/0	2/2
+ 0.5 mg l^{-1} Al^{3+}	0.1	0.018	5.4	2/2	3/3

[a]Based on arbitrary visual evaluation with 4 being apparently
healthy and 0 being visible dead. First number refers to shoots
and second number refers to roots.

of spruce mycorrhizae and the interactions of acid plus metal ions
on the growth of mycorrhizal fungi suggest that this may also be a
factor in the water relations of trees. The observed decline in
red spruce and other forest trees may be due in part to water
stresses resulting from the presence of acidic, metal-polluted soil
solutions.

     Red spruce in not highly tolerant to water stress, and its
growth is decreased by water stresses that may have less effect on
other species. Throughout its growing range, red spruce trees have
undoubtedly been drought stressed on many occasions. Many mature
representatives of this long-lived species have survived until the
past several decades in spite of recurring droughts and have, until
now, shown recovery. It is very unlikely that the presumed
droughts of the 1950-1960 period could, alone, have caused the
extensive mortality that has been reported. The simultaneous
mortality of other, more drought resilient species also argues
against a drought causation of tree decline, although the stress
potential of drought is clear. Spruce trees survive in mountain
environments that include harsh temperature regimes, inadequate
nutrition, poor soil buffering capacity and intermittant droughts
and wet periods. When additionally stressed by acidic depositions
and the concommitant insults imposed by soil solutions containing
reduced levels of divalent cations and increased metallic ions plus
reduced mycorrhizae, increased foliar leaching (Scherbatskoy and

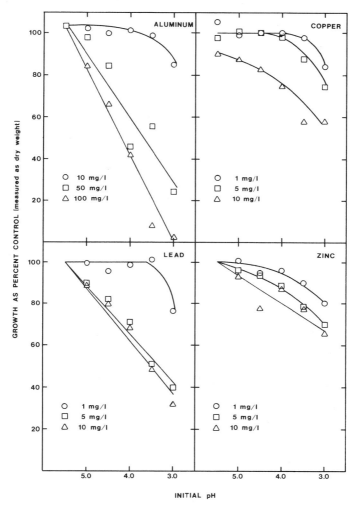

Figure 3. Growth _in vitro_ of axenic cultures of the
mycorrhizal fungus <u>Cenococcum graniforme</u>
at various pH levels and in the presence
of metal ions supplied as chlorides. The
growth period was four weeks at 25°C in
still cultures containing 15 ml of medium.
Results calculated as the mean dry weight
of three cultures.

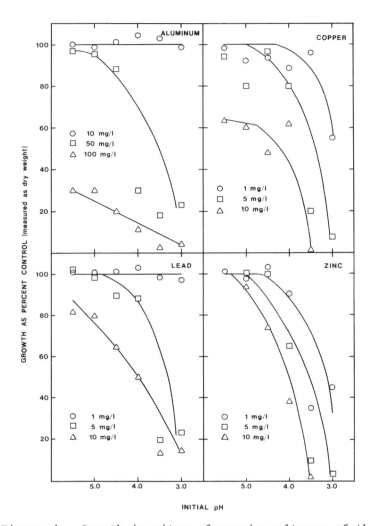

Figure 4.   Growth in vitro of axenic cultures of the
mycorrhizal fungus Polyporus circinatus
at various pH levels and in the presence
of metal ions supplied as chlorides.   The
growth period was four weeks at 25°C in
still cultures containing 15 ml of medium.
Results calculated as the mean dry weight
of three cultures.

Klein, 1983) and other sequelae of acidic deposition, the stress load on trees becomes impossible to withstand.

Johnson et al. (1982) were unable to find significant differences in aluminum concentration of "fine roots" or of foliage from visually healthy and visually unhealthy red spruce. They concluded that the evidence does not support the contention that aluminum accumulation in tissues is an important link between acid precipitation and spruce mortality. Their data, although at variance with those of others (cf. Mayer and Heinrichs, 1981), is not particularly surprising. The roots were taken from soils containing high indigenous levels of aluminum that bind to dead bark cells and cannot be easily removed (Haynes, 1980), and the accumulation of aluminum phosphate in plant roots is known (McCormick and Borden, 1974). Bulk chemical analyses of roots taken from soils cannot provide definitive information on the possible role of aluminum or other elements in destabilizing and stressing forest trees.

Toxic reactions to aluminum and other metallic ions in acidic solutions are due to modifications in the structure and function of cells in very young, actively absorbing root tips (Stoklasa, 1922; Foy and Brown, 1963; Clarkson, 1965; Fleming and Foy, 1968; Rorison, 1972). Böhm-Tuchy (1960) found that aluminum induces plasmolysis and causes the plasmalemma (cell membrane) to "harden" so that uptake of water and minerals is greatly impeded. Calcium can protect membrane integrity under high Al conditions (Schneider, 1952; Foy and Brown, 1963; Lance and Pearson, 1969), but Cronan (1980) and Ulrich and Pankrath (1983) reported that calcium is already in short supply in podzolized soils especially in soil receiving acidic precipitation. The Ca/Al ratio in such soil solutions does not favor membrane protection. The available field chemical data and these laboratory simulation experiments are fully compatable with the concept that water stress, imposed by a variety of environmental and biotic factors centering about acidic depositions, is one of the factors affecting the survival of red spruce.

CONCLUSIONS

Cause and effect relationships are difficult to establish in systems as complex as natural forested ecosystems. The situation is confounded by the diversity of ecosystem components, soils and climatic conditions. There are conflicting reports from laboratories in different parts of the world. Some of the discrepancies are due to differences in site, climate, soil and biota as well as the nature and severity of polluted depositions. The massive tree death in Europe (Ulrich and Pankrath, 1983) reflects the fact that these forests are essentially uniform, monoculture plantations rather than the diverse species and age stands characteristic of mountainous forested areas in North America. In addition, many regions of Europe are receiving much greater levels of acidic deposition and gaseous pollutants than have been experienced by sites such as Camels Hump. Any holistic appraisal and evaluation of research findings must, therefore, be at the level of general ecological principles and patterns of response rather than by point-by-point equivalence of data (Mueller-Dombois et al., 1983).

Acidic depositions plus solubilized endogenous and deposited metals have a reasonably high probability of being solely or partly responsible for the observed alterations in soils and plants growing under natural or near-natural conditions. Spruce decline and the simultaneous declines of hardwood and other conifer species is not likely to be due to a fungal disease or to insect infestation; the simultaneous declines of coniferous and hardwood species militates against an unitary biotic cause. Laboratory simulations have demonstrated that the acid-metal hypothesis is compatable with ecological and physiological principles. The simulations are ecologically realistic and are consistant with the field data although laboratory studies can only indicate possibilities and not necessarily what is actually occurring in the field.

One final point must be made. It is of vital importance to extrapolate from the present situation in an attempt to assess the long-term future of affected ecosystems. Even with an assumption that there will be significant reductions in anthropogenic emissions, biological damage will continue for some years to come, and the possibility of amelioration or reversal of damage should begin to be considered.

Acknowledgements - Research was supported by funding from the Vt. Agric. Exp. Station, the Northeast Forest Experiment Station of the U.S. Dept. of Agriculture Forest Service, The Texaco Foundation, The Jackson Hole Foundation, the A.W. Mellon Foundation, the R.K. Mellon Foundation and the American Electric Power Service Corporation.

Suggestions of Dr. H.W. Vogelmann and Tim Scherbatskoy was greatly appreciated. Technical assistance of Susan C. Adamowicz, Margaret Bliss, Judith Chaves, Tracy Perry, Kathryn Vanmatta and Tim Wilmot is acknowledged with thanks. Journal paper no. 527 from the Vermont Agricultural Experiment Station.

REFERENCES

Aaltonen, V.T., 1948, Boden und Wald, unter besonderer Berück-
    sichtigung des nordeuropäischen Waldbaus. Paul Parey, Berlin,
    Germany. 457 pp.
Alexander, M., 1982, The effects of acid precipitation on microbial
    mineralization of nitrogen in soil, pp. 7 to 21. In: Acidic
    Deposition Ecological Effects Research Peer Review,
    U.S.E.P.A., North Carolina State University, Raleigh, NC.
Baker, D.E., 1974, Copper, soil, water, plant relationships. Fed.
    Proc. 33:1188-1193.
Böhm-Tuchy, E., 1960, Plasmalemma and Aluminiumsalzwirkung.
    Protoplasma 52:108-142.
Carlson, R.W., Bazzaz, F.A., and Rolfe, G.L., 1975, The effects of
    heavy metals on plants. II. Net photosynthesis and
    transpiration of whole corn and sunflower plants treated with
    Pb, Cd, Ni and Tl. Envir. Res. 10:113-120.
Clarkson, D.T., 1965, The effect of aluminum and some other
    trivalent metal cations on cell division in the root apices of
    Allium cepa. Ann. Bot. 29:309-315.
Coleman, D.C., 1983, The impacts of acid deposition on soil biota
    and C cycling. Envir. Exp. Bot. 23:225-233.
Cronan, C.S., 1980, Solution chemistry of a New Hampshire subalpine
    ecosystem: A biogeochemical analyses. Oikos 34:272-281.
Cronan, C.S. and Schofield, C.L., 1979, Aluminum leaching response
    to acid precipitation: Effects on high elevation watersheds.
    Science 204:304-306.
Dixon, R.K., Wright, G.M., Behrs, G.T., Teskey, R.O., and
    Hinckley, T.M., 1980, Water deficits and root growth of
    ectomycorrhizal white oak seedlings. Can. J. For. Sci.
    10:545-548.
Firestone, M.K. and McColl, J.G., 1982, Effects of acid
    precipitation on microbial transformations in soil and
    resulting nutrient availability to plants, pp. 1 to 22. In:
    Acidic Deposition Effects Research Peer Review, U.S.E.P.A.,
    North Carolina State University, Raleigh, NC.
Fleming, A.L. and Foy, C.D., 1968, Root structure reflects
    differential aluminum tolerance in wheat varieties. Agron. J.
    60:172-176.
Foy, C.D. and Brown, J.C., 1963, Toxic factors in acid soil. I.
    Characterization of aluminum toxicity in cotton. Soil Sci.
    Soc. Am. Proc. 27:403-407.
Galloway, J.N., Eisenreich, S.J., and Scott, B.S., eds., 1980,
    Toxic substances in atmospheric deposition. A review and
    assessment. Report No. NC-141, U.S.E.P.A., Fort Collins, CO.
Haynes, R.J., 1980, Ion exchange properties of roots and ionic
    interactions within the root apoplasm: Their role in ion
    accumulation by plants. Bot. Rev. 46:75-99.
Hutchinson, T.C., 1980, Impact of heavy metals on terrestrial and
    aquatic ecosystems, pp. 158 to 164. In: Proceedings of a

Symposium on Effects of Air Pollutants on Mediterranean and
Temperate Ecosystems. Report No. PSW-43, Pacific Southwest
Forest and Range Experiment Station, Riverside, CA.

Ingestad, T., 1959, Studies on the nutrition of forest tree
seedlings. II. Mineral nutrition of spruce. Physiol. Plant.
12:568-593.

Jarvis, S.C., Jones, L.H.P., and Hopper, M.J., 1976, Cadmium uptake
from solutions by plants and its transport from roots to
shoots. Plant Soil 44:191-197.

Jastrow, J.D. and Koeppe, D.E., 1980, Uptake and effects of cadmium
in higher plants, pp. 607 to 638. In: Cadmium in the
Environment. Part I. Ecological Cycling, J.O. Nriagu, ed.,
Wiley-Interscience, New York, NY.

Johannessen, M. and Henriksen, A., 1978, Chemistry of snowmelt
water: changes in concentration during melting. Water Res.
Bull. 14:615-619.

Johnson, A.H. and Siccama, T.G., 1984, Decline of red spruce in the
northern Appalachians: Assessing the possible role of acid
deposition. Tappi J. (in press).

Johnson, A.H., Lord, D.G., and Siccama, T.G., 1982, Red spruce
dieback in Vermont and New Hampshire: Is acid precipitation a
contributing stress? pp. 63 to 67. In: Acid Rain. A Water
Resource Issue for the 80's, R. Hermann and A.I. Johnson,
eds., Am. Water Resources Assoc., Bethesda, MD.

Klein, R.M., 1983, An ecosystem approach to the acid rain problem,
pp. 1 to 11. In: Effects of Acid Rain on Vegetation,
R.A. Linthurst, ed., Ann Arbor Science Press, Ann Arbor, MI.

Klein, R.M. and Bliss, M., 1984, Decline in surface coverage by
mosses on Camels Hump Mountain in Vermont. Bryologist (in
press).

Klein, R.M. and Klein, D.T., 1970, Research Methods in Plant
Science. Natural History Press, Garden City, NY. 756 pp.

Krug, E.D. and Frink, C.R., 1983, Acid rain on acid soil: A new
perspective. Science 221:520-525.

Lamoreaux, R.J. and Chaney, W.R., 1977, Growth and water movement
in silver maple seedlings affected by cadmium. J. Envir.
Qual. 6:201-205.

Lance, J.C. and Pearson, R.W., 1969, Effect of low concentrations
of aluminum on growth and water and nutrient uptake by cotton
roots. Soil Sci. Soc. Am. Proc. 33:95-98.

Ligon, W.S. and Pierre, W.H., 1932, Soluble aluminum studies. II.
Minimum concentration of aluminum found to be toxic to corn,
sorghum, and barley in solution cultures. Soil Sci. 24:
307-318.

Lindberg, S.E. and Harris, R.C., 1981, The role of atmospheric
deposition in an eastern U.S. deciduous forest. Water, Air,
Soil Poll. 16:13-31.

Lovett, G.M., Reiners, W.A., and Olson, R.K., 1982, Cloud droplet
deposition in subalpine balsam fir forests: Hydrological and
chemical inputs. Science 218:1302-1304.

Mayer, R. and Heinrichs, H., 1981, Gehalte von Baumwurzeln an chemischen Elementen einschliesslich Schwermetallen aus Luftvereinreinigung. Zeit. Pflanzenernaehr. 144:637-646.

McCormick, L.H. and Borden, F.V., 1974, The occurrence of aluminum-phosphate precipitates in plant roots. Proc. Soil Sci. Soc. Am. 38:913-934.

Mikola, P., 1948, On the physiology and ecology of Cenococcum-graniforme. Comm. Inst. Forest. Fennica 36:1-104.

Mikola, P. and Laiho, O., 1962, Mycorrhizal relations in the raw humus layer of northern spruce forests. Comm. Instit. Forest. Fennica 55:1-11.

Moloney, K.A., Stratton, L.J., and Klein, R.M., 1983, Effects of simulated acidic, metal-containing precipitation on coniferous litter decomposition. Can. J. Bot. 61:3337-3342.

Mueller-Dombois, D., Canfield, J.E., Holt, R.A., and Buelow, G.P., 1983, Tree-group death in North American and Hawaiian forests: A pathological problem or a new problem for vegetation ecology? Phytocoenology 11:117-137.

Nilsson, S.I., Miller, H.G., and Miller, J.D., 1982, Forest growth as a possible cause of soil and water acidification: An examination of the concepts. Oikos 39:40-49.

Puckett, L.J., 1982, Acid rain, air pollution and tree growth in southeast New York. J. Envir. Qual. 11:376-381.

Raynal, D.J., Leaf, A.L., Manion, P.D., and Wang, G.J.K., 1980, Actual and potential effects of acid precipitation in the Adirondack Mountains. Publ. 56/ES.HS/79 (ERDA 80-28), N.Y. State Energy and Developmental Authority, Albany, NY.

Reeves, F.B., Wagner, D., Moorman, T., and Kiel, J., 1979, The role of endomycorrhizae in revegetation in the semi-arid west. I. A comparison of incidence of mycorrhizae in severely disturbed vs. natural environments. Am. J. Bot. 66:6-13.

Reiners, W.A., Marks, R.H., and Vitousek, P.D., 1975, Heavy metals in subalpine and alpine soils of New Hampshire. Oikos 26: 264-275.

Rorison, I.H., 1972, The effect of extreme soil acidity on the nutrient uptake and physiology of plants. In: Acid Sulphate Soils, Vol. I. H. Dost, ed., Publication 18, Intern. Instit. Land Reclamation and Improvement. Wageningen, The Netherlands.

Rorison, I.H., 1980, The effects of soil acidity on nutrient availability and plant response, pp. 283 to 304. In: Effects of Acid Precipitation on Terrestrial Ecosystems, T.C. Hutchinson and M. Havas, eds., Plenum Press, New York, NY.

Sands, R. and Theodorou, C., 1978, Water uptake by mycorrhizal roots of radiata pine seedlings. Aust. J. Plant Physiol. 5: 301-309.

Schenck, N.C., ed., 1982, Methods and Principles of Mycorrhizal Research. Am. Phytopath. Soc., St. Paul, MN. 244 pp.

Schneider, B., 1952, Erfahrungen mit Kalk und Grunddüngungen auf
    podsolierten Waldböden. Forst Archiv 23:153-158.
Scherbatskoy, T. and Klein, R.M., 1983, Responses of spruce and
    birch foliage to leaching by acidic mists. J. Envir. Qual.
    12: 189-195.
Scherbatskoy, T. and Bliss, M., 1984, Occurrence of acidic rain and
    cloud water in high elevation ecosystems in the Green
    Mountains of Vermont. In: Symposium on the Meteorology of
    Acidic Depositions, Air Pollution Control Association,
    Hartford, CT (in press).
Scott, G.D., 1969, Plant Symbiosis. St. Martin's Press, NY.
    58 pp.
Scott, J.T., unpublished studies, State University of New York,
    Albany, NY.
Siccama, T., Smith, W.H., and Mader, D.L., 1980, Changes in lead,
    zinc, copper, dry weight and organic matter of a forest floor
    of white pine stands in central Massachusetts over 16 years.
    Envir. Sci. Tech. 14:54-56.
Siccama, T.G., Bliss, M., and Vogelmann, H.W., 1982, Decline of red
    spruce in the Green Mountains. Bull. Torrey Bot. Club 109:
    162-168.
Strayer, R.F. and Alexander, M., 1981, Effects of simulated acid
    rain on glucose mineralization and some physicochemical
    properties of forest solids. J. Envir. Qual. 10:460-464.
Stoklasa, J., 1922, Über die Verbreitung des Aluminiums in der
    Natur und seine Bedeutung beim Bau- under Betriebstoffwechsel
    der Pflanzen. G. Fischer, Jena, Germany.
Trappe, J.M., 1962, Fungus associates of ectotropic mycorrhizae.
    Bot. Rev. 28:538-606.
Ulrich, B., 1981, Destabilizierung von Waldökosystemen durch
    akkumulation von Luftverunreinigungen. Forst-Holzwirt. 21:
    525-532.
Ulrich, B. and Pankrath, J., eds., 1983, Effects of Accumulation of
    Air Pollutants in Forest Ecosystems. Kluwer Academic Publ.,
    Hingham, MA. 389 pp.
Volk, B.G., Bitton, G., Byres, G.E., and Graetz, D.A., 1982, Effect
    of acid precipitation on selected soils of the southeastern
    United States, pp. 7 to 21. In: Acidic Deposition Ecological
    Effects Research Peer Review, U.S.E.P.A., North Carolina State
    University, Raleigh, NC.
Witham, F.H., Blades, D.F., and Devlin, R.M., 1971, Experiments in
    Plant Physiology. Van Nostrand Reinhold Co., New York, NY.

PRODUCTIVITY OF FIELD-GROWN SOYBEANS (AMSOY AND WILLIAMS)

EXPOSED TO SIMULATED ACIDIC RAIN

Lance S. Evans,[a,b] Keith F. Lewin,[a] and
Mitchell J. Patti[b]

[a]Department of Applied Science, Brookhaven
National Laboratory, Upton, NY   11973 and
[b]Laboratory of Plant Morphogenesis, Manhattan
College, The Bronx, NY   10471

ABSTRACT

An experiment was performed during the summer of 1982 to determine the effects of simulated acidic rain on seed yields of two commercial cultivars of soybeans grown according to standard agronomic practices. Plants were shielded from all ambient rainfalls automatically by two moveable exclusion covers and exposed to simulated rainfalls in quantities equal to the average amount of rainfall that occurs at the site. Seed yields of Amsoy cultivar exposed twice weekly to simulated rain of pH 4.1, 3.3, and 2.7 were, respectively, 3.0, 9.0, and 12.8% below yields of plants exposed to a pH 5.6 simulated rain. A treatment-response function of seed yield versus rainfall pH for Amsoy was $y = 10.20 + 0.587 x$ and had a correlation coefficient of 0.96 (y is seed mass per plant and x is the pH of the simulated rain). For the Williams cultivar, seed yields of plants exposed to simulated rainfalls of pH 5.6, 4.1, 3.3, and 2.7 were 11.5, 10.5, 11.4, and 11.4 g, respectively. A treatment-response function of seed yield versus rainfall pH for Williams was $y = 11.13 + 0.016 x$ which had a correlation coefficient of 0.038. Plants of Amsoy and Williams grown in plots adjacent to the exclusion shelters and experiencing ambient rain conditions had mean seed yields of 11.4 and 9.8 g per plant, respectively. Seed yield per plant was dependent upon the number of pods per plant because the number of seeds per pod did not vary among treatments of each cultivar.

323

INTRODUCTION

One of the most important objectives of research concerned with the effects of acidic deposition on terrestrial ecosystems is to determine the impact(s) of acidic rain on growth and yield of crops grown under standard agronomic conditions (Galloway et al., 1978; Evans and Hendrey, 1980; Evans et al., 1981b). In such experiments, standard agronomic practices must be employed with large sample sizes, adequate randomization of treatment plots to compensate for local soil variations, and standard statistical analyses in which small differences in yield (10% or less) may be determined.

Rainfall acidity-response functions for crop yield and crop quality are needed to predict the overall impacts on crop yields of ambient and/or anticipated levels of acidic rain. Recently such functions have been determined for field-grown and greenhouse-grown crops (Evans et al., 1980, 1981a, 1981b, 1982, 1983; Evans and Lewin, 1981) as well as for the degree of sexual reproduction of bracken fern grown in an oak-pine forest (Evans and Conway, 1980). Experiments described herein used two commercial varieties of soybeans in an agricultural field situation at Brookhaven National Laboratory, Suffolk County, New York, during the summer of 1982 to determine the effects of simulated acidic rain on seed yields.

Soybeans were used because results of previous experiments suggested that broad-leaved herbaceous plants are sensitive to acidic precipitation (Evans and Curry, 1979; Evans, 1980) and the economic implications of changes in soybean yields could be significant. Moreover, a previous experiment during the summers of 1979 and 1981 demonstrated significant reductions in seed yields (Evans et al., 1981a, 1983). It should be recognized that a 1% reduction in seed yields of soybeans grown in the northeastern United States (Delaware, Illinois, Indiana, Kentucky, Maryland, Michigan, New Jersey, New York, Ohio, Pennsylvania, and Virginia) during 1980 would represent a loss of 53.4 million dollars (ignoring price elasticity; Crop Reporting Board, 1981a; 1981b).

The experiment described herein was performed to determine the effects of simulated rains for pH 5.6 to 2.7 on soybean cultivars Amsoy and Williams throughout the growing season in which all ambient rainfalls were excluded through the use of automatically moveable rainfall exclusion shelters. Seed yields of these cultivars were compared with yields of plants that received only ambient rainfalls from plots adjacent to the exclusion shelters.

MATERIALS AND METHODS

Field plot design

Soybean seed (Glycine max) of cultivars Amsoy 71 and Williams 79 were inoculated with Rhizobium japonicum. Prior to planting, a rye cover crop was plowed to a depth of 23 cm. After plowing, the field was disked with a harrow and fertilized (448 kg ha^{-1} of 5-10-10-3, N-P-K-S), following the recommendations of Sandsted et al. (1980). Seeds were planted on 4 June 1982.

The soil is a Plymouth loamy sand with a silty substratum and the plots have a 0 to 3% slope (Warner et al., 1975). Seeds were planted 6.6 cm apart within each row. Rows were 38.1 cm apart to provide a seeding density of 380,000 seeds per hectare (Ryder and Beuerlein, 1978; Sandsted et al., 1980). Each experimental plot consisted of four adjacent rows, 1.52 m in length. Immediately after planting, the fields were spread with a tank mix of Lasso (Monsanto, St. Louis, MO) and Sencor (Mobey Chemical Corp., Kansas City, MO) at rates of 2.24 and 0.28 kg ha^{-1} of active ingredient, respectively.

Four replicates of four experimental treatments of simulated acidic rainfalls of pH 5.6, 4.0, 3.3, and 2.7 were arranged in latin squares. The field plot design of each experiment included eight latin squares yielding a total of 32 plots per experimental treatment. Each field plot consisted of two rows each of Amsoy and Williams cultivars. The two outside rows of each plot were designated border rows while 13 plants in the central row of each cultivar were harvested. All four rows received simulated rain-falls. The row length of plants harvested for each cultivar within each plot was recorded so that yields could be expressed on a per area basis. For statistical purposes, all plants of each cultivar in each plot were pooled but results are expressed on per plant and per hectare basis. The latin squares used were selected randomly from those given in Table XV of Fisher and Yates (1957).

Simulated rain solutions

The chemistries of the simulated rainfall solutions were derived from several sources. Concentrations of constituents in simulated rain solution with all acidic components removed is shown in Table 1. The nitrate and sulfate concentrations of each treatment solution are shown in Table 2. A 2.5:1.0 equivalent ratio of sulfate to nitrate was used for all simulated rainfall treatments. Simulated rainfalls started on 8 June and ended on 23 September 1982 when almost all leaves were senescent. Harvest began on 27 September 1982 (Table 2).

Table 1.   Concentrations of constituents in simulated
           rain with acidic components removed[a]

Compound	Concentration ($\mu g\ l^{-1}$)
NaCl	2215.26
$(NH_4)_2SO_4$	1242.12
$MgSO_4 \cdot 7H_2O$	788.77
$NaNO_3$	739.59
CaCl	320.76
NH_4Cl	308.70
KCl	276.58
$Zn(NO_3)_2 \cdot 6H_2O$	227.58
$MnSO_4 \cdot H_2O$	39.97
$Pb(NO_3)_2$	28.81

[a]Concentrations of H^+, Na^+, NH_4^+, SO_4^{2-}, Cl^-, and NO_3^- were calculated from data of weighted average of rainfall samples collected by BNL sequential sampler from May through September for the years 1976 through 1980. Concentrations of Mg^{2+}, Ca^{2+}, and K^+ were calculated from data of BNL samples for MAP3S net- work from May through September for the years 1978 through 1980. Concentrations of Zn^{2+}, Mn^{2+}, and Pb^{2+} were kindly provided by Dr. Arthur Johnson from data of rain samples for Lebanon Forest, NJ, for unfiltered, bulk precipitation samples collected from May 1978 through January 1982. Lebanon Forest is located at long. $74°30'W$, lat. $39°50'N$ (approximately 180 km SW of BNL).

Table 2. Experimental design of soybean field experiment of 1982
          at Brookhaven National Laboratory

Duration of the experiment	17 weeks (4 June - 27 September 1982)
Duration of experimental treatments	16 weeks (8 June - 23 September 1982)
Number of rainfalls	32
Frequency of rainfalls	2 wk^{-1}
Number of plots per treatment	32
Individual plot area	2.3 m^2
Number of harvested plants per cultivar per plot	13
Duration of each rainfall	1.17 hr
Total deposition per rainfall	10.4 mm
Total water deposition for all simulated rainfalls	332.8 mm
Total water deposition for all ambient rainfalls	457.2 mm

Rainfall Chemistry

Treatment Solutions	Sulfate (mg l^{-1})	Sulfur Deposition (kg ha^{-1})	Nitrate (mg l^{-1})	Nitrogen Deposition (kg ha^{-1})
Simulated rain of pH 5.6	1.3	4.3	0.67	2.2
Simulated rain of pH 4.1	3.9	13.0	2.0	6.7
Simulated rain of pH 3.3	18	60	9.5	32
Simulated rain of pH 2.7	70	233	36	120
Ambient rainfalls		7.8		1.4

Characterization of ambient rainfall pH levels

Ambient rainfall events from planting to harvest were characterized (Table 3) using methods described by Raynor and McNeil
(1978) and Raynor and Hayes (1981). Data cited were obtained from
G. S. Raynor and J. V. Hayes (unpublished results). Frequency
distributions of ambient rainfall shower durations and volumes
were derived from a rotary rain indicator (Raynor, 1955) and a
tipping bucket gauge, respectively. The rotary rain indicator
records all precipitation except light drizzle, fog, and dew.

Dew formation characteristics

The presence of dew was calculated from temperature, relative
humidity, and rain data. It is assumed that dew formed on the
foliage of soybean plants if dew formation conditions were present
for at least two consecutive hours. Dew-point data were compared
with ambient rainfall data to determine the frequency and duration
of dew conditions in the absence of rain. This procedure has been
used previously (Evans et al., 1981a, 1982).

Table 3. Selected precipitation characteristics for entire growing season in 1982

Time period	Hydrogen (µeq l⁻¹)	Total conductivity (µmhos cm⁻¹)	Nitrate and nitrate (µeq l⁻¹)	Ammonium (µeq l⁻¹)	Sulfate[a] (µeq l⁻¹)	Sodium (µeq l⁻¹)	Chloride (µeq l⁻¹)	N/S[a] ratio	Duration period	Amount event (mm)	Rate (mm ha⁻¹)
4 June to	32.7	20.2	21.1	17.9	53.5	19.9	54.5	0.54	9.0	18.3	2.0
27 September	(25)	(25)	(24)	(24)	(23)	(13)	(15)	(23)	(25)	(25)	(25)

[a]Samples were weighted means of ambient precipitation. See Raynor and McNeil (1978) and Raynor and Hayes (1981) for sampling procedures and methods

Total amount of precipitation = 457.2 mm

Insect infestations

Infestations of Mexican bean beetles (Epilachna varivestis), Japanese beetles (Popilla japonica), and various species of mites were controlled by applications of appropriate insecticides applied at label recommended rates.

Rainfall shielded plots

Soybean plants were shielded from ambient rainfalls with the use of moveable covers. The moveable covers (each cover 30 m x 10 m) were of standard plastic greenhouse construction which were conveyed automatically on rails over the entire crop within 50 sec after the first raindrop impacted on an electronic rain sensor (Aerochem Metrics, Miami, FL). Simulated rainfalls were applied between 1800 and 2400 hours twice weekly. A total of 2.08 cm of rainfall was applied weekly for fifteen weeks. This rate of rainfall was similar to the mean rate calculated from data of 24 consecutive years (Nagle, 1975). The duration of each simulated rainfall event was 1.17 hour. No supplemental irrigation was provided to these shielded plants.

Ambient rainfall plots

Other soybean plots were planted around these field plots. These plots received drip irrigation (Goldberg et al., 1976) to prevent irrigation water from contacting plant foliage.

Statistical analyses

Statistical analyses were performed on all data to determine if treatments within each experiment were significantly different at the 0.05 level of risk. An analysis of variance followed by the Student-Newman-Keuls Test was performed to determine statistical significance (Steel and Torrie, 1960). Weighted least squares regression analysis was performed with means of the 32 plots per treatment averaged over latin squares, rows, and columns. The best fit line and confidence limits around each best fit line are calculated when significant. Correlation coefficients of regression analyses are given when significant (Steel and Torrie, 1960).

RESULTS

The pH distribution of hourly samples of rain events for the 1981 growing season was normal for the region. Over 30% of all ambient rainfalls had a pH below 4.5, 36% had a pH between 4.5 and 5.0, and 34% a pH above pH 5.0. Ambient rainfalls were never below pH 3.0 or above pH 6.5. Ambient concentrations of nitrate plus nitrite, ammonium, sulfate, and hydrogen averaged 21.1, 17.9, 53.5, and 32.7 μeq l^{-1}, respectively (Table 3). The mean nitrogen:sulfur ratio was 0.54.

The simulated rainfall solution of pH 4.1 was similar to the weighted mean hydrogen ion concentration of all ambient rainfalls. The chemistry of simulated rain solution was typical for the region (Table 1) but because the total amount of ambient precipitation was higher (457.2 mm) than the mean (332.8 mm) for the region, mean concentrations of nitrate plus nitrite and sulfate are lower than normal (Tables 2 and 3).

Concentrations of ozone and sulfur dioxide were monitored in Suffolk County, New York. The values obtained were used to estimate the effects of these gases on soybean productivity. Maximum daily and hourly averages on a weekly basis are shown for ozone and sulfur dioxide in Table 4. Except for two days (0.126 and 0.148 nl l^{-1} for 8 and 9 July 1982, respectively) ambient concentrations of ozone did not exceed the current one-hour air

Table 4. Ozone and sulfur dioxide concentrations at Babylon, Suffolk County, New York for the period 6 June through 27 September 1982[a]

Week Beginning	Ozone Concentrations		Sulfur Dioxide Concentrations	
	Maximum daily mean (nl l^{-1})	Maximum hourly mean (nl l^{-1})	Maximum daily mean (nl l^{-1})	Maximum hourly mean (nl l^{-1})
06 June	33	80	17	69
13 June	37	73	13	42
20 June	35	76	11	21
27 June	33	73	12	36
04 July	42	148	11	32
11 July	58	84	9	19
18 July	41	115	13	42
25 July	32	74	12	23
01 August	41	84	7	38
08 August	31	83	8	38
15 August	33	80	17	31
22 August	28	68	8	27
29 August	20	45	12	34
05 September	25	76	12	26
12 September	49	99	12	38
19 September	21	55	13	52
26 September	27	55	5	14

[a]Data obtained from NYSDEC (1982). This air quality station is located 55 km W of Upton, NY.

quality standard. On only 11 days of the 117-day growing season did the maximum hourly average ozone concentration exceed 80 nl l^{-1}. Ambient concentrations of sulfur dioxide did not exceed current air quality standards. On only five days of the growing season did the maximum hourly average SO_4 concentration exceed 40 nl l^{-1}. It is assumed that all plants were exposed to the same ozone and sulfur dioxide concentrations.

The rainfall exclusion shelter system used in this study does not exclude ambient levels of gaseous air pollutants. Plots of both soybean cultivars exposed to simulated rains of pH 5.6 had high yields. Williams cultivar has been shown to be very sensitive to ambient ozone based upon productivity in non-filtered air compared with carbon-filtered air (Heggestad, personal communication). The relatively high sensitivity of Williams to ozone, the high yields of both cultivars in the 5.6 pH experiment, and the relatively low ambient ozone concentrations during the experiment suggest that yields of the two soybeans cultivars were not substantially influenced by this gas.

Plants of Amsoy shielded from ambient rainfalls and exposed to simulated rainfalls of pH 4.1, 3.3, and 2.7 exhibited yields 3.0, 9.0, and 12.8%, respectively, below yields of plants to simulated rainfalls of pH 5.6 (Table 5). A treatment-response function of seed yield versus rainfall pH for Amsoy was y = 10.20 + 0.587 x (Figure 1) and had a correlation coefficient of 0.96 (y is seed mass per plant and x is the pH of the simulated rain). For the Williams cultivar, seed yields of plants exposed to simulated rainfalls of pH 5.6, 4.1, 3.3, and 2.7 were 11.5, 10.5, 11.4, and 11.4 g, respectively. A treatment-response function of seed yield versus rainfall pH for Williams was y = 11.13 + 0.016 x which had a 0.038 correlation coefficient.

The decrease in seed yield per plant in Amsoy was due to a decrease in number of pods per plant (Table 5). The decreased number of pods per plant in high rainfall acidity treatments of Amsoy was not found in Williams. Within each cultivar, the seed number per pod and plant population density did not vary among treatments. Mass per seed varied slightly among treatments for both cultivars. Although statistically significant, mass per seed was only 3.5 and 4.3% higher for plants of pH 2.7 compared with lower acidity treatments for Amsoy and Williams, respectively. Mean seed yield of Amsoy plants was 6.0% higher than Williams under shielded conditions even though seeds of Williams had larger mass (Table 5).

Plants grown under ambient rainfall conditions had lower seed yields than plants (mean of all treatments for each cultivar) grown under exclusion shelters (Table 5). These decreases in yield may be caused by the larger than normal amounts of ambient

Table 5. Seed yield characteristics of field-grown soybeans exposed to simulated acid rain during the 1982 experiment

Cultivar and Treatment	Seed Mass per plant (g)	Pod number per plant	Seed number per plant	Seed number per pod	Mass per seed (g)	Plant population density (plants ha^{-1})	Seed yield (q ha^{-1})
Amsoy cultivar							
Rain pH of 5.6	13.3[a]*	26[a]	68[a]	2.7[a]	0.195[a]	348,000[a]	45.91[a]
Rain pH of 4.1	12.9[a,b]	24[a]	65[a,b]	2.7[a]	0.198[a,b]	343,000[a]	44.29[a,b]
Rain pH of 3.3	12.1[b,c]	24[a,b]	62[b,c]	2.6[a]	0.197[a,b]	352,000[a]	42.57[a,b]
Rain pH of 2.7	11.6[c]	22[b]	58[c]	2.7[a]	0.202[b]	358,000[a]	41.39[b]
Williams cultivar							
Rain pH of 5.6	11.5[a]	20[a]	53[a]	2.6[a]	0.217[a,b]	372,000[a]	42.72[a]
Rain pH of 4.1	10.4[a]	19[a]	49[a]	2.6[a]	0.211[a]	378,000[a]	39.06[a]
Rain pH of 3.3	11.4[a]	20[a]	52[a]	2.6[a]	0.218[b]	363,000[a]	41.24[a]
Rain pH of 2.7	11.4[a]	20[a]	52[a]	2.6[a]	0.220[b]	363,000[a]	41.25[a]
Ambient Plots**							
Cultivar							
Amsoy	11.4	23	60	2.6	0.192	352,000	40.04
Williams	9.8	18	47	2.6	0.209	364,000	35.45

*Mean values followed by a different letter with each parameter are significantly ($p \geq 0.05$) different by analysis of variance (Steel and Torrie, 1960). Values followed by the same letters are not significantly different. For analysis of variance comparisons are made only among treatments within each cultivar.

**Statistical comparisons were not made between cultivars within ambient plots or between exclusion plots and ambient plots.

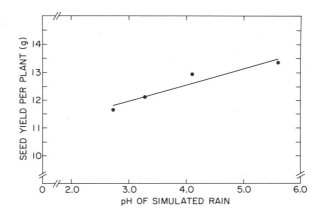

Figure 1.   Relationship between soybean (Amsoy)
seed yield and pH of simulated rainfalls of plants shielded from ambient
rainfalls. By least squares analyses,
the equation fits the relationship as
expressed by y = 10.20 + 0.587 x, where
y is seed mass per plant and x is the
pH of the simulated rain. The linear
least squares equation has a 0.96
correlation coefficient

rainfall that occurred during the experimental period (Table 2).
From visual observations, plants in the ambient rainfall plots had
slightly retarded growth after the 1735 min rainfall that provided
19 cm of water early in the season (several weeks after planting).
However, retarded growth was not visually evident by maturity.
The decreased seed yields in the ambient rainfall plots was
attributed to a decrease in pods per plant since the number of
seeds per pod and mass per seed were both similar to plants grown
under exclusion shelters (within both cultivars). As shown in
Figures 2 and 3, durations and amounts of ambient rainfalls were
usually shorter and smaller than the simulated rainfalls applied.

Dew formation occurred on all test plants. During the
106 day period of the growing season, dew formation occurred for
at least two hours on 38 separate occasions for a total of
236 hours in the absence of rainfalls (Figure 4). Although the
mean duration of dew formation was 6.2 hours, the mode was
2 hours. These data did not consider occasions when conditions
for dew formation were present for less than two consecutive
hours.

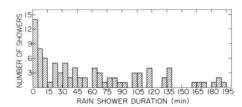

Figure 2.   Frequency distribution of ambient
             rainfall shower durations from
             period of 4 June through 27 Sep-
             tember 1982 at Upton, New York.
             A shower is defined as a precipita-
             tion event in which no additional
             precipitation occurs within 60 min
             before or after the event. Other
             rainfall showers of 160, 180, 186,
             202, 206, 207, 210, 215, 240, 244,
             262, 267, 295, 305, 396, 432, 450,
             516, 576, 1028, and 1735 min were
             not plotted. These higher duration
             rainfalls accounted for 18.3% of the
             total number recorded

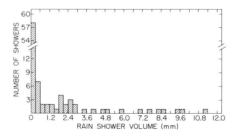

Figure 3.   Frequency distribution of rainfall
             shower amounts from the period of
             4 June through 27 September 1982 at
             Upton, New York. The term "shower"
             is defined in Figure 2. Other rain-
             fall showers of 42.9 and 189.5 mm
             were not plotted

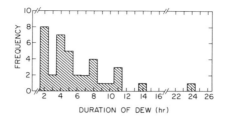

Figure 4.   Frequency distribution of the
              duration of dew formation
              present for at least two hours
              during the period of 12 June
              through 26 September 1982.
              Parameters of the frequency
              distribution are: mean =
              6.2 hours; mode = 2 hours;
              median = 5 hours; total number
              of events = 38; total duration
              of all events = 236 hours

DISCUSSION

     These experimental results show that seed yield of soybeans
of Amsoy cultivar is reduced by exposure to increasingly acidic
simulated rainfalls.   These seed yield reductions resulted from
decrease in the number of pods per plant and are reflected in both
the number of seeds per plant and mass of seeds per plant. Since
the number of seeds per pod and the mass of individual seeds per
plant did not vary among treatments within both experiments, the
decrease in soybean seed yield was attributed to a decrease in the
number of mature pods.   A decrease in number of pods per plant
may result from either a decrease in flower pollination (and
fertilization), a decrease in pod retention, or an inadequate
development of young pods. A decrease in pod development and/or
retention was shown in previous experiments at Brookhaven National
Laboratory in 1978, 1979, and 1981 (Evans et al., 1981a, 1983).

     Qualitative observations on the conditions of plants were
recorded. Foliar injury was observed on young plants shielded
from ambient rainfalls and exposed to simulated rainfalls of
pH 2.7 or 3.3. Similar plants exposed to simulated rainfalls of
pH 4.1 and 5.7 exhibited no visible foliar injury. Visible injury
from the simulated rainfalls was similar to injury to soybeans
grown under growth chamber conditions (Evans and Curry, 1979).
Injury was not observed on such plants after the third true leaf
stage. These results were comparable to those obtained in similar
experiments (Evans et al., 1981a, 1983). No plants in the ambient
rainfall plots exhibited visible foliar injury.

Bean yields were consistent between the 1981 and 1982 growing seasons. When plants of Amsoy were exposed to simulated rainfalls of pH 5.6 under exclusion conditions (1981), seed yields averaged 13.1 g per plant (46.6 q ha^{-1}; population density = 358,000 plants per hectare; Evans et al., 1983). This consistency was also exhibited in plants exposed to ambient precipitation. In 1982, such plots averaged 11.4 g seed per plant, 60 seeds per plant, and 40.0 q ha^{-1} with a plant population of 352,000 plants per hectare. Similar values during the 1981 growing season were 11.7 g per plant, 63.6 seeds per plant, 44.6 q ha^{-1} with a plant population of 383,000 plants per hectare (Evans et al., 1983). These results suggest that year to year variations in plant productivity are slight at the field site when environmental factors are not markedly different.

Yields of soybean plants exposed only to ambient rainfalls may be affected by various shower characteristics. Forty-six and 68 percent of the measured showers had durations less than 10 and 60 min, respectively; 22% of all the showers were longer than 120 min (Figure 2). Fifty-nine and 86% of all showers had amounts less than 0.25 and 3.0 mm, respectively. Moreover, only 7.8% of all showers had amounts over 10 mm (Figure 3). These results show that small showers occurred frequently (mean frequency of one shower per day) but these showers were short in duration with small amounts of total precipitation. In this manner, plants exposed to ambient rainfalls were exposed to different wetting conditions compared with plants exposed to simulated rainfalls under the exclusion shelters.

The rainfall exclusion procedures used in this experiment and previously (Evans et al., 1983) provide an experimental procedure in which crops can be grown with standard agronomic conditions. The rainfall exclusion shelters move over the crop only when ambient rainfalls occur as well as when simulated rainfalls are applied. This experimental procedure minimizes changes in the plant's microclimate. Since the operation of the facilities at Upton, NY, similar rainfall exclusion shelters have been constructed at other research sites in North America.

Our soybean seed yields can be compared with those of crops grown in other areas of the United States. For example, during 1981 the two states with the highest total production (Iowa and Illinois), averaged yields of 2700 and 2600 kg ha^{-1}, respectively (Crop Reporting Board, 1982). Highest county yields in these two states were 3000 kg ha^{-1} (USDA, 1980). Our field plots exposed to ambient rainfalls with supplemental irrigation yielded 35-40 q ha^{-1}, which is comparable to seed yields of the highest yielding counties in Illinois and Iowa.

---

*q ha^{-1} = quintals (100 kg) per hectare

The experimental results obtained at Brookhaven National Laboratory over several growing seasons (e.g., 1978, 1979, 1981, and 1982) indicate that acidic rainfalls which occur during the growing season may decrease soybean seed yields under field conditions. The results also indicate that experiments to accurately determine impacts of acidic rain on growth and yield of field crops must be performed under standard agronomic practices with a large number of plots per treatment, adequate randomization of treatment plots to counteract local soil variations, and appropriate statistical analyses of data.

CONCLUSION

Soybeans were grown under an ambient rainfall exclusion system; results reported in this chapter were for the third growing season. The exclusion system is automated so that two modified plastic covered greenhouses (each 30 m x 10 m) move over the crop within one min after exposure to a single raindrop and are removed after the rain ceases. Data show that cultivars of field-grown soybeans may differ in their responses to simulated acidic rain. Amsoy was shown to be sensitive while Williams was insensitive. Seed yields were dependent upon the number of pods per plant since the number of seeds per pod did not vary among different treatments of simulated acidic rain.

Acknowledgements - This research was supported by the United States Environmental Protection Agency, Office of Research and Development, under IAG 81-DX0533, and by United States Department of Energy, under Contract No. DE-AC02-76CH00016 and in part by Associated Universities, Inc., under Contract No. 550925-S. The authors are grateful for the excellent assistance of Keith Thompson with statistical analyses. Materials that pertain to the chemical composition of rainfall during the summer of 1981 were provided by Gilbert Raynor (Dept. of Energy and Environment, Brookhaven National Laboratory). By acceptance of this article, the publisher and/or recipient acknowledges the U.S. Government's right to retain a non-exclusive, royalty-free license in and to any copyright covering this paper. Although the research described in this article has been funded wholly or in part by the United States Environmental Protection Agency through Contract IAG 81-D-X0533 to Brookhaven National Laboratory, it has not been subjected to the Agency's required peer and policy review and therefore does not necessarily reflect the views of the Agency. No official endorsement should be inferred.

REFERENCES

Crop Reporting Board, 1981a, Crop Production: 1980 Annual
    Summary. CrPr 2-1 (81), Economics, Statistics and Cooper-
    ative Services, United States Department of Agriculture,
    Washington, DC. 26 pp.
Crop Reporting Board, 1981b, Field Crops: Production, Disposi-
    tion, Value--1979-1980. CrPr 1 (81), Economics, Statistics
    and Cooperative Services, United States Department of
    Agriculture, Washington, DC. 35 pp.
Crop Reporting Board, 1982, Crop Production, 1981. Annual
    Summary CrPr 2-1 (82), Statistical Reporting Service, United
    States Department of Agriculture, Washington, DC. 36 pp.
Evans, L.S., 1980, Foliar responses that may determine plant
    injury by simulated acid rain, pp. 239 to 257. In:
    Polluted Rain, T. Y. Toribara, M. W. Miller, and
    P. E. Morrow, eds., Twelfth Annual Rochester Intern. Conf.
    Envir. Toxicity, Plenum Publ. Corp., New York, NY.
Evans, L.S. and Conway, C.A., 1980, Effects of acidic solutions
    on sexual reproduction of Pteridium-aquilinum. Am. J. Bot.
    67:866-875.
Evans, L.S. and Curry, T.M., 1979, Differential responses of
    plant foliage to simulated acid rain. Am. J. Bot.
    66:953-962.
Evans, L.S. and Hendrey, G.R., 1980, Effects of acid precipita-
    tion on vegetation, soils, and terrestrial ecosystems.
    Report from the Intern. Workshop at Brookhaven National
    Laboratory. Report No. 51195, Brookhaven National Labora-
    tory, Upton, NY. 48 pp.
Evans, L.S. and Lewin, K.F., 1981, Growth, development, and yield
    responses of pinto beans and soybeans to hydrogen ion
    concentrations of simulated acid rain. Envir. Exper. Bot.
    21:103-133.
Evans, L.S., Conway, C.A., and Lewin, K.F., 1980, Yield responses
    of field-grown soybeans exposed to simulated acid rain,
    pp. 162 to 163. In: Ecological Impact of Acid
    Precipitation, D. Drabløs and A. Tollan, eds., Proceedings
    International Conference, Sandefjord, Norway, SNSF
    Project, As-NLH, Norway.
Evans, L.S., Lewin, K.F., Conway, C.A., and Patti, M.J., 1981a,
    Seed yield (quantity and quality) of field-growth soybeans
    exposed to simulated acidic rain. New Phytologist
    89:459-470.
Evans, L.S., Lewin, K.F., Cunningham, E.A., and Patti, M.J.,
    1982, Effects of simulated acidic rain on field-grown
    crops. New Phytologist 91:429-441.
Evans, L.S., Lewin, K.F., Cunningham, E.A., and Patti, M.J.,
    1983, Productivity of field-grown soybeans exposed to
    simulated acidic rain. New Phytologist 93:377-388.

Evans, L.S., Hendrey, G.R., Stensland, G.J., Johnson, D.W., and
    Francis, A.J., 1981b, Acidic precipitation considerations of
    an air quality standard. Water, Air, Soil Pollu.
    16:469-509.

Fisher, R.A. and Yates, F., 1957, Statistical Tables for Biolog-
    ical, Agricultural and Medical Research. Hafner Publ. Co.,
    New York, NY. 138 pp.

Goldberg, D., Gornat, B., and Rinon, D., 1976, Drip irrigation:
    Principles, design, and agricultural practices. Drip
    Irrigation Scientific Publications, Kar Shmaryahu, Israel.
    296 pp.

Haggestad, H., personal communication, U.S. Dept. of Agriculture
    Agr. Research Service, Beltsville, MD.

Galloway, J.N., Likens, G.E., and Edgerton, E.S., 1978, Acid
    precipitation in the northeastern United States: pH and
    acidity. Science 194:722.

Johnson, A., personal communication, Geology Dept., Univ.
    Pennsylvania, Philadelphia, PA.

Nagle, C.M., 1975, Climatology of Brookhaven National Laboratory,
    1949 through 1973. Report No. 50466, Brookhaven National
    Laboratory, Upton, NY. 53 pp.

NYSDEC, 1982, Air Quality Surveillance-Summary Report, Eisenhower
    Park (2950-10). Period covered: 1 June 1982 through
    30 September 1982, Monthly Reports. New York State Dept. of
    Environmental Conservation, Albany, NY.

Raynor, G.S., 1955, The rotary rain indicator, an electrical
    precipitation time recorder. Bull. Am. Meteorological Soc.
    36:27-30.

Raynor, G.S. and Hayes, J.V., 1981, Acidity and conductivity of
    precipitation on central Long Island. Water, Air, Soil
    Pollu. 15:229-245.

Raynor, G.S. and McNeil, J.V., 1978, An automatic sequential
    precipitation sampler. Atmos. Envir. 13:149-155.

Ryder, G.J. and Beuerlein, J.E., 1978, A systems approach for
    soybean production. Agronomy Mimeo 221, Cooperative Exten-
    sion Service, The Ohio State University, Columbus, OH.
    14 pp.

Sandsted, R.W., Muka, A.A., Sherf, A.F., Siezcka, J.B., Sweet,
    R.D., and Tingey, W., 1980, Cornell recommendations for
    commercial vegetable production. Cornell Univ., Ithaca, NY.
    51 pp.

Steel, R. and Torrie, J., 1960, Principles and Procedures of
    Statistics. McGraw-Hill, New York, NY. 481 pp.

USDA, 1980, 1972-1980 County Estimates. Economics Statistics and
    Cooperative Service, United States Department of Agricul-
    ture, Washington, DC. n.p.

Warner, J.W., Hanna, W.E., Landry, R.J., Wulforst J.P., Neely,
    J.A., Holmes, R.L., and Rice, C.E., 1975, Soil Survey of
    Suffolk County, New York. Soil Conservation Service, United
    States Department of Agriculture, Washington, DC. 101 pp.

ACID DEPOSITION: A DECISION FRAMEWORK THAT INCLUDES UNCERTAINTY

D. Warner North, William E. Balson, and Dean W. Boyd

Decision Focus Incorporated
4984 El Camino Real, Suite 200
Los Altos, CA   94022

ABSTRACT

This chapter presents a framework for the analysis of decisions on acid deposition; in particular, the decision to impose additional controls on sulfur oxide emissions. The decision framework is intended as a means of summarizing scientific information and uncertainties on the relation between emissions from electric utilities and other sources, acid deposition, and impacts to ecological systems. The methodology for implementing the framework is that of decision analysis, which provides a quantitative means of analyzing decisions under uncertainty. The chapter gives an overview of the decision framework and explains the decision analysis methods with a simplified caricature example. Implementation of the framework has been accomplished in the form of the ADEPT model, available through the Electric Power Research Institute (EPRI).

INTRODUCTION

Acid deposition is a public policy issue characterized by great uncertainty and complexity. This chapter presents an overview of a framework for examining public policy and research strategy decisions on acid deposition. The framework is based on decision analysis, which provides a formal theory for choosing among alternatives whose consequences are uncertain. The key idea in decision analysis is the use of judgmental probability as a general way to quantify uncertainty. Decision analysis has been widely taught and practiced in the business community for several

341

decades (Raiffa, 1968; Brown et al., 1974; Holloway, 1979). It provides a natural way to extend cost-benefit analysis to include uncertainty.

As the debate on acid deposition policy has intensified, the need for an integrating framework for balancing the potential environmental effects with the costs of emission control has grown. Industry and government are faced with the immediate decision of whether to (1) impose additional controls on power plants and other sources, (2) take steps to mitigate the possible effects of acid deposition, or (3) wait until additional understanding can be achieved on the relationship between emissions and ecological effects. The choice involves the careful balancing of very different kinds of risks. Acting now to reduce emissions carries the risk that large expenditures will be made with little or no beneficial effect, while waiting carries the risk that significant ecological damage will be incurred that could have been prevented by prompt action.

If the results of the extensive research programs underway in the United States, Canada, and Europe were available today, the choice might be less difficult. But unfortunately, resolution of critical uncertainties may not occur for five to ten years or longer. Until that time, it will not be possible to predict accurately how changes in emissions will affect the extent of ecological damage from acid deposition. In the absence of perfect foresight, what is needed is a means of reasoning about the best decision based on the information available today.

The objective of the research described in this paper is the development of a framework to summarize current information and uncertainties on acid deposition. A more extensive discussion of the decision framework and the decision analysis methodology on which the framework is based may be found in Balson et al. (1982). The framework is intended to aid decision makers in evaluating strategies for control of anthropogenic emissions and for mitigating the effects of acid deposition. The framework is also intended to aid in the evaluation of research programs for organizations such as Electric Power Research Institute (EPRI) and the U.S. Government, who are spending substantial funds to develop better information as a basis for future decisions.

OVERVIEW OF THE DECISION FRAMEWORK

To understand the effects of alternative control strategies, it is necessary to understand the relation that various levels of emission reduction may have on the impacts of acid deposition. The potential changes in impact must then be balanced with the cost of achieving emission reductions. The comparison of various control strategies is made difficult by several factors:

° There is a large degree of uncertainty about the relationship between emissions and impacts.

° It is difficult to compare the value of changes in
   impacts to the costs of emission reductions.

° People involved in assessing control and mitigation
   strategies have different opinions about the evaluation
   of costs and impacts and may differ in their judgments
   of the uncertainty in costs and impacts.

° The uncertainty in the relationship between emissions
   and impacts will be resolved only over a lengthy period
   of time.

The decision framework is designed to allow explicit treatment
of each of these factors, separating the evaluation of costs and
impacts from consideration of the resolution of uncertainty over
time. In doing this, the framework provides a vehicle for discussion and investigation of sensitive assumptions, which can lead to
the building of consensus.

Three stages can be distinguished in the relationship between
control alternatives and impacts as shown in Figure 1. First,
there is the effect that control strategies will have on emissions.
Then, changes in emissions must be related to changes in acid
deposition. Finally, changes in acid deposition must be related to
changes in the various impacts that can be identified, such as
decreased forest productivity and the loss of sport fisheries.
There is scientific uncertainty about each of these stages.
Relatively little is known about how specific changes in acid
deposition will affect changes in impacts. The range of estimates
given by respected scientists varies over a wide range. There is
somewhat less uncertainty regarding how changes in emissions will
affect changes in deposition; however, the range of uncertainty is
still quite large due primarily to the complex nature of the
chemical transformations that occur in the atmosphere. There is
comparatively little uncertainty about how reduction strategies
would affect changes in emissions. Accordingly, in implementing
the framework, the importance of uncertainty in the other two
stages (emissions to deposition and deposition to impacts) has been
stressed.

At present, the scientific evidence regarding the effects of
emissions is contradictory and subject to different interpretations
by various experts. The decision framework allows for an investigation of the implications of the differing assessments and
evaluates the importance of the disagreements in terms of their

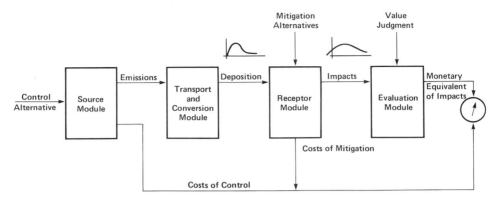

Figure 1. Overview of decision framework

effects on the choice of a control or mitigation strategy. Many experts who disagree about the interpretation of the current state of knowledge, agree that in five to ten years many of those disagreements will be settled (e.g., Interagency Task Force on Acid Deposition, 1982). Thus, in the decision framework, the choice is characterized as one in which we may act now, at a large cost, and accept the possibility that emission reduction will have little beneficial impact. Alternatively, we can wait five to ten years to act on better information that may be available, and accept the possibility that damages may occur during that period. In each case, there is a possibility that the decision will turn out to have been incorrect. From our current state of knowledge, we cannot be sure.

The strategies that are available and the resolution of uncertainty at different points in time are represented as a decision tree (Figure 2). A decision tree is simply an efficient way of describing a set of scenarios. Each particular set of decisions and outcomes representing how uncertainty could be resolved comprises a scenario. Each scenario answers a "what if?" question, corresponding to what if a particular strategy were chosen followed by a particular change in deposition and finally by a particular change in impacts.

The decision tree of Figure 2 provides a generic representation of the time sequence of choices among decision alternatives and the resolution of uncertainty in the areas enumerated in the framework of Figure 1. The first two stages, shown at the far left of the figure, are the decisions within the next few years on control and mitigation options and on a national research program on acid deposition. The next two stages represent resolution of uncertainty on the relation of deposition to emissions and the

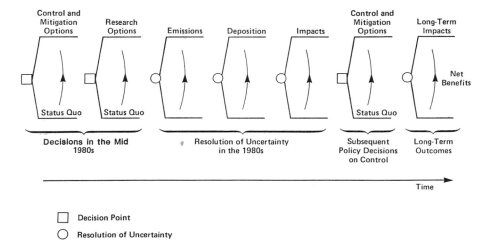

Figure 2.   Decision tree for acid deposition policy

relation of impacts to deposition as the research program is carried out and new scientific knowledge is obtained. Next comes a decision point in the late 1980's or early 1990's when national policy on control and mitigation would be reassessed and an alternative chosen on the basis of the new information that has recently been made available. Further resolution of uncertainty on deposition and on impacts of acid deposition then follows.

The decision tree of Figure 2 provides a rich sequence of scenarios describing the decisions and outcomes characterizing national policy on acid rain. It includes two stages of decision making, one with present information and one with the information that might become available five to ten years hence following an extensive research program. The decision tree explicitly includes the option of taking action now to control emissions or mitigate the effects of acid deposition and the option to wait until better information becomes available in five to ten years. The effect of today's research funding decisions and the choice of emphasis in the research program may strongly affect what information becomes available in the next five to ten years, and this interaction is explicitly considered in the decision tree framework.

The decision tree approach provides a useful separation between value judgments on costs and benefits and judgment about uncertainties in the impacts of acid deposition. Each scenario in the decision tree may be considered as having impacts on a number of concerned parties: consumers who may have to pay more for electricity because of decisions to impose controls on power

plants, fishermen and recreational property owners who stand to
lose if sport fishing in a given lake is degraded by acid deposition, forest products firms and property owners who suffer
economic losses if forest productivity is reduced, and members of
the general public who are concerned about possible ecological
changes from acid deposition. The evaluation of impacts on these
diverse parties is difficult because people see that some parties
bear more of the costs while other parties receive more of the
benefits resulting from a particular decision alternative. People
in Ohio benefit from cheaper electricity because their power plants
burn coal with a higher level of sulfur emissions than would be
allowed in many other eastern states. People in New York may
benefit from reduction in Ohio River Valley sulfur emissions if the
reduction improves the fishing in Adirondack lakes. The political
reality is that government officials must evaluate how tradeoffs
will be made between the costs that one group bears and the benefits that another group receives. Issues of equity and property
rights make such value judgments extremely difficult. It is useful
to separate these value judgments from the uncertainty in the
effects that long-range transport of sulfur and other pollutants
may cause. The decision framework accomplishes this desired
separation between the answer to the question of what will happen
under a given choice of control and mitigation strategies and the
societal evaluation of what each outcome is worth.

A DECISION TREE CARICATURE

     A decision tree of the complexity of Figure 2 can include tens
of thousands of paths or scenarios defined by different combinations of decisions and outcomes at each stage. Carrying out
calculations on decision trees of this size usually involves the
use of digital computers, and our analysis has in fact been implemented in this fashion as a FORTRAN code called ADEPT (Acid DEPosition decision Tree program, which is available from EPRI). The
form of these calculations can be illustrated with the highly
simplified decision tree shown in Figure 3, which has only six
paths or scenarios. This decision tree can be regarded as a
caricature of the complex tree shown in Figure 2. Almost all of
the structure of the decision as described previously has been
eliminated, leaving only two choices: 1) action to further reduce
emissions of sulfur oxides from current levels, and 2) no additional control on sulfur oxides or other emissions. Uncertainty is
represented on the extent to which the emission reduction lowers
acid deposition and on the relationship between acid deposition and
the extent of long-term ecological impacts, which could be minor or
large.

     To some extent, this simplified decision tree illustrates the
current polarized debate about acid deposition. Proponents of

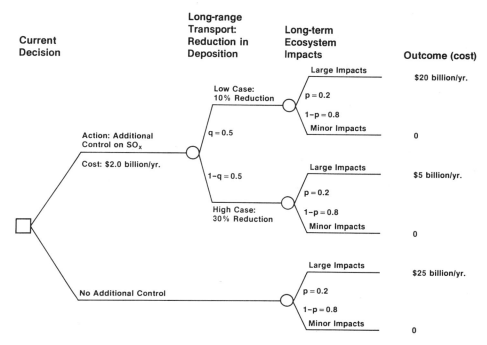

Figure 3.   A simplified decision tree for acid deposition
            control policy

control legislation argue that extensive reduction in sulfur oxide
emissions can only be accomplished if action is taken now, and that
acid deposition resulting from continued emissions could cause far
more extensive damage to lakes, streams, and forests than the
effects observed to date (e.g., National Research Council, 1981).
Their adversaries argue that the links between sulfur emissions,
acid deposition, and ecological damage have not been clearly
established, and that emission reduction would incur large costs
and might achieve little benefit in avoiding damage from the
impacts of acid deposition (e.g., Ruckelshaus, 1984).

These arguments have been represented in the form of six
scenarios, four proceeding from an additional control alternative
and two from an alternative of no additional control.   For the
additional control alternative, assume that a 50-percent reduction
in sulfur emissions from coal-fired power plants in the Ohio River
Valley can be achieved with fuel switching and flue gas desulfurization at a cost of $2 billion per year (in constant dollars;

Balson et al., 1982). Two outcomes are assumed for the reduction in acid deposition in sensitive downwind receptor areas such as the Adirondacks if sulfur emissions are reduced 50 percent by Ohio River Valley power plants. Averaged over the year, acid deposition in the Adirondacks corresponds to about 60 percent sulfate, 30 percent nitrate, and 10 percent other species (Balson et al., op. cit.). If it is assumed that virtually all of the sulfate in the Adirondacks comes from power plants burning coal in the Ohio River Valley and that the relationship between emissions and deposition is linear, then a reduction in sulfur emissions from these sources would lead to a reduction in acid deposition of about 30 percent. However, if local sources of sulfur oxides play an important role or if the emissions-deposition relationship is nonlinear, then the reduction in emissions from Ohio River Valley power plants could result in a much smaller change in the level of acid deposition in the Adirondacks, for example, 10 percent. For illustration, only these two possibilities are assumed, and each is judged to have a probability of 50 percent of being correct.

It is likewise assumed that there are only two outcomes for ecosystem impacts: large impacts and minor impacts. The "minor impacts" outcome implies that the damage to surface waters, soils, forests, and material property will be no worse than the effects that have been observed to date. The "large impacts" outcome assumes that continued high levels of acid deposition will result in damage that is far worse, causing loss of sport fisheries, reduced productivity of forests, and other adverse impacts across much of the northeastern United States and eastern Canada. For this illustrative calculation, the "large impacts" scenario has been assigned a probability of 20 percent, while a probability of 80 percent has been assigned to the "minor impacts" scenario.

It is also assumed that each of the ecosystem impact scenarios can be valued in monetary terms. Conceptually, this monetary value is what decision makers representing the United States would be willing to pay to avoid the ecological damage under each of the scenarios. For simplicity and convenience, the monetary equivalent for the three minor damage scenarios is taken to be $0. For the large impacts, a level of $25 billion per year (in constant dollars) is assumed for the case of no reduction in deposition. In the ADEPT model, such values are calculated based on estimates over time of the extent of forest and surface water acreage damaged by acid deposition. If deposition is reduced by 30 percent, it is assumed that the extent of the damage is reduced more than proportionally so that the equivalent monetary damage is reduced to $5 billion annually. If deposition is reduced only 10 percent, the assumed monetary equivalent of the impact is reduced from $25 billion to $20 Billion.

Once the scenarios resulting have been described in terms of probabilities and values, the decision tree can be used to obtain valuable insights in comparing the decision alternatives.   As a means of evaluating the alternatives, the sum of the control costs plus the expected or probability weighted average of damages are compared.   The expected damage is computed by multiplying the probability of each outcome by the monetary damage if that outcome occurs.   Therefore, the expected damage corresponding to the no additional control alternative is $5 billion per year, which is 20 percent times $25 billion plus 80 percent times $0.   The expected damage for the additional control alternative is computed by multiplying the probability times the monetary value for four cases and summing:

$$(0.5)(0.2)($20) + (0.5)(0.2)($5) + 2(0.5)(0.8)($0)$$
$$= $2.5 \text{ billion per year.}$$

A cost of $2 billion per year for control is added for the additional control alternative:

$$$2.5 + $2 = $4.5 \text{ billion per year.}$$

The least total annual cost is therefore obtained by choosing additional control with an expected total cost of $4.5 billion compared to the $5 billion expected damage with the no additional control alternative.   We can see from this evaluation that the total costs for the two alternatives are close.   Reducing the probability of the large impact scenario from 0.2 to 0.15 would make the no additional control alternative have the lower total expected cost.[1]

    An important concept in decision analysis for evaluating research and other information gathering activities is the expected value of information.   Suppose it is possible to resolve which impact case and which long-range transport case were true before choosing between control and no additional control.   How much would it be worth to gain this information before making the choice? This question can be answered with the probabilities and values in the decision tree of Figure 3 by making another expected value calculation.   This calculation corresponds to reversing the order of the stages in the tree: the expected value is computed assuming that for each possible outcome   on ecosystem impacts, the decision

---

[1]The calculation of the expected value for each alternative is made in the same way as described in the above paragraph, but now using 0.15 instead of 0.20.   The expected damage for the no additional control alternative is (0.15)($25) + (0.85)(0) = $3.75 billion, and the expected damage for the additional control alternative is determined using the first calculation above with 0.15 and 0.85 replacing 0.2 and 0.8 in the terms on the left hand side, yielding a total of $3.875 billion.

alternative will be chosen with the lowest total of control costs
plus damage. If large impacts will occur, the least costly choice
is additional control. If minor impacts will occur, the least
costly choice is no additional control. The probabilities were
given above: 0.2 for large impacts and 0.8 for minor impacts.
Recall that in the absence of information, the least costly
decision was additional control. If the information indicates
large impacts, additional control will remain the least costly
decision alternative. But if the information indicates minor
impacts will occur, no additional control will be the least costly

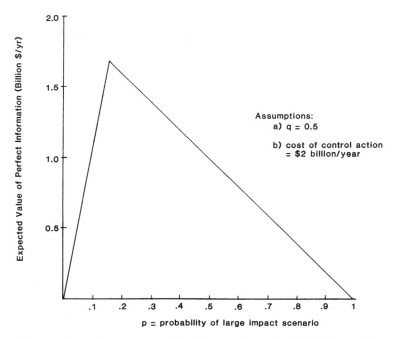

Figure 4. The expected value of perfect information

alternative, and a savings of $2 billion in control cost can be
achieved. The expected (i.e., probability weighted average)
savings from making the decision after the new information is
available is then 0.8 x $2.0 = $1.6 billion per year. The expected
value of perfect information (the expected savings if the decision
could be made knowing which impact case is correct) as a function
of the probability p of large ecosystem impacts is given in
Figure 4; the maximum for the expected value of perfect information
occurs at p = 0.16, the point at which control and no additional
control yield equal total expected costs.

Expanding the Caricature: Act Versus Wait

The decision just examined was a choice between action, additional control on sulfur oxide emissions, and no additional control. A more accurate characterization of the decision facing policy makers is a choice between acting now and waiting. The no additional control alternative is replaced by a wait alternative, where it is assumed that much better information will become available after ten years. The value judgments characterizing scenarios now become more complex, because the incremental damage that occurs by waiting ten years before implementing additional control of emissions as opposed to acting now must be considered. The damage outcome for the scenario in which additional control is implemented after ten years will be less than the damage under no additional control, and more than the damage if additional control is implemented immediately.

Interpreting the Caricature

The caricature decision tree portrays the reason for taking action as the threat of uncertain, but potentially very serious adverse ecological consequences may occur if high levels of acid deposition continue into the future. The damages that have occurred to date may be most important as an indicator, similar to the canary whose death warns miners that the air in the mine can no longer support them.

The choice of whether to take action now to reduce emissions is therefore similar to a decision to buy insurance against a possible future disaster. The large impact outcome is uncertain; in the analysis its likelihood has been described by a probability that summarizes scientific judgment. Establishing this probability will be difficult because scientists differ in their interpretations of the information now available. However, it may be possible to obtain general agreement that the probability lies within a given range, and within this range it may be clear whether purchasing the insurance is or is not a good idea.

If the decision were between additional control and no additional control, it might be expected that the decision would be very sensitive to the probability of the large impact outcome. But the opportunity exists to impose controls at a later time. This time has been rather arbitrarily taken as ten years, but the assumed time can be varied for the future decision point to see if this assumption makes a difference. The consequence of not taking action now is that the ecological effects under the large impact scenario may be made worse by waiting. However, if the wait decision is taken, it might be learned that the large impact scenario is less likely, or even extremely unlikely to happen. It is also possible that alternate less costly ways to reduce emissions or mitigate the impacts of acid deposition may be found.

THE DECISION TREES USED IN THE ADEPT MODEL

    The generic decision tree of Figure 2 is again evaluated to
determine procedures for constructing an implementation of the
decision framework to address the questions discussed above.   There
are various assumptions regarding how fast uncertainty will be
resolved on ecological impacts and long-range transport.   One
limiting assumption is that no resolution of uncertainty will take
place before the second decision point.   The decision tree then has
the form shown in Figure 5.   A major difference between the second
decision point and the first could be the availability of a new
control technology such as the lime injection multistage burner
(LIMB: Balson et al., 1982) that is more effective or less costly
than the technologies now available.

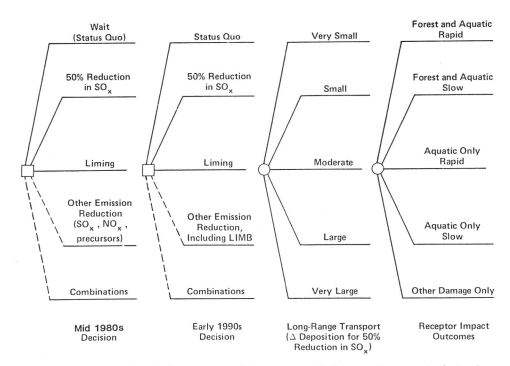

Figure 5.   A decision tree with no resolution of uncertainty by
            the early 1990's

An assumption at the other extreme is that uncertainties on ecological impacts and long-range transport will be essentially resolved within five to ten years, so that the second decision can be assumed to take place under certainty, or with perfect information.[1] This decision tree compares the ecological consequences of waiting before taking action with the value of perfect but delayed information. There is widespread agreement within the scientific community that much will be learned in the next five to ten years regarding the ecological consequences of acid deposition and the long-range transport relationships between emissions at a source region and deposition in a receptor region. It seems that the policy debate is whether the country can afford to wait five or ten years before taking action. Therefore, the decision tree shown in Figure 6 was chosen as the basic decision tree in ADEPT for examining control and mitigation alternatives. The tree is essentially the same form as the expanded caricature described above, except that more decision alternatives have been added. Five, rather than two, cases have been used to represent long-range transport uncertainty; and five, rather than two, cases have been used to represent uncertainty on ecological impacts - a total of 25 combinations instead of four.

A more complex version of the generic decision tree of Figure 2 is given in Figure 7, which incorporates the decision on alternative research programs and the research results obtained from these programs. In this research-emphasis decision tree, it is assumed that better information about long-range transport and ecological effects will become available within ten years. Uncertainty is thus resolved in two stages, a portion in the first time period before the second decision point; the remainder would be resolved afterwards. Larger research efforts give a higher expectation of resolving the uncertainty early, but for all the research options, the research results achieved may be inconclusive or wrong. The research emphasis decision tree in ADEPT permits the evaluation of research programs that should provide a better basis for future decisions on control and mitigation. The research emphasis decision tree is substantially more complex than the basic tree, but the calculations are still easily made on a computer. With the branches represented by solid lines in Figure 7, there are

---

[1]While complete resolution of uncertainty or perfect information is a limiting case, it corresponds practically to the situation where new research results have reduced the uncertainty sufficiently so that the choice among alternatives is clear. Analysis of the value of perfect information is a simple calculation that can be used to guide the evaluation of real, imperfect information-gathering alternatives (Raiffa, 1968; Brown et al., 1974; Holloway, 1979; Balson et al., 1982). The more complex decision tree of Figure 7 described below is used for analysis of imperfect information gathering activities (Balson et al., 1982).

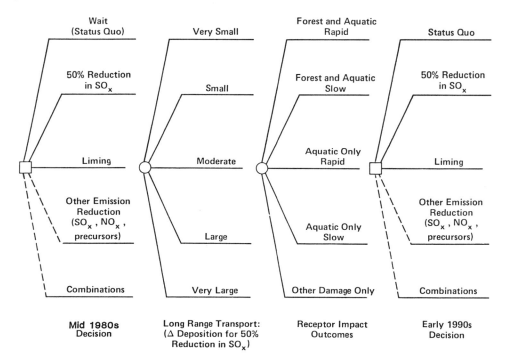

Figure 6.   A decision tree with full resolution of uncertainty
            by the early 1990's

over six thousand scenarios in this tree. The development of
consistent probabilities for the research results and the outcomes
on long-range transport and ecological effects requires a considerable amount of thought about the effect of specific research
findings on the state of scientific knowledge (Balson et al.,
1982).

    Use of the decision framework relies heavily on the assessment
of judgmental probabilities. The ADEPT decision tree models
require that such probabilities be provided as input data. The
assessment process is a difficult and subtle art, especially as in
this application where the judgments concern issues of great
complexity and cut across many scientific specialties (Spetzler and
Stael von Holstein, 1975; Kahneman et al., 1982; Wallsten and
Budescu, 1983). The key to success in using judgmental probability
is the credibility of the analysis process. The expert whose
judgment is being assessed must understand the assessment process
and the way in which his or her judgment is being used. This
requirement implies that substantial time will be needed for
communication between analysts responsible for assessing probabilities and the experts whose judgment is being sought.

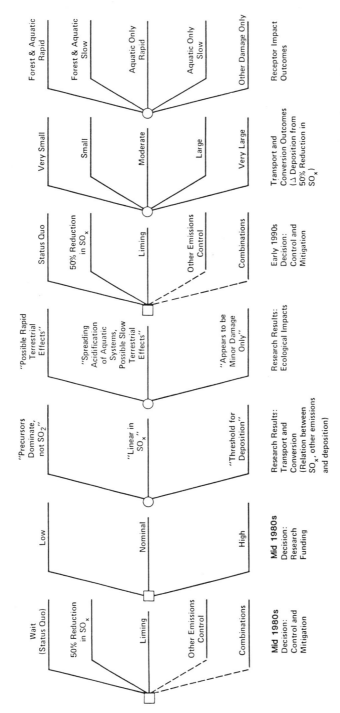

Figure 7. A decision tree with resolution of uncertainity dependent on research funding decisions

The other sensitive aspect for both the basic tree and the research emphasis tree is the characterization of acid deposition impact outcomes. The assumptions and models used in ADEPT are highly modular and easily understood graphically (Balson et al., 1982; Balson and North, 1983). The reduction in emissions resulting from a specific control strategy is phased in over time. Various linear and nonlinear assumptions about the change in deposition which result from a given change in emissions can be utilized. The time pattern of surface water and forest acidification that occurs under the different ecological impact scenarios can be varied. Each of these relationships can be changed within a wide range of possibilities, and each is modular within the ADEPT code so that it can be replaced by an entirely different set of assumptions relatively easily. The monetary equivalents per damaged acre of surface water and forest are important and potentially controversial, but the analysis can easily be repeated over a wide range of value judgments to show the implications of differing opinions for selecting the alternative with the least expected total cost (Balson and North, 1983).

CONCLUSION

In the two years since the acid deposition decision framework and the ADEPT computer implementation were developed, the framework has received considerable attention and acceptance. It has been used within EPRI in planning research program strategy for illustrative calculations of the costs and benefits to individual states for acid deposition control policies (Balson and North, 1983). The ADEPT model has also been used by agencies of the U.S. Government including the Department of Energy and the Environmental Protection Agency.

The fundamental conflicts and differences in perception regarding acid deposition will not be resolved easily, and any attempt at consensus building will be regarded with suspicion by many in the scientific and the policy communities. Nonetheless, it is believed that the decision framework and the ADEPT model offer great potential for consensus building among the electric utility industry, environmental interest groups, government agencies, and other organizations concerned about acid deposition, and about similar complex and divisive environmental issues.

Acknowledgments – The work described in this chapter was carried out through Research Report Project 2156-1 under the sponsorship of the Electric Power Research Institute, Palo Alto, CA.

REFERENCES

Balson, W.E. and North, D.W., 1983, Acid Deposition: Decision
    Framework, Vol. 3, State-Level Application. EPRI Report
    No. 2540, Electric Power Research Institute, Palo Alto, CA.
    51 pp.
Balson, W.E., Boyd, D.W., and North, D.W., 1982, Acid Deposition:
    Decision Framework. Description of Conceptual Framework and
    Decision Tree Models, EPRI Report No. EA-2540, Electric Power
    Research Institute, Palo Alto, CA. 115 pp.
Brown, R.V., Kahr, A.S., and Peterson, C.R., 1974, Decision
    Analysis for the Manager. Holt, Rinehart, and Winston, New
    York, NY. 618 pp.
Holloway, C., 1979, Decision Making Under Uncertainty: Models and
    Choices. Prentice-Hall, Englewood Cliffs, NJ. 522 pp.
Interagency Task Force on Acid Precipitation, 1982, National Acid
    Precipitation Assessment Plan. U.S. Government, Washington,
    DC. 92 pp.
Kahneman, D., 1982, Judgement Under Uncertainty: Heuristics and
    Biases. Cambridge University Press, Cambridge, MA. 555 pp.
National Research Council, 1981, Atmosphere-Biosphere Inter-
    actions: Toward a Better Understanding of the Ecological
    Consequences of Fossil Fuel Combustion. National Academy
    Press, Washington, DC. 263 pp.
Raiffa, H., 1968, Decision Analysis: Introductory Lectures on
    Choices Under Uncertainty. Addison-Wesley, Reading, MA.
    309 pp.
Ruckelshaus, W.D., 1984, Statement before the Committee on Environ-
    ment and Public Works, United States Senate, February 2, 1984.
    21 pp.
Spetzler, C. S. and Stael von Holstein, C. A. S., 1975, Probability
    encoding in decision analysis. Management Science 22:340-358.
Wallsten, T. S. and Budescu, D. V., 1983, Encoding subjective
    probabilities: A psychological and psychometric review.
    Management Science 29:151-173.

ON VALUING ACID DEPOSITION-INDUCED MATERIALS DAMAGES:

A METHODOLOGICAL INQUIRY

Thomas D. Crocker[a] and Ronald G. Cummings[b]

[a]Department of Economics, University of Wyoming
  Laramie, WY and
[b]Department of Ecomonics, University of New Mexico
  Albuquerque, NM

ABSTRACT

Those who study the physical features of pollution impacts upon materials have been granted little guidance about which of these impacts have economic significance. In this paper, an analytical framework is proposed that is appropriate for assessing the economic benefits of reducing the materials damages caused by acid deposition. The structure is intended to provide guidance about the kinds of physical science information that contribute to empirical economic analyses of these damages. Moreover, it supplies a framework that can be used to evaluate the completeness of these efforts. The paper concludes with practical suggestions about means to improve assessments of the economic consequences of materials damages from acid deposition.

INTRODUCTION

Questions about the optimal combination of materials with which to produce one or more types of output have traditionally been attractive to engineers and economists. While drawing upon known physical principles, engineers focus upon the design, construction, and operation of processes capable of producing outputs at minimum pecuniary cost in material inputs. The input prices that compose costs are taken as given. The engineer thus seeks to solve a constrained optimization problem involving detailed knowledge of technological possibilities as well as the expected costs of material inputs. If all nonmaterial inputs such

as labor were free, these engineering results would alone determine optimal process design. However, nonmaterial inputs are
costly, and both material and nonmaterial costs can change.
Engineering results thus identify the least costly set of
materials combinations within the larger set of ultimate technological limits for producing a given type and level of output.

Economists also fix their gaze upon cost minimization objectives, but they assume that the engineer has already solved his
constrained optimization problem. The economic theory of cost and
production describes the effects of varying input prices upon
cost-minimizing combinations of material and nonmaterial inputs.
From the set of cost-minimizing combinations, it identifies those
having the lowest material and nonmaterial costs and specifies how
this "minimum of the minima" will be altered as relative input
prices change. In this fundamental sense, then, economics
portrays the results of engineering reoptimization in terms of the
effects of changes in technology and the relative prices of inputs
on the cost-minimizing combinations of these inputs.

The design engineer considers all inputs consistent with
known physical laws since his objective is to develop and
ultimately implement a detailed plan for the particular production
process. Subsequent efforts to improve upon this plan will center
upon any input or subset of inputs appearing to provide substantial opportunities for cost reduction. In stark contrast, engineering studies of materials damages from air pollution have
concentrated on pollution impacts on a single material in some
production process. A great many have studied the impact independently of any production process. Hence, at best, the focus has
been upon a small set of inputs to the production process. These
highly detailed studies typically either ignore all other inputs
in the production process or implicitly assume that these other
input types and quantities remain constant. They, therefore, fail
to provide the design engineering information the economist
requires if he is to estimate the changes in cost-minimizing input
combinations which a pollution-caused materials impact would
induce.

This separable, piece-by-piece approach to engineering
studies makes the task of covering and synthesizing the materials
impacts of pollution truly awesome. In a prodigious cataloguing
effort, Salmon (1972) identified more than 12,000 distinct
materials having pecuniary value. Each material was, in turn,
embodied in one or more production processes or products which may
appear in a variety of forms and which can be put to a number of
distinctive uses. Moreover, there may be environmental cofactors
such as moisture and temperature that act in concert with pollution to aggravate or to soften its impact. These who study the
engineering features of pollution impacts upon materials have

guidance about which of these embodiments, varieties, and uses take received little, if any, on economic significance. Therefore, most fail to ask whether all the details given in a specific study will have economic content. One purpose of this paper is to provide the engineer and others interested in the materials impacts of acid deposition some guidance about the kinds of engineering information that contribute to empirical economic analyses of these impacts.

Using only information on the value of the capital stock at risk along with exceedingly limited engineering information on acid deposition materials impacts, Crocker (1980) speculated that these impacts constitute the leading source of economic damages from acid deposition. However, this speculation disregarded all the adaptations that economic agents might make when confronted with acid deposition-induced materials impacts. In the following pages, the analytical frameworks are developed for robust empirical support or denial of this speculation.

In order to be meaningful, discussions of how to estimate the economic benefits of reductions in acid deposition-induced materials damages must, at a minimum, suggest a structure that is comprehensive, logical, and theoretically defensible. This structure must serve other than aesthetic purposes: it should provide a basis for assessing the limits and the advantages of alternative methods that might be used to approximate materials damages; further, it would serve to make manifest the types of natural science and economic data required for reasonably comprehensive estimates of these damages.

The purpose of this paper is to provide an analytical structure appropriate for assessing the economic benefits of reduced materials damages caused by acid deposition. To this end, a general framework is described in the next section. Following this, a formal theoretical structure is developed. It is hoped that reader patience with the necessarily ornate theoretical arguments of this section will be rewarded by the conclusions, wherein some prospects for performing empirical research are suggested.

A FRAMEWORK FOR ANALYSIS - OVERVIEW

Research into the economic consequences of air pollution damages to materials has been dominated by investigators mainly interested in physical science dose-response research (e.g., Haynie, 1982). Their economic assessments, most of which appear to have been done as an afterthought, have understandably tended to maximize substantive physical rather than economic content. They have not bothered to establish the degree to which, if at all, these natural science findings are consistent with the

responses that the economic agents, who own and use these
materials, make to the presence of air pollution. Yield changes
for the current level of usage of a particular material multiplied
by an invariant market price have served as the measure of economic losses. Economically relevant changes in input and output
prices and substitutions, producer (consumer) risks, and loca-
tional and temporal redistributions have been overlooked. This
simple procedure involving (service flow yield change) times
(invariant price) will be seriously misleading whenever: (1) the
prices in the input or output markets in which a material is
employed are sensitive to its usage of inputs and to its service
flow; and (2) users of the material have a variety of ways in
which they can adapt to the burden that is imposed or lifted by
any contemplated change in pollution. Heedlessness of the first
criterion implies apathy about the fates of consumers as well as
the fates of inputs, such as labor, that users of the material
employ but do not own. Even more importantly, if the percentage
service flow increases caused by air quality improvements are less
than the consequent percentage output price reductions, an investigator who is inattentive to this first criterion would see
producer gains where there could be producer losses.[1]

The second criterion merely recognizes that economic agents
are observant and purposive creatures. If, by adopting a
different production or consumption pattern, they can reduce the
losses (or increase the gains) they suffer from a pollution
increase, they will do so; similarly, if they can increase the
gains (or reduce the losses) they acquire from a pollution
decrease, they will do so. Failure to account for adaptive
produce and consumer behavior will result in overestimates of the
losses producers suffer and underestimates of the gains they
acquire from changes in pollution levels.

---

[1]Let the producer's total revenue (TR) be pX, where p is the unit
price of the commodity in which the material is embedded, and X
is its quantity. Price is related to quantity by the inverse
demand function $p = p(X)^{-1}$. Marginal revenue (MR) is defined as
the change in total revenue with respect to output quantity, or:

$$MR = \frac{d(TR)}{dX} = p + X\frac{dp}{dX} = p(1 + \frac{X}{p}\frac{dp}{dX}) = p(1 + \frac{1}{\varepsilon}).$$

Since price and quantity are inversely related, ε is always
negative. Thus, MR < 0, when ε < -1. TR thus rises when output
quantity increases and, consequently, price falls. In this
latter case, more-or-less constant marginal costs of production
would be sufficient to cause the inattentive investigator of the
text to be blind to the producer losses resulting from an air
quality improvement.

The objective of efforts to assess the economic consequences of pollution-induced materials damages is to estimate the differences in the sums of consumer surpluses and producer quasi-rents over two or more pollution levels. A consumer surplus is a measure of the net benefit of a service flow or the presence of a stock to a representative consumer. It portrays the difference between the maximum a consumer would be willing to commit himself to pay for a given quantity of the activity output and what he, in fact, has to pay. Similarly, a producer quasi-rent is a measure of the net benefit of the output of an activity to the representative owner of the inputs to the activity. It is the difference between what this owner receives for supplying a particular output and the minimum he must receive in order to be willing to commit himself to that supply. The observable unit prices of other commodities that provide him equal satisfaction set an upper bound to the consumer's maximum willingness-to-pay; the observable earnings his inputs could obtain in other activities set a lower bound on the minimum reward the producer must receive.

Given that producers and consumers have adapted so as to minimize their prospective losses, Figure 1 depicts one example of the changes an air pollution increase can have upon consumer's surplus and producer's quasi-rent. The air pollution increase reduces the desirable physical properties of the output, making smaller the consumer's willingness-to-pay and causing his demand function to shift from D° to D'. Simultaneously, air pollution increases the damages for one or more materials in the production process for the output. The result is an increase in production costs and an upward shift in the supply function from S° to S'. Market price for the output drops from p° to p', partly because of the greater relative magnitude of the shift in the demand function and partly because of the lesser relative slopes of the demand functions. Consumer surplus was the area $ap^\circ b$; it is now the area $dp'e$. Producer quasi-rent was the area $fp^\circ b$; it is now the area $gp'e$. Total social surplus from the production and use of the output in question is thus reduced by the area $fgbh$ plus the area $adeh$.

Of course, differing relative shifts in demand and supply relations, reflecting different relative pollution-induced impacts in the production and consumption sectors, will yield results of a modified character. For example, if the demand curve shifts to D'' rather than D', market price will rise to p''. Qualitative results are unchanged; however, alterations in producer quasi-rents and consumer surplus are obtained from pollution-caused changes in the two sectors. These two examples serve to illustrate the issues of concern: consumers and producers can bear very different economic gains or losses depending on the absolute price elasticities of the demand and supply functions. Moreover, the distribution of

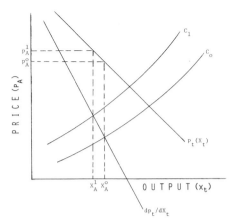

Figure 1.   Changes in producer's quasi-
             rent and consumer's surplus

where economic consequences can differ drastically with the price
elasticity of the demand function relative to the supply function.

The above observations set out the substance of social costs
(benefits) of relevance for pollution-related materials damages in
the most general of terms.   In the following sections these costs
are considered in greater detail, including pollution-related
costs in the production and the household sectors as well as the
social costs from materials damages to public goods.

THE INTERPLAY BETWEEN AGENT DECISIONS AND MATERIALS DAMAGES

In order to perceive the subtleties behind Figure 1, a more
formal analysis must be made.   Let the producer be confronted with
the problem of discovering the least cost method in which to alter
his operating, maintenance, and replacement program for a particular production process as the level of pollution-induced
materials damage changes.   His problem of choosing the optimal
method is conceptually straightforward.   He simply calculates the
optimal time and costs for each method and then chooses that
method having the lowest calculated costs.

Let the producer's decision problem in period t be:

Max: $\sum_{t=1}^{T} [p_t(X_t)X_t - c_t(X_t,K_t) - I_t] \beta^t$ (1)

Subject to: $K_{t+1} = K_t - D_t[X_t, M_t(A_t)] + I_t$ (2)

where: P_t is the unit price of the output in period t,

X_t is the quantity of output in period t,

c_t is the operating cost in period t,

K_t is the stock of nonhuman capital (materials) in period t, and

I_t is additions to capital stock in period t.

T is the end-point of the producer's planning horizon. What happens beyond T is considered to be irrelevant to the producer. D is the rate of deterioration of the capital stock in period t. This deterioration arises from simple wear-and-tear, technological obsolescence, and a lack of maintenance.

M_t is effective maintenance activities during period t. Maintenance activities per se are assumed constant in time -- periodic oiling and greasing of machines -- consistent with the common practice of establishing maintenance programs. A change in air pollution is then assumed to change the effectiveness of a given maintenance program; thus as stated below, it is assumed $\partial M/\partial A \leq 0$, and $\partial D/\partial M \leq 0$. Costs for the fixed maintenance program are therefore subsumed in c_t.

A_t is the level of air pollution in period t.

$\beta^t = (1 + r)^{-t}$, where r is the producer's discount rate.

The objective function in equation (1) is assumed to be continuously differentiable and strictly concave. Various partial derivatives take on the following signs:

$$\frac{\partial P_t}{\partial X_t} \leq 0; \quad \frac{\partial^2 P_t}{\partial X_t^2} \leq 0; \quad \frac{\partial c_t}{\partial X_t} \geq 0; \quad \frac{\partial^2 c_t}{\partial X_t^2} \geq 0; \quad \frac{\partial c_t}{\partial K_t} \leq 0;$$

$$\frac{\partial D_t}{\partial X_t} \geq 0; \quad \frac{\partial D_t}{\partial M_t} \leq 0; \quad \frac{\partial M_t}{\partial A_t} \geq 0; \quad \frac{\partial^2 D_t}{\partial X_t^2} \geq 0; \quad \frac{\partial^2 D_t}{\partial M_t^2} \leq 0. \quad (3)$$

Note that because $\partial c_t/\partial K_t \leq 0$, the producer can increase the present value of his future period net revenues by investing in capital stock in the current period, thereby, from equation (1), reducing his current period net revenues. On the other hand, current production, X_t, implies a user cost inasmuch as X_t depletes capital, $\partial D/\partial X_t \geq 0$, thereby lowering capital stocks available to future periods.

The Lagrangian for the above constrained maximization problem is:

$$L = \sum_{t=0}^{T} \left\{ [p_t(X_t)X_t - c_t(X_t,K_t) - I_t]\, \beta^t \right.$$
$$\left. - \lambda_{t+1}[K_{t+1} - K_t + D(X_t, M_t(A_t)) - I_t] \right\} \qquad (4)$$

Assuming an interior solution, first-order-conditions for a maximum for equation (4) include:

$$\frac{\partial L}{\partial X_t} = \left(\frac{dp_t}{dX_t} + p_t - \frac{\partial c_t}{\partial X_t} \right) \beta^t - \lambda_{t+1}\left(\frac{\partial D}{\partial X_t} \right) = 0 \qquad (5)$$

$$\frac{\partial L}{\partial K_t} = -\beta^t \left(\frac{\partial c_t}{\partial K_t} \right) - \lambda_t + \lambda_{t+1} = 0 \qquad (6)$$

$$\frac{\partial L}{\partial I_t} = -\beta^t + \lambda_{t+1} = 0 \qquad (7)$$

The first two terms in brackets in equation (5) are the producer's marginal net revenues in period t from an additional unit of output during that period; the last term in brackets in equation (5) is the corresponding marginal user cost. If the first bracketed term in equation (5) is defined as marginal net revenues, $M\pi_t$, for period t, then from equation (6):

$$\lambda_t = \sum_{\tau=t}^{T} -\beta^\tau \frac{\partial c}{\partial K_\tau} \qquad (8)$$

and

$$\lambda_{t+1} = \sum_{\tau=t+1}^{T} -\beta^\tau \frac{\partial c}{\partial K_\tau} \qquad (9)$$

From equation 5, discounted marginal net revenues are:

$$(M\pi_t) \beta^T = (\lambda_{t+1})(\partial D/\partial X_t). \tag{10}$$

Expression (7) states that periodic investment, I_t, is set at the level where marginal investment costs equal the marginal value of capital as measured by λ_{t+1}. Equations (8) and (9) define the shadow price of the producer's capital stock in period t, viz., the discounted value of the marginal contribution of the capital stock to reductions in production costs in all future periods. Expression (10) requires the choice of X such that the discounted value of marginal net revenues generated by using the capital stock to produce current output be equated to the opportunity cost (λ_{t+1}) of capital consumed while producing current output $(\partial D/\partial X_t)$.

Given air pollution level A_0 in period t, the profit maximizing firm chooses optimum values I_t^o and X_t^o so as to satisfy (9) and (10). Depreciation in t is $D[X_t^o, M(A_0)]$, where initial stocks K_t and end-of-period stocks K_{t+1} are defined. In what follows, any argument in D other than A is suppressed when referring to the depreciation function. In the market, these conditions result in the market price p_t^o (Figure 2) which yields producer quasi-rents equal to the area under p_t^o and over C_0 evaluated at X_t^o, where

$$C_0 = (\partial c/\partial X_t) + \lambda_{t+1}[\partial D(A_0)/\partial X_t]. \tag{11a}$$

Consumer surplus is that area under the demand curve $p_t(X_t)$ at X_t^o which lies above p_t^o. Net social benefits, the sum of producer quasi-rent and consumer surplus, associated with production at X_t^o is given by:

$$\int_0^{X_0} [p(X)X - C_0(X|A_0)] \, dX. \tag{11b}$$

Suppose now that pollution levels rise to $A_1 > A_0$. According to equation (3), effective maintenance activities decline for all values of X. All else equal[1], capital stocks in all future periods are smaller. Given the concavity of the production function underlying $c(X,K)$, $\partial c/\partial K$ would be larger in all future periods, implying $\bar{\lambda}_{t+1} > \lambda_{t+1}$. Further, for any X, $\partial D(A_1)/\partial X > \partial D(A_0)/\partial X$.[2] All this is to argue that the right-hand side of equation (10) increases. $X_t > 0$ in equation (10) then requires that the left-hand side be correspondingly increased which, from

---

[1] In other words, the future time path of I is being held constant.
[2] Inequality holds when the increase in A changes the slope of the function D.

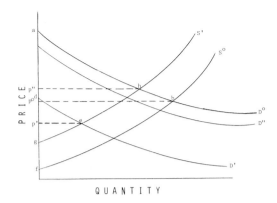

Figure 2.   Prices and output consequences
            of a change in pollution

expression (3), implies a reduction in the level of output X and a
higher price p.   Increases in the right-hand side of equation (10)
are descriptive of a leftward shift in the firm's cost schedule
from C_0 to C_1 as shown in Figure 2, where

$$C_1 = (\partial c / \partial X_t) + \bar{\lambda}_{t+1} [\partial D(A_1)/\partial X_t]. \qquad (11c)$$

With higher production and user costs, output is reduced from X^0
to X^1 and the prices rises from p^0 to p^1.   Social benefits from
production, as defined above, are now given by

$$\int_0^{X_1} [p(X)X - C_1(X|A_1)] \, dX \qquad (12)$$

and the loss in benefits -- social losses -- attributable to the
pollution-caused materials damages associated with the higher
pollution levels A_1 are defined by equation (11b) minus equation (12), or

$$\int_{X_1}^{X_0} [p(X)X - C_0(X|A_0)] \, dX + \int_0^{X_1} [C_1(X|A_1) - C_0(X|A_0)] \, dX \qquad (13)$$

Before continuing these arguments, it will be useful for
later discussions to reflect for a moment on the implications of

the measure for social costs attributable to pollution-related materials damages, given in expression (13), for efforts to measure such damages empirically. Referring to (13), two broad classes of data/information are needed. First, technical/ engineering data are required which describe the physical impact of pollution on that machinery, equipment, buildings, etc., that are inputs to the production process. These physical changes, when weighted by appropriate replacement/maintenance values, describe the shift in the supply curve from C_0 to C_1 in Figure 2. The second class of required data implied by expression (13) is the demand relationship $p(X)$ shown in Figure 2.

Returning to the production model discussed above, a conceptual basis for identifying and measuring social losses was developed associated with materials damages effects on inputs used by a firm (our K_t) in the production of goods and/or services, X. Obviously, however, this is but a part of the potential range of pollution-related materials damages. At least two additional sources for materials damages and their attendant social costs are of interest: damages to the outputs of firms, X, which are used or consumed by households; and damages to nonmarket or public goods, such as national monuments, etc.

Following the standard economic theory of individual behavior, a quasi-concave and differentiable function U is defined which measures the utility, or satisfaction, that an individual derives from consumption or use of goods. Let there be n of these goods, denoted X_1, ..., X_n.

$$U(X_1, \ldots, X_n). \tag{14}$$

Given income, W, and the prices for goods, p_1, ..., p_n, maximization of expression (14) is subject to an obvious constraint: the individual cannot spend more than his income. Thus, the following restriction is imposed:

$$\sum_{i=1}^{n} p_i X_i = W. \tag{15}$$

To get at the problem of concern, some set of goods, Z, must be viewed in a qualitative sense -- that set of goods vulnerable to materials damages, such as lawn furniture, lawn mowers, house paint, etc. These goods are the X's of the production model. The consumer then derives a flow of services through time from this set of goods. These goods, which are not consumed during the time period in which they are purchased but which provide a flow of services through time, are analogous to the firm's stock of capital. Dichotomize the X's such that m of the n goods are capital-type goods denoted as $Y^1, \ldots, Y^m$. The remaining n-m goods are immediately consumed and are denoted as $Z^1, \ldots, Z^s$, where $s = n-m$. Equation (14) can then be rewritten so as to measure the maximized utility at any time as:

$$U\ (Z_t^1, \ldots, Z_t^s, Y_t^1, \ldots, Y_t^m).\tag{16}$$

The goods Y are, by assumption, subject to depreciation, δ, from normal wear and tear as well as from pollution related materials damages. For simplicity, let such deterioration be reflected in the value of Y. Then, at any t, the value of Y is given by:

$$Y_{t+1}\ =\ Y_t + X_t - \delta(Y_t, A_0),\tag{17}$$

where X_t is the periodic purchase (replacement) of the good(s) Y. X_t is, therefore, the purchase of output from firms - the X's of the above-described production model. The function δ measures use-related wear and tear (depreciation) at a given pollution level A_0. Increases (decreases) in A lead to increases (decreases) in δ; i.e., $\partial\delta/\partial A \geq 0$. The consumer's budget constraint, as given in expression (15), now becomes:

$$\sum_{i=1}^{s} p_t^i\ z_t^i\ +\ \sum_{j=1}^{m} p_t^j\ x_t^j\ =\ W\tag{18}$$

In this model, the consumer maximizes

$$L = \sum_{t=1}^{T} [U(Z_t^i, \ldots, Z_t^s, Y_t^1, \ldots, Y_t^m)\ \beta^t] - \psi_t\ (\sum_{i=1}^{n} p_t^i\ z_t^i + \sum_{j=1}^{m} p_t^j\ x_t^j - W)$$

$$-\ \sum_{j=1}^{m} \lambda_{t+1}^j\ [Y_{t+1}^j\ -\ Y_t^j - X_t^j + \delta(Y_t^j, A_0)],\tag{19}$$

where ψ_t, λ_t are Lagrangian multipliers, $\beta^t = (1+r)^{-t}$, and r is the individual's consumption rate of discount. Necessary conditions for a maximum in equation (19) include the following at an interior solution (where superscripts are used only when necessary for expositional clarity):

$$(\partial U/\partial Z_t)\ \beta^t\ =\ p_t^i\ \psi_t\tag{20}$$

$$(\partial U/\partial Y_t)\ \beta^t\ +\ \lambda_{t+1}\ (1 - \partial\delta/\partial Y_t)\ -\ \lambda_t\ =\ 0\tag{21}$$

$$\lambda_{t+1} = \sum_{=t+1}^{T} \beta^{\tau} (\partial U/\partial Y_{\tau}) \prod_{b=t+1}^{\tau-1} (1 - \partial \delta/\partial Y_b) \tag{21a}[1]$$

$$- \Psi_t \, p_t^j + \lambda_{t+1}^j = 0 \tag{22}$$

$$dL/dA = - \sum_{t=1}^{T} \sum_{j=1}^{m} \frac{\partial \delta(Y_t^j, A)}{A} \tag{23}$$

Conditions shown in expressions (20) to (23) describe the familiar situation for the maximization of individual utility. With equation (20), the consumer equalizes the ratio of marginal utility to price for all Z^i purchased. Major concern is with the purchase of X, the rule for which is given by equation (22). To appreciate this rule, consider the interpretation of λ_{t+1} given in (21a). A unit of goods Y at t produces the flow of services, $\partial U/\partial Y$, in future periods t+1, ..., T. At any future period, say t* > t, that unit, as measured in t, is subject to depreciation during years t+1, t+2, ..., t*. Thus, the amount of Y available to produce $\partial U/\partial Y_{t*}$ in year t* is a unit of Y discounted by the product of $(1 - (\partial \delta/\partial Y_{t+1}))$, $(1 - \partial \delta/\partial Y_{t+2})$, ..., $(1 - \partial \delta/\partial Y_{t*})$. This, of course, is the measure given on the right-hand side of (21a). λ_{t+1} is the present value of utility derived from a unit of Y acquired in t, where that unit is adjusted to reflect depreciation. λ_{t+1} can then be interpreted as the marginal utility value of a unit of Y acquired (via X_t) in time t. Defining Ψ_t as the marginal utility of income, the logic of the decision rule given in expression (22) is apparent: acquire X in t to the point where the utility value (opportunity cost) of the price of $X(\Psi_t p_t^j)$ is equated to the marginal utility value of Y.

Of particular importance are the equi-marginal conditions implied by equations (20) and (22):

$$\Psi_t = \frac{\partial U/\partial Z_t^i}{p_t^i} = \frac{\lambda_{t+1}^j}{p_t^j} , \quad \forall_{i,j} \tag{24}$$

The condition given in equation (24) alerts us to the interdependence of the consumer's purchases of all goods and services, Z and Y.

Use of the Implicit Function Theorem (Silberberg, 1978) allows for S + 3M + 1 equations in S + M + Z variables to be

---

[1]For this expression, $\prod_{b=t+1}^{t} \approx 1$; $\prod_{b=t+1}^{t+1} (1 - \partial \delta/\partial Y_b) \approx (1 - \partial \delta/\partial Y_{t+1})$.

derived from L in equation (19).   These numbers are then expanded
by time periods. Of interest here are the m functions:

$$x_t^j = x_t^j \, (p_1^t, \; \ldots, \; p_s^t, \; p_{s+1}^t, \; \ldots, \; p_{s+m}^t, \; W, \; A_0), \; j = 1, \; \ldots, \; m. \quad (25)$$

The demand functions in expression (25) are for goods x^j in t.
For a given x^j, holding constant all prices other than p_t^j, incomes
(W) and the pollution level A_0, the relation $x^j(p_j)$, where
$\partial x/\partial p \leq = 0$, is the demand function illustrated in Figure 2 and is
of central importance in our derivation of social costs in the
production sector (equation 13).

The issue as to the impacts of a change in pollution in the
household sector can now be addressed.   Consider any good y^j and
periodic purchases of this good, x_t^j.   The individual's demand for
x^j is given in equation (25).   An increase in pollution from A_0 to
A_1 will shift upwards[1] the depreciation function δ in equation (17), thereby reducing the value of λ_{t+1}; in equation (24)
the term to the right becomes less than the left-hand term.
Satisfaction of expressions (20) to (24) first requires a reduction in X_t, implying lower future values for Y_t and (with
diminishing marginal utility) higher values for the $\partial U/\partial Y_\tau$ terms
in λ_{t+1} (see equation 21a).   Secondly, concomitant with reductions
in X_t, expenditures are reallocated to other goods, thereby
reducing the $\partial U/\partial z^i$ terms in equation (24).

These simultaneous responses to the change in A are made
clear by stating that A in the demand function (expression 25)
acts as a shift parameter.   An increase in A "shifts" the demand
function to the left as illustrated by the shifts from D^0 to D'
and D'' in Figure 1.   Of interest here are the social costs
associated with this shift relative to the consumer's initial
position.   In general, these costs will be given by:[2]

$$\int_0^{X_2} [p(X,A_0) - p(X,A_1)] \, dX + \int_{X_2}^{X_1} [p(X,A_0) - C_1(X,A_0)] \, dX \quad (26a)$$

where X_2 is the equilibrium value of X after the pollution-caused
shift in demand.   Total social costs (TSC) from pollution-related
materials damages to market goods and services, which includes
costs to producers and consumers, is derived by combining expressions (13) and (26):

---

[1] It can be assumed that the slope of $\delta(y,A)$ would also be
affected.

[2] The functions $p(X)$ are the inverse functions of $X(p)$ given in
equation (25).

$$TSC = \int_0^{X_2} [p(X,A_0) - p(X,A_1) - C(X,A_0) + C(X,A_1)] dX$$

$$+ \int_{X_2}^{X_1} [p(X,A_0) - C(X,A_0)] dX \qquad \text{(26b)}$$

As in the previous inquiry concerning the empirical implications of social cost measures for the production sector, it
remains useful to inquire about empirical measures of the terms
given in equation (26). Basic to any effort to derive these
empirical estimates are two classes of data: First, as before,
technical/engineering data which describes the physical impacts of
pollution on consumer goods Y -- the $\delta(Y,A)$. Secondly, measures
are required for consumer responses to shifts in $\delta(Y,A)$ as they
relate to changes in the valuation of X, i.e., the shift in the
demand curve $p(X)$.

The final types of goods that may be seriously affected by
materials damages are commonly described as "public goods." The
economic consequences of damages to this class of goods are developed with the following example. Consider a public commission
charged with identifying and nominating for preservation those
man-made structures in an urban region considered to be noteworthy
for historical and/or aesthetic reasons. The commission's budget-
ing and political resources do not allow it to preserve every
structure considered noteworthy on historical and aesthetic
grounds. It must make choices. Presumedly, the relative values
that the representative citizen of the region attaches to alternative combinations of these structures would play some part in the
commission's deliberations. If this representative consumer
possessed the powers of the commission, the following formulation
could be used to describe the behavior he would adopt. To
simplify the exposition, this individual's concerns for the
welfare of his progeny has been disregarded; also it is assumed
that there is only a single decision period, the individual's
expected lifetime. The complications introduced by discounting
and marginal-rates-of-substitution among contingent temporal
states are therefore suppressed.

Assume that our citizen obtains utility from three classes of
goods and services. The first class of goods consists of the
purchased goods and services, Z and Y, of concern immediately
above. To avoid repeating the complex analysis of intertemporal
stock-flow relationships associated with Y, Y-type goods are
treated as Z-type goods in what follows. Thus, our first class of
goods is dubbed Z. The second class of goods from which our

consumer derives utility is that stock of historical structures of
concern to the commission. This is to say that the consumer may
acquire satisfaction from the mere existence of some or all of
these structures, reflecting the fact that he too appreciates
their historical values and their abilities to stimulate the
senses. It is emphasized that these values accrue to the
individual whether or not he/she actually sees, touches, visits
the structures -- the simple existence of the structure(s) serves
as the source of values. If $\bar{K}$ (still another capital stock)
measures the stock of these structures at the beginning of time
t*, these arguments imply that, like Z, $\bar{K}$ is an argument in the
individual's utility function.[1] Finally, a third class of goods,
S, are also derived from the stock of structures $\bar{K}$. Utility from
these goods obtains, however, from the use -- e.g., visitation --
of $\bar{K}$.

Within this simplified, timeless context, it is assumed that
the quality or value of $\bar{K}$ is instantaneously affected by the level
of air quality, A. Thus, a value for A determines the average
value of $\bar{K}$ as perceived by the individual. For an initial air
quality level, A_0, there is by definition $\bar{K}(A_0)$, $\partial\bar{K}/\partial A \leq 0$, as an
argument in the consumer's utility function, given by:

$$U = U [Z, K(A), S]. \tag{27}$$

In expression (27), $U(\cdot)$ is assumed to be weakly separable in Z
and K, and in Z and S. The structural service flows available to
the individual from S are affected by the size of the structural
capital stock, K, and by other goods, q, such as shoes, subways,
and automobiles, that aid his access to these flows. This relationship can be represented by a production function assumed to be
homogeneous of degree one:[2]

$$S = S [K(A), q] \tag{28}$$

The marginal products of K and q are assumed positive, i.e.,
$\partial S/\partial K$, $\partial S/\partial q \geq 0$, for a given value of A.

The citizen's decision problem is to maximize equation (27),
subject to (28), and the following budget constraint:

$$W = p_1 Z + p_2 q, \tag{29}$$

---

[1]The quality dimensions of K are discussed below. The problem in
using a single measure for $\bar{K}$ to aggregate across many unique
historical structures (e.g., "adding" the Statue of Liberty to
the Alamo, etc.) is acknowledged.

[2]Pollak and Wachter (1975), among others, show that this assumption is necessary in order to distinguish between changes in
behavior due to relative price changes and shifts in technology.

where p_1 and p_2 are, respectively, the exogenous unit prices for Z and q. W is the citizen's income or budget.

The problem specified in equations (27) to (29) has features that reflect earlier comments and warrant a bit more emphasis. First, the flow of services the structural capital provides cannot be directly acquired; instead, other goods must be purchased or used in order to influence these service flows in the manner described by expression (28). One of the goods which must be used is the structural capital stock. Whether it be due to their ambiences or to the peculiar configurations of the structure, equation (28) implies that there exist few adequate substitutes for these structures. Second, apart from their contributions as settings for the consumption of commodities, the appearance of the K term in equation (27) means that the structures are valued in their own right. This existence value could originate, for example, from the sense of wonder or historical continuity and permanence the structures provide, or perhaps they serve as physical symbols (e.g., the Statue of Liberty) of abstract principles woven into the social fabric (Roth, 1982). Further, the problems illustrated in equations (27) to (29) imply that the citizen is unwilling to sacrifice everything to ensure the existence of these structures, regardless of whether it be for the sake of art, ancestor worship, or the regional social fabric. More will be said about existence and existence values at the end of this section.

Upon substitution of equation (28) into (27) and (29), setting up the appropriate Lagrangian expression, and solving for the first-order-conditions, the following conditions for an interior solution are obtained:

$$\frac{\partial L}{\partial Z} = 0 = \frac{\partial U}{\partial Z} - \lambda p_1 \tag{30}$$

$$\frac{\partial L}{\partial q} = 0 = \frac{\partial U}{\partial S} \frac{\partial S}{\partial q} - \lambda p_2 \tag{31}$$

$$\frac{\partial L}{\partial K} = 0 = \frac{\partial U}{\partial K} + \frac{\partial U}{\partial S} \frac{\partial S}{\partial K} \tag{32}$$

The interpretation of equations (30) to (32) differs a bit from the earlier analyses inasmuch as the consumer, in his role as the commission, chooses the stock $\bar{K}$, and the choice of any stock K would be considered costless to him/her. It will be shown, however, that his/her choice of K is still bounded.

From expressions (30) and (31), the familiar conditions can now be written as:

$$\frac{\partial U/\partial Z}{P_1} = \frac{(\partial U/\partial S)\,(\partial S/\partial q)}{P_2}, \tag{33}$$

which describes the individual's allocation of income: the marginal utility of Z, per dollar spent on Z, equals the marginal utility of the consumption (visitation) of K via trips q, per dollar spent on q.

The condition described in equation (32) allows for the choice of K when U in expression (27) is maximized and invariant at some U*, and when Z and q are fixed. This follows from taking the total differential of equation (27) and setting $dU = dq = dZ = 0$. K is chosen such that

$$\frac{\partial U}{\partial K} = -\frac{\partial U}{\partial S}\,\frac{\partial S}{\partial K}, \tag{34}$$

which, by substituting $\partial U/\partial S$ from (31), becomes

$$\frac{\partial U}{\partial K} = -\lambda P_2 \frac{\partial S/\partial K}{\partial S/\partial q} \tag{35}$$

Thus, from equation (35) a level of $\bar{K}$ is chosen so as to equate the marginal utility derived from the existence of K with the marginal S-contributions of K, per marginal S-contribution of q, times λP_2, the utility cost of q. Combining equations (35) and (30) yields the expression:

$$\frac{\partial U/\partial K}{\partial U/\partial Z} \cdot \frac{\partial S/\partial q}{\partial S/\partial \bar{K}} = -\frac{P_2}{P_1} \tag{36}$$

where the weighted ratio of marginal utilities for K and Z equals the ratio of prices. Note that if K is not consumed by the individual, $q = 0$ and $S(K,0) = 0$. Equation (34) then becomes:

$$\partial U/\partial K = 0, \tag{37}$$

and the individual chooses a (larger) $\bar{K}$ such that the marginal value to him of more K is zero. The consumer is then satiated with K. Of course, if a cost for K, say t-dollars per unit (a tax?) is added, the marginal utility associated with a dollar, the λ, would appear on the right-hand side of equation (37) and fewer units of K would be consumed (in the existence sense) by the individual.

What of the effect on the consumer of an increase in pollution? From earlier discussions concerning the household sector, it is known that a demand relationship can be defined for K of the form:

$$\bar{K} = \bar{K} \; (p_1, \; p_2, \; W, \; A). \tag{38}$$

The function on the right side lacks the interpretative neatness of the earlier demand function, however, for one simple, but important, reason: there exists no market price $\bar{p}$ for K which, through the market, manifests the individual's preferences for K vis-a-vis other goods. The price p_2, the price paid for goods (travel, etc.), used in consuming K is known. Given the many goods of the q-type which may be involved in using K, and the fact that such goods are also likely used for other purposes, looking to the p_2's for consumer valuation of K would likely be a futile task; further, the p_2's would not reflect the existence value dimensions relevant for the valuation of K.

Of course, this is the nature of public goods. A public good is one for which any one person's consumption of the good leaves the good unaffected and for which one cannot deny its consumption or use to any individual. Since it is not traded in the market place - property rights to the good are vested in no one individual but to society as a whole - there exists no market value (price) for the good. This Commission for Public Structures wished to include individual valuations of K in decisions as to whether or not, and to what extent, to spend monies in maintaining that stock of public goods susceptible to pollution-related materials damages. Given market determined demand and cost relationships in the production and household sectors, one could, at least conceptually, draw these valuations from market observations.

At A_0, let the consumer satisfy the conditions described in expressions (30) to (32), in which case Z^*, q^*, and K^* are optimal utility-maximizing values for these variables. Holding $U^*(Z^*, S[K^*(A_0), q^*], K^*(A_0))$ constant, the following relationship can be defined:

$$\frac{dW}{dA} = p_1 \; \frac{\dfrac{\partial U}{\partial S}\dfrac{\partial S}{\partial K} + \dfrac{\partial U}{\partial K}\dfrac{\partial K}{\partial A}}{\partial U / \partial Z} \tag{39}[1]$$

The expression (36) shows, for a given change in A, the amount by which the consumers income would have to be increased for the consumer to be as well off at the higher pollution level

---

[1] Set $dU^* = 0$, and then substitute the expression for dZ in expression (29) into dU^*. Upon substituting (33) into the resulting expression, equation (39) is obtained.

as he was at the previous lower pollution level. Alternatively, dW/dA can be interpreted as what the consumer would be willing to pay for a decrease in pollution level by the amount dA.

A conceptual measure for pollution-related materials damages, as they might impact on public goods, is also defined in equation (39). The issue remains, of course, as to how this measure might be estimated. Travel cost methods might be employed as was done by Cicchetti et al. (1976) to approximate lower bounds for dW/dA through information related to q and p_2. Aside from methodological problems associated with such approximations, it is known from the above that p_2 will not capture the "existence" dimensions of the conceptually correct measures. In the end, it may be that an attempt has to be made to elicit this information directly from individuals, perhaps by employing the contingent valuation survey methods originated by Davis (1963). In any case, dW/dA will include a technical component and a behavioral, value component. The technical/ engineering component is the weighting term $\partial K/\partial A$ in expression (39), i.e., the physical effect on structures attributable to a change in pollution levels. The value term has three components: behavioral adaptations in the consumption of goods other than the public good, Z; valuation of K in terms of consumption, use, or visitation of K; and the existence value of K.

The presence of K in the individual's objective function, and therefore in equation (39), has not thus far been well-motivated. Its presence is said to be due to something called "existence value"; substituting a similar set of words rather than an extraction of more primitive terms has been done to avoid having to shape the expression. There are several plausible alternative constructs that can be used to pump analytical content into "existence values". Perhaps the most familiar of these plausible constructs is Krutilla's (1967) treatment of the amenity features of natural environments. Krutilla (op. cit.) and many subsequent commentators argue that the use of an undeveloped natural setting as an amenity must be distinguished in economic analysis from its use as a provider of commodity or service flows. Moreover, economic growth, if interpreted as increases in commodity and service flows, has quite different effects on the demand and supply of natural environments. In particular, the stock of more-or-less unique natural environments is fixed and even diminishing; if demand for them increases with economic growth, their value will also increase. Thus, to the extent that the flow of commodity goods and services is greater in the future than currently, the current values of these environments serve as negatively biased indicators of their future values. Second, decisions to develop a natural environment often tend to have irreversibility features. A development decision is irreversible, whereas if the environment is preserved, both preservation and development remain available as options. Irreversible conversion thus includes a user cost in that the set of opportunities available for adapting to altered future circumstances has been diminished.   If unique and unusual man-made structures are substituted for natural environments in the above discussion, the arguments seem to lose none of their force. For the consumer as well as the producer, materials damages must be examined not only in view of foregone present commodity and service flows, but also in view of negative impacts on future commodity and service flows and future opportunities.

An analytical basis for existence values might be discovered by exploring the degree of complementarity between K and assorted private goods. Paraphrasing an example from Hart (1980), suppose that the structures surrounding and the statuary within an urban public square serve as an aesthetically pleasing addition to the site-specific production of private musical performances and artwork. The structures and statuary substantially reduce the artists' costs of attracting an audience to view and perhaps purchase their work. However, when removed from the square, the unit costs of attracting an audience exceed any artist's reservation price for producing art. Yet the minimum value for a combination of the square and the art may be much greater than the sum of reservation prices for them when treated separately. The setting that the public square provides must be available at some necessary minimal scale in order for the art to be supplied. A benefit-cost analysis of artwork based upon the price structure present in the absence of the square would give very different results than an analysis done with the square present. The absence of the square causes the removal (from existence) of another good from the commodity space.[1]

Endogenous preferences offer yet another plausible means of analytically motivating existence values. As Elster (1979) and Thaler and Shefrin (1981) suggested, certain kinds of current choices such as more education and locating in handsome urban or rural environments may cause a person later to be what he or she now wants to be later. For example, one may now be indifferent among modern art forms but nevertheless attend art shows so that one may later have discriminating tastes. This personal transformation can arise through deliberate acquisition of information or through the practice of learning by doing. Minor modifications of any one of numerous extent habit formation models (e.g., Pollak, 1976) seem likely to capture the essence of the problem. In these models, an individual's current preferences are made to depend on his past consumption. Additional modifications could allow one to explain how the current provision of certain environ-

_____
[1] It is noted that the basic theorems of welfare economics presume the commodity space to be exogenously determined.

ments may cause you later to be what somewhat now wants you to be later. Again, one may take the lifelong urban dweller into the mountains so that he or she may experience and come to like the natural settings that another likes. This obviously raises the ancient spectre in economic analysis of interdependent preference orderings. Nevertheless, as long as individual X's preference ordering in any period is made to depend upon his past consumption, in principle, no problem arises. Individual Y, by his manipulation of X's current equilibrium consumption pattern, can then be represented as producing individual X's future preferences. Does this not describe what many responsible parents attempt to do? Unfortunately, the discipline of economics has practically no formal understanding of the forms and parameters central to this production process.[1]

CONCLUSIONS

The potential economic benefits from reducing acid deposition-induced materials damages are plausibly large. Limited insight as to the potential magnitude of such damages is given by an admittedly crude estimate for one of many damage components. According to the U.S. Department of Commerce (1980), the 1979 values of stocks of equipment and structures in the U.S. manufacturing sector were $365 billion and $160 billion, respectively. Reported replacement life was 12 years, on the average. Equipment purchases were $46 billion in 1979, and investments in new structures amounted to $14 billion. Assuming a ten percent discount rate and a reduction in acid deposition levels that extends average replacement life by six months, the present value of the reduction in acid deposition-induced materials damages to the manufacturing sector's 1979 capital stock would be $5.7 billion. Assuming that the 1979 investment magnitudes in capital stock would be continued, the benefits caused by reductions in materials damages to this annual investment would be on the order of $0.9 billion per year, or, in present value terms, $8.9 billion over 50 years in the manufacturing sector alone.[2]

---

[1]An excellent recent paper by Kreps (1979) perhaps provides the axiomatic basis for overcoming this problem.

[2]For example, suppose a machinery/equipment item with replacement cost C-dollars would be replaced in 10 years with pollution level A_1 and in 12 years with a lower pollution level A_0. The present value of economic benefits attributable to a policy which lowers pollution levels from A_1 to A_0 is $C(1+r)^{-10} - C(1+r)^{-12}$, where r is the discount rate. In general, given a replacement life of t-years for an article, a pollution-level change of ΔA which increases replacement life by Δt years causes economic benefits (B) attributable to ΔA to be:

$$B(\Delta A) = C \left[\frac{1 - (1+r)^{-\Delta t}}{(1+r)^t} \right].$$

Current prospects are poor for improving these estimates by extrapolating across materials damage studies that are particularized across goods at risk, and across time and place. There are several reasons for this. Perhaps the most important is simple lack of information about physical materials damage relations and the adaptations which economic agents can make to these physical relations.[1] The physical information on materials impacts is limited because of the many complex interactions with which the researcher must contend in his efforts to estimate cause-effect relations. There are many materials grades, types, and surface coatings, and the impact of a particular pollutant upon each may be conditioned by other pollutants and by the environment.

The difficulties for benefits estimation that these physical complexities introduce are accentuated by the propensity of physical science researchers to concentrate upon materials rather than upon groups of materials in their manifestations as machinery, furniture, buildings, etc. A necessary condition for performing economically meaningful studies of pollution-induced materials damages is the acquisition of information on the covariation of pollution and maintenance/replacement policies defined in terms of their economic (e.g., machines or tires) rather than in their physical (e.g., steel or rubber) manifestations. Many of the analytical issues and trade-offs that must be confronted in constructing units for these economic manifestations are presented in Bongers (1980) and Lau (1982). If the adaptations which economic agents can make are to be captured, one must know the cause-effect relations between acid deposition and physical damages to goods and processes; these must be defined in dimensions corresponding to those in which economic agents rather than natural scientists choose to define them.

Although the example is drawn from agricultural rather than materials damages, the reader may find it useful to have some feel for the degree by which one can be misled by control benefits estimates that ignore price effects and agent adaptations. Crocker (1982), while drawing upon original work of Adams et al. (1982), studied the economic gains of ambient oxidant reductions to consumers and producers of 14 crops grown in four southern

---

[1] In principle, the duality methods employed in Manuel et al. (1982) did not require explicit investigator knowledge of underlying physical damage relations. However, to be fully credible, some attention to consistency with physical realities is necessary. Moreover, because their empirical implementation involves observed behavior, the reliability of duality methods when applied to circumstances outside the realm of historical experience is an open question.

California subregions. Of the 56 (14 x 4) cases, the estimated percentage changes in yields, after accounting for consumer and producer adaptations, exceeded by a factor of at least two in 29 cases the triggering oxidant induced percentage yield changes.

Given current inabilities to aggregate from particularized materials damage studies, the ordering of the question might be reversed; this is by starting with an aggregate statement similar to an earlier calculation for the U.S. manufacturing sector. This national benefits estimate is consistent with one or more allocations of benefits across types of goods, locations, and times. Any allocation must be based upon some set of particularized economic opinions and/or formal studies, each of which is specific in type of good, location, and time. It can be asked what will be gained in terms of national control benefits estimates by improving an allocation; that is, if A and B are two allocations but B is more accurate, what is it worth to policymakers to have B rather than A available? An answer to this question requires an understanding of the accuracy of the data (the particularized studies) on which a national benefits estimate is based. Bayesian statistical techniques, such as those found in Leamer (1978), allowed assessment of the precision and accuracy of data as well as improvements in precision and accuracy likely to result from alternative kinds of additional data (particularized studies). Since the allocation of a national control benefits estimate across times, locations, and types of goods is exactly determined by the data, the precision and the accuracy of the allocations can, in principle, be determined. Having answered these questions, it can be determined whether the costs of improving the data (the particularized control benefits studies on which the national estimates are based) would be justified by the resulting improvements in the allocations. Of course, formation of this last judgement would require a further specification of the tradeoffs policymakers are willing to make between the costs of data improvements and the benefits of more precision and accuracy in allocations.

The suggested procedure need not be limited to the pecuniary features of the particularized control benefits studies that, when extrapolated, yield national benefits estimates. As in Adams et al. (in press), similar general procedures may be applied to evaluations of the precision and accuracy of the natural science informational features of these particularized benefits studies.

Acknowledgments - This research was partly supported by USEPA Grant #A808893010. Bruce Forster and Donald Adams have provided helpful comments.

REFERENCES

Adams, R.M., Crocker, T.D., and Katz, R.W. (in press), On the
    adequacy of natural science information: A Bayesian frame-
    work. Review Economics Statistics.
Adams, R.M., Crocker, T.D., and Thanvibulchai, N., 1982, An
    economic assessment of air pollution damages to selected
    annual crops in southern California. J. Envir. Economics
    Management 9:42-58.
Bongers, C., 1980. Standardization: Mathematical Methods in
    Assortment Determination. Martinus Nijhoff Publ., Boston,
    MA. 248 pp.
Cicchetti, C.J., Fisher, A.C., and Smith, V.K., 1976, An econom-
    etric evaluation of a generalized consumer surplus measure:
    The mineral king controversy. Econometrica 44:1259-1276.
Crocker, T.D., 1980, Testimony on September 23 to the Select
    Committee on Small Business and the Committee on Environment
    and Public Works, pp. 103 to 111. In: Economic Impact of
    Acid Rain, U.S. Senate, 96th Congress, 2nd Session, U.S.
    Government Printing Office, Washington, DC.
Crocker, T.D., 1982, Pollution damages to managed ecosystems:
    Economic assessments, pp. 103 to 124. In: Effects of Air
    Pollution on Farm Commodities, J.S. Jacobsen and A.A. Millen,
    eds., Izaak Walton League of America, Arlington, VA.
Davis, R.K., 1963, Recreation planning as an economic problem.
    Natural Resources J. 3:239-249.
Elster, J., 1979, Ulysses and the Sirens: Studies in Rationality
    and Irrationality. Cambridge University Press, New York, NY.
    264 pp..
Hart, O.D., 1980, Perfect competition and optimal product differ-
    entiation. J. Economic Theory 22:279-312.
Haynie, F.H., 1982, Economic assessments of pollution - related
    corrosion damage, pp. 214 to 231. In: Atmospheric Corro-
    sion, W.H. Ailor, ed., John Wiley and Sons, New York, NY.
Kreps, D.M., 1979, A representation theorem for "preference for
    flexibility". Econometrica 47:565-577.
Krutilla, J.V., 1967, Conservation reconsidered. The American
    Economic Review 57:777-786.
Lau, L.J., 1982, The measurement of raw material inputs, pp. 167
    to 200. In: Exploration in Natural Resource Economics,
    V.K. Smith and J.V. Krutilla, eds., The John Hopkins Univer-
    sity Press, Baltimore, MD.
Leamer, E.E., 1978, Specification Searches: Ad Hoc Inference
    with Nonexperimental Data. John Wiley, New York, NY.
    370 pp.
Manuel, E.H., Jr., Horst, R.L., Berennan, K.M., Lanen, W.N.,
    Duff, M.C., and Tapiero, J.K., 1982, Benefits analysis of
    alternative secondary national ambient air quality standards
    for sulfur dioxide and total suspended particulates,

    Vol. III. Final Report to EPA Contract No. 68-02-3392,
    Mathtech, Inc., Princeton, NJ. 277 pp.

Pollak, R.A. and Wachter, M.L., 1975, The relevance of the house-
    hold production function and its implications for the alloca-
    tion of time. J. Political Economy 83:255-277.

Pollak, R.A., 1976, Habit formation and long-run utility func-
    tions. J. Economic Theory 13:272-297.

Roth, J.W., 1982, Some illustrative preservation problems and
    treatments in Washington, D.C., pp. 31 to 48. In: Conserva-
    tion of Historic Stone Buildings and Monuments. Committee on
    Conservation of Historic Stone Buildings and Monuments,
    National Academy Press, Washington, DC.

Salmon, R.L., 1972, Systems analysis of the effects of air pollu-
    tion on materials. Publ. No. 1 PB209-192, National Technical
    Information Service, Washington, DC. 342 pp.

Silverberg, E., 1978, The Structure of Economics. McGraw-Hill
    Book Co., New York, NY. 543 pp.

Thaler, R.H. and Shefrin H.M., 1981, An economic theory of self-
    control. J. Political Economy 89:392-406.

U.S. Department of Commerce, 1980, Statistical abstract of the
    U.S., U.S. Govern. Printing Office, Washington, DC. 1156 pp.

THE AGRICULTURAL SECTOR, AIRBORNE RESIDUALS, AND POTENTIAL

ECONOMIC LOSSES

Walter P. Page

Center for Business and Economics
State University of New York
Plattsburgh, NY   12901

ABSTRACT

This article summarizes the results of two U.S.E.P.A.-funded studies dealing with potential economic losses to the agricultural sector from airborne residuals. Results are reported for the Ohio River Basin and the northeast United States by selected crop and pollutant and for state shares in total regional losses. For each region, losses appear to be relatively large and concentrated by crop and state area. Policy implications of these results are discussed with particular emphasis given to NO_x standards in existing U.S.E.P.A. regulations.

INTRODUCTION

The purpose of this paper is to present results of two research efforts dealing with agricultural losses from airborne residuals. One research project was conducted in connection with the U.S.E.P.A.-funded study, the Ohio River Basin Energy Study (ORBES; Page et al., 1980a). The second study was again done for the U.S.E.P.A. and represented an exploratory regional study to assess possible damages in the northeast United States from acid rain (Page, 1982). The first study dealt with monetarized losses to producers of corn, soybeans, and wheat from SO_2 and O_3 concentrations in the ORBES area (Figure 1). The second study dealt with damages to corn and hay in the northeast from acidic deposition and NO_x concentrations. The study region consisted of the states of Maine, Massachusetts, New Hampshire, New York,

385

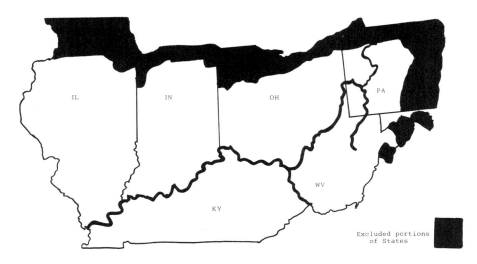

Figure 1:   Ohio River Basin Energy Study Region, Phase II, for the
            U.S.E.P.A. (see text for details).

and Vermont.   In the case of the ORBES study, the purpose was to
estimate potential losses from O_3 and SO_2 during the years 1976 to
2000 under a variety of scenarios concerning energy and fuel use in
the basin.   The northeast U.S. study focused on monetary losses
only for the year 1979.   In addition, the study of monetarized
agricultural losses in the northeast included a sensitivity
analysis; both physical and economic parameters were varied in
order to assess the sensitivity of final results to physical versus
economic estimates.   Further, ORBES study results were obtained for
utility contribution to agricultural losses versus other background
concentrations.   It was not possible to perform a similar analysis
in the study of the northeast because of the lack of information
concerning the transport of airborne residuals.   The northeast U.S.
study is therefore based on total measurements of rainfall acidity
and concentrations of NO_x.

     In the case of the ORBES study, the principal input to the
analysis consisted of physical crop losses, by scenario (Loucks et
al., 1980).   These estimates, in turn, relied upon several other
research outputs concerning future energy and fuel use in the
region, estimates of economic growth in the region, a siting model
for generating facilities, and a transport model (Fowler et al.,
1980; Page and Gowdy, 1980; Page et al., 1980b; Teknekron Research,
1980).   Results were obtained for three different energy and
fuel-use scenarios.   Scenario 2 represented a base-case or
business-as-usual set of future economic and energy fuel-use
characteristics.   The scenario assumed compliance with State
Implementation Plans (SIPs) for existing units and New Source

Performance Standards (NSPS) and Revised New Source Performance Standards (RNSPS) for additions to generating capacity. Scenario 2d was identical to scenario 2 with respect to NSPS and RNSPS standards being met, but differed in that there was no compliance with SIP requirements. Hence, scenario 2d was more lax with respect to air quality than scenario 2. The last scenario examined, scenario 7, was identical to scenario 2 except the growth rate for electric capacity, 1976-2000, was greater and utility plant life was assumed to be 45 years (35 years in scenarios 2 and 2d). All other assumptions were identical to those in scenario 2. Comparing scenarios 2 and 2d, then, provides information concerning the difference in monetarized welfare losses to agricultural producers from alternative policy assumptions concerning compliance with SIPs. Differences found between scenarios 2 and 7 reflect alternative assumptions concerning anticipated regional growth in electric demand as well as generating facilities plant life. Further, two different analyses were performed based on alternative assumptions concerning emissions. One analysis was based upon nominal-load emissions from utilities and the second was based on peak-load. The principal results reported from the ORBES study concern damages from peak-load emissions. It is unclear whether nominal or peak-load emissions is more appropriate; this issue is presently unresolved. Economic losses to the agricultural sector based upon probable crop losses from peak-load emissions, however, will be discussed.

The study of economic losses to the agricultural sector in the northeast was carried out in response to the growing awareness of potential damages due to acid deposition in the northeastern U.S. and eastern Canada. These areas are major recipients of airborne residuals from distant point sources. The project focused on two concerns:

° Does existing information allow investigation of economic damages to the agricultural sector from acid deposition or other airborne residuals in the northeast U.S.? If so, what are the characteristics of the region that warrant separate treatment?

° How sensitive are estimated monetary losses from acid deposition or other airborne residuals to variations in physical and economic parameters.

As noted earlier, the study focused only on damages in a single year (1979) and did so only for those crops where there is some evidence of physical damage from residuals. Results are reported in both studies with respect to losses as a percent of potential clean-air production in the agricultural sector. The reason for doing so is to put the absolute value of dollar losses

in some meaningful perspective: it is much easier to understand
the statement that losses are five or six percent, or whatever
percent, of potential clean-air production than it is to understand
the statement that losses are X dollars.

ECONOMIC ANALYSIS OF AGRICULTURAL LOSSES

     The appropriate economic assessment of losses to agricultural
producers from airborne residuals has been extensively discussed
(Leung et al., 1978; Adams et al., 1979; Benedict and Jaksh, 1979;
Jaksh, 1979; Leung et al., 1979; Forster, 1984).   In general, the
kinds of losses which would be experienced are illustrated in
Figure 2, where the reduction in productivity due to airborne
residuals can be represented by a leftward shift of the supply
curve from S_0 to S_1.   The appropriate measure of losses to the
agricultural producer is the difference in producer surplus between
the initial pollution-free curve (S_0) and the curve in the presence
of airborne residuals (S_1).   Producer surplus in the case of
clean-air production would consist of the area "ade" whereas
producer surplus in the presence of contaminants would consist of
the area "bce."   The effect of airborne residuals is to raise the
equilibrium market price for the crop from a to b.   Finally, in
addition to the loss in producer surplus, there is an associated
loss in consumer surplus.   In the case of the clean-air supply
curve (S_0), consumer surplus consists of the area "agd" whereas
with the dirty-air supply curve (S_1) consumer surplus consists of
the area "bgc".   A complete assessment of welfare losses to
producers and consumers would consist of looking at the differences

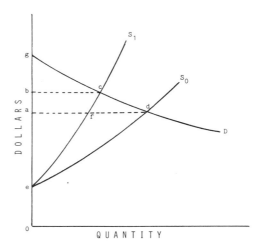

Figure 2:   Producer and consumer surplus with
            shifting supply curve.

in producer surplus, consumer surplus and the increase in price due
to dirty air.

Estimation of these losses as well as the change in product
price would require estimating both supply and demand curves for
the crop in question. A more limited analysis, the one reported in
this paper, is the estimation of losses to producers only as
illustrated in Figure 3. Estimation of such losses is based on the
assumption that there is a fixed price, a, faced by each producer.
This assumption may be justified on the grounds that only a single
region was examined and in both studies the regions do not account
for a large share of total United States production. Hence the
regions would tend to be price takers[1]. Only if all producing
regions for these crops experienced the same concentrations of
airborne residuals would there be an associated price effect. The
producer surplus would consist of the dollar value of the area
"Oac" (Figure 3) in a pollution-free environment. It consists of
the area "Oab" given contaminants which reduce agricultural
productivity. The difference, "Obc", is the estimated welfare loss
to agricultural producers from airborne residuals. This loss
estimate does not include potential losses to those supplying
commodities to the agricultural sector. An estimate of these
indirect losses in the ORBES region can be found in Miller (1982).
If productivity in the production of agricultural goods is reduced,
there may be associated reductions in other sectors which provide
resources for agricultural production. An example would be the
reduction in earnings in the transport sector associated with lower
levels of agricultural productivity. A complete analysis of the
social losses from airborne residuals would include estimating the
potential direct and indirect losses to agricultural producers as
well as consumer surplus losses.

The problem, however, is even more complex than suggested in
the above statements. An air pollution damage function is a
statement relating the harmful physical effects that result from
various levels of contaminants introduced into the air due to human
and non-human activities or processes. Air pollution is costly
because it reduces the capacity for proper functioning of human and
natural processes and activities. The cost of pollution may be
defined as the value that people place on reducing those damages
suffered because of airborne residuals. The greater the reduction
of damage, the greater will be the value attached by people to
damage reduction. A host of problems stand in the way of measuring
the cost of air pollution: a market does not exist which permits

---

[1]Price takers are firms in a competitive industry faced with the
existing market price. Revenues for such firms are only a function
of quantity.

Figure 3:   Producer surplus with fixed price
and shifting supply curve.

people to make actual payments based on individual valuation; many
pollution costs are unknown or only vaguely perceived; certain
types of pollution damages, although real, are not understood; some
damages do not effect people directly, although society still
places a value on their elimination; and some costs are recognized
and experienced directly but individuals do not know the valuation
they would place on their reduction.   Ideally, an economic analysis
of air pollution damages would entail a comparison of the schedule
of benefits of pollution reduction with a schedule of cost of
pollution abatement.   Since optimal abatement levels require a
comparison of incremental costs and benefits, it would be necessary
to develop a schedule of incremental benefits.   Because it is
difficult to develop entire schedules of abatement benefits, at a
minimum it is desirable to estimate the benefits which would result
from marginal reductions in concentrations from current levels.

The approach taken in the two studies (Page et al., 1980a, b,
1982) followed the suggestions of Ridker (1967), where three
different approaches for the measurement of the cost of air
pollution are discussed.   The measure adopted in these studies was
restricted to the estimation of the direct effects to agricultural
producers in the absence of adjustments.   The analyses, then,
focused only on the direct losses to agricultural producers in the
absence of any assumed adjustments to enhance productivity and
without including estimates of the indirect losses experienced by
suppliers to the agricultural sector.   Finally, no effort was made
to estimate the optimal level of pollution.   Such an estimate would
require, as noted earlier, a complete schedule of benefits and

costs associated with the agricultural sector and potential damages from airborne residuals. Because this analysis did not include consumer surplus losses nor losses to industries supplying inputs to the agricultural sector, damage estimates clearly understated the true value of social losses from airborne residuals.

DATA BASES AND PARAMETER ESTIMATES

Several parameter estimates are required to conduct this type of analysis (Figure 3): estimates of the annual real price and production of affected agricultural goods, supply elasticities for each of the goods, and the size of the agricultural sector with respect to each crop. Potential clean-air production for each crop in the absence of contaminants also needs to be estimated.

The ORBES study (Page et al., 1980a) evaluated corn, soybeans, and wheat, while corn and hay were the crops used in the northeast U.S. study (Page, 1982). Corn was thought to be influenced by atmospheric ozone and hays by acid precipitation. Fixed prices were used in both studies as weighted average prices received by farmers over the historic period. Weighted averages were computed by weighting state prices in each year by the percent that years' production was of total production over the historic period. The reason for using weighted averages was to avoid the influence on average prices associated with a few years of unusual crop conditions. In the case of the ORBES study it was assumed that the weighted price for each crop and state was unchanged over the period 1976-2000; the implication of assuming a fixed weighted price for the period 1976-2000 is that agricultural prices would increase at the overall inflation rate for the entire period. In the study of northeast agriculture the weighted price had to be fixed for 1979.

The calculation of producer surplus losses also relies upon estimates of supply elasticities for each crop. Secondary source information on supply elasticities for these crops was not adequate. The principal problem was the large range in values from extant studies. The variation in literature estimates was particularly pronounced when considering regional, as contrasted with national, studies (Askari and Cummings, 1977). Finally, it was not possible to select elasticity estimates from the literature for each crop based upon similar specifications, geographic areas, or years covered by data sets. Elasticity estimates for both studies, with the exception of hay for the northeast study, were calculated separately using a distributed lag model for the estimation of supply elasticities (Nerlove, 1968). In general, the model used for the estimation of supply elasticities incorporated past as well as future information concerning expectations of agricultural producers. The expectations were formulated on the

basis of past information. Yet all past information did not have
equal influence on producer expectations, for recent values are
more indicative of future price expectations than earlier values.
Hence, future expectations can be expressed as a weighted moving
average of past values in which the weights decline going back in
time. This is a widely used and accepted method for modeling
producer expectations of future prices. In the case of the ORBES
study, the mean elasticities for soybeans, wheat, and corn,
respectively, were 0.263, 0.56, and 0.187. The same elasticity for
corn (0.187) was used for the northeast agriculture study while
elasticity for hays was assumed to be the same as for corn.

Finally, it was necessary to identify current levels of
production associated with existing levels of airborne residuals.
The calculation of potential clean-air production for 1979 was a
simple matter. It consisted of observed production in 1979 (in the
presence of dirty air) divided by one minus the physical damage coefficient from observed airborne residuals. In the case of the
ORBES study, the physical damage coefficients were provided by
Loucks et al. (1980) for minimum, maximum, and probable estimates
and for nominal and peak-load emissions. Physical damage
coefficients were also based upon this information as well as
recent studies concerning damages to hay crops from acid deposition
for the northeast agriculture study. A six percent physical damage
loss to corn production from ozone concentrations was selected.
The reason for this decision was related to the similarity in ozone
readings for the five northeast states (246 $\mu g \, m^{-3}$) and the ORBES
region (270 $\mu g \, m^{-3}$). This value was roughly midway between the
minimum and probable estimates used in the ORBES study. The
physical damage coefficient from acid precipitation for hay in the
northeast was based upon experimental evidence on yield of
marketable portions of Fescue and Timothy in 1979 suggesting a
ten percent reduction in productivity at pH levels of
approximately 4 (Lee et al., 1981). Such levels in atmospheric
acidity have been commonly experienced in the northeast. Given
these physical damage coefficients, it was possible to derive a
potential clean-air production level in order to estimate losses in
producer surplus between clean-air and dirty-air cases. Knowing
potential clean-air production, the real price of agricultural
goods and supply elasticities, the difference in producer surplus
under clean air conditions and in the presence of airborne
residuals was calculated.

INTERPRETATIONS OF RESULTS

Major findings from each study (Page, 1980, 1982) are reported
and discussed as well as implications for public policy. With
reference to the study of damages in the northeast, results are
found in Tables 1 to 4. The dollar value of producer surplus for

Table 1. Distribution of clean and dirty air producer surplus by crop and area for the northeastern United States during 1979.

AREA	Producer Surplus, Clean Air (thousands of dollars)			Producer Surplus, Dirty Air (thousands of dollars)		
	Corn	Hay	Total	Corn	Hay	Total
Maine	12,261	23,277	35,538	11,525	20,950	32,474
Massachusetts	13,370	16,869	30,238	12,567	15,182	27,749
New Hampshire	7,259	11,367	18,626	6,824	10,230	17,054
New York	284,664	334,891	619,555	267,584	301,402	568,986
Vermont	31,760	51,392	83,152	29,855	46,252	76,107
Five-State Region	349,314	437,795	787,109	328,355	394,015	722,370

Table 2.   Losses to agricultural producers from airborne residuals
           in the northeastern United States in 1979 for corn and
           hay.

AREA	Producer Surplus, Clean Air (thousands of dollars)	Absolute Differnce in Producer Surplus with Airborne Residuals (thousands of dollars)	Percent Loss For Each State
Maine	35,538	3,063	4.7
Massachusetts	30,238	2,489	3.8
New Hampshire	18,626	1,572	2.4
New York	619,555	50,569	78.1
Vermont	83,152	7,045	10.9
Five-State Region	787,109	64,738	100.0

clean and dirty air by crop and region is given in Table 1.
Table 2 contains the total producer surplus for clean air over both
crops by region, the difference between clean and dirty air in
producer surplus for both crops, and the percent of loss for each
state.   The distribution of losses from dirty air by region and
crop is shown in Table 3, where the third column shows the
percentage losses from dirty air for potential producer surplus in
each region.   Finally, Table 4 contains sensitivity results from
varying price, elasticity of supply, and damage co-efficients.   The
second column indicates the percent difference for each case in the
total five-state region.   Table 1 is included to give the basic
results concerning clean and dirty-air producer surplus by crop and
region.   The principal interpretation of the results is based on
data listed in Tables 2-4.

From Table 2 it is seen that the five-state region potential
clean-air producer surplus is $787 million, and the difference
between producer surplus with clean versus dirty air is about
$65 million.   Similar information is found in the table for each of
the five states.   The five-state region potential losses during
1979 were on the order of 8.2% of producer surplus if clean air
were present (see also Table 3).   The losses to the five-state
region shown in Table 2 were highly concentrated in certain areas.
Of the total losses of $65 million, approximately 78.1% was
accounted for by New York alone.   In decending order of importance,
the rank was: Vermont (10.9%), Maine (4.7%), Massachusetts (3.8%),
and New Hampshire (2.4%).   As a matter of public policy, these
results suggested that the states of New York and Vermont should
surely be highly sensitive to potential agricultural losses for
corn and hay crops due to airborne residuals.

Table 3.   Distribution of losses from dirty air by area and crop in
           the northeastern United States during 1979[a].

AREA	Corn (%)	Hay (%)	Total (%)
Maine	6.0	10.0	8.6
Massachusetts	6.0	10.0	8.2
New Hampshire	6.0	10.0	8.4
New York	6.0	10.0	8.2
Vermont	6.0	10.0	8.5
Five-State Region	6.0	10.0	8.2

[a]Entries in columns 1 and 2 reflect the percent reduction from
potential clean air producer surplus associated with corn and hay
in the specified areas.   Column 3 reflects the total reduction in
producer surplus from both crops for each area.

The percent reduction from potential clean-air producer
surplus for both crops was constant; corn was 6% and hay 10%
(Table 3).   Each of these columns contain numbers calculated by
dividing the losses to the particular crop by the potential
clean-air producer surplus for that crop.   Even though the five-
state region losses were highly concentrated for New York and
Vermont, it is nonetheless true that for each of the states, as
well as the five-state region, the percent of crop losses as a
proportion of potential clean-air producer surplus was quite
similar; the highest percentage was 9.6% in the case of Maine and
the lowest 8.2% in the case of New York.   A second perspective,
then, would be to observe that while the five-state region losses
are highly concentrated in New York and Vermont, nonetheless for
each state the percent losses for corn and hay production are quite
similar.

The reader should be cautioned when interpreting the results
for percentage variations in critical parameters (price, elasticity
of supply or physical damage coefficients) shown in Table 4.   For
the base-case values of the three parameters, the five-state
regional difference in producer surplus with clean versus dirty air
was $65 million.   The procedure was to increase the price by
five percent and re-estimate potential clean-air producer surplus
and dirty-air producer surplus.   A new difference was then

Table 4.   Sensitivity of regional losses to variations in physical
           and economic parameters.

Parameter Variations (%)	Sensitivity Results[a] (%)
PRICE:	
+ 5	+ 05.0
+ 10	+ 10.0
+ 20	+ 20.0
− 5	− 05.0
− 10	− 10.0
− 20	− 20.0
ELASTICITY:	
+ 5	− 00.8
+ 10	− 01.6
+ 20	− 03.1
− 5	+ 00.8
− 10	+ 01.6
− 20	+ 03.3
DAMAGE COEFFICIENT:	
+ 5	+ 05.5
+ 10	+ 11.1
+ 20	+ 22.4
− 5	− 05.5
− 10	− 09.3
− 20	− 21.5

[a]In each case, the percentage is calculated by dividing the
difference between potential clean-air producer surplus and dirty-
air surplus with a plus or minus change in only that parameter by
the corresponding difference from the base case.

obtained. The last column in Table 4 consists of percentage figures derived by dividing the difference in producer surplus between clean and dirty air, i.e., as if the price were five percent higher, by the difference between the base-case price and other base-case parameters. Only one parameter was changed at a time. Part of the reason for examining the sensitivity of results to such variations is to ascertain whether or not monetarized losses are most sensitive to physical damage coefficients or economic parameters. A second reason, however, relates to studies which may be conducted under various scenarios concerning acid rain or ozone concentrations over time. If, for instance, damages to crops from acid rain over a 25-year period are examined, then price projections will be required. If current period results are quite sensitive to this parameter value, then it becomes critical that such projections be state-of-the-art and as believable as possible. If results are not particularly sensitive to variations in price, one would be somewhat less responsive to arguments over price forecast.

The results of the sensitivity analysis are particularly interesting when evaluating price variations. Changes in damages to corn and hay production for the region are almost identical to the parameter variations themselves. In large measure the elasticities of supply for the crops were quite inelastic (Table 4). This is typical of agricultural crops which are used as inputs for the production of other agricultural goods, in this case milk and beef production which are themselves subject to inelastic demand. In the case of the present work, weighted average prices were derived from extant data so that the point estimates were acceptable. If prices were forecast through time, however, these sensitivity results suggested that such forecasts were important because resulting monetary damages were quite sensitive to price forecasting errors. As expected, the results were not particularly sensitive to variations in elasticities. Even a 20 percent increase in elasticity would only lower total damages to the five-state region by 3.05 percent while a 20 percent decrease would raise the total damages by 3.25 percent (Table 4).

The sensitivity of the results to variations in the damage coefficients were quite similar to those observed with regard to price, although somewhat higher by a percentage point or two. The effect of increasing the damage coefficient was to shift the potential clean-air production point to a larger value and hence to increase the area calculated as producer surplus losses. It appears that results were quite sensitive to the estimated physical damage coefficients. From these calculations, the most sensitive parameters for estimating producer surplus losses were agricultural prices and estimates of physical damage coefficients. As observed in Table 4, if the price forecast or the damage coefficient were off by ±20%, results could well be off by the same amount.

Results from the ORBES study are given in Tables 5-9. The first three tables contain the net present value of probable total and utility-related crop losses from sulfur dioxide and ozone for the years 1976 to 2000 for each scenario (2, 2d, and 7). Calculations appearing in Tables 8 and 9 provide information on the distribution of losses by crop as well as total and utility losses by crop. Data listed in Tables 5 to 7 were based upon estimates using peak-load emissions from utilities. Tables 8 and 9, on the other hand, were based on nominal-load emissions from utilities. The percent distributions by crop are invariant whether peak or nominal-load emissions were used.

The value of monetarized agricultural losses for scenario 2, in millions of (1975) dollars, is given in Table 5. Calculations are also provided for pollution-free output for the 1976-2000 period as well as the percent losses by ORBES-related state areas. In the case of scenario 2 (base-case scenario), total regional losses constituted approximately 10.3% of the discounted value of pollution-free output over the 1976-2000 period. Those losses uniquely attributable to utilities constituted approximately 4.3 percent, or about 40 percent of the total losses. These losses were unevenly distributed with respect to the ORBES-related portion of each state. Illinois, which accounted for 53.6% of total regional losses, had the single largest loss within the region. This was followed by Indiana with 25.1%, and then by Ohio with 13.9%. The remaining states constituted only a small fraction of the total regional losses from airborne residuals: 6.1% for Kentucky, 0.4% for Pennsylvania, and 0.1% for West Virginia.

Results for scenario 2d, which differs from scenario 2 only with respect to SIP (State Implementation Plans) standards, are provided in Table 6. The loss pattern was almost identical to that of scenario 2. The total regional losses constituted 10.4% of the present discounted value of clean-air production, with utility related losses being 4.3%. As was the case with scenario 2, the distribution of losses was uneven with respect to the ORBES-related portions of states and was almost identical in its distribution. Compliance or noncompliance with SIPs did not significantly influence the degree of agricultural losses within the region nor within the ORBES-related portions of states within the region.

As noted earlier, scenario 7 differed from scenario 2 only with respect to the anticipated rate of growth in electric generating capacity and plant life: Respective values for scenario 7 were 3.9% and 45 years while scenario 2 values were 3.13% and 35 years. The effect of these two differences was to raise the percent losses of pollution-free output from 10.3% in scenario 2 to 12.3% in scenario 7 (Table 7). The losses uniquely attributable to utilities also increased from 4.3 to 5.5 percent. The distribution of losses with ORBES-related portions of each

Table 5. Net present value of probable total and utility-related cumulative crop loss from 1976 to 2000 using Scenario 2[a].

ORBES area	Losses[b] (millions of dollars)	Percent losses are of pollution-free output (%)	Percent losses are of ORBES total losses[c] (%)
IL	3755.56 (1565.78)	10.5 (4.4)	53.6 (53.8)
IN	1758.51 (727.60)	10.0 (4.1)	25.1 (25.0)
KY	479.76 (201.18)	10.6 (4.4)	6.1 (6.9)
OH	976.00 (403.18)	10.1 (4.2)	13.9 (13.8)
PA	27.71 (11.58)	7.6 (3.2)	0.4 (0.4)
WV	4.48 (1.91)	7.7 (3.3)	0.1 (0.1)
ORBES TOTAL	7002.03[d] (2911.22)	10.3 (4.3)	100.0 (100.0)

[a] Assumes a 10% discount rate (numbers in parentheses are for utility-related losses)
[b] Crops are corn, soybeans, and wheat
[c] Sum may not be 100% due to rounding
[d] Sulfur dioxide losses are 0.7% of total and ozone is 99.3%

state was comparable in magnitude to that observed for scenarios 2 and 2d. The higher growth rate in capacity for scenario 7 produced the only case where agricultural losses, and hence monetary losses, increased over the entire 1976-2000 period. In scenarios 2 and 2d, losses increased until 1984 and then decreased from 1985 to the year 2000. In the case of scenario 7, however, the higher growth rate in electric generation capacity would elevate the physical losses, and hence economic losses; these would rise throughout the period.

The percent of individual crop losses to total ORBES losses (Table 8) as well as the percent of individual crop losses for a

Table 6.   Net present value of probable total and utility-related
           cumulative crop loss from sulfur dioxide and ozone for
           1976 to 2000 using Scenario 2d[a].

ORBES area	Losses[b] (millions of dollars)	Percent losses are of pollution-free output (%)	Percent losses are of ORBES total losses[c] (%)
IL	3767.02 (1577.24)	10.6 (4.4)	53.6 (53.6)
IN	1771.41 (740.50)	10.0 (4.2)	25.2 (25.2)
KY	482.59 (204.01)	10.6 (4.5)	6.9 (6.9)
OH	977.96 (405.11)	10.1 (4.2)	13.9 (13.8)
PA	27.95 (11.82)	7.6 (3.2)	0.4 (0.4)
WV	4.51 (1.93)	7.8 (3.3)	0.1 (0.1)
ORBES TOTAL	7031.41[d] (2940.61)	10.4 (4.3)	100.0 (100.0)

[a] Assumes a 10% discount rate (numbers in parentheses are for
   utility-related losses)
[b] Crops are corn, soybeans, and wheat
[c] Sum may not be 100% due to rounding
[d] Sulfur dioxide losses are 1.1% of total and ozone 98.9%

given area (Table 9) are also provided. Both tables are based on
nominal-load emissions. For any given state within the ORBES
region, losses were heavily concentrated for soybeans and corn,
with soybeans being the worst (Table 8). This was calculated as a
percent of ORBES total losses. Losses to wheat production were
insignificant. The same statement was true with respect to
utility-related losses, with the percentage being almost identical.
This reflects, of course, the distribution of total acreage across
crops in the Ohio Basin area as well as the parametric values of
the loss coefficients for corn, soybeans, and wheat.

Table 7.   Net present value of probable total and utility-related
           cumulative crop loss from sulfur dioxide and ozone for
           1976 to 2000 using Scenario 7[a].

ORBES area	Losses[b] (millions of dollars)	Percent losses are of pollution-free output (%)	Percent losses are of ORBES total losses[c] (%)
IL	4491.960 (1998.010)	12.6 (5.6)	53.7 (53.8)
IN	2090.910 (924.048)	11.8 (5.2)	25.0 (24.9)
KY	574.604 (256.996)	12.7 (5.7)	6.7 (6.9)
OH	1168.870 (516.399)	12.0 (5.3)	14.0 (13.9)
PA	30.909 (13.649)	8.4 (3.7)	0.4 (0.4)
WV	4.978 (2.231)	8.6 (3.9)	0.1 (0.1)
ORBES TOTAL	8362.221[d] (3711.345)	12.3 (5.5)	100.0 (100.0)

[a] Assumes a 10% discount rate (numbers in parentheses are for
   utility-related losses)
[b] Crops are corn, soybeans, and wheat
[c] Sum may not be 100% due to rounding
[d] Sulfur dioxide losses are 0.6% of total and ozone 99.4%

     Percent distribution of losses across crops for each of the
individual states within the ORBES region is listed in Table 9,
where the extent of losses within each state crop is noted.
Soybean losses for Illinois, Indiana, Kentucky, and Ohio
constituted upward to 60% of their total losses. Similarly, corn
losses were very large, ranging in the vicinity of 37-43% for
Illinois, Indiana, Kentucky, and Ohio and constituting on the order
of 95% in the case of Pennsylvania and West Virginia. It is clear,
then, that the principal losses in the region are related to
soybean and corn losses.

Table 8. Distribution of ORBES total and utility-related
          agricultural losses by Crop for the years 1976-2000 using
          Scenario 2[a].

ORBES area	Losses to crops (millions of dollars)	Losses as a Percent of Total ORBES Losses[b]		
		Corn	Soybeans	Wheat
IL	4349.089 (1747.129)	20.6 (20.6)	31.7 (31.8)	1.3 (1.3)
IN	2032.809 (814.236)	11.0 (10.9)	13.1 (13.1)	1.0 (1.0)
KY	557.455 (226.088)	2.6 (2.6)	4.1 (4.1)	0.2 (0.2)
OH	1136.639 (455.411)	5.3 (5.3)	7.8 (7.8)	0.9 (0.9)
PA	30.324 (12.179)	0.4 (0.4)	0.0 (0.0)	0.0 (0.0)
WV	4.858 (1.972)	0.0 (0.0)	0.0 (0.0)	0.0 (0.0)
ORBES Total	8111.174 (3257.015)	39.9 (39.9)	56.7 (56.7)	3.4 (3.4)

[a] Assumes a 10% discount rate and nominal load emissions (numbers
in parentheses are for utility-related losses)
[b] Each percentage was calculated by dividing the dollar losses for
each crop by ORBES total losses for all three crops. The same
procedure was followed for calculating utility-related
percentages. Zero percent indicates a percentage less than 0.05.

Table 9. Distribution of area total and utility-related agricultural losses by Crop for the years 1976-2000 using Scenario 2[a].

ORBES area	Losses to crops (millions of dollars)	Losses as a Percent of Total Area Losses[b]		
		Corn	Soybeans	Wheat
IL	4349.089	38.5	59.1	2.4
	(1747.129)	(38.4)	(59.2)	(2.5)
IN	2032.809	43.8	52.3	4.0
	(814.236)	(43.8)	(52.3)	(4.0)
KY	557.455	37.4	59.5	3.0
	(226.088)	(37.3)	(59.7)	(3.0)
OH	1136.639	38.1	55.7	6.2
	(455.411)	(38.2)	(55.6)	(6.2)
PA	30.324	93.9	0.0	6.1
	(12.179)	(93.9)	(0.0)	(6.1)
WV	4.858	96.2	0.0	3.9
	(1.972)	(96.0)	(0.0)	(4.0)
ORBES Total	8111.174	39.9	56.7	3.4
	(3257.015)	(39.9)	(56.7)	(3.4)

[a] Assumes a 10% discount rate and nominal load emissions (numbers in parentheses are for utility-related losses)
[b] Each percentage was calculated by dividing the dollar losses for each crop by the area total losses for all three crops. The same procedure was followed for calculating utility-related percentages. Zero percent indicates a percentage loss less than 0.05.

CONCLUSIONS

The Ohio River Basin Energy Study (ORBES) supported the following general conclusions for the period 1976 to 2000:

o   monetary losses to agricultural producers in the ORBES six-state region were 12 percent of the present discounted value of clean-air production;

o   similarly defined, losses from utilities were 4.8%;

o   losses were highly concentrated in the ORBES portions of Illinois, Indiana, and Ohio;

o   the overwhelming monetary losses were attributable to ozone concentrations in the region;

o   high growth in electric demand (scenario 7) produced annual losses which increased to the year 2000, while losses for scenarios 2 and 2d leveled off after 1985; and

o   losses were highly concentrated in soybeans and corn as a percent of ORBES total losses as well as a percent of a given loss for each state.

A study of northeastern five-state agriculture economic losses supports the following general conclusions:

o   using base period parameter estimates, potential losses to producers of hay and corn in the five-state region, were about $65 million for 1979;

o   losses in producer surplus constituted approximately 8.2% of potential clean-air producer surplus values for the five-state region during 1979;

o   the five-state region losses were highly concentrated in the two states of New York (78.1%) and Vermont (10.9%);

o   for each of the states taken individually, losses to corn and hay were approximately 8.2 percent of producer surplus as related to a clean air situation;

o   in each state, as well as for the five-state region, losses to corn production in the presence of dirty air was 6% of potential clean-air producer surplus, while for hay it was 10%; and

o   regional losses were most sensitive to variations in the prices of crops and the estimated physical damage coefficients. Losses appeared not to be particularly sensitive to the estimated supply elasticities for the crops.

These two studies supported several general observations. First, monetarized agricultural losses were heavily attributable to ozone concentrations in at least one of the regions. Yet NO_x emissions, the principal precursor of ozone, did not figure prominently in SIPs nor in NSPS or RNSPS standards. With regard to provisions of the amended Clean Air Act, secondary standard provisions have not been adequately emphasized in establishing NO_x standards for mobile and point sources. Secondly, losses to the agricultural sector from airborne residuals in these two regions tend to be quite large relative to potential clean air production. In addition, losses are very unevenly distributed both spatially and by agricultural crop. The third observation concerns sensitivity of monetarized agricultural losses to various physical and economic parameters. It appears that economic analysis of losses is most sensitive to the estimated price of the agricultural goods and the estimated physical damage coefficient relating particular airborne residuals to particular crops. Finally, both studies strongly suggested the need for a regional perspective in examining the relationships between airborne residuals and crops. This is seemingly due to several reasons: the nature of cropping practices varies from one region to another and individual crops vary in their sensitivity to concentrations of residuals. The sensitivity issue relates to the concentrations of airborne residuals themselves, where there are large variations from one region of the country to another. Also, the existence of air corridors or patterns of acidic deposition distributions, which account for various areas of concentrations across the United States, must be considered. To put the matter differently, it appears reasonable to think, for both technical and economic reasons, that economic damages suffered by agricultural producers are spatially very unevenly distributed. It behooves policy makers and researchers to recognize this fundamental fact.

Acknowledgements - Kern O. Kymn, West Virginia University, is thanked for helpful comments on earlier drafts of this paper.

REFERENCES

Adams, R.M., Thanaribulchai, N., and Crocker, T.D., 1979, Preliminary Assessment of Air Pollution Damages for Selected Crops within Southern California, Vol. III. Method Developments for Assessing Air Pollution Control Benefits. U.S.E.P.A., Laramie, WY. 99 pp.

Askari, H. and Cummings, J.T., 1977, Estimating agricultural supply
    response with the Nerlove model: A survey. Internat.
    Economic Review 8:257-292.
Benedict, H.M. and Jaksh, J.A., 1979, Protocol for economic
    assessment of damage to vegetation by air pollution, pp. 23 to
    42. In: Proceedings of an APCA Technical Conference,
    Methodologies for Assessment of Air Pollution Effects on
    Vegetation. Minneapolis, MN.
Forster, B.A., 1984, Economic impact of acid precipitation: A
    Canadian perspective, pp. 97 to 122. In: Economic
    Perspectives on Acid Deposition Control, T.D. Crocker, ed.,
    Butterworth Publ., Stoneham, MA.
Fowler, G.L., Jansen, R.J.C., Jones, W.W., Bailey, R.E., and
    Gordon, S.I., 1980, The Ohio River Basin Energy Facility
    Siting Model, Vol. I. Grant No. EPA R805588 and Subcontract
    to University of Illinois at Chicago Circle under Prime
    Contract EPA R805588, U.S.E.P.A., Chicago, IL. 196 pp.
Jaksh, J.A., 1979, Economic evaluation of air pollution damage to
    crops: The essentials for a proper analysis. Abstract, 72nd
    Annual Meeting of the Air Pollution Control Association,
    Pittsburgh, PA.
Lee, J.L., Neely, G.E., Perrigan, S.C., and Grothaus, L.C., 1981,
    Effect of simulated sulfuric acid rain on yield, growth and
    foliar injury of several crops. Envir. Exper. Botany
    21:171-185.
Leung, S., Johnson, R., Ling, T., Noorbakhsh, M., Reed, W., and
    Walthall, R., 1979, The Economic Effect of Air Pollution on
    Agricultural Crops, Application and Evaluation of Methodology:
    A Case Study. Interim Report, Eureka Laboratories, Inc.,
    Sacramento, CA. 58 pp.
Leung, A., Reed, W., Cauchois, S., and Howitt, R., 1978,
    Methodologies for Evaluation of Agricultural Crop Yield
    Changes: A Review. Corvallis Environmental Research
    Laboratory, U.S.E.P.A., Washington, DC. 168 pp.
Loucks, O.L., Armentano, T.V., Usher, R.W., Williams, W.T.,
    Miller, R.W., and Wong, L., 1980, Crop and Forest Losses Due
    to Current and Projected Emissions from Coal-Fired Power
    Plants in the Ohio River Basin. Prepared for Ohio River Basin
    Energy Study (ORBES), Subcontract under Prime Contract
    U.S.E.P.A. R805588, The Institute of Energy, Indianopolis, IN.
    282 pp.
Miller, J., 1982, An input-output study of agricultural losses in
    the Ohio River Basin due to airborne residuals. Ph.D. Thesis,
    West Virginia Univ., Morgantown, WV. 121 pp.
Nerlove, M., 1968, Dynamics of Supply: Estimation of Farmers
    Response to Price. Johns Hopkins University Press, Baltimore,
    MD. 267 pp.
Page, W.P., 1982, Economic Losses to the Agricultural Sector in the
    Northeast Due to Airborne Residuals. Subcontract No. 81-140
    between West Virginia Univ., Morgantown, WV, and the

University of Illinois under Cooperative Agreement No.
CR809461010 between the University of Illinois and the
U.S.E.P.A., Washington, DC. 101 pp.

Page, W.P. and Gowdy, J., 1980, Gross Regional Product in the Ohio
River Basin Energy Study Region, 1960-1975. Subcontract under
Prime Contract EPA R805588 to West Virginia University,
Morgantown, WV. 74 pp.

Page, W.P., Gilmore, D., and Hewing, G., 1980b, An Energy and Fuel
Demand Model for the Ohio River Basin Energy Study Region.
Grant No. EPA R805585 and Subcontract under Prime Contract EPA
R805588 to West Virginia University, Morgantown, WV. 115 pp.

Page, W.P., Ciecka, J., Rabian, R.B., and Arbogast, G., 1980a,
Estimating Regional Losses to Agricultural Producers from
Airborne Residuals in the Ohio River Basin Energy Study
Region, 1976-2000. Prepared for Ohio River Basin Energy Study
(ORBES), Phase II, Grant No. EPA R805585 and Subcontract under
Prime Contract EPA R805588 to West Virginia University,
Morgantown, WV. 144 pp.

Ridker, R.G., 1967, Economic Costs of Air Pollution. Praeger
Publ., New York, NY. 214 pp.

Teknekron Research, Inc., 1980, Air Quality and Meterology in the
Ohio River Basin: Baseline and Future Impacts. Subcontract
under Prime Contract EPA R4833530, Teknekron Inc., Berkeley,
CA. 162 pp.

ECONOMIC IMPACT OF ACID DEPOSITION IN THE CANADIAN AQUATIC SECTOR

Bruce A. Forster

Department of Economics
University of Guelph
Guelph, Ontario, Canada   N1G 2W1

ABSTRACT

Water quality and aquatic life are important variables for various economic activities, which include commercial fisheries, sport-fishing and other aquatic-based recreational endeavors such as boating, swimming, and SCUBA-diving. The impacts of acidification on some of these activities may be valued by examining market data. The paper will discuss appropriate methods and provide approximate estimates of the value of these impacts where possible. For other markets, such as prices, participation or visitation, data may not exist. When such data does exist, it may not provide complete economic valuations of the relevant activities. Techniques that have evolved during the last decade that permit economists to estimate monetary values for activities that are not appropriately revealed in market data are discussed in this paper. These techniques will be related to a major study conducted for the Ontario Ministry of the Environment on impact of acid deposition on environmental amenities.

INTRODUCTION

The purpose of this chapter is to discuss the economic impacts of acid deposition on the Canadian aquatic sector. Even though acid deposition is an important problem that requires action, the economist cautions that what is at issue is how much control at what cost is socially desirable. In order to answer this question, the economist needs to know what benefits will result from varying degrees of control. The benefits of controlling pollution are the

409

values to society of the adverse physical impacts that are avoided as a result of the control measures. The economist generally attempts to calculate monetary estimates of these damages so that they may be compared with the costs of control which are generally expressed in monetary terms.

It is frequently this use of monetary estimates that disturbs the general public. They may feel that some environmental attributes are above monetary measurement and are somehow on a different plane than "common" goods that have money prices attached to them. However, underlying these monetary values are real economic resources. In order to improve environmental quality or prevent a deterioration, it is generally necessary to reorganize the use of society's productive economic resources. This usually means producing less of some commodities desired by society. The economist asks how much society is prepared to pay in foregone commodities to achieve improved environmental quality. In general, society will not be willing to pay more than the benefits that accrue from improving environmental quality.

Crocker (1982) cautioned that various features of the acid deposition phenomenon may result in traditional benefit-cost analyses producing misleading answers. In particular, there are reasons to expect downward biases in benefit estimates of control and upward biases in control cost estimates. Provided that this is kept in mind, the exercise of attempting to estimate benefits is still useful for as Brady (1983) stated:

> "The very process of identifying effects, arraying them, and attempting quantification produces valuable information which will aid policy makers in making rational choices. It is fine and good if quantification and monetization can be obtained; but if it is not possible, this is equally valuable information for the policy maker. The process of undertaking a benefit-cost analysis makes explicit the effects and assumptions used in quantification. The value of the exercise lies in the precise specifications of some variables, while indicating the limitations with reasonable assurance concerning other variables. For others, it is useful to be specific that valuation is simply not attempted."

In areas in which there are high degrees of uncertainty concerning benefit estimation, Brady (op.cit.) suggested that information be arrayed as follows:

$$\begin{pmatrix} \text{Social valuation} \\ \text{of environmental} \\ \text{effects} \end{pmatrix} = \begin{pmatrix} \text{Effects valued} \\ \text{in} \\ \text{economic terms} \end{pmatrix} + \begin{pmatrix} \text{Effects valued} \\ \text{in} \\ \text{"environmental} \\ \text{quality" terms} \end{pmatrix} \qquad (1)$$

Economists would replace the word "economic" in the second bracket by "monetary," which is what Brady (op. cit.) presumably meant. For the third bracketed item, he is suggesting a qualitative assessment be made of environmental amenities for which monetary estimation currently may not be feasible. Good economics should consider both of these categories.

DISCUSSION

The following discussion will consider economic assessments of the aquatic impacts of acid deposition. Where possible, monetary estimates of the impacts will be offered using various techniques from economic analysis. An assessment of the information available or lacking in the sciences and other disciplines essential for such an analysis will be made. It is ironic that while the aquatic impacts of acid deposition are best understood by the science community, there are special features of the problem and the information collected that make it a more difficult sector for the economist to assess.

The Aquatic Sector at Risk

In order to determine the possible extent of aquatic damage from acidic deposition, it is necessary to determine the area that is a) sensitive to acidification, and b) receiving acidic deposition in amounts likely to cause acidification of aquatic ecosystems. A waterway is classified as sensitive to acidification if it has a low buffering capacity, i.e., a low ability to neutralize acidity. One common measure of a waterway's ability to neutralize acidity is its alkalinity. Waterways for which alkalinity measures less than 200 μeq l^{-1} are commonly classified as sensitive (Linthurst, 1983). Another common measure of buffering ability is the Calcite Saturation Index (CSI) which is a relative measure combining information on pH, alkalinity and calcium concentrations (Jeans, 1982). Waterways with a CSI greater than 3 are sensitive while those with a CSI less than 3 are insensitive.

For severity of deposition, there are also two indicators in the literature of which one is the rate of sulphate deposition. The Memorandum of Intent Final Report (MOI, 1983) stated that "there have been no reported chemical or biological effects for regions currently receiving loadings of sulphate in precipitation at rates less than about 20 kg ha^{-1} yr^{-1}." The other indicator is the pH of precipitation and a pH <4.5 is believed necessary to produce acidification (Linthurst, 1983).

While most inland waters of Newfoundland are sensitive to acidification, current depositions are low enough that these waters are unlikely to be affected (Jeans, 1982). Increases in deposition rates would, however, put these waters at risk. For the remaining areas of eastern Canada, the MOI (1983) determined areas of sensitivity and associated deposition rates. A subset of their data showing only sensitive regions receiving deposition rates in excess of 20 kg ha^{-1} yr^{-1} is given in Table 1. The Maritimes has 8,719 km^2 of surface water area that is classified as highly sensitive and receiving 20-40 kg ha^{-1} yr^{-1} of sulphate deposition. A further 13,447 km^2 is classified as moderately sensitive and receiving 20-40 kg ha^{-1} yr^{-1}. However, an increase in pH has been noted in Nova Scotian rivers since 1973; without supporting evidence, this is attributed to changing weather patterns (Linthurst, 1983).

Estimates from the Quebec Ministry of Environment suggested that there are 100,000 Quebec lakes that would be classified as sensitive (Hansen, 1982); however, many of these lakes are not in zones of high deposition. About 10,137 km^2 of Quebec surface water is classified as highly sensitive and receiving more than 20 kg ha^{-1} yr^{-1}.

Early estimates from the Ontario Ministry of the Environment (OME) indicated that 48,000 lakes in Ontario were considered to be sensitive to acidification and 140 lakes in the LaCloche Mountain region were fishless as a result of acidification (OME, 1980). The sensitive lakes cover an area of roughly 16,540 km^2. Not all of these lakes are in regions receiving high deposition. Less than 9,000 km^2 is highly sensitive and receiving high deposition rates. Slightly less than an additional 2,000 km^2 is moderately sensitive and receiving more than 20 kg ha^{-1} yr^{-1} of sulphate deposition.

Using rainfall with a pH <4.5 as the criterion of high deposition and conductivity as a measure of sensitivity, Minns (1981)

Table 1. Impacted regions at risk from acid deposition in
          eastern Canada (from MOI, 1983)

	Sulphate Deposition (kg ha^{-1} yr^{-1})	Area (km^2)	
		Highly Sensitive	Moderately Sensitive
Ontario	20-40	8,452	1,890
	>40	408	98
Quebec	20-40	10,137	730
	>40	--	456
Maritimes	20-40	8,719	13,447

calculated that there were 11,434 lakes that might become acidified. Of these 1,213 lakes were classified as acidic already,
3,442 were defined as impacted, and 4,186 were at high risk. A
further 7,248 were classified as of moderate risk. Minns (op.cit.)
stated that the majority of these 11,434 lakes were less than
10 hectares[1] in area; however, this quantitative information is not
as useful for economic analysis as the area measures provided by
the MOI (1983) inventory discussed above.

The Economics of Mitigative Action

     If it is possible to neutralize the acidity of waterways
through chemical means, then the cost of this neutralization
provides a measure of benefits to be derived from reducing acid
deposition rates since these costs could be avoided. The major
chemicals that have been used in neutralization are lime and
limestone, and the process is simply referred to as liming.
Various procedures are discussed by Linthurst (1983).

     The objection that "it is not economically feasible" or that
"it is prohibitively expensive" to lime all sensitive lakes is
often raised. Such assessments are either bad economics or at best
uninformed judgments. It is only uneconomic if the flow of all
relevant benefits accruing to society from the neutralization
process are not worth the cost. In this case, the cost of liming
at least shows an upper bound on the benefits. Those who object to
the liming approach would not readily admit such a concession.

     It is not necessary to lime all sensitive lakes, but rather
those that are at risk given current deposition rates and their own
buffering capacities. The Canadian members of the aquatic Work
Group for the Memorandum of Intent (1983) believed that deposition
rates of 20 kg ha^{-1} yr^{-1} would protect all but the most sensitive
aquatic ecosystems in Canada. Considering those highly sensitive
areas receiving sulphate deposition loadings in excess of
20 kg ha^{-1} yr^{-1}, this amounts to a surface water area of 27,716 km²
or roughly 2.8 million hectares.

     Studies conducted by the OME on lake liming suggest that the
costs range from $50 to $500 per hectare with the average dosage
being 0.5 ton per hectare (Lucyk, pers. comm.). These costs are
comparable (monetary exchange rates aside) to those quoted by
Haines (1981) and Driscoll and Menz (1983) for the United States.
The cost of application depends upon the ease of access, the target
alkalinity level, and the limnological characteristics of the
waterway. The neutralization is good for three years of buffering
action. The average cost of liming an easily accessible 70 ha lake
in Nova Scotia was $160 per ha; however, the application rate was

---

[1]hectare (ha) = 2.47 acres = 10,000 m² = 0.01 km² (100 ha = 1 km²)

about double the Ontario rate (MOI, 1983).

Assuming that $100 per ha is a reasonable average cost of liming the sector at risk given the relative ease of access for much of it, then the cost of liming is $280 million. This does not include the costs of having adequate personnel available to monitor the results between applications and determine appropriate applications for each body of water which will generally have its own unique chemical and biological attributes. Suppose that these costs amount to an additional 20 percent. This results in a total cost of $336 million for a three-year program producing an annualized cost figure of $112 million. If deposition rates remain constant, then this cost must be borne in perpetuity. Assuming a real discount rate of five percent, the present discounted value for this continuous stream of annualized costs is $2.24 billion. Highly sensitive areas receiving 10-20 kg ha^{-1} yr^{-1} may also be susceptible. This would add a further 30,000 km^2 to be limed. It is likely that these areas will be less accessible and hence more expensive to lime. More detail would be needed to calculate liming costs for this group of lakes.

While neutralization may be technically feasible if only surface waters are acidified, it may be less successful if the acidification process has proceeded too far. Problems that may arise include the threat of aluminum toxicity to the remaining fish and the potential for re-acidification of the lake. At the Mersey salmon hatchery in Nova Scotia, treating the water with limestone resulted in reducing the mortality of Atlantic salmon fry from 30 percent to three percent (MOI, 1983). On the other hand, there are examples in which liming has raised the pH of the water but aluminum did not precipitate; this resulted in fish mortality (Haines, 1981).

In cases where lakes have lost fish due to the acidification process, then restocking of fish will need to accompany neutralization efforts. Saunders (1981) reported that a major Atlantic salmon enhancement program was planned for Atlantic Canada. There may be problems for the viability of restocking programs, for it needs to be done carefully to avoid the toxicity of heavy metals remaining in the water column. Haines (1981) reported that fish stocked in several Ontario lakes after neutralization did not survive. If the food chain has been disrupted, then the restocking program could fail. Problems may also result from introducing domestic or semidomestic strains into wild conditions. The original stocks could be harmed by the domestic strains or the domestic strains may simply fail as a result of genetic inbreeding leading to a loss of wildness and adaptability (Fraser, 1981; Saunders, 1981). Fraser (1981) identified two wild strains of trout in Ontario which may serve as better stocking fish than domestic strains.

In those cases where the lost fish stocks had evolved unique gene characteristics, then it is impossible to recover these by restocking with alternatives unless society places little value on these unique characteristics. Scientists argue that "a naturally evolved complex of stocks appears essential to utilize fully the productive capacity of waters" (MOI, 1983). It is not clear to the economist just how to link this concept of species diversity to the welfare of members of society. It is not likely to be revealed fully in market data. Thus, nonmarket techniques for assessment are relevant, but such a discussion will be postponed until later. Goodin (1983) argued from an ethical point of view that society should never knowingly destroy natural assets that are irreplaceable since by definition there were no substitutes. Genetically unique fish stocks may qualify for this category. This view is echoed in the MOI (1983) where it is stated:

"As the extinction of any species can never be remedied, the threat of extinction of any species by acid rain would justify lake liming or virtually any other feasible protective measure regardless of immediate costs or benefits."

Further,

"Liming of an acidified habitat would also be justified if it were inhabited by a population which was genetically unique and consequently for which no replacement could be found if it were exterminated."

Where required and viable, the costs of restocking must be added to the costs of neutralization. When using neutralization and restocking as a mitigative action, the policy maker must realize that costs could be incurred without a completely successful outcome. In this case, the benefit of reducing acidic deposition is larger than the cost of neutralization plus restocking.

Economics of Impacts to Fisheries

According to Linthurst (1983), "the clearest evidence for impacts of acidification on aquatic biota is adverse effects on fish." The concern of Canadian government officials in this area is expressed by a former Minister of the Environment, John Roberts (1981), "even a slight lowering of the pH balance. . . . .a slight increase in acidity. . . . .can wipe out the fish and with them a major tourist, sport, or commercial fishing industry."

The best known studies of lake acidification in North America are those conducted by Beamish and Harvey (1972) for the LaCloche Mountain region in Ontario. While more is known about the adverse

effects of acidification on fish than other receptor categories,
the information produced by the scientific community thus far is
generally not in a form that readily submits to economic analysis.

The scientific literature is replete with threshold concepts
where an entire fish stock or species is lost.   As Evans et al.
(1981) summarized:

> "Concern has centered around the complete elimination of
> fish from impacted waters and this elimination seems to be
> taken as the de facto definition of injury. . . . .   Most
> of the available information concerning effects on aquatic
> biota is qualitative."

As an example, the pH thresholds at which fish in the LaCloche
Mountain lakes stopped reproduction are shown in Table 2, where it
is illustrated that sensitivity of fish to acidity is species
specific.   The literature also points out that the loss of fish
species need not be correlated with large declines in annual pH,
but rather changes in aluminum concentrations or perhaps episodic
or pulses of acidification.

The economist usually attempts to find a dose-response curve
that smoothly relates outputs to varying levels of the offensive
inputs.   According to Linthurst (1983) "unfortunately, loss of fish
populations from acidified surface waters is not a simple process
and cannot be accurately summarized as "X" pH (or aluminum concentration) yields "Y" response.   The mechanism by which fish are lost
seems to vary between aquatic systems and probably within a given
system from year to year."   This makes it very difficult for the
economist to determine the value of fish likely to be impacted.
Even if the thresholds can be applied cross other lakes and at
different points in time, the literature usually does not supply
the quantity of fish that would be lost as the pH levels decline

Table 2.   Approximate pH at which fish in the LaCloche Mountain
           Lakes ceased reproduction (from Beamish, 1976)

pH	Species
6.0-5.5	smallmouth bass, walleye, burbot
5.5-5.2	lake trout, trout perch
5.2-4.7	brown bullhead, white sucker, rock bass
4.7-4.5	lake herring, yellow perch, lake chub

through the various thresholds. There seems to be little information about the possible losses of fish productivity prior to
threshold values. Further, there seems to be little information
about the magnitude of damage caused by deformities or uptake of
heavy metals as a result of acidification. These effects may
reduce the marketability of fish even if yield measures have not
been affected. This may be especially relevant for mercury uptake
by fish in acidified areas.

An interesting exception to this general threshold study
approach is the work of Hough, Stansbury, and Michalski, Ltd.
conducted in conjunction with J. E. Hanna Associated, Inc. (Hough,
et al., 1982) for the Canadian Department of Fisheries and Oceans.
This report, referred to as the HSMHA study, was aimed at determining fish productivity responses in the sports fishing sector.
Ironically, for reasons outlined below, the framework is possibly
more relevant for commercial fisheries. Their framework was
apparently used by Victor and Burrell (1982) in their study of
commercial fisheries in Ontario conducted for the OME. The Victor
and Burrell study had not been released to the public at the time
of this writing. Comments must be restricted to the HSMHA framework and its potential relevance for economic and biological
assessment of acidification.

The HSMHA study tried to quantify fish productivity losses
prior to threshold pH values as well as extinction losses. In
order to determine fish yields, HSMHA used the morphoedaphic index
(MEI) proposed by Ryder (1965) which is used frequently in
fisheries management to predict fish yields. The MEI is calculated
as the ratio of the concentration of total dissolved solids, TDS,
to the mean depth of the lake, $\bar{z}$, or

$$MEI = TDS/\bar{z}. \tag{2}$$

Noting that in most natural water TDS is comprised largely of
bicarbonates, HSMHA replaced TDS by alkalinity. Thus, fish yield
is determined as a function of alkalinity and mean depth (in the
form of the adjusted MEI). As acidification proceeds, lake alkalinity is reduced and fish losses can be predicted.

In principle, this approach has appeal for economists since
they need to know fish yield responses to varying degrees of
acidification. Unfortunately, the MEI is not accepted universally
as the appropriate predictor of fish yields. Young and Heimbuch
(1982) argued that TDS has little predictive power after lake
surface area is considered. Since mean depth and fish yields are
both correlated with surface area, the relationship between fish
yield and the MEI indicator involving TDS may be somewhat spurious.
Hanson and Leggett (1982) found that total phosphorus concentration
and macrobenthos biomass/mean depth were better predictors of fish

yields and biomass than MEI, TDS, or mean depth for the same data
set. Prepas (1983) concluded that mean depth was as good a
predictor of fish yield as MEI; and hence in his view, TDS was not
a useful parameter. These observations seriously question the
usefulness of the HSMHA (1982)approach since the crucial variable
is alkalinity as a surrogate for TDS. If TDS is not relevant, then
neither is alkalinity; and hence the link between acidification and
fish yields is severed and the projections become meaningless.
This may, in fact, explain the extremely low yield response
obtained by HSMHA. A similar criticism will likely apply to the
Victor and Burrell (1982) study.

The criticisms of MEI have a disconcerting aspect to them.
The mean depth seems to be a crucial variable according to some
researchers (e.g., Prepas, 1983). Clearly, acidification does not
alter mean depth; hence statistical models of fish yields that
depend upon mean depth or surface area will be of little use in
assessing impacts of acidification for those waters. While mean
depth and surface area may be well-known biogeographic factors
limiting species diversity and productivity, these must set upper
bounds on the environmental carrying capacity in the absence of
stress and would become less relevant for systems under stress such
as from acidification. Schlesinger and Reiger (1982) estimated
theoretical upper bounds for fish productivity in North America
using MEI and temperature. Much of the impacted regions of Ontario
and Quebec have upper limits of productivity between 5 and
10 kg ha^{-1} yr^{-1}, while southern Ontario and Nova Scotia regions
support more than 10 kg ha^{-1} yr^{-1}.

Harvey and Lee (1982) calculated the number of species of fish
that might have been lost in the LaCloche Mountain lakes due to
acidification. They derived a regression equation that specified
the number of fish species in a given lake as a function of the
logarithm of lake surface area for lakes in the LaCloche region
with pH >6.0. Given the surface area, the obtained equation was
then used to predict the expected number of species in lakes with
pH <6.0. The difference between the predicted and the observed
number of species would be the number of species lost due to
acidification. According to Harvey and Lee (op.cit.), the predictive power of the model was not very reliable for lakes in the
pH 5.2-6.0 range. For lakes with a pH <5.2, the loss in species
ranged from 20 to 100 percent with this large range being
attributed to differential species tolerance toward acidity. This
procedure provides for an expost determination of losses, and it
may also be useful in assessing potential losses from acidification.

While this exercise is intriguing, it is difficult to know
just how to assess the economic impact of changes in the number of
species. Society seems to place different implicit values on

various species. Trout, for example, are a greater prized species than yellow perch. This approach does not yield information on the population sizes of the lost species. If the disappearance of some species permits an increase in the population size of the remaining fish, this may lessen the loss depending upon the desirability of the remaining species. It may be possible to combine this approach with a modified version of Schlesinger and Reiger (1982) to provide estimates of population size and species loss as acidification proceeds. In a study of Adirondack lakes in New York and White Mountain lakes in New Hampshire, Confer et al. (1983) determined lossed of 2.4 species of zooplankton and 22.6 mg dry wt per m² of lake surface area per unit decrease in pH. Until such a link is established for fisheries, quantitative economic assessments will be elusive.

In 1979, the commercial fish harvest in Ontario was worth roughly $26 million. Most of this harvest was from the Great Lakes region which is well buffered against acidification except for some nearshore spawning areas that could be adversely affected by acid pulses from spring snowmelt (Giles, 1982). The harvest in the northern inland waters was $2.2 million in 1979. Presuming that these are the impacted waters and that a total loss of these stocks could result from continued atmospheric deposition of acids, then this value is the cost of acidification to the Ontario commercial fishing industry provided that the loss in catch is sufficiently small so as to leave fish prices unchanged. Given the size of the northern inland waters catch to Ontario and Canadian catch sizes, this seems like a reasonable assumption. However, behavior responses by fishermen could mean that the 2.2 million dollar annual loss estimate is at best an upper bound to the true damage estimate. As acidification proceeds and stocks diminish, fishermen could increase fishing effort and hence maintain yields (until extinction occurs). Schlesinger and Reiger (1982) noted that fishing effort is a major variable influencing fish yields, and Peterman and Steer (1981) found an inverse relationship between catchability and fish abundance. These responses, of course, reduce losses only in the short-run prior to extinction; however, there is also a feasible response to reduce long-run losses. Once a lake is fishless, the fishermen may move their efforts to substitute lakes and maintain fish yields. The cost incurred in such a move becomes the relevant cost of acidification to the commercial fishery.

Economic Impacts on Recreational Fishing

There is more concern for the impact of acidification on recreational fishing opportunities than for the commercial fishing sector. Mitchell (1980) concluded "that in terms of economic importance, the recreational fisheries make a greater contribution to Canada's GNP than the commercial fisheries." For comparison, Mitchell (op.cit.) pointed out that in 1975, the market value of the commercial fishery was $694.3 million while expenditures and investment of sports anglers totalled $1,021.6 million. This is particularly true for Ontario where the recreational fish catch in 1975 was 63 percent of the total Ontario catch. In fact, Mitchell's (op.cit.) estimates indicate that the Ontario sport-fishing sector dominated the value of commercial fisheries in the Atlantic region. Apart from the losses of sportfish in the LaCloche Mountain lakes, "there is limited documentation of sport-fishery losses in Canadian waters coincident with acidification, primarily because no concerted effort has been made to collect such information" (Harvey and Lee, 1982). In the Muskoka-Haliburton region of Ontario where concern for the sportfishery impacts is quite high, no adverse effects on fish populations have been documented to date. Even though the region is sensitive and receives high deposition rates, the lake pH values have not dropped to harmful levels (Linthurst, 1983).

According to Watt et al. (1983), there are seven rivers in Nova Scotia with a pH <4.7 that previously supported salmon popula-tions but no longer do so. In rivers with 4.7 < pH < 5.0, angling returns have declined 2.8 percent per year since 1954. Given the current acid loadings in the southern and central regions of Ontario, Minns (1981) reported that Canada could ". . . .lose significant proportions of brook trout, ...bass and walleye communities during the 1980's. Further, we might expect to lose the lake trout associations with the above species over the next two decades." He offers the assessment that "the nonlake trout communities most at risk must be considered in more immediate danger simply because they occur in smaller lakes. However, the overall picture indicates that lake trout communities represent a greater potential loss."

The HSMHA (1982) study determined yield responses for Ontario's sportfishery as a result of acidification. The projec-tions seem very low even for a lake in a sensitive region receiving high deposition. At current deposition rates, it estimated a 0.27 percent loss in yield over a 40-year period. If a reduction in acidity of rainfall is assumed, there is a 2.7 percent increase in yield over a 40-year period. Since impacted areas are a small fraction of total surface water area, this figure extrapolated over impacted lakes produces a smaller aggregate yield change than three percent. However, the HMSHA (op. cit.) study cautioned against such extrapolations at this time. The comments made concerning the relevance of the HMSHA framework for the commercial fishery may explain why the projections are so low. The short-comings of the MEI (Ryder, 1965) as a predictor of fish yields mean that the link between acidification and fish yields is weak.

The relevant output of recreational fishing is not the "fish caught" but rather the "fishing (or recreation) experience." Recreation studies generally focus on the number of fishing days. It is not immediately clear how "fishing days" vary with fish abundance. Lee and Kreutzweiser (1982) reported that sportfishing was enjoyed by 38 percent of the Ontario population. Mitchell's (1980) figures, in fact, showed that slightly more than half of Canada's anglers reside in Ontario. The Ontario Recreation Survey determined that up to 30 percent of participants fish less than they desired as a result of lack of opportunities or good facilities resulting in a short-fall of roughly 24.2 million angler-days (Lee and Kreutzweiser, op. cit.).

Forster (1983) reported that in 1977-1978, nonresident fishermen were willing to pay $6.1 million for licenses to fish in Ontario waters; six million dollars of this was paid by non-Canadians. Ontario does not require its residents to purchase fishing licenses. In Canada, one out of six anglers is a nonresident. If this ratio is true for Ontario as well and if residents would be willing to pay similar amounts as nonresidents, then a lower bound for the willingness to pay for fishing (in 1977-1978) in Ontario was $36.6 million. It may well be that Ontario has a higher proportion of foreign anglers, indicating that this estimate may be too high. In the late 1970's, the Ontario population was slightly less than nine million. Assuming that at least 66 percent were aged over 18 and that 38 percent of this group would have been prepared to purchase the $10.75 annual license, this would yield roughly $25 million in possible license revenues from Ontario residents. The combined revenue would be roughly $31 million.

Other provinces require licenses of both residents and nonresidents. Quebec revenues amounted to almost $2.8 million. Sales in the Atlantic provinces amounted to almost another million dollars. An educated guess would suggest that if Ontario residents had to purchase licenses, the total value of license revenues for eastern Canada might be in the order of $35-40 million. This figure has limited value in economic assessments of acidification impacts. First, the numbers do not reveal maximum willingness to pay for the licenses and hence they provide lower bounds at best. Furthermore, regional participation rates would be needed; i.e., how much of the fishing activity takes place in regions at risk to acidification.

A more sophisticated approach to valuing recreational fishing is to calculate the consumer surplus - this is roughly the difference between what the fisherman would be willing to pay, at a maximum, to fish and what he actually pays. In this approach, the analyst must determine the days of fishing lost as a result of acidification. It is unlikely that fishermen perceive variations in pH directly, hence variations in fishing will be in response to the quality of fishing. It is also not clear how fishermen will

vary their activity in response to variations in fish biomass. It
is quite possible that fishing quality could improve in the short-
run as a result of acidification. This could happen as the
remaining fish grow larger from decreased competition for food
supplies. Thus, acidification may paradoxically yield short-run
benefits to sportfishing with sudden losses in the longer term as
repeated recruitment failures lead to extinction.

Menz and Mullen (1983) examined the impact of acidification on
the Adirondack fishery by considering the reduction in visitation
days that resulted from a loss of water acreage supporting fish due
to acidification. In 1978 in Canada, the number of recreational
fishing days was roughly 87 million with 78 million being resident
and nine million being nonresident. Mitchell (1980) assumed an
annual growth rate in fishing activity of 6.25 percent. This
resulted in roughly 104 million fishing days by 1981. Suppose that
the process of acidification would lead to a one percent decrease
in total fishing activity given the small area that is impacted.
Crocker et al. (1980) used a consumer surplus value of $32.30 per
fishing day in 1978 to calculate the loss associated with acidification to the American sportfishery. Martin et al. (1982) determined
the 1978 value per fishing day at the Lake Mead fishery in the
southwestern United States was $60 for general fishermen and $100
for largemouth bass club members. Given this range of estimates
and considering the rate of inflation and the U.S./Canada exchange
rate, it is assumed that a consumer surplus range of $50-$100 per
fishing day might be appropriate for Canada in 1981. This provides
a range of $52-$104 million for the estimated loss in consumer
surplus for a one percent reduction in fishing activity.

The limitations of the foregoing exercise are obvious. In the
absence of Canadian estimates for the value of a fishing day,
inferences must be made from available American sources. Fishing
activity responses to acidification cannot be observed since the
effects may not yet be occurring, but rather will take place in the
future. Another issue of concern is the availability of lakes that
substitute for impacted lakes that lose their fisheries. As
previously noted, the impacted surface area at risk is a small
fraction (13 percent) of total surface water area. Mitchell (1980)
stated:

> ". . . the scale of Canada's fresh waters is so vast that
> the frontier fringe of intensive sport fishery has not yet
> been pushed back to the stage where productivity per se has
> emerged as a problem (largely because of costs and means of
> mass access). This is not to say that some trophy fish
> species, e.g., lake trout have not been creamed off or that
> problems due to overfishing or other factors (e.g., acid
> rain) do not abound within the settled areas of Canada."

To the extent that recreational fishermen can migrate to alternative sites at negligible cost, the impact of acidification on the value of the experience may likewise be negligible. However, if the move is not possible without substantial cost, then the costs that the recreationist is prepared to incur in order to continue a given quality of fishing activity is a measure of the acidification damage. To the extent that individuals can substitute alternate sites or alternate activities at a given site, the damages are lessened in the sense that the fishermen (recreationist) will be willing to pay less to preserve environmental quality at a given site. In their study of the Adirondack sportfishery, Menz and Mullen (1983) found that the economic losses from acidification dropped from $1.96 million, with no assumptions for interfishery substitution, to $1.66 million when interfishery substitution was assumed to exist.

Economic Impacts on Amenity Values

In the previous analyses, impacts of acidification were assumed to result in a reduction in angling days. This resulted in a reduction in the consumer surplus of fishermen and this reduction is taken as the welfare loss of acidification to recreational fishing. This approach may be too narrow in scope since for many people the fishing activity is part of a broader recreational experience. As Buchanan (1983) argued ". . . results indicate that current research may be erroneous in attributing satisfaction scores to one main activity while failing to consider the influence of secondary activities." Buchanan (op. cit.) found that while catching fish was the most important characteristic associated with the experience of central Illinois recreational fishing activity, several other characteristics were ranked not far behind (Table 3). Adams (1979) found that catching fish actually rated lower than other characteristics in the fishing experience of residents and nonresidents in Wyoming. Some selected mean scores obtained by Adams (op.cit.) are presented in Table 4.

Bryan (1977) argued that "catching a fish, any fish" was likely to be the prime concern of the occasional fisherman while for "the most specialized fisherman, the fish are not as important as the experience of fishing as an end in itself." Bryan (op.cit.) determined the following propositions concerning the behavior of fishermen:

1. Fishermen tend to become more specialized over time.

2. The most specialized fishermen comprise a leisure subculture with unique minority recreationist values.

Table 3. Seven important characteristics of recreational fishing
          (from Buchanan, 1983)

Characteristic	$\bar{x}^a$
Catching Fish	7.49
Physical Rest	6.80
Escaping Personal/Social Pressures	6.68
Being With Friends	6.62
Family Togetherness	6.45
Escaping Physical Pressures	6.40
Experiencing Nature	6.36

[a]mean score on a nine-point scale (nine is the highest)

Table 4. Characteristics and satisfaction of recreational fishing
          in Wyoming (from Adams, 1979)

Characteristics	Wyoming Residents	Non-Residents
Being able to relax	1.39	1.47
Just being outdoors	1.48	1.62
Getting away from people	1.76	1.90
Getting out with friends	1.80	2.00
Getting good eating fish	1.97	1.98
Fishing water surrounded by pleasant scenery	2.22	2.43
Seeing wildlife	2.25	2.25
Catching some fish	2.50	2.35
Catching native or wildfish	2.89	2.57
Chance to catch large fish	3.28	2.72

Scale: 1 = extremely important
         2 = very important
         3 = moderately important
         4 = a little important
         5 = not at all important

3.   Increased specialization results in a shift from fish consumption to preservation and emphasis on the activity's nature and setting.

4.   As specialization increased, dependency upon particular resource types increases.

These observations pointed out that even recreational fishing has various characteristics that affect the satisfaction derived from the experience. Some of these characteristics would not be affected by acidification. These include, for example, those factors that promote relaxation, being outdoors, escaping various pressures, and being together with friends or family. On the other hand, other aspects of the aquatic ecosystem, in addition to fish, are also likely to be adversely affected by acidification, and these impacts would reduce the satisfaction derived from the experience. Bryan's (op. cit.) propositions suggested that highly specialized fishermen will be more concerned with the preservation of fish species and the natural setting. Adverse impacts on fish species and the ecosystem would result in welfare losses for this group that was not reflected in consumer surplus associated with fishing days. The reliance of this specialized group on particular species would mean that their welfare losses would be more substantial than for the occasional angler if these species were lost as a result of acidification.

It is unlikely that the full impact of acidification on the aquatic sector will be a simple direct relationship with fish biomass changes and/or changes in fishing days. Recreational fishing should be considered as part of the broader recreational experience based upon aquatic resources. Such an approach would allow recreationists to specify the relative importance of various components in the experience. In general, the impact of acidification on all recreational amenities should be considered. The general recreationist places value on living things in the ecosystem. They may derive enjoyment from merely watching fish swim in brooks or lakes or seeing it jump with no intention of fishing. The SCUBA diver, for example, derives satisfaction from observing the fish in their own habitat. Disappearance of fish and other biota changes will reduce the enjoyment of diving experiences. Recreationists derive satisfaction from observing or hearing wildlife which depend upon the aquatic resource. Frequently, concern is expressed for the fate of the loon and other waterfowl.

Recreationists not only use lakes for fishing but also for swimming and boating activities. Acidification of waterways is accompanied by an increase in transparency that may improve the quality of swimming or boating. On the other hand, MOI (1983) reported that acidified waters often have an increased growth of

benthic filamentous algae that may depreciate shoreline recreational values and activities such as swimming. Further, in some
poorly buffered lakes, the growth of certain phytoplankton species
can create obnoxious odors and hence reduce recreational satisfaction. These offensive conditions have already been observed in
Ontario waters.

It is not clear what the net change in perceived quality will
be when these various physical traits and the health concerns of
individuals swimming in acidified waters are considered. Robert R.
Nathans Associates (1969) was unable to detect a response in
swimming or boating activity with respect to varying pH levels in
the Appalachian region. This result is consistent with the suggested dilemma facing the recreationist. A further dilemma may
face the SCUBA diver - fantastic visibility but nothing to look at!

As mentioned above, many of these feared adverse impacts have
not yet occurred on a large scale, if at all. Thus the current
damage in an economic sense is probably minimal, but the process
may be irreversible once started. What is relevant in terms of
economic value is the "willingness to pay" by society to ensure
that the process does not continue unabated. The only way to
determine these values is to survey people directly to determine
the value they place on those items that are likely to be impacted
by acidification.

A major study of acidification impacts on amenity values was
undertaken by ARA Consultants (1982) for the OME. The objectives
of the study were:

1.   to determine the monetary value that members of society
      place on changes in the quality of environmental
      amenities caused by acid deposition or other forms of
      pollution;

2.   to determine the socioeconomic factors that might account
      for the variation in monetary values specified by members
      of society;

3.   to determine the level of awareness regarding acid
      deposition as well as individual attitudes and beliefs
      concerning pollution and acid deposition; and

4.   to determine the substitution of activities that
      individuals would make as a result of changes in environ-
      mental quality.

At the time of this writing, the ARA study had not been released to
the public, so a detailed presentation and analysis of the results
is not possible. The following information, however, can be
provided.

The survey was conducted using face-to-face interviews with 920 individuals aged 18 years and older. In order to determine the attitudes of foreign tourists which are an important part of Ontario tourism, 100 visitors from the U.S. were included. It is also important to determine the values that possible nonusers of the resources in the impacted areas would place on these resources. These values may reflect "option values" for their descendents to use the resource. Or, they may be "existence values" if the individual would be willing to pay to preserve the aquatic resource with no intention or expectation that any of his/her family would ever use it. In order to assess these values, 206 urban residents distant from the impacted areas were surveyed. The remainder were interviewed in Ontario's "Cottage Country", namely the Muskoka-Haliburton region and the Kawarthas district around Peterborough. Muskoka-Haliburton is an impacted region at risk while the Kawartha district is believed to be not sensitive to acidification.

The questionnaire was quite long with 58 basic questions, many of which had several parts. The questionnaire obtained information on distance travelled for recreational experiences, most frequent recreation area, and time spent in the cottage country "this year" and in the "previous three years." Information was obtained on the proportion of time spent on a variety of activities. Respondents were asked how far they would be prepared to travel in order to swim or fish if these activities were no longer possible at their current site. They were also asked what activities they would substitute for swimming or fishing if these were no longer possible at a given site. Respondents were asked to assign importance levels to protection from pollution for fish, small animals, water birds, water plants and flowers, and land plants and flowers. These various questions attempted to obtain information which techniques for recreational fishing assessment discussed above did not yield.

The major objective of the study was to determine monetary values which individuals attach to variations in environmental quality that might be caused by acidification. In order to obtain these values, a "contingent valuation" procedure was employed. These basic procedures evolved over the last decade and were proven quite useful in assessing recreational values in areas of the southwestern U.S. adversely affected by energy developments (Schulze et al., 1981).

In order to represent the process and impacts of acidification as accurately as possible for the respondents, the following "Environmental Quality (EQ) Ladder" in Table 5 was shown to the respondent with colored illustrations. These descriptions were developed by scientists at the Ontario Ministry of the Environment. In order to help the respondents think in monetary terms about environmental quality, they were shown a table of individual tax

Table 5.   Environmental quality ladder used for ARA consultant's
           survey (1982).

---

10   "Unpolluted" environment - all fish, wildlife and plants
      healthy and abundant. Restoration of some fish species
      such as trout and salmon to waters where they used to be
      naturally, but declined.

9   Environment with a wide variety of wildlife, plants, etc.
      In some parts of the Province, the best sports fish no
      longer exist, such as walleye.

8   Ontario Environmental Quality Level - Good fishermen would
      catch 10 fish in 2 days. Forests, wildflowers, and wild-
      life healthy and in abundance.

7   Generally healthy and varied wildlife and vegetation.
      Some decline in sports fish numbers and types (bass).
      Average fishermen would catch 7 fish in 2 days.

6   Wildlife and vegetation still relatively healthy and
      varied. Aquatic life not as abundant - fewer frogs, less
      waterplants and loss of some sports fish such as trout.
      Fishermen would catch 3 fish in 2 days.

5   Wildlife and vegetation declining - fewer others, ducks
      and loons. Very few fish types - yellow perch, chub and
      suckers left in lake.

4   Only one kind of fish - yellow perch left. Frogs are rare
      and the small vegetation growth around the lake is rare.
      Water seems clear, and there are very few water birds
      (ducks, loons) and fish-eating wildlife (raccoons, otters).

3   No fish left in the lake and wildlife which depend on
      fish for food are gone. Water plants, except for moss,
      have disappeared and the lake water is crystal clear
      except for a green film in water.

2   The lake water is very clear, but there is virtually no
      sign of fish-eating wildlife in the form of ducks or small
      animals, and very few birds. The trees and shrubs seem
      thin and the leaves of some are spotted with brown.

1   Polluted environment - fish, wildlife, plants seem less
      healthy and less abundant. Some species no longer exist
      in the Province.

0

payments by income level for various public services such as police and fire, highways, hospitals, public education, defense, and unemployment benefits. The respondents were then asked:

> "How much, if anything, would you pay in taxes and prices annually to protect the Ontario environment from declining from level 8 to level 4?"

The response was recorded and the respondent was then asked "Why did you choose this amount." The question was repeated for changes from 8 to 7, 8 to 2, and 8 to 6. The order of the changes was jumbled in order to prevent the respondent using simple rules of thumb to generate additional responses. The respondents were also asked how much they would be willing to pay to improve the quality of the environment from 8 to 9 and from 8 to 10. Underlying these questions is the presumption that individuals receive satisfaction, or utility U, from income Y (which will enable individuals to purchase goods and services), and environmental quality EQ (see equation 3). Further, it is presumed that variations in income levels will substitute for variations in environmental quality in terms of satisfaction.

The questions relating to willingness to pay, WTP, to prevent a decline in EQ (a negative ΔEQ) are what economists term "equivalent surplus" measures of the value of ΔEQ. The maximum WTP bid results in an equivalent reduction in welfare or utility as would a decline in EQ which the WTP bid is being offered as prevention. In mathematical shorthand, the equivalent surplus WTP value is determined by:

$$U(EQ^0, Y - WTP) = U(EQ', Y) \tag{3}$$

where $EQ^0 = 8$ and $EQ' = 7, 6, 4, 2$ in the specific questions posed.

The WTP bids obtained in the questions relating to environmental improvement from level 8 to levels 9 or 10 are what economists refer to as "compensating surplus" measures of value. The maximum WTP in this case is just large enough to offset (or compensate) for the welfare gain that would otherwise result from an increase in environmental quality:

$$U(EQ^0, Y) = U(EQ^1, Y - WTP), \quad EQ^0 = 8, \; EQ^1 = 9, 10. \tag{4}$$

In order to see if the WTP bids obtained through the survey were consistent with the above theoretical considerations, a simple econometric analysis of the equivalent surplus data was performed. As a first approximation, the following linear bid equation was estimated

$$WTP = \alpha_0 + \alpha_1 \Delta EQ + \alpha_2 Y \tag{5}$$

This bid equation is a "demand curve-like" construct that relates the WTP bid to the variation in EQ and the respondents' income level (Bradford, 1970). The linear form is restrictive in terms of the structure which it imposes upon the preferences of the respondent. More flexible forms should be examined with this data set in the future. It must be assumed that the respondents WTP bids are sufficiently close to their true maximum values to be consistent with the theory. A ΔEQ variable must be constructed. In order to do this, it was presumed that the information contained in the EQ descriptions could be aggregated in a meaningful sense and indicated by a numerical value on the ladder. The questions as posed in fact assume that such aggregation is meaningful. Conditions for consistent aggregation in environmental economics were outlined by Forster (1981). The respondents' income was assumed to occur at the center of the relevant income interval. In estimating the WTP equation, one is combining a large number of observations and deriving a single common "inverse demand" equation for environmental quality. This aggregation is legitimate provided the various individuals are sufficiently similar in regard to their attitudes in preserving environmental quality. In order to allow for potential differences, separate demand equations were estimated for the various subgroups. The resulting parameter estimates are presented in Table 6.

Table 6.   Inverse demand equations by sample group (see text).

Sample Group	Const α_0	ΔEQ α_1	Y[a] α_2	R^2	F	# of obser.
All Respondents	-149.23	32.14 (99.38)[b]	10.31 (548.17)	0.16	319.95	3257
Muskoka/Haliburton	-115.85	32.01 (62.51)	8.88 (298.67)	0.20	179.41	1394
Kawartha District	-104.80	25.95 (15.86)	9.00 (92.08)	0.12	53.87	748
Urban Residents	-267.87	39.65 (19.81)	16.61 (167.62)	0.19	93.51	756
U.S.A. Visitors	-192.17	30.54 (8.87)	10.15 (54.32)	0.15	31.51	348

[a]Y is the respondents income level expressed in thousands of dollars

[b]Numbers in parentheses are F-statistics

All parameter estimates are significant at the 0.01 level indicating that variations in ΔEQ and Y income play a role in explaining variations in WTP values. However, the R^2 statistics vary from 0.12 to 0.20 indicating that these two variables explain only a small fraction of the variance of the WTP values. The low R^2 values shown in Table 6 are common when working with cross-section micro-data referring to short time periods (Theil, 1971). It is possible that the WTP values were influenced by a variety of personal respondent characteristics. While these characteristics may be important, some may not be easy to identify, let alone quantify. If bids are averaged within income groups, the resultant equation explains 80 percent of the variation. The averaging obviously removes some personal variation.

From the equation for All Respondents in Table 6, the average WTP value increased by $32.14 for each unit decrease away from level 8 on the ladder. The estimates for incremental bids from the individual groups varied from $25.95 to $39.65; however, using simple t-tests these estimates were not significantly different from $32 suggesting that responses to variations in EQ were similar across the income groups. The WTP values were also dependent upon the respondents income level. For the All Respondents Group, the average bid for a given ΔEQ rose by $10.31 for every $1000 increase in respondent income. The estimates for the individual groups ranged from $8.88 for the Muskoka-Haliburton group to $16.61 for the urban area. Using simple t-tests, these estimates are significantly different from $10 suggesting that the income influences on WTP values do differ across the groups. The parameter estimates for the Kawarthas group and the U.S.A. Visitors are not statistically different from $10. This strong relationship between income and willingness to pay to protect the environment is consistent with the suggestion that environmental quality is a "normal" good - one for which demand rises as income rises.

The ARA (1982) study did not calculate the aggregate willingness-to-pay; however, in principle, such an aggregate could be calculated with the data. However, there are some problems in using the results in this fashion. It would first be necessary to determine whether the sample groups were representative of the broader population groups. For example, were the 100 U.S. visitors typical of all U.S. visitors? Were the 206 urban residents typical of all Ontario residents? Were the recreationists typical? Care would also be needed to ensure that there were no double-counting - urban residents may also be recreationists. It is not clear whether the target population should be "individuals" or "households." It may be that the respondent when stating WTP was stating his/her own willingness to pay. Or, they may have reported a "household" WTP which included all household members and not just their own WTP. This could cause quite divergent results when determining aggregate bids. Both aggregates should be calculated to determine a possible range of estimates.

The more important issues concern the accuracy of the individual WTP bids themselves. There are two issues here. One results in a downward bias; the other results in an upward bias. The WTP bids were simple first responses to the individual questions posed by the interviewer. Studies done in the southwestern U.S. used an "iterative bidding technique" which encouraged the respondent to reveal his maximum WTP value (Schulze et al., 1981). The respondent was asked if he would pay $X for a given result. If he answered "yes," then he was asked if he would pay $(X + 1), and so on until a negative answer was reached. The highest WTP receiving a "yes" was recorded. The ARA (1982) study did not use the iterative bidding approach because of the already lengthy survey instrument. Schulze et al. (1983) found that the iterative bidding approach led to significantly higher WTP values than the first offered bid of the respondent. Coursey and Schulze (1983) suggested that the simple first response of the respondent might be interpreted as the "opening bid" in an iterative bidding process which one would expect to be lower than the ultimate WTP bid. Cox et al. (1982) demonstrated that in laboratory results, maximum WTP values were obtained only after a number of iterative learning periods even when it was in the immediate best interest of the respondent. For these reasons, the WTP values were likely to be lower than the true WTP values to protect the level of environmental quality in Ontario.

However, the impact of acidification is not to lower the level of environmental quality across all of Ontario, but rather the impacts are more regional. The questions posed to the respondent, however, referred specifically to the "Ontario environment" suggesting widespread deterioration which will minimize the availability of close substitutes. To the extent that the respondent believed the deterioration in question applied province wide, the WTP offer may have been larger than if a more specific and restricted geographic area had been suggested such as the impacted region or perhaps just the Muskoka-Haliburton region. One would expect the recreationist who goes only to the Kawarthas district to offer a higher WTP if he feels the Kawarthas will also suffer than if he believes only the Muskoka region will suffer. Since it is not possible to state unambiguously the ultimate bias in the WTP values, they should probably not be used for aggregation.

Greenley et al. (1981) used a contingent valuation procedure to estimate recreational values, option values, bequest values, and existence values for aquatic-based recreation activities during 1976 in the South Platte River Basin, Colorado, where water quality could suffer irreversible degradation due to mining activity. This has some similarity to the acid deposition problem. The researchers found WTP measures quite sensitive to the proposed method of payment with the sales tax vehicle producing bids roughly three times as large as a water services fee. This discrepancy was attributed to feelings of equity. Everyone using the services would pay sales taxes while only residents would pay the water fees. Average household WTP measures are given in Table 7. The total recreation plus preservation annual values ranged from $38.25 with water fees to $121.23 with the sales tax vehicle. The $121.23 figure was close to the estimated maximum WTP to reduce environmental risk due to nontoxic wastes, where Burness et al. (1983) found that in a contingent valuation procedure households in Albuquerque would pay roughly $10 per month to hedge against environmental risk. For Houston households, the estimates were $28.30 and $20.35 per month depending upon income class ($40,000 and $25,000, respectively). The range of estimates was quite large.

Suppose that households in eastern Canada would be willing to pay at least $100 per year on the average, to preserve environmental quality in the impacted regions. This would seem to be in the relevant range given for estimates in the U.S. The 1981 population for the provinces east of Manitoba was roughly 16.8 million. Assuming an average household size of three, this would produce 5.6 million households. Thus, if these households are willing to pay at least $100 annually to preserve their sensitive aquatic regions from acid deposition degredation, then the value of the aquatic region is at least $560 million. Assuming a real rate of interest of five percent, the present value of benefits of environmental preservation is $11.2 billion.

Table 7. Resident household annual WTP for water quality preservation in the South Platte River Basis, 1976 (adapted from Greenley et al., 1981)

	Water Fee	Sales Tax
Recreation Value	18.60	56.68
Option Value	7.68	34.05
Bequest Value	5.40	16.97
Existence Value	6.60	24.98
Total Preservation and Recreational Value	38.25	121.23

CONCLUSIONS

   The preceding sections have provided an assessment of the
economic impacts of acid deposition on the aquatic ecosystem in
eastern Canada.   The estimates provided are very crude at best.
More detailed examinations of the Canadian situation will be needed
to further refine these values.   Nevertheless, given the available
scientific evidence, it seems likely that the current economic
impacts are minimal.   This assessment is made because the region at
risk is small compared to the total surface water area.   This means
that there are substantial substitution possibilities.   For
example, of the 4655 lakes in Ontario which are currently acidic
or impacted, most are small and in the Sudbury region.   There are,
by comparison, 181,450 lakes in Ontario.

   Given the availability of substitute sites and activities, it
is possible that future economic impacts will be lessened.   In
order to assess these impacts, it is important to determine how
individuals will respond to the biological impacts of acidification.   These behavior adaptations are as important in determining
the ultimate economic impacts as are the biological impacts and in
some cases more so.   Research on the value that Canadians place on
preserving the impacted region itself for recreational and
aesthetic purposes is required.

Acknowledgments – This paper was completed while the author was a
visiting professor of economics at the University of Wyoming.
Thanks are due to Tom Crocker for reading a draft manuscript and
offering his insight and criticisms. Responsibility for errors in
the final product remains with the author.

REFERENCES

Adams, S.W., 1979, Segmentation of a recreational fishing market:
      A canonical analysis of fishing attributes and party compo-
      sition. J. Leisure Res. 11:82-91.
ARA Consultants, 1982, Value Awareness and Attitudes Associated
      with Acid Precipitation in Ontario: The Amenity Value Survey.
      Report to the Ontario Ministry of the Environment, Toronto,
      Ontario, Canada. 190 pp.
Beamish, R.J., 1976, Acidification of lakes in Canada by acid
      precipitation and the resulting effects on fish. Water, Air,
      Soil Poll. 6:501-514.
Beamish, R.J. and Harvey, H.H., 1972, Acidification of the
      LaCloche Mountain Lakes, Ontario and resulting fish mortal-
      ities. J. Fish. Res. Board Can. 29:1131-1143.
Bradford, D.F., 1970, Benefit-cost analysis and demand curves for
      public goods. Kyklos 23:775-791.

Brady, G.L., 1983, The benefit/cost analytical framework: Handling uncertainties in the Clean Air Act. J. Environmental Management 16:335-346.

Bryan, H., 1977, Leisure value systems and recreational specialization: The case of trout fishermen. J. Leisure Res. 9:174-187.

Buchanan, T., 1983, Toward an understanding of variability in satisfactions within activities. J. Leisure Res. 15:39-51.

Burness, H.S., Cummings, R.G., Mehr, A.F., and Walbert, M.S., 1983, Valuing policies which reduce risk. Nat. Resources J. 23:675-682.

Confer, J.L., Kaaret, T., and Likens, G.E., 1983, Zooplankton diversity and biomass in recently acidified lakes. Canadian J. Fish. Aquatic Sci. 40:36-42.

Coursey, D.L. and Schulze, W.D., 1983, The application of laboratory experimental economics to the contingent valuation of public goods. Unpublished manuscript, Economics Department, University of Wyoming, Laramie, WY. 40 pp.

Cox, J.L., Roberson, B., and Smith, V.L., 1982, Theory and behavior of single price auctions, pp. 21 to 45. In: Research in Experimental Economics, Vol. 2, V.L. Smith, ed., JAI Press, Greenwich, CT.

Crocker, T.D., 1982, Conventional benefit-cost analyses of acid deposition control are likely to be misleading, pp. 76 to 91. In: Acid Rain: A Transjurisdictional Problem in Search of Solution, P.S. Gold, ed., Canadian-American Center, State Univ. New York, Buffalo, NY.

Crocker, T.D., Tschirhart, J.T., Adams, R.M., and Forster, B.A. 1980, Methods development for assessing acid deposition control benefits. Report for Grant No. R806972010, U.S.E.P.A., Washington, DC. 201 pp.

Driscoll, C.T. and Menz, F.C., 1983, An assessment of the costs of liming to neutralize acidic Adirondack surface waters. Water Resources Res. 19:1139-1149.

Evans, L.S., Hendrey, G.R., Stensland, G.J., Johnson, D.W., and Francis, A.J., 1981, Acidic precipitation: Considerations for an air quality standard. Water, Soil, Air Poll. 16:469-509.

Forster, B.A., 1981, Separability, functional structure, and aggregation for a class of models in environmental economics. J. Environmental Econ. Mgt. 8:118-133.

Forster, B.A., 1983, Economic impact of acid precipitation: A Canadian perspective, pp. 97 to 121. In: Economic Perspectives on Acid Deposition Control, T.D. Crocker, ed., Butterworths, Stoneham, MA.

Fraser, J.M., 1981, Comparative survival and growth of planted, wild, hybrid, and domestic strains of brook trout in Ontario lakes. Canadian J. Fish. Aquatic Sci. 38:1672-1684.

Giles, W., 1982, Acid rain: An Ontario perspective, pp. 24 to 29. In: Acid Rain: A Transjurisdictional Problem in Search of Solution, P.S. Gold, ed., Canadian-American Center, State Univ. New York, Buffalo, NY.

Goodin, R.D., 1983, The ethics of destroying irreplaceable assets. International J. Environmental Studies 21:55-66.

Greenley, D.A., Walsh, R.G., and Young, R.A., 1981, Option value: Empirical evidence from a study of recreation and water quality. Quarterly J. Economics 657-673.

Haines, T.A., 1981, Acidic precipitation and its consequences for aquatic ecosystems: A review. Transactions American Fisheries Soc. 110:669-707.

Hansen, P., 1982, Bitter rain. Outdoor America 47:6-8.

Hanson, J.M. and Leggett, W.C., 1982, Empirical prediction of fish biomass and yield. Canadian J. Fish. Aquatic Sci. 39:257-263.

Harvey, H.H. and Lee, C., 1982, Historical fisheries changes related to surface water pH changes in Canada, pp. 45 to 54. In: Int. Symp. Acidic Precipitation and Fisheries Impacts in Northeastern North America Proceedings, Cornell University, Ithaca, NY.

Hough, Stansbury and Michalski, Ltd. and J. E. Hanna Associates, Inc., 1982, An approach to assessing the effects of acid rain on Ontario's inland sports fisheries. Canadian Department of Fisheries and Oceans, Rexdale, Ontario, Canada. 177 pp.

Jeans, D., 1982, Acid rain: A Newfoundland and Labrador perspective, pp. 30 to 39. In: Acid Rain: A Transjurisdictional Problem in Search of Solution, P.S. Gold, ed., Canadian-American Center, State Univ. New York, Buffalo, NY.

Lee, A.G. and Kreutzweiser, R., 1982, Rural landowner attitudes toward sport fishing access along the Saugeen and Credit Rivers, Southern Ontario. Recreation Research Review 9:7-14.

Linthurst, R.A., ed., 1983, The Acidic Deposition Phenomenon and its Effects, Critical Assessment Review Papers, Vol. 2. Effects Sciences. EPA-600/8-83-016B, U.S.E.P.A., Washington, DC. 655 pp.

Lucyk, C., personal communication, Ontario Ministry of the Environment, Toronto, Ontario, Canada.

Martin, W.E., Bollman, F.H., and Fum, R.K., 1982, Economic value of Lake Mead fishery. Fisheries 7:20-24.

Menz, F.C. and Mullen, J.K., 1983, Acidification impact on fisheries: Substitution and the valuation of recreation resources, pp. 135 to 155. In: Economic Perspectives on Acid Deposition Control, T.D. Crocker, ed., Butterworths, Stoneham, MA.

Memorandum of Intent on Transboundary Air Pollution (MOI), 1983, Final Report of the Impact Assessment Work Group, Ottawa, Canada and Washington, DC.

Minns, C.K., 1981, Acid rain: A preliminary estimate of the risk to Ontario's inland fisheries. Report No. MA 1622, Department of Fisheries and Oceans, Government of Canada, Ottawa, Canada. 36 pp.

Mitchell, C.L., 1980, Canada's fishing industry: A sectoral analysis. Canadian special publication of fisheries and aquatic sciences, No. 52, Dept. Fisheries and Oceans, Ottawa, Canada. 49 pp.

Ontario Ministry of the Environment (OME), 1980, The case against the rain. Toronto, Ontario, Canada. 24 pp.

Peterman, R.M. and Steer, G.J., 1981, Relation between sportfishing catchability coefficients and salmon abundance. Transactions American Fisheries Soc. 110:585-593.

Prepas, E.E., 1983, Total dissolved solids as a predictor of lake biomass and productivity. Canadian J. Fish. Aquatic Sci. 40:92-95.

Robert R. Nathans Associates, 1969, Impact of acid mine drainage on recreation and stream ecology. Appalachian Regional Commission, Washington, DC. 183 pp.

Roberts, J., 1981, Acid rain: Solving a border problem. Catalyst for Environment/Energy 7:14-17.

Ryder, R.A., 1965, A method of estimating the potential fish production of north temperate lakes. Transactions American Fisheries Soc. 94:214-218.

Saunders, R.L., 1981, Atlantic salmon stocks and management implications in the Canadian Atlantic provinces and New England, U.S.A. Canadian J. Fish. Aquatic Sci. 38:1612-1625.

Schlesinger, D.A. and Reiger, H.A., 1982, Climatic and morphoedaphic indices of fish yields from natural lakes. Transactions American Fisheries Soc. 111:141-150.

Schulze, W.D., d'Arge, R.C., and Brookshire, D.S., 1981, Valuing Environmental commodities: Some recent experiments. Land Economics 57:151-172.

Schulze, W.D., Cummings, R.G., Brookshire, D.S., Thayer, M., Whitworth, R., and Rahmatian, M., 1983, Experimental approaches for valuing environmental commodities. Grant No. CR808-893-01, U.S.E.P.A., Washington, DC.

Theil, H., 1971, Principles of Econometrics. John Wiley and Sons, New York, NY. 736 pp.

Victor and Burrell Research Consultancy, 1982, A methodology for estimating the impacts of acid deposition in Ontario and their economic value. Ontario Ministry of the Environment, Toronto, Ontario. 180 pp.

Watt, W.D., Scott, C.D., and White, W.J., 1983, Evidence of acidification of some Nova Scotia rivers and its impacts on Atlantic salmon. Canadian J. Fish. Aquatic Sci. 40:462-473.

Youngs, W.D. and Heimbuch, D.G., 1982, Another consideration of the morphoedaphic index. Transactions American Fisheries Soc. 111:151-153.

ACID RAIN IN NORTH AMERICA: CONCEPTS AND STRATEGIES

Volker A. Mohnen and Jerre W. Wilson

Atmospheric Sciences Research Center, State University
of New York at Albany, Albany, New York

ABSTRACT

The major scientific issues and policy options are reviewed in
an attempt to present a comprehensive picture of the acid rain
problem in North America. It is found that several scientific
problems still remain to be resolved, but there is a relatively
good understanding of emissions of acidifying substances, their
transport and chemical transformation, and their ultimate
deposition.

INTRODUCTION

Acid rain occurs at many places in North America but exhibits
a significant maximum in the northeastern United States and
adjacent Canadian provinces. A pH contour map and precipitation
weighted concentrations of sulfate and nitrate ions derived from
the NADP and CANSAP networks from 1979 through the end of 1981
(Semonin, personal communication, 1983) are given in Figures 1-3.

There are large regions in North America containing lakes that
are potentially sensitive, based on bedrock geology, to
acidification by acid precipitation (Figure 4, from Galloway,
personal communication). The places marked by a star indicate
lakes located within such sensitive areas and with clearly
demonstrated "damages." The solid circles also are locators for
lakes within sensitive regions but where there are no noticeable
"damages." The main difference between these two aquatic systems

439

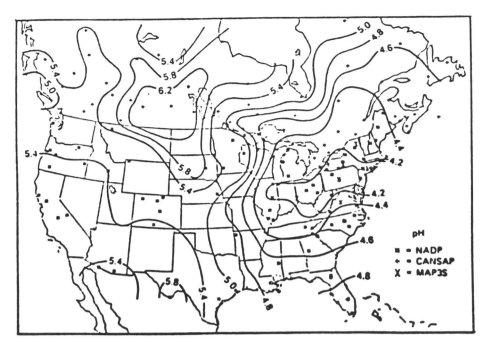

Figure 1.   Precipitation pH contour map of North America for the
             period 1979-1982 (Semonin, personal communication).

sensitive to acidic deposition is that one set of lakes received
precipitation with pH levels well below 4.6 while the precipitation
acidity received by the other lakes was well above the 4.6 level.

    Identification   of   aquatic   systems   sensitive   to   acidic
deposition should take into account not only the bedrock geology,
but include the vegetation cover, the litter layer, and the organic
and inorganic soil layers.   However, as a general guideline, the
fact remains that aquatic damage has been documented in areas where
increased depositional loading of pollution-related ions (hydrogen,
sulfate, nitrate, etc.) occurs on sensitive systems.   It is the
belief of the scientific community that anthropogenic emissions of
sulfur and nitrogen compounds are responsible for the observed high
concentration   levels   of   hydrogen   ions,   sulfate   and   nitrate   in
precipitation.   In the eastern United States and adjacent Canadian
provinces, these emissions of sulfur and nitrogen compounds are in
excess of 90 percent of the natural emissions.   The anthropogenic
sources are related to fossil fuel combustion of oil and coal, with
coal and oil-fired power plants and the transportation sector
accounting for the major portion of the emissions. SO_2 and NO_x
emissions for the eastern United States are shown in Figures 5 and
6, respectively.

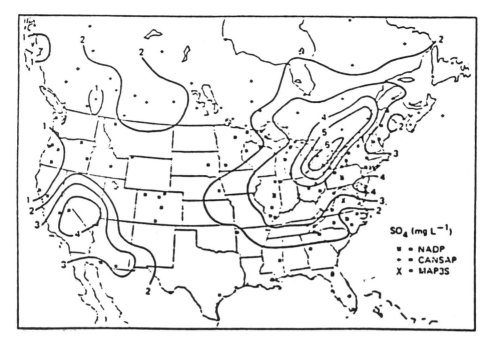

Figure 2.   Precipitation weighted sulfate concentration contour map
             of North America for the period 1979-1982 (Semonin,
             personal communication).

DISCUSSION

     The two major areas of uncertainty in establishing source
receptor relationship are horizontal and vertical transport over
long distances and incorporation of sulfur and nitrogen compounds
into cloud droplets, followed by aqueous phase reactions leading to
the wet deposition of acidifying substances.   These will be
reviewed separately as transport and transformations.   Lastly,
different control strategies will be evaluated.

Transport

     One of the basic observations about acid rain is the
approximate coincidence of the region of highest deposition with
the areas of greatest sulfur dioxide and nitrogen oxide emissions.
This circumstance represents some of the strongest evidence linking
acid deposition with man-made emissions of SO_2 and NO_x.   These
two pollutants reach the surface of the earth as dry acidic
deposition and are transformed, at least partially, to sulfuric and
nitric acids, which are removed from the atmosphere in both dry and

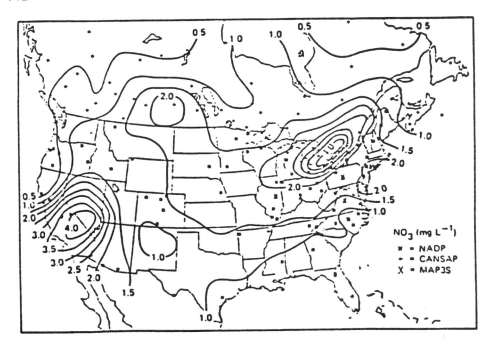

Figure 3.   Precipitation weighted nitrate concentration contour map
             of North America for the period 1979-1982 (Semonin,
             personal communication).

wet form.   The role and depositional pathways of all the precursors
in the formation and fallout of acidic compounds are being
extensively studied.   The speed and direction of the winds at the
level of emission exert a major influence in the movement of these
pollutants.   Their transport distance will depend upon whether the
given pollutants are gases or particles, vertical movement during
their lifetime, the types of atmospheric chemical reactions, and
the efficiency whereby they are removed (scavenged) from the
atmosphere.   The consensus in the scientific community is that the
long range transport of pollutants leads to acid deposition.   It is
also recognized that both long range and local sources may
contribute to acidity at a receptor.

    Regional models have been developed in an attempt to establish
source receptor relationships between acidic deposition precursors
and wet and dry deposition.   These have been operated as part of
joint efforts organized under the United States-Canada Memorandum
of Intent on Transboundary Air Pollution (1980).   Two distinct
types of models are in use today for studying long range transport
of air pollutants: Eulerian grid models and Lagrangian trajectory
models.   Existing acid deposition models have already contributed
to our understanding of annual wet deposition over a large

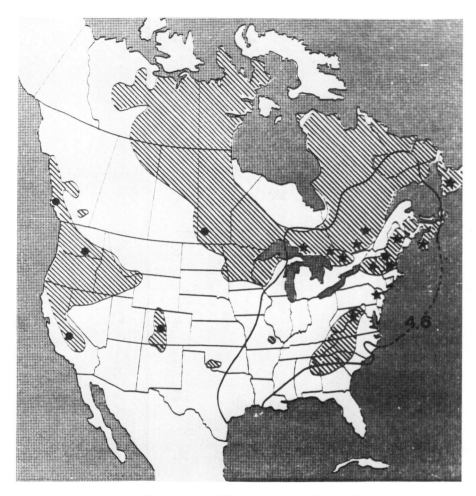

★ Lakes and Streams Acidified by Acid Deposition

● Lakes and Streams Not Acidified by Acid Deposition

Figure 4.   Sensitivity map of North America based on bedrock
            geology. The most sensitive areas are shown as dashed
            zones (Galloway, personal communication), along with the
            precipitation pH 4.6 contour line.

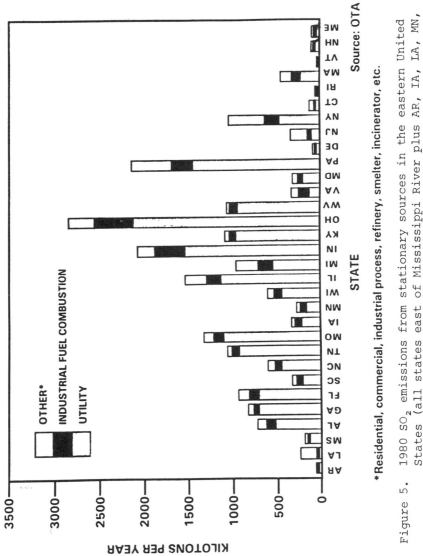

Figure 5. 1980 SO$_2$ emissions from stationary sources in the eastern United
States (all states east of Mississippi River plus AR, IA, LA, MN,
and MO; Office of Technology Assessment, 1980).

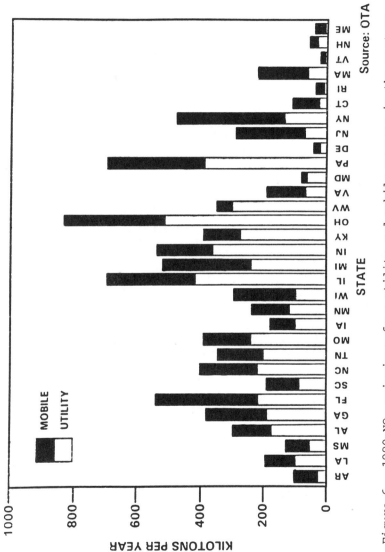

Figure 6.   1980 NO$_x$ emissions from utility and mobile sources in the eastern
United States (all states east of Mississippi River plus AR, IA,
LA, and MO; Office of Technology Assessment, 1980).

geographic area such as the eastern United States. But similar
efforts for shorter time periods are not thought to be feasible at
this time. Similarly, efforts to model dry deposition are hampered
by the almost complete lack of field measurements against which to
judge the model results. Obviously, uncertainties in our knowledge
of transport and chemical transformations are two of the basic
reasons for the limited faith that the community has in current
acid deposition models.

Verification of any transport model is critical to its
acceptance both as a description of scientific understanding and as
a tool for analyzing policy choices. Two methods include constant
altitude balloons (Pack et al., 1978) and chemical tracers. In the
past, the first method did not yield conclusive answers over longer
transport distances, and the latter method has only very recently
been successfully employed in the field of long range transport
(Cross Appalachian Tracer Experiment, carried out in 1983). While
techniques for computing trajectories have advanced markedly over
the past decade, they are still plagued by a host of additional
uncertainties; most notably are the sparsity of meteorological
data, both in time (upper air soundings are made only twice daily)
and space (stations are widely separated in the United States).

It is not possible at this time to unequivocally associate an
uncertainty value to current acid deposition models resulting from
their simplified treatment of transport processes. It is
understandable why a healthy degree of skepticism should be applied
regarding the usefulness of these current models in accurately
predicting acid deposition over eastern North America. In
particular, those models are considered less reliable when the time
averaging period is shortened to an event-by-event basis or when
the geographic region of interest (receptor) is relatively small.

By necessity, the deterministic models that are available
employ approximations to the atmospheric transformation processes
that are hypothesized to be important to acid deposition.
Reflecting a lack of knowledge of specific mechanisms, the chemical
transformations are treated parametrically in all existing models.
It becomes apparent that none of the currently available models
contain an aqueous phase submodule capable of realistically
treating SO_2 transformations. Furthermore, none of the present
models contain any detailed nitrogen chemistry.

Transformations

Considerable insight has been obtained over the past years
into the chemical transformations that occur during transport. For
example, it is now known that certain cloud and precipitation
elements play a major role in the overall conversion of sulfur and

nitrogen compounds to sulfates and nitrates. For example, rates of SO_2 oxidation through gas phase reactions are relatively slow (a few percent per hour during daylight), whereas in theory those for the aqueous phase pathways may be as high as 100 percent per hour for seemingly realistic concentrations of the reactants in cloud water. A comparison of reaction pathways for aqueous phase oxidation of SO_2 is shown in Figure 7 (Martin, 1984). H_2O_2 is the only oxidant for which the rate dependence on $[H^+]$ compensates for the decreased solubility of SO_2 with increased $[H^+]$. The concentrations of reactants used in deriving Figure 7 are representative of those that might be anticipated in the atmosphere. Oxidation by H_2O_2 dominates all reactions for conditions of low pH. Oxidation rates can be greater than 100 percent per hour but this situation cannot persist for more than 1-3 hours, which is the average lifetime of clouds.

There is some evidence of the formation of HNO_3 in cloud water and precipitation. Both theory and experiments suggest that HNO_3 may be formed rapidly from a combined gas phase-aqueous phase process via H_2O_5 generated by O_3-NO_2 reactions (Gertler et al., 1982). Although significant uncertainty remains concerning the source of HNO_3 in clouds and precipitation water, the limited evidence currently available favors the probably importance of the formation of N_2O_5 followed by its reaction in cloud droplets to form HNO_3.

This does not mean, however, that aqueous phase reactions dominate the overall production of sulfates. According to current understanding, the SO_2 gas phase conversion rate is of the order of 16 percent per 24-hour period for polluted summer sunny skies and three percent per 24-hour period during winter sunny weather (Calvert and Mohnen, 1983). These rates are sufficiently large to compete with the aqueous phase processes when averaged over longer periods of time. The relative importance of either mechanism varies, of course, depending on a variety of meteorological conditions such as the extent of cloud cover, relative humidity, presence and concentrations of various pollutants, intensity of solar radiation, and amount of precipitation.

Recent laboratory results (Calvert and Mohnen, 1983) strongly suggested that the SO_2 gas phase conversion is approximately linear. This finding restored some of the severe doubts concerning current models containing "linear" transformation chemistry. Of course, there still exists the possibility that the aqueous phase conversion of SO_2 may be oxidant limited (H_2O_2 limited). Results from the Atmospheric Sciences Research Center's field station at Whiteface Mountain, New York, strongly suggested that during the winter months the H_2O_2 levels in clouds were very low, sometimes below 1 ppb (cloud water), while measurements of over 1000 ppb occurred in the summer (Camarota et al., 1983). Since

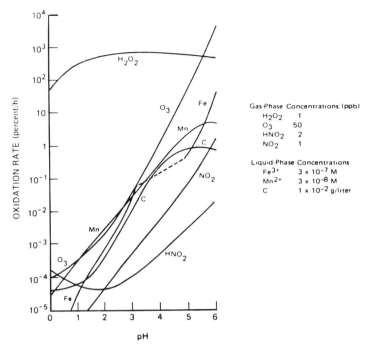

Figure 7.   Theoretical rates of liquid phase oxidation of SO_2
             assuming 5 ppb of SO_2, 1 ml m^{-3} of water in air, and
             concentrations of impurities as shown (from Martin,
             1984).

approximately 65 percent of the annual sulfate in precipitation is
deposited during the summer months, the oxidant limitation that
prevails during the winter months might be masked. Hence, on an
annual basis the aqueous phase oxidation might be closer to
linearity. As a result, the chemistry of gas and aqueous phase
sulfur might be "not significantly non-linear", as was recently
assessed by the National Academy of Sciences report on "Acid
Deposition" (1983).

STRATEGIES FOR CONTROL

     A high level of public pressure is demanding some Federal
action to reduce the deposition of acidifying substances onto
ecologically sensitive receptor areas. The National Academy of
Sciences report (1983) on acid deposition expressed reservation and
conservatively stated: "We do not believe it is practical at this
time to rely upon currently available models to distinguish among
alternative control strategies."

However, decisions on almost all issues of public concern are routinely made in light of uncertainties in scientific knowledge. It is therefore no surprise that 12 bills have been introduced in the United States Congress to address the issue of "acid rain." These range from research only and lake liming approaches (Byrd-Rahall) to major SO_2 reductions as high as 50 percent (Mitchell/Stafford and Waxman/Sikorski). The research bills are supported by the Administration, most coal interests, the utility industry, the mining industry and the mine workers.

During June, 1983, three major scientific reports on acid rain were released. It was concluded that over 90 percent of acid rain in the northeast and Canada comes from man-made sources (National Acid Precipitation Assessment Program, 1983); that over a very broad region and over a long period of time a reduction in sulfur dioxide emissions will result in a proportional sulfur deposition reduction, but that current information does not allow us to predict the effect of those reductions on sensitive receptors (National Academy of Sciences, 1983); and that some further reductions in emissions be undertaken now to avoid the potential for irreversible damage later (White House Science Advisor's Panel, 1983).

The fundamental question that is being decided by the United States Government at this time (September, 1983) is whether or not further steps are warranted in light of our present scientific understanding and available data on damage. Options available are as follows:

Minimal change options: These options specifically recognize that the science supporting the need for acid deposition controls is inadequate. Proponents recognize that the arbitrary regional SO_2 rollbacks contained in most congressional proposals are not based on sound science and are likely to be inefficient at solving the acid deposition problem. These are divided into two broad areas:

1. Accelerated research and lake liming. This approach would accelerate the acid rain research program under the National Acid Precipitation Assessment Program. Current plans call for assessments of current and anticipated future damage (1985), a methodology for evaluating costs and benefits of alternative control strategies (1987), and a final evaluation of optimal control strategies (1989). The results could enhance policy makers' ability to choose among control and mitigation options and greatly aid in efforts to effectively implement those decisions.

2. Application of current and augmented rule-making authorities. This option would provide the Environmental Protection Agency with additional legislative authority to improve the Clean Air Act's provisions for controlling interstate air pollution and set regionally-variable secondary air quality standards, as opposed to uniform standards, for acid deposition and precursor pollutants. This would also improve EPA's ability to enforce the attainment of secondary standards for air pollution.

Major emissions reductions options. These options call for a major rollback in SO_2 emissions. Proponents of these approaches believe that, in spite of the scientific uncertainty, the weight of the evidence suggests that action should be taken now to control the dominant source of acid deposition - SO_2 emissions. One of these options (number 1 below) focuses on the near-term need to protect aquatic resources in the northeast while learning more from accelerated research before deciding to expand a control program to other parts of the United States. Another option (number 2 below) accepts the notion of a need to control SO_2 emissions on a broad regional scale to address concerns about both known damages in the northeast and potential damages in other parts of the country.

Either of these options could prove to be very costly and could have significant impacts on miners of high sulfur coal and utility rate payers in the affected states. Financing schemes or subsidies to mitigate these impacts could take the form of a generation tax on fossil fuel electricity production, an emissions fee on each ton of SO_2 produced, transferrable emission permits, or direct Federal budget outlays. Furthermore, prohibitions against fuel switching, while protecting the miners of high sulfur coal, would increase the cost of these options.

1. Experimental emission reduction option. This option would require by 1990 a reduction of three million tons of SO_2 below 1980 levels in a 13-state targeted area which includes New England, New York, Pennsylvania, Ohio, West Virginia, Maryland, Delaware, and New Jersey. This option was designed to reduce acid deposition in the most sensitive receptor regions, the Adirondacks and New England, while providing an opportunity for further research before controls are imposed on a wider area.

Emissions reductions would be allocated to the 13 states based on a formula which requires reductions proportional to the average emissions rate of each state in 1980. SO_2 emissions in the other 18 states outside the 13-state region would be capped at 1980 levels (allowing some growth in areas that are predominantly gas-fired).

These reductions represent roughly a 25 percent lowering in emissions from states in the 1000 km zone of influence of the sensitive region of Algoma-Muskoka region of Canada. As such, Canada should reduce SO_2 emissions in those areas of Canada within 1000 km of the Adirondacks by a comparable amount.

2.   Large scale regional rollback over either: a) 31 states (Stafford/Mitchell), or b) 48 states (Waxman/Sikorski)

Option 2(a) would mandate an SO_2 emissions reduction target of 8 million tons below 1980 levels in 31 states by 1993-1995. Owners seeking to build new plants after 1995 would have to secure offsets. Each state would be allocated an emissions reduction target through a formula attempting to simulate a least cost approach. The Stafford/Mitchell bill uses a formula based on excess emissions from the utility sector. EPA has devised a formula based on a rate per Btu of fossil fuel consumed, as discussed in the experimental emissions reduction option listed above.

Option 2(b) would mandate a reduction of 14 million tons (up to 10 million tons of SO_2 reduction from existing power plants and no additional emissions from new sources) over 48 states, with control technologies funded by a 1 mil/kilowatt hour fee on electricity generation within the 48 states region. Also included would be a 4 million ton reduction in NO_x emissions.

CONCLUSIONS

It was not the purpose of this paper to choose among possible emission control strategies should they be deemed necessary. Unfortunately, the overall scientific knowledge is not yet at a level that would lend itself to a logical course of action. The issue of an acid rain control program in light of limited scientific knowledge should therefore not be made by atmospheric scientists alone but must include all sectors that determine public policy.

REFERENCES

Calvert, J.G. and Mohnen, V.A., 1983, The chemistry of acid
    formation, Appendix A in NAS Report, Atmospheric processes in

acid deposition. National Academy Press, Washington, DC.
60 pp.

Camarota, N.A., Kadlecek, J.A., McLaren, S.E., and Mohnen, V.A.,
1983, 1983 winter cloud study, ASRC Whiteface Mountain Field
Station. Pub. No. 981, Atmospheric Sciences Research Center,
State University of New York, Albany, NY. 31 pp.

Galloway, J.N., personal communication, 1983, Dept. Environmental
Sciences, University of Virginia, Charlottesville, VA.

Gertler, A.W., Miller, D.F., Lamb, D., and Katz, U., 1982, SO_2 and
NO_2 reactions in cloud droplets, pp. 131 to 160. In:
Chemistry of Particles, Fogs, and Rain, J.L. Durham, ed., Ann
Arbor Science, Ann Arbor, MI.

Martin, L.R., 1984, Kinetic studies of sulfate oxidation in aqueous
solutions, pp. 62 to 100. In: Acid Precipitation: SO_2, NO
and NO_2 Oxidation Mechanisms: Atmospheric Considerations,
J. Calvert, ed., Ann Arbor Science Publ., Ann Arbor, MI.

National Academy of Sciences, 1983, Acid Deposition: Atmospheric
Processes in Eastern North America. National Academy Press,
Washington, DC. 375 pp.

National Acid Precipitation Assessment Program, 1983, 1982 Annual
Report to the President and Congress. U.S. Government
Printing Office, Washington, DC. 69 pp.

Office of Technology Assessment, 1980, United States Congress,
Washington, DC. Adapted from: G. Gschwandtner and K.
Eldridge, 1983, Historic Emissions of Sulfur and Nitrogen
Oxides in the United States from 1900 to 1980. Draft report
under EPA Contract No. 68-02-3311, Pacific Environmental
Services, Inc., Durham, NC. 350 pp.

Pack, D.H., Ferber, G.F., Heffter, J.L., Telegardad, K., Angell,
J.K., Hoecker, W.H., and Machta, L., 1978, Meteorology of long
range transport. Tellus 33:132-141.

Semonin, R.G., personal communication, 1983, Illinois State Water
Survey, P.O. Box 5050, Station A, Champaign, IL.

United States-Canada Memorandum of Intent on Transboundary Air
Pollution, 1980, Government Printing Office, Washington, DC.

White House Science Advisor's Panel, 1983, Panel appointed in 1982
by science advisor George A. Keyworth II. Title of Panel:
Office of Science and Technology Policy - Acid Rain Peer
Review Panel. Chairman, William Nirenberg.

ACID RAIN: LEGISLATIVE PERSPECTIVE

Frances F. McPoland[a]

Legislative Assistant to
Congressman Richard L. Ottinger (D-N.Y.)
Washington, D.C.

ABSTRACT

Moving legislation through Congress can be a somewhat painful process. The Clean Air Act of 1970, which was amended in 1977, did not include any provisions directly and comprehensively addressing the need for acid rain control, so Congress has been unsuccessfully trying to resolve this problem for the past three years. The problems for Congress have been regional differences ranging from economic need to environmental awareness. Congress has, so far, found it impossible to balance the economic needs of the Midwest with the environmental needs of the Northeastern United States. A number of solutions have been offered to deal with the problem of acid rain, but so far they have not appealed to all of the regions of the country. In time, when the American people become aware of acid rain damages, Congress will be forced to find a solution to the problem.

INTRODUCTION

It is appropriate that within this discussion of acid deposition some mention be made of the law which struggle to control it and attempts to improve that law. Other chapters in this volume discuss the scientific aspects of acid rain control, but it must be realized that the eventual outcome of acid rain control is likely to be determined by Congress. This chapter provides an overview of the acid rain issue as it currently (January, 1985) stands in Congress. But the reader should be

---

[a]Present address: 2800 Wisconsin Ave., N.W. #907, Washington, DC 20007

warned that while Congress is a ponderous body it can sometimes move with lightning speed, and legislation on this issue could occur at any time, and will most likely happen when the public becomes aroused. For a proper overview, this chapter needs to combine basic information on the various pieces of legislation that were under consideration in the 98th Congress with a discussion of the legislative process that goes beyond the kinds of information normally provided in civics text books.

DISCUSSION

The eventual fate of acid rain legislation is likely to depend both on scientific aspects of the debate and on Congressional factors. Those Congressional factors, in turn, can be divided into three general categories: the process that the legislation must go through, the clash of special interests, and the personalities of the individuals that move the legislation.

Background: The legislative process

At some point, most adults come to realize that the real process of government is not exactly the one they learned about in their eighth grade civics classes. The nice, neat "checks and balances system" emphasized in school often has very little to do with the process of legislation. This realization, while important, is only the first step in coming to understand how complicated the legislative process really is.

A vast amount of legislation is proposed during every session of Congress. In the 97th Congress (January 1982/December 1983) a total of 13,240 bills were introduced; of those 529 became public law (U.S. Congress, 1983). While most of this legislation is serious, some of it is introduced solely for its public relations impact and some for its educational value; in any case, very little passes. If we are to understand what is and is not happening with legislation on acid rain, it would be worthwhile as background to understand some of the reasons why many rolls are called but few bills are chosen. Some of the relevant information is the sort that may be familiar from civics textbooks, but hopefully this chapter will go beynod basic civics in order to help the reader understand the problems of the modern legislative process.

Members of Congress propose specific items of legislation (most of which are actually written by staffers or others, such as constituents or lobbyists) that begin winding their way through the legislative process. In the House of Representatives these bills are referred by the Parliamentarian to the appropriate committee or, as is often the case, two or more committees. Before the legislation can become law, it must first be approved by the subcommittees, then the full committees, and then the full chamber in both houses of Congress - the Senate and the House of Representatives. But even if the bill passes all these hurdles, it may wind up having little resemblance to what was originally proposed.

At the earliest stages, the chairs of the subcommittee schedule hearings on the bill so that all the subcommittee members have an opportunity to learn about the proposed legislation. This is not as simple or as automatic as it sounds: in the 98th Congress, for example, acid rain hearings were conducted in five cities around the country over a period of four months before a single hearing was held in Washington (U.S. Congress, 1984a,b).

If it appears that the chair has sufficient support in the form of votes for the legislation to proceed, the chair then schedules a "mark-up" session to amend the bill accordingly and sends it to the full committee where the process begins again. The chair can also intentionally not bring the legislation up at all, if he or she so chooses, thus defeating the bill simply by ignoring it. Often this leverage is used by full committee chairs to get subcommittee chairs to move or delay legislation that is important to them.

Very few bills get as far as full committee approval in even one house of Congress. If a full committee approves a bill, i.e. "reports" it, a "rule" is requested from the Rules Committee to establish the procedure for floor debate and vote. The bill is then sent to the House floor for debate, amendments and final passage or rejection by the entire House of Representatives. A somewhat different procedure would be followed in the case of bills being reported out of committee in the Senate.

One of the first keys to a deeper understanding of this process is that the chairs of the committees have almost absolute control over the movement of legislation that has been referred to them.[1] It is perhaps a reflection of this control that the only

---

[1] One example of the power of the chair came in late 1983 during the Energy and Commerce committee mark-up of natural gas decontrol legislation. Since this was a full committee mark-up, the powerful John Dingell (D-Michigan) was in the chair. The debate was acrimonious on both sides and gradually it became clear that the members might not approve the legislation as Dingell wanted. When the opposition gloated that they had the votes, Dingell is reported to have said, "yes, but I have the gavel." At that he slammed it down and adjourned the session...effectively ending the mark-up on that legislation until he chose to reopen it many months later. Such is the power of the chair.

acid rain legislation that had hearings, let alone a mark-up, in the House of Representatives during the 98th Congress was a bill co-authored by Representative Gerry Sikorski (D-Minn.) along with Rep. Henry Waxman of California, the chair of the Health and Environment Subcommittee of the House Energy and Commerce Committee. Even in that case, it took almost a year from the time that the bill was introduced until it was scheduled for mark-up in subcommittee, where the chair was still unable to muster sufficient votes for passage. Similar actions occurred in the Senate.

The legislation actually reported out of committee continues to follow a very rocky road, for a bill can be defeated in many different ways. It can be amended beyond recognition at any point in the process, the Rules Committee can give it an unfavorable rule, or the leadership in the House can simply neglect to place the bill on the calendar for floor consideration. There is room for a great deal of personal discretion on the part of key individuals.

All of this is to say that the process is not simple. The procedures in the Senate are totally independent from the House and are just as complicated. Even if the companion bill -- the Senate bill introduced at the same time and for the same purpose as the House legislation -- is identical to its House counterpart at introduction, the Senate may produce a bill that is drastically different from the one produced in the House. If both the House and Senate pass their respective pieces of legislation, the bill then goes to a "Conference Committee" to iron out the differences; this is a committee made up of key legislators on the particular issue from both chambers. Sometimes the Conference Committee receives instructions from either House that some part of that House's version of the legislation is inviolable, and sometimes similar feelings exist about the legislation passed by the other chamber. This could lead to a deadlock and no possible hope of legislation if the issue cannot be resolved.

Remarkably, some legislation does survive this process. The Clean Air Act Reauthorization did not survive in the 98th Congress. Before turning to specific reasons for its lack of passage, it would be useful to provide a brief background on the Clean Air Act.

History of clean air legislation

The 1970 Clean Air Act was preceeded by legislation in the early 1960's[1], none of which provided a comprehensive control

---

[1] The Clean Air Act of 1970 was preceeded in 1966 by PL 89-675, the Clean Air Act Amendments of 1966, and in 1967 by PL 90-148, the Air Quality Act of 1967. It was also amended in 1977 by PL 95-95, the Clean Air Act Amendments of 1977.

program. These legislative efforts were generally geared towards reducing local air pollution in order to protect public health. The 1970 Clean Air Act, although amended extensively in 1977, is still largely intact. It has six major functions: it sets ambient air pollution levels through National Ambient Air Quality Standards, and monitors the pollution control programs for each state through State Implementation Plans. It establishes National Emission Standards for motor vehicles, mandates the Prevention of Significant Deterioration of air quality conditions in selected areas, and uses New Source Performance Standards to impose technology-based control requirements on emissions from stationary sources. Lastly, it limits new emissions in non-attainment areas.

    While the Clean Air Act was a very significant piece of legislation, it was also a compromise. Persons with varying interests in the clean air debate have made a number of efforts to amend it, although generally those efforts have been unsuccessful. Because of the so called energy crisis, for example, an unsuccessful attempt was made in 1974 to permit temporary waivers for industry in order to encourage the increased use of coal. There are also a number of issues that were not specifically addressed in the 1970 Act, and many aspects of acid rain fall into this category. The long-distance transportation of air pollutants that are the precursors of acid rain, for example, were not specifically dealt with in the 1970 effort; legislative language was ambiguous as to congressional intent on this issue. In addition, the enforcement efforts of the Environmental Protection Agency (EPA) during the administration of President Reagan have been questionable. The approach EPA has taken during the first four years of the Reagan administration has been to drastically relax the limits for sulfur dioxide emissions in the high density industrial regions of the Midwest. The EPA has also consistently refused to analyze the cumulative impacts of massive relaxations of emission limits which they approved. Their position has been not to evaluate the cumulative interstate impact of all sources but only look at the short range impact. Yet, most of the damage to New York State's Adirondack region can be directly traced to the cumulative impact of emissions from the Midwest. In short, the issue of acid rain is not separately addressed in the 1970 Act. It has become a major factor in the debate over re-authorization of the Clean Air Act, which was supposed to occur in 1980. There is another major issue concerning the Clean Air Act which must also be resolved: the need for stronger legislation on hazardous air pollutants - carcinogens, mutagens, and other extraordinarily dangerous substances like EDB and dioxins. Since 1971, EPA has been studying 37 of these pollutants and has so far only take action on seven.

Proposals for acid rain controls in the 98th Congress

These have run the gamut from plans to study the problem to
ambitious projects to reduce acid deposition by fifty percent
within ten years. Recent legislative initiatives concerning reauthorization of the Clean Air Act have been some of the most
complicated and contentious ever to face Congress. Regional
differences abound and party loyalty has been less evident on this
than on almost any other issue in recent memory.

The reality of Congress can be viewed as being different from
the reality of the rest of the world. It is important to remember
that the primary concern that many members of Congress face is
employment - their own. Therefore, their actions on issues of
national importance tend to be reduced to the lowest common
denominator - how an issue effects their constituency and their
chances for re-election. Happily, and ironically, this personal
concern and the resulting delay can actually contribute to
carefully worked-out legislation.

Scientific evidence indicates that acid deposition occurs when
sulfur and nitrogen oxides (SO_2 and NO_x) and their transformation
products return to the earth from the atmosphere. If this is
accepted, then the argument arises over where these sulfur and
nitrogen oxides come from. Again, differences of opinion abound,
but most of the evidence indicates that the major source of acid
deposition, at least in the eastern United States, is SO_2 emissions
from utilities in the Midwestern United States which burn a high
sulfur grade of coal. This generally refers to coal with a sulfur
content of 1% or higher. In order to control emissions from these
utilities they must either switch to coal with a lower sulfur
content or install scrubbers to remove the pollutants before they
are emitted into the atmosphere.

There are major problems with both of these policy options.
The Midwest supplies much of the coal it burns. Most of this coal,
which provides thousands of jobs, has a high sulfur content. If
the utilities are forced to reduce the amount of SO_2 that they
emit, it might be more cost effective for them to switch to low
sulfur coal rather then install costly scrubbing equipment. The
cost for any of these actions would be directly passed on to the
utility customer. Much of the low-sulfur coal in the United States
is from the West and is "stripped mined," which does not provide
many jobs because surface mining is performed by large equipment.
So, it is believed that allowing utilities to switch to low-sulfur
coal would cost jobs to the Midwest and not create new ones where
the low-sulfur coal is being mined. This information is disputed
by the low-sulfur coal industries lobbying arm, the Alliance for
Clean Energy (ACE), which contends that jobs would simply shift
from high-sulfur to low-sulfur coal fields. So the issues of

concern are: the jobs in the Midwest, the cost of electric utility rates primarily in the Midwest and the ecological damage that is being done to the Northeastern portion of the United States.

The arguments, while far from simple, can be simply presented. While the scientific evidence indicates that the majority of sulfur dioxide emissions come from the Midwest, it is by no means the case that a majority of the concern over the issue can be found in the same region. Many of the emissions pour out of tall stacks that send their emissions elsewhere, and most Midwestern soils have a high buffering capacity while Midwestern industries have sometimes shown a lower capacity to buffer job losses in the recent past. In addition, the science of acid rain remains somewhat uncertain. There is, however, little doubt in the minds of Northeasterners that Midwestern emissions substantially contribute to the problem of acid rain.

People in the Western United States, meanwhile, have tended not to be interested in what has been viewed as primarily an Eastern problem, although concern has been increasing in the recent past (NWF, 1984). Their problems are much more related to nitrous oxide (NO_x) emissions from automobiles than SO_2 emissions, primarily because the West has newer more efficient utility systems, tends to mine and use a lower sulfur coal, has a higher reliance on oil fired plants, and has a good deal of hydroelectric power. On the whole, therefore, members of Congress from this region tend to question why they should be involved in control programs that focus on utility emissions, and they are reluctant to help pay the bills.

In the Northeast, however, the problem is seen as more pressing. In a recent survey conducted in New York (Reeb, 1983) more people identified acid rain as a major problem than knew the name of their representative in Congress. This kind of information is very real to New England members and makes them realize that they must at least take a stand on the issues. In fact, some members of Congress have made this the primary issue of their campaigns, running for re-election on it every two years.

An additional issue that Congress has dealt with for the last four years is the Reagan administration's concern that environmental problems can only be solved with an approach that reflects the direct economic benefit of the control program. The previous administrator of the E.P.A. (William Ruckelshaus) has spend considerable time and effort attempting to convince President Reagan's cabinet council that acid rain was an issue that should be dealt with at some level. The Reagan administration simply cannot see that the bottom line numbers reflect sound economic policy, although evidence indicates that it does (Menz and Muller, 1983; Crocker in press). So, in spite of mounting evidence to the contrary this administration continues to propose a "study only" strategy.

Persons who feel that members of Congress pay little attention to the perspectives of their states or districts would have a difficult time explaining congressional proposals to deal with the acid rain issue. Various solution have been proposed; most of them reflect the concerns of the regions of the members making the proposals. Among them are major proposals by Senator Durenberger (R-MN), Senators Mitchell (D-ME) and Stafford (R-VT) and Representatives Sikorski (D-MN) and Waxman (D-CA) (Table 1). In addition, there have been proposals by Senator Glenn (D-OH), Representative Rinaldo (R-NJ), Senator Byrd (D-WV) and Senator Randolph (D-WV), which have added to the discussion. These bills, however, have not been a major factor in the debate. Finally, the New England Congressional Caucus [Congressmen D'Amours (D-NH) and Conte (R-MA)] introduced legislation supported by the national environmental community, as represented by the National Clean Air Coalition[1] which seemed to influence the scope of discussion in congress and in other political spheres as well.[2]

The debate in the Senate had a different character than the debate in the House. There are only two Senators from each state, no matter how many people they represent, and the Senators are elected for six year terms. This is unlike the House, which holds its elections every two years and in which the members represent districts of approximately equal size. Senators may not feel as much intense political pressure to respond to local issues as do representatives. It also means that the Senate does not have the heavy concentration of representatives from the depressed Midwest or the acid-rain-conscious Northeast. It allows the Western

---

[1]The National Clean Air Coalition is made up of the following national organizations: American Lung Association, Center for Auto Safety, Citizens for a Better Environment, Environmental Defense Fund, Environmental Policy Center, Environmentalists for Full Employment, Friends of the Earth, International Association of Machinist and Aerospace Workers, Izaak Walton League of America, League of Women Voters of the United States, National Audubon Society, National Consumers League, National Farmers Union, National Parks and Conservation Association, National Wildlife Federation, Natural Resources Defense Council, Oil, Chemical and Atomic Workers, Sierra Club, United Steelworkers of America, The Wilderness Society, and Western Organization of Resource Councils.
[2] The democratic platform committee included language in its 1984 platform calling for a fifty percent reduction in acid rain emissions from the 1980 levels as well as additional NO_x reductions.

Senators to have a stronger influence on preventing legislation which will cost their region money, since they do not believe that they should pay for the problems created by the Midwest and affecting the Northeast. Also, the Western Senators know that most of the coal burned in power plants in the Western United States is of the low-sulfur variety, and the Western states would benefit financially from an increased demand for low-sulfur fuel.

What this has meant in the Senate is that the only legislation considered by the Committee on Environment and Public Works was S.768, introduced by the Chair, Senator Stafford of Vermont (Table 1). This is a strong New England oriented bill which would require a ten million ton reduction of sulfur dioxide emissions in the eastern 31 states within ten years of its passage. Funding for the program would be provided entirely by generators of the pollution, the utility customer. This would mean that electric utilities would either switch to cheaper low-sulfur Western coal, at the possible cost of local Midwestern coal mining jobs, or would install expensive scrubbing equipment paid for by the local electric consumer through their utility bills. In either event this might prove to be a costly approach for the Midwest.

Another approach, advocated by Senator Durenberger of Minnesota (S.2001) also calls for the reduction of 10 million tons of sulfur dioxide emissions in the Eastern 31 states. The Environmental Protection Agency would be directed to establish a priority list of sources for control. Failure of the EPA to establish and act upon this priority list would automatically activate a standard of reductions directed at sources emitting more than 1.2 pounds of SO_2 per million BTUs over a 30 day period. This approach would allow the utilities to decide how best to achieve their share of the reductions by installing scrubbers or switching to low-sulfur fuel. However, the EPA would be directed to monitor the economic impacts of fuel switching and would be authorized to limit its use if these impacts were detrimental. The funding mechanism for this program is designed to spread out the costs to those industries thought to be responsible for the pollution. All stationary sources of sulfur dioxide will contribute 2/3 of the fund, all stationary sources of NO_x will contribute 1/6, and mobil NO_x sources will contribute 1/6 of the fund. Ultimately, this means that the major portion of the costs of this reduction program would be born by electric utility consumers in the six states emitting the most pollution (Ohio, Illinois, Kentucky, West Virginia, Indiana, and Pennsylvania).

House Resolution 3400, the primary legislation in the House of Representatives, was introduced by Representatives Sikorski (D-MN) and Waxman (D-CA). It also required a ten million ton reduction of SO_2 but added a requirement for a four million ton reduction in NO_x emissions over a ten-year period (Table 1). In addition, H.R. 3400

Table 1.   Possible acid rain legislation during the
           98th Congress.

Bill Number	Major Sponsor	Compliance Date	Emissions Reductions (Million Tons)	Achieved by	Funding
HR 3400	Waxman-Sikorski	1995	10 SO_2 4 NO_x	Retrofit 50 largest emitters (utilities) for SO_2 reductions tighter NO_x controls	1 mil kwh fee on all non-nuclear electricity
S 2001	Durenberger	1995	10 SO_2	All sources SO_2 & NO_x	Emissions tax
S 768	Stafford-Mitchell	1995	10 SO_2	State by state on the basis of utility emissions	None
HR 4404	D'Amours-Conte	1995	12 SO_2 4 NO_x	Based on HR 3400 for NO_x and the first 10 mil tons of SO_2 reductions. Remainder of the reductions from utility and industrial boilers based on emissions	1.5 mil kwh fee on all non-nuclear and non-hydro electricity

takes a very different approach to the reductions: the 50 utilities that produce the greatest amount of sulfur dioxide would be required to install scrubbers. All of these plants are located in the Midwest as is most of the high-sulfur coal they currently burn. The requirement for the installation of scrubbers was written into the legislation to prevent utilities from switching to cheaper low-sulfur coal. Such a switch, which could be less expensive than the scrubbers for the industries, would create severe economic hardships in the Eastern United States where most high sulfur coal is produced. Installing scrubbers on the 50 largest emitters would account for 6 1/2 million tons of SO_2 reductions, with the other 3 1/2 million tons of reductions in the second phase of the program being allocated to the states as they determine. The program would be funded by a 1 mil per kilowatt hour "user fee" imposed on all electrical consumption throughout the country except for the electricity generated from nuclear power plants. This would spread out the costs to all utility customers, not just those in the Midwest, and would cost the average family fifty cents a month. The program would run until 1997.

As mentioned earlier, the New England Congressional Caucus also entered the fray against acid rain. Although most Northeastern political leaders have consistantly advocated a "polluter pay" philosophy, with most of the cost being in the Midwest, Representatives D'Amours (D-NH) and Conte (R-MA) introduced legislation (H.R. 4404), cosponsored by the entire New England delegation, that was based on the Waxman-Sikorski "user fee" concept. H.R. 4404 went even further then the Waxman-Sikorski legislation; it required an additional two million tons of SO_2 reductions. These additional reductions were to be obtained from sources such as industrial boilers and process emitters[1] which also produce sulfur dioxide. The additional reductions were to be allocated on a state-by-state basis determined by each of their emissions. They also redesigned the fee structure to collect more money (1.5 mils per kilowatt hour from all non-nuclear utility customers) and to distribute the collected funds based on the same formula as the reductions. On the other hand, the Waxman-Sikorski bill would have the funds distributed in the second phase of the program on a first-come-first-served basis. The 12 millions ton bill was endorsed by the National Clean Air Coalition and was offered in the Subcommittee on Health and the Environment as an

---

[1] Process emitters include oil refineries, smelters, iron and steel mills, pulp plants, concrete factories, natural gas plants and sulfuric acid plants.

amendment to the Waxman-Sikorski legislation by Congressman
Richard Ottinger (D-NY).[1]

Much of the other legislation introduced in the 98th Congress
concerning acid rain dealt with either mitigating the effects of
the damage (Senator Randolph, S.766) or simply studying the
feasibility of acid rain controls (Senator Byrd, S.454). This
legislation has been generally promoted by the Midwestern
delegation to Congress, the Republican members who agreed with the
market-oriented philosophy of the Reagan Administration, and the
high-sulfur coal producers in the Midwest.

PROSPECTS FOR A RESOLUTION

Acid rain control legislation did not come out of the 98th
Congress which ended in December 1984. The legislation which had
the greatest chance in the House of Representatives was H.R. 3400.
In subcommittee "mark-up" a coalition of Republicans and Midwestern
Democrats were able to defeat the acid rain portion of the
reauthorization vehicle.[2] The chair, Mr. Waxman (D-CA), then

---

[1] The amendment failed to be attached to the Waxman-Sikorski acid
rain legislation. The 12 million ton proposal was offered by
Mr. Ottinger (D-NY) and during the debate word reached Mr. Waxman
(D-CA), the chair, that a Democrat from Ohio, Mr. Eckart (D-OH),
felt that he would be forced to vote against final passage of the
entire bill if this amendment was attached to it. Mr. Waxman also
came to understand that the Republican minority, which opposed
final passage of the bill, planned to vote for the amendment in
order to force Mr. Eckart to vote against final passage. He would
have been the swing vote. Mr. Waxman asked Mr. Ottinger to
withdraw his amendment, which he was not permitted to do by the
minority (unanimous consent is required). A vote was taken and all
the Democrats who favored the legislation, many of whom wanted to
see 12 million tons of reductions achieved, voted against the
amendment. Mr. Ottinger abstained until he was sure that his
amendment would be defeated and then he voted for it.

[2] During the third day of the Health and the Environment Subcommittee mark-up of this legislation, a "motion to strike" the
entire acid rain section of the bill was offered by the
Republicans. Mr. Eckart (D-OH), who was the swing vote, was
feeling the pressure of abnormally high levels of unemployment in
his district. He voted in favor of the "motion to strike" and
caused the acid rain legislation to be defeated for 1984 on a
10-9 vote. Since that time he has offered his own version of acid
rain control legislation (H.R. 5794) which he feels better protects
the economic interests of the Midwest.

indicated that since the prospects for passing good acid rain legislation was out of question, no attempt would be made to pass a straight reauthorization of the Clean Air Act. Since each Congress lasts two years and is an entity onto itself, the process begins all over again in January, 1985 with the 99th Congress. It does seem clear, however, that the same or similar legislation will be introduced again.

In the Senate, however, legislation proceeded further then it did in the House. Mr. Stafford's S.768 bill was voted out of committee by a coalition of Western and Eastern Senators, but it was not permitted to come to the floor for a vote because the majority leader of the Senate was Howard Baker from Tennessee who does not believe in the "polluter pay" philosophy. Mr. Baker, along with a coalition of Western Senators who threatened to filibuster if the bill was brought to the Senate floor, effectively killed the legislation for the 98th Congress.

It must be understood that even if both of these pieces of legislation passed their respective Houses there would have been major unresolved issues. The legislation from the Senate was strictly a "polluter pay" program with the entire burden of the costs of pollution controls to fall on the Midwestern utilities. S.768 also contained no protection for jobs in the high-sulfur coal mining regions. The version that the House was most likely to pass placed the financial burden on all utility customers throughout the country and tried to establish provisions that would have prevented significant job losses; however, it did nothing to encourage further development of low-sulfur coal mining in the Western United States, which some interests would like to see. These issues would have been very difficult to resolve in a Conference Committee.

The acid rain issue will remain contentious in Washington for a long time to come. Resolution of the issues will come, not necessarily when the science gets better, but when the people of the United States demand a solution. Inevitably Congress will then act. It may appear from this discussion that democracy is messy and unwieldy, and probably it is. But one reason that it is so messy is that members of Congress really do try to represent their constituents, and we, the people, want different things in different sections of the country, or different sections of society. It may take a long time, but ultimately it will be accomplished. The question remains if it will be accomplished in time.

Acknowledgements - The author acknowledges help and contributions from numerous individuals, especially David R. Wooley, Assistant Attorney General for the State of New York, Dr. William R. Freuden-burg of Washington State University, Pullman, Washington, and

Edward Schillinger, Yale University, New Haven, CT. Survey data
were kindly made available by Mr. Anthony Taverni, Department of
Environmental Conservation, Albany, NY.

REFERENCES

Crocker, T.D., in press, What economics currently say about acid
    rain control. In: Adjusting to Regulatory, Pricing and
    Marketing Realities, H.R. Trebing, ed., Institute of Public
    Utilities, East Lansing, MI.
Menz, F. and Muller, J., 1983, Acidification impact of fisheries:
    Substitution and the valuation of recreation resources,
    pp. 135 to 155. In: Economic Perspectives of Acid Rain
    Control, T.D. Crocker, ed., Ann Arbor Science Press, Ann
    Arbor, MI.
National Wildlife Federation, 1984, Acid Rain: State by State
    Impacts. 102 pp.
Reeb, D.J., 1983, Economic Impact Study of Acid Precipitation.
    Center for Financial Management, Institute for Government and
    Policy Studies, Rockerfeller College of Public Affairs, SUNY,
    Albany, NY. 250 pp.
U.S. Congress, 1983, Congressional Record. V. 129, No. 3, D-17
    (Jan. 25 daily edition).
U.S. Congress, 1984a, House Committee on Energy and Commerce,
    Subcommittee on Health and the Environment. Acid Rain Control
    Hearings. 98th Congress., 1st and 2nd Sessions, Dec. 1, 9,
    1983 and Feb. 10, 1984. Serial No. 98-114, Washington, DC.
    855 pp.
U.S. Congress, 1984b, House Committee on Energy and Commerce,
    Subcommittee on Health and the Environment. Acid Rain Control
    Hearings. 98th Congress, 2nd Session, Feb. 17 and
    March 5, 1984. Serial No. 98-115, Washington, DC. 833 pp.

VIEWPOINTS FROM SCIENCE, INDUSTRY AND THE

PUBLIC IN THE STATE OF NEW YORK

Roman R. Hedges[a] and Donald J. Reeb[b]

[a]Department of Politicial Science, Nelson A. Rockefeller
College, State University of New York, Albany, NY 12222
and [b]Department of Economics and Center for Financial
Management, State University of New York
Albany, NY 12222

ABSTRACT

Surveys of the acid rain physical science literature are
available. A few aggregate economic estimates of environmental
damage and a number of economic cost studies of reducing emissions
and assessing acid rain damages to micro-environments have been
made. Yet, little has been done in evaluating public opinion.
This study deals with three aspects:

° Interviews with natural resource based businesses in New York
  where it was found that they are cognizant of the acid rain
  literature but are waiting for the university community to
  document the acid rain damage to their industry.

° A survey of New York households concerning acid rain where it
  is believed that acid rain is a serious problem. Those
  interviewed were willing to spend $300 per year on its
  reduction.

° Lastly, a proposal was developed to tie Canadian surplus
  hydro-electric power exports to New York sulfur emission
  reductions. The hydro-electric power sales would guarantee a
  market for Canada and reduce electricity costs in New York
  while permitting manufacturers and public utilities to
  minimize the cost of reducing sulfur emissions.

INTRODUCTION

Government and industry policies surrounding acid rain will be examined in this paper. This will include a perusal of the existing knowledge of acid rain found in the scientific community, the business community, and the public at large. In addition, a rough assessment of the economic impact of acid rain in New York State will be provided. Finally, an attempt will be made to draw these diverse elements together to assess the nature of current policy and make some recommendations to policy makers.

THE ACID DEPOSITION PROCESS

Acid rain refers to the deposition of acidic material in the atmosphere on the natural and man-made surfaces of the earth. The term should be acid deposition because the phenomenon includes all forms of wet, acidic deposition (rain, snow, and fog) as well as the deposition of dry, acidic substances. In addition acid deposition carries the connotation that the acidic material has been transported over some long distance before deposition takes place. This further distinguishes acidic deposition from local air pollution (which may or may not be acidic) and has the effect of making scientific assessments of the impact of acidic deposition complex (Interagency Task Force on Acid Precipitation, 1982).

The processes which produce acidic deposition are not well understood. Primary pollutants such as nitrogen and sulfur oxides are released into the atmosphere by burning of fossil fuels, whether from the production of electric power, from factories and smelters, or from automobiles, and by occasional natural sources such as volcanoes. Once in the air, these primary pollutants are transformed by chemical processes into several different acidic substances, including sulfuric acid, nitric acid, and hydrochloric acid. The specific substances created and the site of deposition are dependent upon atmospheric conditions and location and content of emissions. As the acidic substances are deposited on various materials and biotic surfaces, the acids react and new chemical substances are created. These reactions in turn alter subsequent chemical reactions between the acidic substances and other materials and biota (Mohnen and Geis, 1981).

The cloud chemistry, long-range transport systems, and deposition processes would probably remain in the realm of physical scientists, however, the chemical and biological reactions are believed to be of considerable importance to other disciplines (Mohnen and Geis, op. cit.). There are abundant claims that acidic deposition damages aquatic life, reduces forest productivity, injures crops, destroys buildings, adversely affects human health, and degrades the quality of life. Such consequences are bound to be scrutinized outside the laboratory and field experiments of the physical scientist and involve the use of methodologies of the social scientists. Examination of govenment and business policy concerning acid deposition and scientific attempts to scrutinize the economic and social issues surrounding acid deposition is growing. Before the various possible effects of acid deposition can be enumerated, the hypotheses and estimates of the physical scientist must be examined.

VIEWPOINTS ON THE EFFECTS OF ACID DEPOSITION

The study of acid deposition is a new area of inquiry. It was believed, albeit naively, that acid deposition effects might have been observed by others than those in the traditional academic disciplines. A survey was therefore undertaken to evaluate the perceptions of other segments of society.

The Scientific Viewpoint

Most materials and biota react to the acidity levels of the immediate environment. Some of the reactions are quite mild, others are quite dramatic. Many of the reactions can only be assessed by direct observation before and after changes in the acidity levels are made. The reactions are contingent upon a long list of attendant conditions, including the initial acidity level and the buffering capacity of the environment, making direct assessments difficult (Glass et al., 1982). Local air pollution is seen as a separate matter meaning that the effects of acid deposition are confined to situations where the doses of acid are rather more diluted. Since even seemingly simple measurements of the acidity level itself have proven to be suspect, this further complicates the situation. As a result, the state of knowledge concerning the effects of acidic deposition is primitive by traditional standards. In many instances the evidence is circumstantial. Laboratory results are often too constrained to permit generalizations. Field studies of sufficient duration to detect effects which might not appear for years have just begun. Reliable historical data are practically non-existent, making archival research all but impossible.

The effects of acid deposition on aquatic life should be considered. As the water gets more acidic there is a reduction in the variety of plant and animal species which can be sustained. Many fish of interest to sport and commercial fisheries (e.g., trout) are particularly sensitive to the acidity of the water. Fry seem to be especially vulnerable. Increased acidification is also associated with a rise in the levels of biologically available forms of frequently toxic metals such as aluminum, mercury, cadmium, and zinc, which may play a role in further undermining the fish stock (U.S.-Canada MOI, 1981; CNR, 1981).

The response of a specific body of water to the introduction of a given amount of acid will depend on many factors. A larger volume of water will show a smaller response because the acid will be diluted. A body of water located in soils which contain relatively large amounts of materials which can buffer the acid will be less affected. The buffering capacity of the soil is so variable that some areas have essentially an unlimited capacity to absorb acid while others do not. The Adirondack Mountains, portions of the eastern side of the central Hudson Valley, and Long Island all have soils with little in the way of buffering capacity (Glass et al., 1982). Studies by the New York State Department of Environmental Conservation showed that approximately one-fifth of the lakes in the Adirondack Mountains, which are concentrated on its western edge, have waters too acidic to support game fish such as trout (Pfeiffer, 1982). This is due to the combined effects of their small, shallow size, relative high levels of acidic depositions, and surrounding soils with little or no buffering capacity. Moreover, it appears that the waters of the Adirondacks have become less supportive of fish life over the years which suggest that increased acidification due to acidic deposition could be the cause.

The effects of acidic deposition on the forests of the state are less substantial. This is partially so because both the direct effects of acidic deposition upon trees and the indirect effects on forest plants attributable to the alterations of soil chemistry produced by acidic deposition must be considered. Highly acidic soils cause aluminum to be made available for absorption by trees. Aluminum decreases the ability of trees, such as spruce, to draw water from the soil and thereby limits their growth (Vogelmann, 1982). More importantly, the effects of acidic deposition on trees are species and site specific.

Acid deposition is suspected as a cause in the relative decline in the growth of high altitude red spruce in New York, and evidence exists suggesting the deleterious effect of high levels of acid on maple seedlings and seed germination (Raynal et al., 1982). It is also suspected that forest microbe species diversity is reduced where soils are acidic, and that the surviving microorganisms are pathogenic (Wang et al., 1980). On the other hand, some laboratory evidence suggested that increased acidity might be good for some tree species under certain conditions (Mohnen and Geis, 1981).

Knowledge of the effects of acidic deposition on agricultural production is also sparse. As with forest, the effects of acidity are both direct and indirect. The indirect effects of acidity through changes in soil chemistry seem to be significant. Soil acidity is a well understood and an important factor in agriculture. Tomatoes, green peppers, alfalfa and timothy like an acidic soil while carrots and melons are adversely affected. In each species the mechanisms differ, but often the effects of pH will relate to certain vital trace minerals which can be tied up or freed for plant use by changing the soil acidity. A farmer will raise only those crops which match the soil conditions or will alter the soil conditions through the use of fertilizers, lime, and other common farm chemicals. Few farmers probably stop to think about what causes longterm variations since soil acidity will be systematically altered anyway. The large-scale application of farm chemicals so common to modern agriculture probably serves to overwhelm any and all evidence of acidic deposition from air pollution. In addition, much of the agricultural activity of New York State takes place in areas which have a large natural buffering capacity; these areas are less vulnerable to acid deposition. Specific agricultural effects attributable to airborne acidic materials are not well known (Forsline and Kender, 1982).

Little is known about the effects of acid deposition on the built environment. It is difficult to separate these effects from those due to general air pollution, particularly in urban areas. Baer and Berman (1983) studied the deterioration of marble tombstones in rural and urban areas and tentatively concluded that local pollution accounts for two-thirds of the damage. Crocker (ms) argued that the major costs of acid deposition were damages to man-made materials.

The effects of acidic deposition on human health are also poorly understood. There is currently no evidence about the effects from direct deposition of acids on people. But acid precipitation is thought to make toxic metals such as aluminum and lead available in some water systems at levels in excess of drinking water safety standards (Fuhs and Olsen, 1979). Large-scale studies of the presence of such potentially dangerous substances are relatively recent with first results being inconclusive (Taylor, 1983).

The final area of concern centers on issues of the quality of life in New York. The scientific community has not evaluated this area. This is undoubtedly because the quality of life is seen as a derivative of the problem and dependent upon both the perceptions of people and the scientific assessment of acid deposition effects. The lack of scientific evidence makes assessment of the quality of life premature.

This brief review of scientific findings has painted a picture of considerable research activity and little in the way of definitive conclusions. While it could be argued that this is not unusual for any substantive pathogenic social science research area (e.g., alcoholism, drugs, prisons), it seems less defined because the implicit comparison is physics or physical chemistry. Additionally, legislators are not aware that science is inherently conservative. Evidence in science is supportive, not conclusive. In addition, much of the research is focused on complex chemical and atmospheric processes in an effort to pinpoint the origins of acidic deposition. While this research is necessary, such a focus serves to divert attention from the study of effects. Scientific research is constrained by time and money where artificial laboratory research and widely spaced monitoring efforts are the norm; a norm which confines and confounds an assessment of effects.

The normal division of scientific labor further confines the issues. Local air pollution, which is highly acidic, does not count in the calculations. By defining acid deposition in terms of the longrange transport of acids, the maximum effect is limited to small, dilute quantities of acid spread over time and across vast areas far removed from urban concentrations. Since local pollution undoubtedly overwhelms any attempt to analyze acidic deposition in the more polluted, urban areas, i.e., in precisely those areas where society's resources are most concentrated and valuable, analyses of effects are biased against discovering significant deleterious effects.

The Viewpoint of Industries

An attempt was made to directly contact a select group of industry representatives in an effort to determine whether affected industries had already identified problems which had not appeared in the scientific literature. In addition, the current state of industry knowledge concerning acid deposition was assessed.

Using existing literature as a primary guide, industries which could be affected by acid deposition were identified. Attempts were made to contact knowledgeable individuals in those organizations. Interviews were arranged and conducted. These were quite lengthy, often exceeding two hours. The respondents were frank in their comments and usually displayed a level of knowledge about their industry and acid deposition which was beyond expectation. In all, more than four dozen interviews from all over the state of New York were completed.

Respondents were unwilling to translate their general concern, either personal or professional, into detailed estimates of the effects of acid deposition. This reluctance was based upon the general observation that the issues surrounding acid deposition were complex scientific matters and that scientists were not unified in their views. While this view serves the short-term interests of their industry, the respondents were of the opinion that an economic assessment was premature, that scientists must be the source of basic knowledge, and that relevant scientific knowledge would be available to their industry through existing channels.

Lest there be any doubt about the sophistication of the respondents, many knew the names of one or more researchers in the area of acid precipitation. Many had read scientific articles on the subject and virtually all knew the name of a scientist they could contact for more information on scientific matters related to their industry. It was not unusual for the respondents to think in terms of a specific college, university or department chairman as a source of information. These representatives do not generate their own studies. They do not see acid precipitation as an issue with a ground swell of concern from the membership. They did not believe that there was knowledge about acid deposition in the field which was not available or already a part of the public body of knowledge on the subject.

Responses to inquiries about acid deposition varied by each industry. In those industries which have large companies and/or capital requirements, the responses were more sophisticated. These industries seemed to be more attentive to the scientific community and more knowledgeable about acid deposition. Similarly, those industries which have strong ties to colleges and universities (e.g., agriculture and forestry) have a great deal more knowledge. Indeed, the extent to which research and personnel of the New York agricultural and forestry colleges permeate industries which center on agriculture and forestry is remarkable. When scientists at those well-connected schools indicated that acid deposition is or is not a problem in New York, the affected industry will know immediately. And, if the current state of knowledge is any indication, most industries will accept the university judgement.

The forest products industry in New York consists of two major economic segments: wood for fiber and wood for furniture, construction lumber, etc. The Christmas tree portion is much smaller as is the maple syrup portion of this industry. The value of hardwoods is substantially higher than that of the softer woods, and hardwood accounts for about three-fourths of the wood value harvested in the state. The forests of the state have more biomass now than they did fifteen years ago, although the quality of the wood is somewhat less than it once was.

Acid deposition is an issue on the minds of most everyone connected with the forest products industry, but the concern is markedly reduced among those whose primary economic interest are hardwoods outside of the Adirondack region. The concern about the Adirondacks centered on the low buffering capacity of Adirondack soils and the lack of specific research findings in regards to the effects of acid deposition on hardwood forest stands which are often found on sweeter soils outside the region. In effect this means that acid deposition is not thought to be a major problem for the large sawmill operators in the southwestern portion of the state, nor is it a major concern for the economically smaller segments of the industry such as maple syrup producers and Christmas tree growers. The judgment of industry representatives was such that it made no sense to expend resources interviewing the hardwood segment of the industry.

Representatives from the smaller segments of the industry, e.g. the maple producers and Christmas tree growers, were interviewed because the larger industry-wide organizations seemed to know little about them. These respondents were unlike those in the larger, better-financed segments of the industry. They were part-timers and indicated that many of their members were also part-time workers in the forest products industry. They were less knowledgeable about scientific findings, and they were less inclined to act on their own to acquire additional knowledge. Their knowledge was fundamentally dependent upon the schools of forestry and agriculture. It should be noted, however, that these respondents were quite receptive to information on acid deposition, but more immediate problems such as rainfall, insects, and deer browsing dominated their list of concerns.

The best understanding of and most concern about acid deposition within the forest products industry comes from the pulp and paper producers. These companies usually have extensive holdings, particularly in those parts of the state which exhibit vulnerability because of the poor buffering capacity of Adirondack soils. Pulp and paper mills are expensive and have a long payback period. Decisions about fiber availability vis-a-vis a specific mill site are of paramount concern in this segment of the industry. Many of the companies have large-scale forest management programs which include company holdings, leased holdings, and fiber purchase on privately held land. Scientific forest management designed to produce maximum long-term yield seems to be the basic policy through this portion of the industry. This means that a great deal of attention can be devoted to issues which will not be fully addressed in the immediate future. Moreover, the slow growth of the forest means that environmental effects are known only by careful, well-designed research. Hence, there is a heavy reliance on the research done by professional foresters at schools such as Syracuse, NY.

The pulp and paper company representatives were acutely aware of the scientific examination of acid deposition. They cited numerous laboratory studies, published and unpublished field research, conferences attended, conversations with academic and industry researchers, and personal experience on the subject. The professional judgement was that there is reason for concern but that scientific knowledge permitted no reasonable conclusion. These industry representatives talked in specific terms of the problems of acid deposition on species, soils, and locations. They identified the same conditions for placing trees at risk as did search of the scientific literature. They believed that high altitude red spruce are a good candidate for acid deposition effects. Even though the pulp and paper mills often use other trees, the possibility that red spruce were giving an early warning was not disregarded. Still, the overall judgment was that there was no documentable damage of economic importance at this time.

The concern should not be minimized. Some of the companies have embarked upon lengthy research projects to assess the impact of acid deposition on the various species found in their holdings. For example, Finch-Pruyn and International Paper have a new, joint project designed to study tree growth for a variety of species over an extended period of time. The project is based upon observations of individual trees in an existing forest ecosystem. The project is broader in scope than a simple assessment of acid deposition and clearly indicates that the effects of this problem should be placed on the table. Results from this research are not expected for many years, however.

Those interviewed in the forest products industry expressed concern and skepticism simultaneously. They seemed to appreciate the potential for harm and the lack of a firm scientific basis for action. Since the entire industry is suffering from the current recession, it is not clear that an economic analysis based upon scientific evidence could provide sound estimates of these effects on the industry. The slack is simply too great. The forest is growing at a rate faster than it is being harvested, and industrial production is systematically undervalued.

Agriculture is a major industry in New York. Dairy farms, fruit orchards and vineyards, and various grain and vegetable farms abound outside of the New York City metropolitan area. Modern farming practice is dependent upon scientific research. Every state, including New York, has its land grant college devoted to the promotion of sound agricultural practice based on the latest scientific research. A network of information is created by county extension workers who attempt to bridge the gap between scientific norms and the day-to-day life of the farmer. Extension workers and agricultural researchers are acquainted with the details of disease resistance, nutrient needs, soil chemistry, and species responses. A great deal of this knowledge centers on the effects of soil acidity on the release of trace minerals and plant nutrients.

Because of this emphasis on science, and especially on the effects of soil acidity, it could be expected that a great deal of

knowledge about the effects of acid deposition on agriculture could
be found. Literature reviews and interviews did not confirm such
an expectation. It is essentially irrelevant where soil sweetness
or sourness comes from as long as its effects are easily
controlled. Since acidity is directly responsive to the use of
readily available and moderately priced lime, the agricultural
community has practiced soil liming with great regularity since
before the 1930's. Wherever the soil acidity gets too high, no
matter whether the source is acid deposition or the farmer's own
applications of fertilizers, liming is an inexpensive and
traditional remedy. In addition, it must be noted that the bulk of
New York agriculture occurs in those areas of the state which have
a high buffering capacity. This means that effects would be
especially hard for farmers to discern.

Farm organizations are generally unwilling and unable to
embark on extensive independent research. Individual farmers may
have   anecdotal   evidence   about   the   particular   features   of
agriculture, but the overwhelming fact is that agricultural
knowledge is based upon academic research. Again, concern exists.
The New York Farm Bureau has attempted to get its parent
organization, the American Farm Bureau, to take an official
position of concern and support for further research. But the
limits of that concern are well defined by the academic research
which has not, as yet, produced definitive answers.

Acid is known to affect building materials. But if the
distinction between local air pollution and acid deposition is
considered along with the fact that acid washes have been common
practice in urban building restoration, the effects of acid
deposition will be all but overwhelmed within the urban areas.
Very little about the effects of acid deposition from home building
industry representatives was determined. Acid deposition does not
appear to be of major concern, and documentation of possible
effects or systematic research was not known to them. The home
builders of New York do not have a policy on construction which
takes into account possible effects of acid deposition nor were
they aware of research which would cause them to adopt such a
policy. On the other hand the roofing industry now recommends a
minimum pitch on all roofs to avoid ponding and associated
materials damage. Acid deposition was cited as one of the reasons
for this recommendation.

Because of what was found in the scientific literature, human
health effects were confined to an examination of acid deposition
on water supplies. The acidity of water supplies has the potential
for affecting the availability of minerals and metals in the water
as well as being an issue of water quality in its own right. There
is simply no administrative concern in New York about the effects
of acid deposition on water supplies of the State outside the

Department of Health (Fuhs and Olsen, 1979). Neither those concerned with water management, water supply to municipalities, or private residential and commercial wells displayed organizational concern. This might well be a reflection of the source of most of the State's water supply; most commercial and public water is drawn from those areas of the state which have rather highly buffered soils.

Leadership interviews were not useful in discovering the effects of acid deposition on the quality of life. A good deal of what was found in these areas stemmed from analysis of citizen surveys. The effects of acid deposition seemed to be related to the public's perception of acid deposition rather than decline in tree growth. A decline in recreational activity needs to be evaluated in terms of the subjective judgment of citizens as well as in more formal econometric terms. Because much of what is heard about the effects of environmental issues related to the quality of life comes from environmental groups, their views were examined briefly. These groups are quite concerned about acid deposition. They see the issue in terms of general air quality and argue on esthetic, economic, and moral grounds that clean air is a national priority of considerable magnitude.

Fishing groups also see the issue as having both esthetic and economic importance. Unless there are fish, dollars will go elsewhere. Economic documentation does not come from the fishermen themselves. Better information is available from the research community than from those who are engaged in sport and recreational fishing.

Apart from the specific views of groups which have an interest in the environment, if the quality of life of an area is diminished, it might be expected that population growth and demand for real estate and recreation would be affected. Interviews were therefore conducted with representatives of the real estate industry. Inquiries were limited to the effects of acid deposition on the price of real estate sold in the Adirondack region. The decision to confine the inquiries to the Adirondack region was also based upon the assumption that effects to real estate prices for other areas of the state could not be meaningfully tied to acid deposition. Pending the outcome of the citizen survey of public opinion, acid deposition was a local problem only for the Adirondack region.

Industry representatives expressed personal concern about the possible effects of acid deposition on real estate sales, yet neither believed that it had any noticeable impact upon sales or prices. The explanations offered for this differed. The limits on land use established by the Adirondack Park Agency were offered as reasons why the going price was already low, making additional adverse effects difficult to measure. The effects of an economic boom surrounding the Lake Placid region, which was attributed to the 1980 Winter Olympic Games and attendant facilities development, was cited as a reason why any downward pressure in prices which might exist would not be apparent at this time. In any case, the real estate industry has no internally generated reports or systematic knowledge of the effects of acid deposition. The industry consists of small operators in a relatively competitive marketplace. They do not have the internal resources to engage in organized research, and the educational efforts of the industry center on increasing the skill and sophistication levels of realtors in regards to house construction, home financing, mortgages, and the like.

One set of interviews was conducted with representatives from the electric utilities sector, since this industry is usually singled out as the major source of substances which produce acid deposition. It is also the source of a great deal of knowledge about acid deposition. The electric utilities have engaged in a full program of research on the subject of acid deposition. Utility-sponsored research in New York seems to be pointed toward atmospheric chemistry rather than the effects of acid deposition. The major exception is a study focused on the costs of implementing proposed federal legislation, such as U.S. Senate Bill S. 3041, on air quality.

These interviews with representatives from New York industry revealed a substantial concern about acid deposition in a wide range of potentially affected industries. Representatives were knowledgeable and sincere. They simply did not have information beyond that found in the community of researchers. There is little doubt that the industries do not see evidence which warrants direct action at this time, but neither is there any doubt that virtually no one is willing to deny the existence of a potentially serious problem whose full dimensions are not yet known. There is obviously widespread support in private industry, government, and universities for more research on an identifiable problem.

The Public Viewpoint

Crucial elements of acid deposition implications have thus far not been adequately addressed by the public. The lack of a firm scientific footing for building a full model of effects means that attempts to make policy at this time will necessarily be contingent on non-scientific factors. Public perceptions and knowledge play a significant role in policy making in any democratic society.

A survey of the public's perception and knowledge of acid deposition was done by a telephone survey of 601 New York State residents. This survey was conducted by Gordon S. Black Associates of Rochester, NY, based upon a sample design and survey instrument provided to them. The sample design was created to study the public's perception throughout the state and to enable a separate examination of the perceptions and understanding of acid deposition in the Adirondack region. This was accomplished with a stratified probability sample of the State. The Adirondack region of 300 individuals was defined as a separate sample strata. A separately defined rest of the State strata was allocated for 301 interviews. These two representative strata can, therefore, be analyzed separately or, with the application of proper statistical weightings, can be combined to produce a complete, representative State-wide picture of public perception. Each of the two sample strata will produce results which are within six percent of the true population for 95 percent of the time. The combined sample has a slightly smaller margin of error. The Adirondack region was singled out for special attention because it seems to be the one place in the state where significant effects of acidic deposition are likely to be observable.

The questions included in the final survey instrument were designed to serve several purposes. Because there were few studies which have examined the public perceptions of acid deposition in any manner, several items were included in the instrument simply for purposes of providing context. These were items which tap the concerns of the citizenry. Two items were designed to place the concerns about the environment into a more realistic focus. They asked the respondents to consider trade-offs between the economy and the environment. A series of direct questions were also included about acid deposition. To avoid confusing the respondents the term acid rain was used in the interviews. Current awareness of and knowledge about the issue, the level of concern which exists, and the respondents preferences for ways to address the issue were examined.

Fully 93 percent of the respondents from the Adirondack region said that they had heard of acid rain and more than three people in four said that acid rain is a problem in New York State. Over one-third of the respondents believed that acid rain is a serious problem in the State. Three in four residents from elsewhere in the State said that they had heard of acid rain and almost 60 percent said that it was a problem in New York State. To place these findings into context, only one in three indicated that the shortage of oil, gasoline, and other fuels was cause for a great deal of concern and three in four indicated that either "too little" or "about the right amount on improving and protecting the environment" was spent in the State. Indeed, a general concern for environmental protection was the norm, and acid rain was just one of several specifics. Three-fourths of the respondents indicated that the disposal of hazardous industrial chemical wastes is of great concern while two of three expressed a similar level of concern about toxic chemicals such as PCB's and pesticides. About 75% did not want to relax the current environmental regulations to produce additional electricity and three in five New York residents believed that strict environmental controls could be instituted without hurting business.

As might be expected, the respondents were unanimous that if cost were not a factor additional pollution-control devices to reduce the amount of acid rain should be considered. Only four percent rejected cost-free reductions in acid rain, and the Adirondack and non-Adirondack regions did not differ from one another. Four out of five respondents said that acid rain was a problem for the national government, or State and national governments together. Fewer than one person in fifty said that acid rain was not a government problem, despite the fact that everyone was given the option to indicate if they agreed with the statement that acid rain is not the government's problem, and the government should stay out.

One finding stands out: over 55 percent of the people said that severe damage to the (Adirondack) region would affect them personally. To be sure, those who live in the region said this more frequently (65 percent), but this form of concern even outstriped the number who lived or have vacationed there in the last five years (42 percent). Pending the results of further investigation, the respondents presented a sophisticated version of the general argument that the diminuition of the Adirondack region is an issue of personal concern above and beyond that of a narrow definition of self-interest. It is possible that once some information about recreational interests are included in the calculations, the disjuncture will be minimized. In the interim, it is felt that the creation of a special place in the Adirondack Mountains for all New York residents was central to the framing of several provisions of a State constitution almost one hundred years ago. These preliminary results suggest that New Yorkers see the region in terms larger than their own immediate self-interest.

Even more indicative of the depth of feeling on the subject of acid deposition was the response to the one item in the survey which directly confronted the question of cost:

"Experts estimate that the required pollution controls [needed to reduce acid rain] could be quite costly for companies to install. These costs, passed along to consumers, could mean that the electric companies might have to raise their average bill $25 per month. In your opinion, is it worth the cost of clean-up, or is it just too costly?"

Almost half (49.5 percent) indicated that it is worth the cost of clean-up while less than one-third said that it was not. More than 15 percent said that it depends or that they didn't know. Residents of the Adirondacks were statistically indistinguishable from those from the rest of the state.

The $25 per month increase is far greater than what the actual electricity costs would be for New Yorkers and is far from negligible when compared to the size of other possible expenditures thought large enough to alter basic, economically motivated behavior. It is large enough to speak forcefully for the claim that New York residents see acid rain as a pressing problem which they would like to see addressed by their government even if it costs money to solve the problem.

AN ASSESSMENT OF THE PROBLEMS

Taken together, the scientific, industry, and public knowledge of acid deposition is not complete, but society should move beyond a cataloguing of effects into the realm of an assessment. There is a small body of knowledge concerning the economic effects of acid deposition. This literature is based upon crude estimates of the effects of acid deposition. The usual strategy is to examine the costs and benefits which accrue to proposed solutions.

These proposed solutions take the form of prevention, diversion, or amelioration. Prevention would focus on limiting the generation of primary pollutants thought to produce acidic deposition, e.g., sulfur oxides. A diversion would center on having the acid deposition take place where it would do little harm. Acid-making industries could be placed so that the acids fell over the North Atlantic, for example. An ameliorative solution would center on treating the effects of acidic deposition. If acidic deposition alters the acidity of a lake and if it causes fish to die, an ameliorative solution would be to chemically lower the acidity and, thereby, reduce the negative effects.

The costs of the various solutions are thought to be easier to assess than are the benefits which result. It is not known whether the fish in some remote Adirondack lake would be saved by reducing the production of sulfur oxides in a specific coal-fired electric-generating plant in Cleveland because of the scientific complexities (discussed above) or even the worth of those fish to society. It is calculable how much it would cost to wash the coal or put scrubbers on the smokestack to accomplish a specified level of sulfur reduction.[1]

   This results in a bias against any specific actions, which is exacerbated in other ways. Many of the effects of acidic deposition are not directly proportional to increases in acidification: initial, small effects do more damage than later large increases, i.e., there are non-linear diminishing marginal effects. Such effects can serve to produce erroneous prices in calculating benefits for traditional benefit/cost analyses. If the present environment is less than pristine (coal-fired electricity production is just one hundred years old), this produces special problems in the analysis. A large reduction in acidic deposition may have very little effect in an already dirty environment. Only after major efforts are made and costs are incurred will large benefits begin to appear.

   Of greater consequence is the possibility that some damage might be irreversible. Conventional analyses will also point away from action if reversibility is assumed when it is not warranted. The usual techniques assume that benefits given up today can be recaptured (at some identifiable cost) some time in the future. If

---

[1] The New York Power Pool (1983), consisting of seven investor-owned electric utilities, made estimates of the cost to New York utilities of U.S. Senate bill S.768 (a modified S.3041 of 1982) to reduce sulfur dioxide. If implemented, the minimum cost to New York was $99.7 million (1982 dollars) for a 109 thousand ton reduction in sulfur emissions; this is about 1.2 percent of total (gas, electric, and steam) sales of the seven combination utilities and about 6.3 percent of 1983 capital expenditures. The maximum cost was estimated at $719.1 million, a seven-fold increase over the minimum cost. One of the major differences between the minimum and maximum cost is the reduction in sulfur emissions -- a 396.6 thousand ton reduction versus a 109 thousand ton reduction from 1982 levels. The caveats for these estimates are many: assumptions were made about Canadian hydro purchases, nuclear plants coming on line, the estimated cost of fuel, coal conversions, and many others. The major weakness of the study methodology is an assumption of increased electricity output of 1.4 percent annually. While this is the assumption made in the State energy plan, it is some 4.2 times greater than the 1972-1980 growth in energy sales. If fuel sales increase more slowly, then this legislation (S.768), which is based on a percentage reduction from 1980 emission levels, will be much different in impact, because plant construction would be slower and fuel demand would be less, partly from the increased contribution of Canadian hydro to total sales.

acid deposition damages agricultural crops and the problem is not addressed immediately, the shortfall can be made up next year by planting more seed, using more fertilizer, and cutting back on sulfur emissions. But acid deposition may not work that way. For example, treating a lake with lime will lower the acidity levels, but such treatment may not reduce the toxicity of metals made biologically available through the original increase in acidity. If the effects of acidic deposition are not reversible, then the analysis is properly adjusted to lower the tolerance to the risks attributable to making a mistake. This places a higher cost on uncertainty. If present action (or inaction) forecloses future attempts to improve the environment, the society might proceed differently than if the environment is only harmed temporarily.

All of this is made more confusing because of low rates of increases in electricity demand. A deflection in the growth curve for electrical demand may mean that the environment will begin to clean itself. Given these biases and the limited scientific and field knowledge, a solid estimate of the economic benefits available from an attempt to solve the acid deposition problem is not possible. Nevertheless, attempts have been made.

The Organization for Economic Co-operation and Development (1981) of western Europe calculated that a 37 percent reduction in sulfur dioxide emissions would produce $9 to $85 per person reduced damages. At $9, New York might avoid $158.4 million in damages from such reductions. The EPA (1982 draft) study contained no dollar estimates for acid rain damage to crops or for forests. It gave estimates of increased morbidity and excess deaths, but these were not related specifically to acid rain (rather to SO_2 emissions), and dollar estimates were not provided. The EPA study also cited the Fuhs and Olsen (1979) data on drinking water contamination by toxic metals. Again, dollar estimates were not given.

Renshaw et al. (1983a) attempted a macro-type study. Their hypotheses - if the dollar damages cannot with present knowledge be assessed, are people behaving as if they know of negative impacts from acid rain or as if they are risk adverse and presume that there are high probabilities of negative effects from acid rain. By studying the population growth of towns and counties, the results were: 1) greater population losses were occurring in "acid lake counties" in New York's western Adirondacks even after other macro variables were statistically accounted for and 2) there were indicators that out-migration, induced by acid rain, was occurring. The dollar loss in property for acid rain-induced population migration was not assessed. Vrooman and Brown (1982) studied property value effects near acidified lakes. They found little effect, but the data base was sparse. The study showed great promise and needs to be done again.

Page (1982) estimated hay, corn, and other agricultural product losses related to SO_2 and NO_x. The dollar loss for New York was 78 percent of the five-state (Maine, Massachusetts, New Hampshire, New York, and Vermont) regional total, and the New York loss was equal to $51 million (1979 values). The study estimated that six percent of the corn crop and ten percent of the hay crop were lost in each state.

Menz and Mullen (1982) calculated losses in recreational fishing in New York from acid rain: they were valued at $2 to $8 million annually (including regional multiplier effects). The estimate did not include the possibility of substitutions for recreational fishing.

The EPA (1982 draft) estimated the maximum annual benefit realized through control of sulfur oxides to be as high as $5 billion for Minnesota and all of the states east of the Mississippi River. This consisted of estimated damages of $3.5 billion for materials, $0.6 billion for forest ecosystems, $16 to $160 million in crop loses, and $60 million for human health effects. Damages attributable to nitrate precursors to acid deposition were estimated by the Commission on Natural Resources (1981) at $125 million for sport fishing, $15 million for forest loses, $135 to $542 million for materials, $135 million for crops, and $160 to $900 million for human health.

The costs of damage are not necessarily absorbed at the damaged site. Consider the possible effects of acidic deposition on New York forests. The forest products industry (lumber, wood, furniture, and paper) contributes roughly $2 billion to the economy of the State. A five percent reduction in the forest product industry due to acidic deposition would cost the state economy $100 million annually. Lewis County in the wood-producing western Adirondacks would lose $0.8 million, but Kings County (Brooklyn) would lose $4.5 million and Queens would lose $5.6 million. The reason for this is that furniture, paper, and other products derived from the forest are sometimes manufactured at plants near their markets and the metropolitan New York market dominates the state economy (Renshaw et al., 1983b).

Even if given governmental priority, it is not completely clear how to proceed. If areas such as the Adirondacks display more apparent effects from acidic deposition, it is not even clear if it is because there is more damage in this region than others or if it is because the region is more pristine that a small amount of damage which would go unnoticed elsewhere is perceived as having major effects. If it is the former, policies which are ameliorative or diversionary are in order. If it is the latter, prevention should be of higher priority.

Moreover, a decision must be made about whether the Adirondack region is uniquely vulnerable or simply uniquely visible. The choice will affect who is to bear the burden and obtain the benefits associated with proposed solutions. Those who are vulnerable are often entitled to societal compensation while the merely visible often must give way to harder political and economic realities. In the national debate on acid deposition, the proposed solutions concentrate on the reduction of emission (prevention) and presume that acidic deposition is the result of long-range transportation of chemically transformed pollutants. Appalachian coal regions will be asked to incur costs to satisfy the preferences of those in the northeast. The Appalachian coal regions will see jobs given over to those who mine western coal which is generally lower in sulfur. If the assumption about long-range transportation is incorrect, the northeast does not gain despite the losses incurred by others.

Residents of New York, and especially those from the Adirondacks, should not permit themselves to believe that public pressure will all be in their favor. Costs are not willingly borne without compensating benefits. Careful analyses of the costs of reducing acid deposition are easier to construct than are forceful analyses of the importance of popular concern and hard-to-measure benefits. The introduction of costs in the survey questions reduced support for dealing with acid rain from over 90 percent to about 50 percent, though this was related to an unrealistically high estimate of the marginal cost for reducing acid rain.

A PROPOSAL

Acid rain is much more of a controversy in academic methodology than it is a subject for sharp disagreements in public opinion. Researchers focus on the costs of acid precipitation control and whether these are less than the benefits to be realized from its reduction. Public opinion focuses on the presumption that acid precipitation cannot be good for the state and that they are going to pay for it. While further studies on the cost of damages from acid rain might be illuminating, a more relevant study could be one relating Canada and New York's acid rain problems to Canadian hydro-electricity sales to New York. Such a study might show that it is wise for New York to have a reduction in acid rain and Canada to sell hydro-electricity to New York, the northeast and midwest at reduced prices. Canada would thereby save its lakes and sell its electricity at a guaranteed price, while the U.S. would pay part of the capital costs of the hydropower, reduce its own sulfur dioxide emissions, and purchase electricity at a lower price.

The basis for an efficient trade may be that New York, because of its geography, would be willing to transport electricity into the grids of other states (Vermont, Massachusetts, New Jersey, Pennsylvania, and Ohio). Canada would require that a reduced rate in electricity be joined by reduced SO_2 emissions. The greater hydro-capacity of Canada and the increased levels of acid rain deteriorations could be re-positioned such that both lower electricity prices and less acid rain could be envisioned.

The concept of working with Canada to reduce acid rain and to change the price of hydro-electricity sales to the U.S. combines what are now two separate activities into one transaction. And, since it is presumed to be a transaction (without coercion), it must be to the benefit of both parties--the U.S. and Canada--or it would not occur. The proposal would involve:

- ° study of technical feasibility (transmission lines, excess capacity of Hydro-Quebec, implementation time);

- ° study of legal feasibility (agreements with Canada and with neighboring states for transmission and sale of electricity) and controls (cutbacks of about 25 percent in SO_2 and NO_x);

- ° economic feasibility (effects on costs of electricity in each state, the profitability to each public utility, the security of the Hydro-Quebec bonds).

The three steps are as one: the economic feasibility requires intimate knowledge of the prior two parts such that technical and legal components can be adjusted to achieve maximum economic benefits.

CONCLUSION

Substantial disagreement exists over such basic items as the definition of acid deposition. It invites controversy in the policy arena. Without agreement on the basic science, assessments of economic and social impact are made more problematical. And herein lies the basic point concerning acid deposition: scientific certainty is not available. This transforms the problem from a question of science to a question of preference. If nothing is done, the current sensibilities of New Yorkers will be violated. They stand ready to pay substantial costs to solve what they see as a serious problem while at the same time believe that many of the usual trade-offs between the economy and the environment are illusionary.

The very fact that acid deposition is of public interest means that armed with appropriate facts and scientific uncertainty, it may prove easy for those who have different interests on the issue

to construct counter arguments. At the very least, current
scientific uncertainty makes it feasible to prolong the debate and
delay action. To accept such a delay is to ignore the concerns of
New Yorkers who fear that scientific uncertainty might help create
an acidified world in which the quality of life will be reduced.

REFERENCES

Baer, N.S. and Berman, S.M., 1983, Marble tombstones in national
    cemeteries as indicators of stone damage: General methods.
    Air Pollution Control Association, 76th Annual Meeting,
    Atlanta, GA. unpublished paper. 21 pp.
Commission on Natural Resources, 1981, Atmosphere - Biosphere
    Interactions: Towards a Better Understanding of the
    Ecological Consequences of Fossil Fuel Combustion. National
    Academy of Science, Washington, DC. 363 pp.
Crocker, T.D., Manuscript, Scientific truths and policy truths in
    acid deposition research. Department of Economics, University
    of Wyoming, Laramie, WY. 27 pp.
Crocker, T.D. and Forster, B.L., 1981, Decision problems in the
    control of acid precipitation: Non-convexities and
    irreversibilities. J. Air Pollu. Control Assoc. 31:32-37.
Environmental Protection Agency, 1982, Critical Assessment
    Document, The Acidic Deposition Phenomenon and It's Effects,
    Chapter E-8, Draft document, U.S.E.P.A., Washington, DC.
    29 pp.
Forsline, P.L. and Kender, W.J., 1982, The effects of acid rain on
    fruit crops, pp. 47-99. In: New York Symposium on Acid
    Deposition, Proceedings. Center for Environmental Research,
    Cornell University, Ithaca, NY.
Fuhs, G.W. and Olsen, R.A., 1979, Acid precipitation effects on
    drinking water in the Adirondack mountains of New York State.
    New York State Department of Health, Albany, NY. 7 pp.
Glass, N.R., Arnold, D.E., Galloway, J.N., Hendrey, G.R.,
    Lee, J.J., McFee, W.W., Norton, S.A., Powers, C.F.,
    Rambo, D.L., and Schofield, C.L., 1982, Effects of acid
    precipitation. Envir. Sci. Tech. 16:163-169.
Interagency Task Force on Acid Precipitation, 1982, First Annual
    Report to the President and the Congress of the United States.
    U.S.E.P.A., Washington, DC. 26 pp.
Menz, F.C. and Mullen, J.K., 1982, Acidification impact on
    fisheries: Substitution and the valuation of recreation
    resources. American Chemical Society, 1982 Annual Meeting,
    Las Vegas, NV. 29 pp.
Mohnen, V.A. and Geis, J.W., eds., 1981, Acid Precipitation
    Research Needs Conference, Proceedings. College of
    Environmental Science and Forestry, State University of New
    York, Syracuse, NY. 88 pp.

New York Power Pool, 1983, Cost Impact of Acid Rain Legislation on
    the Member Electric Systems of the New York Power Pool.   Draft
    No. 3, Albany, NY.   38 pp.
Organization for Economic Co-Operation and Development, 1981, The
    Costs and Benefits of Sulphur Oxide Control.   Paris, France.
    164 pp.
Page, W.P., 1982, Electricity generation, acid rain, and regional
    economic agricultural losses.   Fourteenth Annual Conference of
    the Institute of Public Utilities, Williamsburg, VA.   38 pp.
Pfeiffer, M.H., 1982, Acidity Status Summary Memorandum.   New York
    Department of Environmental Conservation, Albany, NY.   20 pp.
Raynal, D.J., Roman, J.R., and Eichenlaub, W.H., 1982, Response of
    tree seedlings to acid precipitation, I and II, Effect of
    substrate acidity on seed germination, and effect of simulated
    acidified canopy throughfall on sugar maple seedling growth.
    Envir. Exper. Botany 22:377-392.
Reeb, D.J., ed., 1983, The Economic Impact of Acid Precipitation.
    Center for Financial Management, State University of New York,
    Albany, NY.   227 pp.
Renshaw, E.F., Kamya, M.M., and Mann, M.A., 1983a, Acid lakes and
    population growth, Exhibit 4b, pp. 1-18.   In: The Economic
    Impact Study of Acid Precipitation, D.J. Reeb, ed., Center for
    Financial Management, State University of New York, Albany,
    NY.
Renshaw, E.F., Kamya, M.M., and Mann, M.A., 1983b, The possible
    economic impact in New York State of acid precipitation on the
    forest ecosystem, Exhibit 4c, pp. 1-17.   In: The Economic
    Impact Study of Acid Precipitation, D.J. Reeb, ed., Center for
    Financial Management, State University of New York, Albany,
    NY.
Taylor, F.B., 1983, A Cooperative Study of the Effects of Acid Rain
    on Water Supplies.   New England Water Works Association,
    Dedham, MA.   8 pp.
Vogelmann, H.W., 1982, Catastrophe on Camel's Hump.   Natural
    History 91:8-14.
Vrooman, D.H. and Brown, W.V., 1982, Acidity and the value of
    waterfront properties in the Adirondack Park region.   St.
    Lawrence University, Canton, NY.   17 pp.
United States-Canada, 1981, Memorandum of Intent on Transboundary
    Air Pollution, Impact Assessment, Working Group I.
    U.S.E.P.A., Washington, DC.   39 pp.
Wang, C.J.K., Manion, P.D., Leaf, A.L., and Raynal, D.J., 1980,
    Actual and Potential Effects of Acid Precipitation on an
    Adirondack Forest.   ERDA 80-28, New York State Energy Research
    and Development Authority, Albany, NY.   80 pp.

ACID DEPOSITION - THE CANADIAN PERSPECTIVE

Alexander N. Manson

Environmental Protection Service
Environment Canada
Ottawa, Ontario   K1A 1C8

ABSTRACT

This paper provides a summary of the Canadian position on acid rain abatement programs. The problem is defined and relavent scientific findings and environmental effects highlighted. The main causes of acid deposition in Canada, particularly in the eastern portion of the country, are identified. Canadian air pollution control programs in general, and the Canadian acid rain abatement program in particular, are discussed. International legal principles and obligations have also been addressed. The paper concludes by outlining the desired elements of a future Canada-United States agreement on acid rain.

INTRODUCTION

Acid deposition, commonly termed acid rain, is causing serious environmental and economic damage in eastern Canada. An area of more than one million square miles is vulnerable to its effects. The extent and intensity of the damage will increase unless action is taken to reduce precursors, which are mainly man-made emissions of sulphur dioxide and nitrogen oxides. When the rate of acid deposition exceeds the rate at which the receiving environment produces neutralizing substances, there is a loss of the ability of soils and waters to neutralize the deposited materials.

DISCUSSION

The long term, cummulative effects of acid deposition on the environment and its ecosystems are of great concern to Canada. The resulting damage to aquatic life, forest resources, buildings, and materials has important economic ramifications.

Environmental Impacts and Objectives

Over the long term, acid deposition will gradually reduce the receiving environment's neutralizing ability (i.e., the neutralizing material is gradually removed). Ecosystems then become acidified. These systems will remain acidified until deposition is reduced and the neutralizing ability is restored. The latter aspect is essentially a function of the rate of weathering of soil and rock material. The longer term acidification will also cause extensive alteration and damage to biological systems. In eastern Canada, this type of acidification is primarily caused by sulphur deposition (MOI, 1983).

During spring snow melt, 50-80 percent of the acidifying materials that have accumulated in the snow-pack are released with the first 30 percent of melt-water (Elder, 1984). This surge of very acid melt-water can exceed the rate at which the receiving environment can provide neutralizing material. Lakes which normally have enough neutral buffer to resist long-term depressions in pH may suffer damage to fish populations and biota as a result of the spring pH depression following snow melt. Substantial biological damage can be done to amphibians and other aquatic life. In general, sulphuric acid is the major contributor to acidity during spring run off, but nitric acid can be a major contributor in some instances. Of the lakes surveyed in Ontario, about 45 percent are categorized as vulnerable to acidification (Ontario MOE, 1982). In many of them, there are already critical signs of a reduction in neutralizing capacity, and a similar situation exists in the province of Quebec. In Nova Scotia, salmon runs no longer exist in about ten percent of the former salmon rivers, and acidification is showing its effects in another 20 percent (DFO, 1984).

Recently, there has been much discussion about whether aquatic ecosystems respond in a delayed or direct manner to acid deposition. In Canada, biological degradation is continuing to be observed except for those lakes and streams that have already been severely damaged. Therefore, with respect to biological damage, most aquatic systems are responding in a delayed manner. With respect to chemical change, both types of responses are occurring. Thus, because of the cumulative nature of the problem, it appears that the damage is becoming worse with time (Elder, 1984).

Much of the new scientific information on acid rain deals with forestry. Almost 50 percent of the forest growth in eastern Canada occurs in areas which are receiving sufficient· deposition to damage nearby aquatic ecosystems. At the Maritime Forest Research Centre in Fredericton, New Brunswick, scientists have carried out controlled experiments on the effects of acid rain on germination and early growth of several key tree species. They found rain with a pH of 4.6, and constituent make-up similar to that which falls in the Atlantic region of Canada, reduced the germination rate and growth of some species (Percy, 1982). In another controlled environment at the University of Toronto, workers found that the fertility of some conifer spores is affected by rain with an acidity similar to that which falls over much of eastern Canada (Cox, 1982).

Some very disconcerting observations have also been emerging in other countries. Damage to a wide variety of tree species has been noted in virtually every state along the Eastern seaboard of the United States from North Carolina northward. Researchers are not completely sure about the cause, but air pollution, including sulphur dioxide, sulphates, acid deposition and ozone, seems to be strongly implicated. In the Federal Republic of Germany, forests have deteriorated dramatically in the past year - in contrast to 1982 findings which showed a forest damage area of eight percent, a 1983 survey reported 34 percent (and a 1984 survey 50 percent). In issuing this report, it was stated, "in relation to a single factor, the clues indicate that air pollutants and their conversion products are the fundamental cause; all clues indicate that without air pollutants the forest damage would not have occurred" (FRG, 1983). While the deposition rates and ambient air concentrations of pollutants are considerably higher in West Germany than in eastern Canada, this information cannot be ignored.

Acid deposition and precusor pollutants also cause damage to some agricultural crops as well as to buildings and materials. Sulphates can cause a significant reduction in visibility and are implicated with respiratory problems. Acidification of untreated drinking water supplies can cause increased levels of toxic metals such as lead and copper (Fuhs et al., 1985) which are leached from distribution systems; however, the magnitude of this problem is not well known at present (MOI, 1983).

The natural resource base at risk sustains vital components of the economy and life-style in many areas of eastern Canada. For example, the gross economic activity generated by sport fishing in eastern Canada in 1981 exceeded $1.1 billion. Tourism revenues were $10.4 billion. Shipments of forest products from eastern Canada amounted to $14.6 billion. Together these revenues accounted for about eight percent of the gross national product for the whole of Canada (CFS, 1984; DFO, 1984).

Canadian scientists concluded that a reduction of wet sulphate deposition to a level less than 20 kg ha^{-1} yr^{-1} would protect moderately sensitive lakes and streams from long-term acidification, leaving only the most vulnerable ecosystems open to damage (MOI, 1983). Above this amount, environmental damage occurs and can become increasingly severe at higher levels of acidic deposition. Further scientific studies are underway to develop environmental objectives to avert long-term acidification effects for the limited number of very vulnerable areas, for forest ecosystems and other sectors. Short-term deposition effects are also being examined. Should these objectives be more restrictive than the 20 kg ha^{-1} yr^{-1} proposal, abatement programs could be adjusted accordingly.

Emissions within Canada

The main cause of acid deposition is the oxidation of sulphur and nitrogen compounds emitted into the atmosphere from natural and man-made sources. Man-made emissions contribute over 90 percent of the deposition in northeastern North America (Nierenberg, 1983). Natural sources of sulphur and nitrogen emissions such as swamps, marshes, lightning, etc., exist but are estimated to be only a few percent of the total. Base case sulfur dioxide emissions in 1980 for the eastern part of Canada (Manitoba and the eastern provinces), at regulated levels, were:

Source	Kilotonnes
Non-ferrous smelters	2,720
Thermal power	726
Non-utility fuel use	580
Other sources	490
TOTAL	4,516

A detailed listing is given in Table 1.

Deposition within eastern Canada

Two main factors must be considered in tracing the sources of deposition and determining the nature and magnitude of abatement programs. First, deposition of sulphur and nitrogen compounds is occurring at relatively uniform rates over large areas. However, emissions are not uniformly distributed. In most of the vulnerable areas in eastern Canada, deposition rates seem to be associated with regional levels of emissions rather than with local source/receptor relationships. Second, environmental damage is for the most part caused by cumulative winter or annual deposition rates; shorter term variations can be important in some instances.

Table 1. Sulfur dioxide emissions (thousand metric tonnes) for
          eastern Canada (Sheffield, pers. comm.)

Province		1980 Base Case
Manitoba	Inco[a]	414
	HMBS[a]	293
	All other sources	31
	TOTAL	738
Ontario	Inco (Copper Cliff Smelter)[a]	1,072
	Inco (IORP)[a]	83
	Falconbridge[a]	154
	Ontario Hydro[b]	452
	Algoma Steel - Wawa	161
	All other sources	272
	TOTAL	2,194
Quebec	Noranda	552
	Gaspe Copper	91
	All other sources	442
	TOTAL	1,085
New Brunswick	N.B. Electric Power	122
	All other sources	93
	TOTAL	215
Nova Scotia	N.S. Power	124
	All other sources	95
	TOTAL	219
Prince Edward Island	All sources	6
Newfoundland	All sources	59
TOTAL		4,516

[a]Owing to decreased production in the smelting industry, the actual
1980 emissions were lower by approximately 600,000 tonnes
[b]1980 emissions from Ontario Hydro were lower than projected levels
by approximately 56,000 tonnes

Sulphur compounds can remain in the atmosphere two to five days before being deposited up to 1,000 km downwind. Nitrogen compounds typically remain in the atmosphere for shorter periods of up to two days, thus limiting their transport distances to several hundred km (MOI, 1982). Annual wet deposition of sulphate has been linked by computer simulation models to large source areas (Table 2). Even though there is some debate on the accuracy of these models, they have been verified within reasonable limits by the correlation of results with field measurements (MOI, op. cit.). These models can be used to estimate changes in regional emission rates that would be required to achieve desired changes in long-term rates of wet sulphate deposition (Shaw, 1985). As shown in Table 3, a reduction of about 50 percent from the 1980 base case SO_2 emission level together with compatible reductions in U.S. emissions would be required to achieve the environmental objective of 20 kg ha^{-1} yr^{-1} for wet sulphate in deposition.

In 1983, the final reports of the Canada/U.S. Work Groups, established under the 1980 Memorandum of Intent, released their final reports (MOI, 1983)[1]. The U.S. and Canadian scientists on the Work Groups agreed on a number of important points:

° damage in both the short and long-term is occurring in areas vulnerable to acid rain as a result of sulphur deposition;

° wet sulphate deposition above 20 kg ha^{-1} yr^{-1} (annual rate of 18 pounds per acre per year) in vulnerable areas is associated with damage while areas with deposition less than this level have no recorded damage;

° the damage is caused by sulphur deposition;

° the solution for curtailing these damages is to reduce emmissions;

° acid deposition occurs in eastern North America in areas both within and downwind from major industrial regions;

° technology exists to reduce emissions by substantial amounts; and

° if there are no changes in abatement programs, emissions are forecast to increase through the remainder of this century.

These reports were subjected to peer reviews. This was conducted by the Royal Society of Canada (1984) and the President's Office of Science and Technology Policy (Nierenberg, 1984). In

---

[1]Work Groups 1, 2, and 3B reports are available through Environment Canada, Ottawa, Ontario, Canada.

Table 2.   The estimated percentage contribution to wet sulphate
           deposition in several areas in eastern Canada (Shaw,
           pers. comm.)

Source Area	North Central Ontario	South Central Ontario	Southern Quebec	Southern Nova Scotia
Canada	29-32	32-46	35-58	30-34
United States	68-71	54-68	42-67	66-70

both cases, the reviewers concluded that abatement programs were
needed and that sufficient information was available for their
design.   In addition, the U.S. peer review noted that one of the
most important long-term impacts of acid deposition was to
unmanaged forest soils.

Canadian Air Pollution Control Programs

     Air pollution control in Canada is a shared responsibility of
the federal and provincial governments.   The provinces are
primarily responsible for controlling emissions which impact within
their borders while the federal government is responsible for
interprovincial and international air pollution.

     Except for pollutants dangerous to human health, Canadian air
pollution control programs are generally designed to achieve
specified environmental quality objectives rather than prescribing
the use of specific technologies.   Air quality objectives for
sulfur dioxide have been published by the federal government[1] for
the guidance of the provinces. Comparison of Canadian and U.S. air
quality requirements in Table 4 indicates that the Canadian acceptable levels are slightly more stringent than the U.S. primary
standards while the Canadian desirable levels are more than twice
as stringent as U.S. standards.   Most provinces base their control
programs on the acceptable level; two use the desirable level.   As
in the United States, the provinces use a variety of regulatory
mechanisms to achieve their air quality standards.

     As a result of Canadian air pollution control efforts, levels
of SO_2 in urban areas have decreased by about 50 percent in the
last decade.   The annual acceptable level for Canada has not been

---

[1]Ambient Air Quality Objectives Order No. 1, C.R.C. 1978, c.403
and Ambient Air Quality Objectives No. 3, SOR/78-74 (as am. by
SOR/78-812); both issued under the Clean Air Act, S.C. 1970-
71-72, c.47.

Table 3. Model projections of overall Canadian and United States emission reductions needed for predicted wet sulphate deposition. Values are provided by Shaw (pers. comm.)

Receptor Area	Projected Deposition (in kg ha^{-1}yr^{-1}) from Model Calculations				
	Current Deposition	25% Canadian Reduction	50% Canadian Reduction	100% Canadian Reduction	50% Canadian and U.S. Reduction
North Central Ontario	18-22	17-21	16-20	14-18	11-15
South Central Ontario	29-35	27-33	24-30	20-26	13-19
Southern Quebec	27-33	25-31	23-29	19-25	15-21
Southern Nova Scotia	17-23	17-23	16-22	14-20	12-18

Table 4. Comparison of sulphur dioxide ambient air quality
          requirements ($\mu g\ m^{-3}$)

Averaging Period	Canadian Objectives		U.S. Standards	
	Desirable	Acceptable	Primary	Secondary
Annual	30	60	80	-
24 hours	150	300	365	-
1 hour	450	900	-	-
3 hours	-	-	-	1,300

exceeded since 1972 (Environment Canada, 1981). The 1-hour and 24-hour desirable levels were exceeded only one percent of the time. Sulphur dioxide, as a local air quality problem, has thus been successfully controlled in Canada. This was achieved through a variety of actions including emission containment, use of tall stacks, use of low sulphur fuels, intermittent control procedures and fuel regulations. These actions resulted in a 27 percent reduction in total Canadian sulphur dioxide emissions between 1970 and 1980; 30 percent in eastern Canada and eight percent in the west.

Canadian non-ferrous smelters have used selective mining and mineral separation preparation practices, sulphur containment and changes in process technology to reduce emissions by 49 percent. Regulations limiting the sulphur content of fuels have reduced nonutility fuel use emissions by 18 percent. Canadian utilities have used natural gas, low sulphur coal, coal-blending, coal-washing, nuclear and hydraulic generation and load management to limit increases in emissions to 71 percent while increasing total generation by 91 percent.

International Obligations

Canada's approach to transboundary aspects of acid rain has been based in large measure on existing and evolving international law and principle. While the body of international law in matters of transboundary air pollution is relatively small, there are some significant precedents. One of these is the Trail Smelter Arbitration of 1941[1]. This established an important international legal principle:

"...no State has the right to use or to permit the use of its territory in such a manner as to cause injury by

---

[1]3/R. Intl. Arb. Awards 1905, 1965 (1941).

fumes in or to the territory of another or the properties
or persons therein when the case is of serious consequence and the injury is established by clear and convincing evidence."

Another internationally recognized principle which is important and
which both the United States and Canada have accepted is Principle 21 of the Stockholm Declaration on the Human Environment[1].
It states:

"States have, in accordance with the Charter of the
United Nations, and the principles of international law,
the sovereign right to exploit their own resources
pursuant to their own environmental policies, and the
responsibility to ensure that activities within their
jurisdiction or control do not cause damage to the
environment of other States or areas beyond the limits of
national jurisdiction."

Article IV of the Boundary Waters Treaty of 1909, which has guided
Canada/U.S. transboundary relations in the field of water quality,
also provides a useful principle. It reads:

".....waters herein defined as boundary waters and waters
flowing across the boundary shall not be polluted on
either side to the injury of health or property on the
other."

These several principles have been used as an approach to Canadian
"international obligations" to the United States as regards transboundary air pollution.

The United States and Canada have a long standing tradition of
resolving transboundary environmental quality issues. An example
is the Great Lakes Water Quality Agreement[2]. Despite some
recognized limitations in the available scientific data a consensus
was reached on the necessity of taking mutual agreed environmental
water quality objectives. Each country was then in a position to
design and introduce measures to achieve those objectives. Anala-
gous in the acid deposition issue would be an agreement on the wet
sulphate deposition objective of 20 kg ha^{-1} yr^{-1} and commitment to
the development of compatible programs to achieve that objective.

---

[1]U.N. Conference on the Human Environment, Stockholm, Sweden;
U.N. Doc., A/Conf. 48/14, at 7 (June 5-16, 1972).

[2]Both 1972 (entered into force 15 April 1972) and 1978 (entered
into force 22 November 1978) Agreements Between The United States
And Canada On Great Lakes Water Quality.

The Canadian approach to transboundary air pollution also recognizes that, together with shared environmental goals, Canada and the United States have different socio-economic conditions and constitutional frameworks which can result in different approaches to similar problems. These can in turn produce different domestic standards and control strategies. It is for this reason that Canada has turned to the common denominator of international legal principles and scientific criteria for guidance in the resolution of potential disputes, rather than the applicability of domestic law and regulation on one side of the border or the other.

On August 5, 1980, the Canadian and U.S. governments signed a Memorandum of Intent on Transboundary Air Pollution (MOI, 1980). Under the terms of the MOI, both countries acknowledged the already serious problem of acid deposition and agreed:

°   to negotiate a transboundary air pollution agreement as soon as possible;

°   to establish working groups to develop the scientific and technical basis for an agreement; and

°   to vigorously enforce existing laws and regulations in the interim.

Canadian Acid Deposition Abatement Program

Using the federal/provincial cooperative approach, Canada is now designing and initiating a number of actions to combat acid deposition. An acceptable rate of deposition in selected receiving areas is determined and then the ranges of reductions in emissions for contributing source areas that would achieve the environmental objective are estimated. Alternative programs are then developed and technically assessed for those ranges with respect to each emission area; these are then reviewed in relation to social, economic, energy and other factors.

On February 15, 1982, federal and provincial Environment Ministers[1] agreed that:

°   wet sulphate deposition should be reduced to less than 20 kg ha^{-1} yr^{-1} by 1990 to protect moderately sensitive lakes and streams;

°   to achieve this objective, emissions in Canada east of the Manitoba/Saskatchewan border and in the United States east of

---

[1]Federal/Provincial Ministerial Committee on Long Range Transport of Airborne Pollutants, Toronto, Ontario, Canada, 15 Feb 1982.

the Mississippi needed to be reduced by up to 50 percent; and

° Canada would reduce emissions by up to 50 percent contingent
   on parallel and compatible action in the United States.

On February 24, 1982, this proposal was introduced to U.S. negotia-
tions under the MOI. On June 22, 1982, the U.S. rejected the
Canadian proposal as premature.

   Regulations, programs and commitments were accepted which
would reduce the 1980 base case sulfur dioxide emissions by about
25 percent by 1990 (Table 5). These initiatives include the
following reductions[1]: 37 percent at the Inco, Sudbury smelter,
40 percent at the Noranda smelter, 43 percent at Ontario Hydro, and
up to 40 percent in nonutility fuel use emissions through natural
gas conversion and fuel upgrading.

   On March 6, 1984, federal and provincial Environment Ministers
agreed to proceed independently with the development and implementation of further Canadian abatement actions. They reaffirmed
their commitment to the 20 kg ha^{-1} yr^{-1} objective to protect
moderately sensitive areas in Canada. They agreed to reduce the
1980 base case sulphur dioxide emissions by 50 percent by 1994 (or
a residual emission rate of about 2.25 million tonnes in eastern
Canada; see Table 1).

   Utilities contribute less than 15 percent to total eastern
Canadian sulphur dioxide emissions. To reduce these emissions
further, Canadian utilities are relying, where possible, on load
management and increased use of hydraulic and nuclear capacity,
including the advancement of schedules for new construction. They
are also considering further application of some of the techniques
successfully used to combat local air pollution. Projections of
load-growth for some eastern utilities indicate that their coal-
fired steam capacity will be more cost-effective when used for peak
loads rather than base loads; they are therefore finding that
operational changes to these generating stations, particularly
through the use of low sulphur and washed coals, are likely to be
less expensive than capital expenditures. Scrubbers will be
necessary and economically sensible for Canadian utilities only if
long-term needs call for increased use of coal-fired generation to
meet base-load requirements.

---

[1]Copper Cliff Smelter Complex Regulation R.R.O. 1980, Reg. 301,
and Ontario Hydro Emissions Regulation O. Reg. 7/82; both enacted
under the Environmental Protection Act of Ontario; Atmosphere
Quality (Acid Rain Amendment) Regulation, a draft regulation
enacted under the Environmental Quality Act of Quebec; and
pp. 82-88 of Work Group 3B report.

Table 5.   Emission reductions for Canada from the 1980 base case

Source	Controls	Thousand metric tonnes
Ontario	– Ontario Hydro, regulation requires phased reduction from 452 to 260 kilotonnes by 1990	192
	– INCO (Copper Cliff Smelter). Control Orders required cut from 1072 to 646 kilotonnes by December 31, 1982	426
Quebec	– Noranda. Cut of 40% on 1980 emissions of 552 kilotonnes announced by provincial government	221
New Brunswick	– Emission reductions from electrical generation, forest industry and other sources	30
National Energy Program	– Reductions in non-utility fuel use emissions from 745 to 445 kilotonnes by 1990 as a result of gas conversions and reductions in heavy oil use (includes 30 kilotonnes shown above)	270
	TOTAL	1,139

While non-utility fuel use accounts for about 13 percent of eastern Canadian sulphur dioxide emissions, abatement programs applicable to the non-utility use of fuel tend to be relatively more expensive than for the non-ferrous smelting and thermal-power sectors. Controls in this sector are therefore directed to such things as energy conservation, fuel substitution and fuel upgrading.

Achieving the 2.25 million tonne emission ceiling by 1994 will be expensive. As one might expect, the costs vary depending on how and where the reductions are achieved. The annualized costs of the necessary actions are estimated to be between $600 million and $1 billion per year. Capital expenditures of $1.1 to $1.5 billion are required in the non-ferrous mineral industry. Capital expendi-

tures of $1 billion or more could be required in the utilities
sector. This depends on a number of factors including the availability of nuclear power (EMR, 1984).

    The non-ferrous smelting industry is the largest single source
of sulfur emissions, representing almost 60 percent of the eastern
Canadian total. To reduce their emissions further, eastern Canadian smelters are looking at process modernization, increased
production of sulphuric acid, greater selectivity in mining operations, and enhanced mineral-separation during milling. Non-ferrous
smelters are responsible for about 60 percent of the SO_2 emissions
in the eastern part of Canada. A major study of the Canadian
non-ferrous mineral sector stated that it is unlikely that Canadian
nickel and copper producers will be in a financial position to
proceed with major modernization and SO_2 abatement programs before
the late 1980's (EMR, op. cit.). At some smelters a number of
actions can be taken using existing technology. At others there
are attractive technology demonstration possibilities that could
significantly increase plant productivity and/or make pollution
control more affordable. Most of this technology demonstration
work can be done in three to five years, allowing it to be applied
by 1994. The plan will include a mix of technology demonstration
projects together with some initiatives using existing technology.

United States Acid Deposition Abatement Programs

    Programs to abate sulphur dioxide emissions in the United
States will differ from those in Canada because of the overwhelming
differences in the nature and importance of various emission
sources. Policies, priorities, and regulatory mechanisms may also
make the U.S. approach different. The only aspect in which the
programs of the two countries need be compatible would be in
achieving the agreed environmental objective.

    Similarly, it is a matter of judgment by the Government of the
United States as to the most appropriate mix of technologies, fuel
uses, policies, and regulatory mechanisms to be introduced with the
object of reducing sulphur dioxide emissions. Many studies (i.e.,
OTA, 1984) and analyses have been done in recent years on the
feasibility and associated costs of different abatement strategies
and programs for the United States. The cost estimates differ
widely because of differing assumptions about fuel use, control
technologies, economic growth and regulatory approaches.

CONCLUSIONS

    Scientific evidence already available provides ample information on the causes, effects and solutions to the acid rain problem. Further information will emerge as the scientific efforts continue. Recommendations for a course of action must be based on an imperfect, always increasing, body of pertinent data whose quality and completeness can be expected to improve. Actions based on imperfect data run the risk of being in error. Inaction, pending collection of all the desirable information, can possibly entail even greater risk of damage.

From the Canadian perspective, the adequacy of United States abatement programs will be judged in terms of the extent to which, in concert with the Canadian abatement program, the 20 kg ha^{-1} yr^{-1} objective is achieved in vulnerable areas in Canada. Since a tonne of emission reductions in the U.S. midwest is far more important to Canada than a tonne of reductions in the southeast, Canadian objectives vis-a-vis U.S. emission reductions cannot be stated in terms of tonnes or percentage emission reductions.

Basically, what Canada is seeking from the United States is an agreement which includes:

°   the necessity to achieve the 20 kg ha^{-1}yr^{-1} environmental objective;

°   significant emission reductions aimed at achieving the objective; and

°   a bilateral mechanism to monitor progress and recommend changes to abatement programs.

REFERENCES

CFS, 1984, Oral testimony and committee documents. Forest Research and Development Branch, Canadian Forestry Service, Environment Canada, Ottawa, Canada. 6 pp.
Cox, R.M., 1982, Sensitivity of forest plant production to acid rain. Dept. Botany and Inst. Environ. Studies, Univ. Toronto, Toronto, Ontario, Canada. 7 pp.
DFO, 1984, National program - acid rain: Project summaries and progress achieved. Dept. Fisheries and Oceans, Burlington, Ontario, Canada. 132 pp.
Elder, F.C., 1984, Effects of nitrate on the acidification of the aquatic system. Report No. EPS 2/TS/1, Environment Canada, Ottawa, Canada. 14 pp.

EMR, 1984, Canada's non-ferrous metals industry: Nickel and
    copper. Catalogue No. M37-26/1984E, Energy, Mines and
    Resources Canada, Ottawa, Ontario, Canada. 102 pp.
Environment Canada, 1981, Urban air quality trends in Canada,
    1970-1979. Report No. EPS 5-AP-81-14, Inventory Management
    Division, Environment Canada, Ottawa, Ontario, Canada. 56 pp.
Federal Minister for Food, Agriculture and Forests, 1983, Bericht
    des Bundesministers für Ernährung, Landwirtschaft und Forsten
    anlässlich der Waldschädenserhebung, 1983, Report of the
    Federal Minister for Food, Agriculture and Forest on the
    Forest Damage Level, Bonn, Republic of Germany. 17 pp.
Fuhs, G.W., Olsen, R.A., and Bucciferro, A., 1985, Alkalinity and
    trace metal content of drinking water in areas of New York
    State susceptible to acidic deposition, pp. 143 to 162. In:
    Acid Deposition - Environmental, Economic and Policy Issues,
    D.D. Adams, ed., Plenum Publ., New York, NY.
MOI, 1980, Memorandum of Intent Between the Government of Canada
    and the Government of the United States of America Concerning
    Transboundary Air Pollution (signed 5 August 1980).
MOI, 1982, Canada-United States Memorandum of Intent on Trans-
    boundary Air Pollution, Work Group 2 Report, Atmospheric
    Services and Analysis. Environment Canada, Ottawa, Ontario,
    Canada. 148 pp.
MOI, 1983, Canada-United States Memorandum of Intent on Trans-
    boundary Air Pollution, Work Group 1 Report, Impact Assess-
    ments. Environment Canada, Ottawa, Ontario, Canada. 626 pp.
Nierenberg, W.A., Chairman, 1983, Interim report - acid rain peer
    review panel. Office of Science and Technology Policy,
    Washington, DC.
Nierenberg, W.A., Chairman, 1984, Report of the acid rain peer
    review panel. Office of Science and Technology Policy,
    Washington, DC. 63 pp.
Ontario MOE, 1982, Acid sensitivity survey of lakes in Ontario.
    APIOS Report No. 003/82, Ontario Ministry of the Environment,
    Toronto, Ontario, Canada. 18 pp.
OTA, 1984, Acid rain and transported air pollutants: Implications
    for public policy. Office of Technology Assessment, Washing-
    ton, DC. 323 pp.
Percy, K.E., 1982, Sensitivity of Eastern Canadian forest tree
    species to simulated acid precipitation. Canadian Forestry
    Service, Maritimes Forest Research Centre, Fredericton, New
    Brunswick, Canada. 8 pp.
Royal Society of Canada, 1984, Long-range transport of airborne
    pollutants in North America: A peer review of Canadian
    federal research. Toronto, Canada. 115 pp.
Shaw, R.W., 1985, The use of long-range transport models in deter-
    mining emission control strategies for acid deposition, pp. 75
    to 94. In: Acid Deposition - Environmental, Economic and
    Policy Issues, D.D. Adams, ed., Plenum Publ., New York, NY.

Shaw, R.W., Personal communication, Atmospheric Environment
    Service, Environment Canada, Downsview, Ontario, Canada.
Sheffield, A., Personal communication, Environmental Protection
    Service, Ottawa, Canada.

TRANSBOUNDARY AIR POLLUTION: THE INTERNATIONAL EXPERIENCE

John E. Carroll

Department of Forest Resources
University of New Hampshire
Durham, NH   03824

ABSTRACT

Acid rain is a serious problem in Canada-U.S. relations. A divergence exists between the Canadian and U.S. positions on acid rain, for reasons of geography, economics, and political ideology. The United States, whose position has been reactive to Canada, has taken such an uncompromising stance that treaty and agreement violations can be alleged. The issue has risen to the multilateral as well as bilateral level. Canada has called for a 50% reduction in pollution emissions causing acid rain. The United States could respond with a credible compromise if it chose to do so. This could then be readily implemented. The position of the United States seems to be an unwillingness to compromise, resulting in environmental, economic, and political damage.

INTRODUCTION

Acid rain is now accepted by many as a serious environmental and ecological problem. It is obvious to many that it is a serious interstate and interregional problem, pitting, for example, the emitting industrial Midwest against the receiving and vulnerable Northeast in the United States. And, society is coming to realize that acid rain is a serious international problem with significant foreign policy dimensions; not only has it had a history of creating difficulties between Scandinavian receptors and British, German and east European emitters, but indeed it now shows signs of causing a substantial rupture in Canadian-U.S. relations, the

507

damage from which all North America will pay for in the coming
years (Carroll, 1982, 1983).

DIPLOMATIC ISSUES ASSOCIATED WITH ACID RAIN

     The problem has become serious for two basic reasons. Canada
has unilaterally raised the acid rain issue and given it a high
place on the diplomatic agenda, and the United States has thus far
failed to respond substantively to this concern. This has sent a
negative signal back to Canada and its people.

The Canadian Concern

     Why has Canada raised the issue to such a high diplomatic
level and placed such a high political stake on its resolution?
The four imbalances described below establish the foundation for a
fundamental assymetry which should alone be sufficient to cause a
serious bilateral problem.

     First, at least 50% of all Canada's acid deposition (and
perhaps as much as 60%) comes from U.S. sources over which Canada
exerts no influence, except through diplomacy. Conversely, only a
very small percentage of U.S. acid deposition (perhaps 15%) comes
from Canadian sources (U.S.-Canada Research Consultation Group,
1979, 1980). This fact has been determined by a group of highly
qualified government scientists from both the Canadian and U.S.
Federal governments - the bilateral Research Consultation Group.
Thus, there is international movement of pollutants across the
border in both directions, but the national contributions to that
transboundary movement are not close to being balanced. All
interests accept this as fact. Hence, no matter what sacrifices
Canada makes to reduce its own emissions, it still faces very
substantial pollution over which it has no control.

     Second, greater proportions of Canada are geologically
vulnerable to the impacts of acid deposition than are the U.S.
Specifically, almost all of Atlantic Canada, Quebec and Ontario,
most of Manitoba and Saskatchewan, and smaller portions of Alberta,
British Columbia and the northern territories of Canada are highly
vulnerable. This vast region, much of which is known as the
Canadian Shield, is composed of exposed granite bedrock, already
containing highly acidic soils, numerous small lakes, and vast
coniferous forests, none of which can take any additional incoming
acid without experiencing serious ecological change. Not the least
of these changes is the death, for all intents and purposes, of
thousands of small lakes. These bodies of water can no longer
support fish or other aquatic life. In other words, only small
areas of Canada have buffering capacity sufficient to accept this

acidic deposition without showing signs of damage, at least in the short-term. Areas such as the prairie grasslands, the farmlands of southwestern Ontario and Prince Edward Island, and a few other Canadian areas are in this buffered category. Further exacerbating the situation, however, is the fact that some of Canada's most cherished lake and recreational resort country, a country which has been described in the literature and paintings of Canada's most revered writers and artists, a country located just north of Toronto and other major cities and heavily used by thousands of Canadians-- the Muskoka Lakes and Haliburton Highlands--is in the zone of highest vulnerability (Howard and Perley, 1980). Thus, the proximity of acid rain damage (or alleged acid rain damage) to large numbers of people, including the most influential decision-makers of Canadian society, has contributed to the seriousness of the bilateral problem.

Third, Canada is, perhaps more than any other nation, dependent upon the forest products industry for export to much of the world (Curtis and Carroll, 1983). Many jobs are dependent on Canada's lumber, pulp and paper, and newsprint exports and on the country's ability to sustain a competitive position on world markets. What does this have to do with acid rain? The data are by no means complete, but there is increasing suspicion among government and industry foresters and academic scientists that acid deposition may be inhibiting the growth of commercially valuable softwood timber. The full story is yet to unfold, but Canadian forestry officials believe that a decline in growth rate of only one percent will be sufficient to drive Canada off the world markets--especially when combined with Canada's other disadvantages of a short growing season, high labor cost, and increasingly sharp competition from the American southeast. Anything which threatens or appears to threaten Canada's world position in forestry exports must be viewed critically by any Canadian government. Many are now awaiting research results in this area with great anticipation, not the least being the forest industry and those commercial activities dependent upon it.

A final factor contributing to the seriousness of acid rain as a bilateral issue is the great disparity between the peoples vis-a-vis their knowledge of acid rain, and concern over its effects (Howard and Perley, 1980). Most Canadians have been saturated with media attention on this subject for at least five or six years, and polls indicate that as many as 85% of Canadians nationwide are well aware of the issue. Torontonians, residents of Canada's largest city, are especially aware and have heavily influenced their politicians in Ottawa to negotiate a hard line with the United States. Americans have had significantly less exposure to the issue in general and to its impact on bilateral relations in particular. Polls indicate that American awareness until rather recently has been the adverse of Canadian awareness--in other words, only

15-20 percent. It has been climbing, however, and may be as high as 50 percent today. This differential goes a long way toward explaining such an obvious lack of concern south of the border. The disparity in awareness is dangerous in the bilateral sense, for the image conveyed to Canadians is that Americans simply don't care and that they willingly reap the material benefits of uncontrolled pollution emissions while assuming no responsibility for serious damages to Canada. Further, as signatories to a joint Memorandum of Intent (1980) to resolve the problems, Americans do not seem to be taking such obligations seriously. The diplomatic damage which results from this perception, whether justified or not, is a threat to future U.S. interests in Canada as well as to Canada's long-term well being and relationships towards the United States.

The United States Response

Why has the United States failed so far to respond substantively to this concern, even at the risk of sending such a negative signal north of the border and creating a diplomatic problem? Among those in the United States who are well informed about acid rain and perhaps even of its diplomatic consequences, the knowledge of the enormous cost of regulation, the cost of emissions control, particularly for SO_2, is all too real. Environmentalists from both counties should not kid themselves into thinking that the costs of significant sulfur dioxide emissions reduction of the order mandated by some of the proposed federal legislation is inconsequential. It is not. Such costs, generally involving the costly construction and operation of wet sulfur scrubbers, is indeed very significant. This is not to say the costs of enduring the consequences of acid rain might not be even greater. They may be. But it is to say that the costs of many of the proposed solutions are also very high and must be examined carefully before such decisions can be made. Current sulfur scrubber technology is proven technology, is commercially available, and can remove as much as 90% of the SO_2 pollutant from utility stacks (Wetstone and Rosencranz, 1983). But does this mean we should go the sulfur scrubber route? Aside from the great capital investment to build such scrubbers where they are in fact a separate manufacturing operation, they are very costly to keep in operation, are easy to turn off, and create environmental problems of their own. These problems are associated with limestone quarrying and transporting to those related to safe storage of great quantities of a byproduct sludge for which there is little or no market.

Perhaps one of the biggest disadvantages of scrubbers, however, is that being such a large investment they preclude the availability of dollars for many other good purposes, represent the burning of a greater quantity of coal in order to produce the same amount of electricity, and postpone the day when promising new technologies such as fluid bed combustion and limestone injection might become commercialized. Further, for all their cost and other disadvantages, they do nothing to get at the other side of the problem - nitrogen oxides (NO_x), now fully one third of acid deposition and growing in proportion to SO_2. Indeed, it is time for those concerned about acid rain to start focusing on the many sources of NO_x in both our societies. At present rates of growth, these sources are getting out of control and may well represent a more critical problem than sulfur gases.

Contributing to the lack of a U.S. substantive response is the regional decline of the American industrial heartland in Ohio, Michigan, Indiana, Illinois, and adjacent states; the trauma of the steel and automotive industries; the regional dependence on locally mined high-sulfur coal and the employment that coal represents; the state of antipathy of the Reagan Administration against government regulation coupled with the low priority which has been given to environmental concerns; and the U.S. drive toward coal-based energy independence as an element in the nation's security. The Canadian aspect of the issue is more concerned with sulfur dioxide emissions from metal smelters and the export viability of Canada's minerals and metals industry. Therefore, Canada represents a different though not necessarily less complex situation.

Role of the International Joint Commission

It is to the International Joint Commission (IJC), and particularly to its Great Lakes Water Quality Committee, that credit must be given for first bringing to the attention of Ottawa and Washington the increasing threat of acid rain as a bilateral problem. Although addressing acid rain in the context of the Great Lakes, the Committee--and later the Commission itself--made clear its belief in the broad and complex nature of the phenomenon and the likely imbalance in precursor acidic emissions between the two countries. Students of acid rain and bilateral relations will be quick to realize, however, that the IJC role subsequent to its early warnings has been non-existent, and for one basic reason: the two Federal governments have not chosen to use the IJC to monitor or to help achieve resolution of the issue and, in fact, have steered a path which carefully avoids IJC involvement. Given the IJC's expertise in this area, as well as its historic early warning role, one can only speculate as to the reasons for this avoidance. High on the list, however, must be the suspicion that Ottawa and Washington diplomats do not wish to lose direct control over such a sensitive and consequential issue, as they most assuredly would do if the dispute were referred to the independent IJC. Thus, the IJC, under its Great Lakes water quality and general transboundary air pollution warning responsibility,

continues to watch the evolution of the acid rain issue, but
clearly as an observer and not as a player.

VIOLATIONS AND RESOLUTIONS

     The bilateral conflict over acid rain and the large export of
U.S.-based acidic pollution to a very vulnerable Canada may
eventually come under a degree of control, indeed of containment as
an international problem, when the U.S. government finally comes to
realize the seriousness of the issue and initiates a positive
reaction, however humble. The acid rain problem as an issue in
public policy will not be so readily resolved, however, and will be
with us ecologically and economically for a long time to come, both
domestically and internationally. When positive movement comes
from Washington, the issue will necessarily take a lower bilateral
profile as Canada, the complainant, begins to assess that U.S.
response. To leave the matter at that, however, would be to
grossly understate the reality of the damage that this dispute, and
many other international environmental disputes, has done and is
doing to American integrity, in North America and in the estimation
of our friends on other continents. Our response to acid rain is
symptomatic of a much larger problem, both in our relations with
Canada and also in response to our international environmental
obligations and responsibilities. U.S. administrations, both
Democratic and Republican, have at the least demonstrated great
carelessness vis-a-vis the fulfillment of such obligations. At
worst, they've shown blatant disregard for their commitments and
responsibilities and for the impact of their actions on the lives
and livelihoods of other peoples who are called friends and
partners.

     Acid rain presents a highly complex as well as a serious case.
Not only is transborder air and water pollution at issue in both
directions, albeit most of the damage is done by the U.S. to
Canada, but there has been a notable failure by both governments to
use the appropriate machinery available to rectify the situation -
the International Joint Commission.

     Although there has been some contradiction and uncertainty
over the scientific facts surrounding the acid rain phenomenon, it
now appears that the concerns raised by Scandinavian scientists a
decade ago and by Canadian and some U.S. scientists and
environmentalists more recently, have validity (Howard and Perley,
1980; Wetstone and Rosencranz, 1983). Continuing studies,
including those of the U.S. National Academy of Sciences (1983)
and even the White House's own Interagency Task Force on Acid Rain
(1983) now reach similar conclusions to those by Environment Canada
and many academic scientists in both the U.S. and Canada. The
issue now is less one of debate over scientific validity and more

one of how best to reduce the causative emissions and allocate
costs. The diplomatic damage between Canada and the U.S. over the
issue of acid rain has, however, already been done, and future
generations of Americans will pay that cost.

Treaty Violations

For many good reasons, including differences in geologic
vulnerability, levels of awareness, imbalances in transboundary
movement of pollutants and even differences in dependence on the
timber industry, and on coal as an energy source, the U.S. and
Canada started from different positions on acid rain. These
differences have spurred Canada to action and have contributed to
foot-dragging in the U.S. There may now be an increasing
willingness to act by those south of the border. What is less
excusable, however, is the callous disregard that too many in
positions of responsibility and leadership in the U.S. have shown
toward the Canadian complaint, to the nature and intensity of that
complaint, and to the U.S.'s own obligations under the Boundary
Waters Treaty, under international law and custom, and, simply,
under tenets of good neighborliness and plain human decency. Such
is not meant to exempt Canada from its own sources of transborder
pollution or its own unwillingness to employ the treaty mechanism,
although it is obvious that Canada had to take, and has taken,
strong steps internally to get its own house in order. In any
event, the U.S. is not responsible for what Canada does or does not
do. It is responsible for its own actions. And, given its great
power, its great ability to do both harm and good, this
responsibility is a heavy one. Its "head in the sand" attitude
from 1978 to 1984 has not done this nation or its people justice.

A major failing on the part of both the U.S. and Canadian
governments has been their unwillingness to invoke the Boundary
Waters Treaty of 1909 and use the International Joint Commission.
It was the IJC which first alerted the two governments to the
bilateral threat arising from this pollution. The Commission did
so as part of its normal duties under the Great Lakes Water Quality
Agreement and also under the authority of its standing charge to
alert governments to new transboundary problems arising in the area
of air pollution. The Commission has continued to remind both
governments each year of the bilateral seriousness of the acid rain
problem, an unnecessary reminder at this point of great public
clamor but an important one for the Commission under its duties and
responsibilities.

It is now known that in 1980, the International Joint
Commission went further. In a letter dated April 17, 1980, from
both sections of the Commission and addressed to Sharon Ahmad, then
Deputy Assistant Secretary of State for European Affairs (including
Canada), the Commission notes: "... acid precipitation is one of

the most serious environmental problems facing the United States
and Canada." Further, "The Commission believes that in line with
emerging state practice prior restraint should be observed for
activities having potentially serious transboundary impacts until
the impacts of those actions .are adequately addressed. Prior
restraint for such new activities should be practiced until there
are firm assurances of no additional transboundary harm arising
from such activities... (S)uch assurances may be possible by
employing best available technology." Concluding, "The Commission
believes that vigorous domestic initiatives in both countries must
be pursued and that a potentially very serious problem between the
two countries can be avoided by a cooperative undertaking to
protect the shared environment. Now is not the time for the
relaxation of efforts against atmospheric pollution. The
Commission believes that... (G)overnments can put in place in a
timely manner measures to control and reduce atmospheric pollution.
These measures...will enable both countries to protect their shared
environment" (IJC, 1980).

   These are not the words of one complaining nation acting to
protect its self-interest. They are the words and findings of a
highly respected bilateral body appointed by both governments to
act in the collective best interests of both nations. And yet, how
did the governments respond? They didn't, and, going a step
further, the U.S. government, to its discredit, took steps to see
that the letter was not made public. This letter, representing as
it does the concern and the foresight of the IJC, constitutes
strong evidence that the United States Government is not willing to
use the pre-ordained means at its disposal to avoid and/or resolve
bilateral disputes. The letter also demonstrates the unwillingness
of the Canadian Government under the circumstances of the time to
raise the Commission's concerns publically to the U.S. Government
and the American people and to use the letter to support its case
as a complainant. It also shows the reluctance of Canada to
unilaterally refer the acid rain dispute to the IJC for
recommendation, even though it is fully within the purview of
either signatory under the terms of the Treaty to unilaterally
refer a matter of concern to the Commission. That acid rain was
not referred to the Commission in 1980, therefore, when it could
readily have been so referred in the U.S.-Canada Memorandum of
Intent on Transboundary Air Pollution (1980), is the fault of both
Federal governments, each in turn refusing to use a highly
appropriate mechanism available to them.

   The Convention on Long-Range Transport of Air Pollutants
(1979), conducted by the United Nations Economic Commission for
Europe with the U.S. and Canada as signatories, presents further
evidence of U.S. recalcitrance in fulfilling its international
environmental obligations in general and acid rain in particular.
In this Convention, the contracting parties endeavor "... to limit

... and gradually reduce and prevent air pollution including long-range transboundary air pollution" (Article II). Not only has the United States failed to reduce such pollution or to control such pollution sources in the intervening years, but, far more importantly, and far more damagingly, has clearly increased such pollution through its relaxation of Clean Air Act controls over sulfur dioxide emissions from coal-fired power plants in the Midwest. That action, taken in the spring of 1981, though legal under conditions of the U.S. Clean Air Act of 1977, is at the least a violation in spirit of the Convention. A case might be made that this action, along with severe reductions in Federal energy conservation and new coal technology program budgets, actions contradictory to Article VII, may represent a violation of the letter of the Convention as well - or, at the least, give the perception of such violation.

The Convention did not enter into force until March of 1983, upon the signature of the 24th signatory nation, so one might argue that the twenty-three prior signatories were not bound under international law to observe its provisions until the accession of the twenty-fourth signatory. This is technically correct. However, the United States, Canada and many other nations agreed to a supplemental resolution to the Convention stating a commitment to observe the requirements of that Convention effective immediately (in November 1979). Such a resolution does not carry the force of international law, but its violation certainly does constitute a violation of spirit. The United States was not prepared to honor its voluntary obligations under this supplemental resolution. Since the Convention itself only entered into full force in mid-1983, it remains to be seen if the U.S. is now willing to honor its more formal commitments under the ECE Convention, both in letter and in spirit. It also remains to be seen if the U.S. is willing to do so convincingly, giving no opportunity to other nations, whether friends or detractors, to perceive otherwise.

It is often said that acid rain is a most complex issue and that its remedy is or at least can be extremely expensive (Harrison, 1984). This is true, and it is, therefore, understandable that an acid-emitting nation is not quick to embrace the idea of putting on controls and reducing emissions. However, the tragedy of acid rain diplomacy is that so much could have been rectified diplomatically without the extremely costly, capital intensive, inflexible kind of measures which are called for by some. The United States, for example, witnessing the serious concerns of so many highly qualified scientists from so many different disciplines, should have recognized the foolishness of trying to maintain an ultimately untenable "head in the sand" posture, one which crudely denied the evidence, suppressed good scientists, exposed itself to allegations of data manipulation, forced its diplomats to represent indefensible positions, damaged

the credibility of American scientists, and made itself look
foolish before the world in addition to damaging Canadian–American
relations to an as yet unassessed but unquestionably serious
degree.

What Should be Done

    The Canadian–U.S. acid rain dispute must be resolved if
serious long-term damages to bilateral relationships are to be
avoided. Regardless of what one thinks of the validity of the data
or the seriousness of the issue as an environmental one, there can
be no denial of the issue's seriousness to diplomatic or bilateral
relations. It is serious, and the search for resolution demands
the best of good faith efforts - not rhetoric - from both
governments.

    The first step in achieving resolution must be a recognition
that there are essentially two sides to the issue, not without
their varying shades of differences, but essentially two sides.
The leading protagonists on one side are the Canadian Federal
government, the Ontario provincial government, Canadian and
American citizen environmentalists, and a small number of acid-
vulnerable U.S. states (in New England, New York and the upper
Midwest). The opposing side is best represented by coal interests
and the coal-dependent utility industry, the U.S. Government (and
particularly the Reagan Administration), U.S. states heavily
dependent on coal-fired power generation and the jobs such
generation represents (centered in the U.S. Midwest), and the coal-
based province of Nova Scotia. These, then, are the forces which
must be brought together.

    Granted that the stakes are likely very high regardless of
which decision is made (i.e., to control and reduce or not to
control and reduce SO_2 and NO_x emissions), and granted that all
interests agree that much more research is needed before much
scientifically conclusive information is available to
satisfactorily prove cause and effect linkages. Therefore, a
compromise is in order. But can the issue be compromised? It can
be and, in fact it must be if serious diplomatic as well as
environmental costs are to be avoided. That compromise must yield
some reasonably significant decrease in SO_2 and NO_x reductions, but
through the avoidance of any decision to build large and costly
sulfur scrubbers.

    Canada's diplomats have offered to reduce Canadian sulfur
dioxide emissions by 50% by 1990 if the United States will do
likewise. Canada suggested in 1983 that, lacking such bilateral
agreement, Canada might be willing to reduce these emissions by 40%
by 1990 in any event (Brandt, 1983; Canadian Embassy, 1984). In

response to this, the United States should be willing to offer, and Canada should be willing to accept, a compromise U.S. sulfur dioxide emission reduction of 40% in that same time period, or a combination SO_2 and NO_x reduction which would equal Canada's 40% SO_2 reduction. It is possible to achieve this without scrubber technology.[1]

Achievement of These Goals

Such a goal for reducing emissions of sulfur dioxide might be achieved in the following four steps (Carroll, 1982):

° Through application of universal and mandatory coal-prewashing at the mine mouth. This is capable of removing at least 15% of the sulfur.

° Through the widespread use of the operational technique known as Least Emissions Dispatching (LED), wherein reliance is placed on new lower emission power plants for base-load generation and on older more polluting equipment to answer peak power demand. This, in effect, would be the reverse of the present procedure, which is called least-cost dispatching. This would remove at least another 15%.

° Through somewhat greater reliance on low sulfur coal in the fuel mix, while avoiding especially serious disruption of that coal mining industry based on high-sulfur coal. This could easily remove another 10% without threatening or displacing the high sulfur coal industry. (This could be anything up to about 60%, though achieving such a percentage would seriously damage the U.S. high sulfur coal industry and regions dependent upon it).

° Through more innovative use of regulatory legislation which permits trade-offs between NO_x and SO_2 reductions, perhaps on a 2-1 basis.

In addition, maximum resources should be devoted in both countries, but particularly in the United States, toward research and demonstration necessary to speed the introduction of new coal-combustion technology, especially fluid bed combustion and limestone injection techniques. There is especially great value to be gained by a bilateral pooling of jointly managed research efforts in this area.

---

[1]Canada has now gone further and announced (February 5, 1984) a unilateral 50% reduction of SO_2 by 1994 regardless of U.S. action. This will increase pressure on the United States to at least support a 30% reduction.

In the Aftermath: Implementation

     The ratification of an agreement or treaty as such would not be the end of these deliberations. Critical attention must be given to the nature of the vehicle designed to implement the agreement or treaty, essentially in perpetuity. Such a vehicle could be an existing organization, such as the International Joint Commission (IJC), or it could be a wholly new body designed expressly for that purpose, and perhaps also designed to address local air pollution problems at various points along the border.

     Because of the magnitude and complexity of acid rain and the fact that the phenomenon affects major populations and regions far removed from the border, it is felt that a new vehicle may be needed. Such a vehicle, perhaps to be called the Canada-U.S. Bilateral Air Quality Commission, must involve private sector as well as public sector interests if it is to do a credible and effective job. Industries which emit SO_2 and NO_x, those who are not emitters but which have a stake in healthy bilateral relations and regional economic interests, financiers, labor unions, and citizen environmental groups -- these are all examples of private sector interests that have much expertise in the acid rain phenomenon and/or have a major stake in the continued resolution of the issue domestically and internationally. They must be willing to contribute if any agreement is to be effective. Hence, the commission should have a permanent composition of two federal officials, diplomatic and environmental, from both governments; one state and one provincial official; a business executive from an emitter concern in each country; a business executive from a non-emitter concern in each country involved in Canada-U.S. bilateral trade; a financier from each country with an institution involved in bilateral activity; a labor union representative from each country; and at least one citizen environmentalist for air and one for water from each country. Various other combinations are possible as long as there is national balance, balance by interest type, and assured participation by the private sector. Federal diplomatic officials from each nation should co-chair the commissiion at all times to insure acceptance of findings by the governments.

     In its organization and rules of procedure, such a commission could well rely on the IJC model. This would include small national staffs in Ottawa and Washington, various technical advisory boards and committees, appointments of commissioners to terms of office, following the Canadian IJC model, and semi-annual meetings of the commissioners alternating between the two national capitals. Unlike the IJC, the commission should have full initiatory authority under the guidelines laid out in the treaty or agreement. Like the IJC, however, the commission should be advisory to government and should be devoid of any power to make

binding decisions.   Such power could be threatening and would
ultimately damage the effectiveness of the commission.   Other
behavior could undoubtedly be adopted or modified from the valuable
IJC model.   Whether or not commissioners should attempt to serve
collegially, however, as they are expected to serve on the IJC, or
whether they should be representing their nation's position and
their functional area of interest deserves investigation (Carroll
and Mack, 1982; Curtis and Carroll, 1983).

CONCLUSION

     Most importantly in this period of impasse over a critical
issue, our two nations must move now, must be as innovative and
creative as possible, and must embrace the challenge openly.   Any
benefit/cost assessment of acid rain must be careful to add to the
cost side of the calculation increasingly serious foreign relations
costs, costs which might be avoided if the will were there to do
something (Carroll, 1983a,b, 1984).

     Sadly for the United States and for the health of Canadian-
U.S. relations, the Government of the United States, both Executive
and Legislative branches, knows the problem is real and that acid
rain will create further costs.   But the United States still
refuses to act, a refusal based upon a lack of concern for the
environment, for the implementation of necessary regulatory control
to protect people and property, for the health of the bilateral
Canada-U.S. relationship and for U.S. obligations in that relationship, and a determination to protect the economic status quo in the
coal-producing and coal-dependent regions of the country.
Ironically, failure to act now poses an even greater long-term
threat to the economic viability of those same regions.   The time
for compromise is, unfortunately, passing, as are the opportunities
for compromise.   As evidence accumulates as to the threat acid rain
may pose to forest ecosystems, the woods products sector of the
economy and perhaps even to human health, the political pressure to
take significant and perhaps precipitous action will become
irresistable.   The regions that will feel that shock most directly
are those coal-producing and coal-dependent regions which the U.S.
Federal Government, in its policy of inaction, now seeks to
protect.   Americans in all regions will suffer the increasingly
justifiable devolution of Canada from U.S. collaboration and
cooperation.   The U.S. will have only itself to blame for this
cost, a far greater cost than most Americans now realize.   Acid
rain's significant place in North American history will be
cemented.

Acknowledgments - The basis for part of this chapter is the book
Acid Rain: An Issue in Canadian-American Relations (1982) by

John E. Carroll. Publication of this book was sponsored by the Canadian-American Committee (C.D. Howe Institute, Toronto and the National Planning Association, Washington, DC). Portions previously appeared in Alternatives Vol. II, No. 2, Winter, 1983, and the Bulletin of the Atomic Scientists, Vol. 40, No. 1, January, 1984.

REFERENCES

Brandt, A., 1983, Letter to Rep. John Dingell, U.S. House
     Representatives, from Ontario Environment Minister, Toronto,
     Canada. November 1, 1983, as reported.
Canadian Embassy, 1984, Canada to go it alone in acid rain control
     program. Press Release, Washington, DC. (March 7, 1984).
Carroll, J.E., 1982, Acid Rain: An Issue in Canadian American
     Relations. Report No. 194, National Planning Association,
     Washington, DC. 80 pp.
Carroll, J.E., 1983a, Environmental Diplomacy: An Examination and
     a Prospective of Canadian U.S. Transboundary Environmental
     Relations. Univ. Michigan Press, Ann Arbor, MI. 382 pp.
Carroll, J.E., 1983b, Acid rain diplomacy: The need for a
     bilateral resolution. Alternatives, Perspectives on Society
     and Environment 11:9-12.
Carroll, J.E., 1984, Water dampens U.S.-Canadian relations. Bull.
     Atomic Scientists 40:20-25.
Carroll, J.E. and Mack, N.B., 1982, On living together in North
     America: Canada, the United States and international
     environmental relations. Denver J. Intern. Law Policy
     12:35-50.
Convention on Long-Range Transboundary Air Pollution, 1979,
     UN/ECE/GE 79-42960. Environ. Law Policy 6:37-40.
Curtis, K.M. and Carroll, J.E., 1983, Canadian-American Relations:
     The Promise and the Challenge. D.C. Heath and Co., Lexington,
     MA. 128 pp.
Harrison, W.B., 1984, An explantation of the Southern Electric
     System position on acid rain legislation. Southern Co.
     Services, Inc., Birmingham, AL. 8 pp.
Howard, R. and Perley, M., 1980, Acid Rain: The North American
     Forecast. Anansi, Toronto, Canada. 207 pp.
IJC, 1980, Letter from International Joint Commission to
     Sharon Ahmad, U.S. Deputy Assistant Secretary of State,
     Washington, DC. April 17, 1980, as reported.
Interagency Task Force on Acid Rain, 1983, General Comments on Acid
     Rain: A Summary by the Acid Rain Peer Review Panel for the
     Office of Science and Technology Policy, Executive Office of
     the President, Washington, DC. 6 pp.

M.O.I., 1980, Memorandum of Intent on Transboundary Air Pollution, United States-Canada. T.I.A.S. No. 9856, Govern. Printing Office, Washington, DC and Queens Printer, Ottawa, Canada. 4 pp.

U.S.-Canada Research Consultation Group on the Long-Range Transport of Air Pollutants, 1979, The LRTAP problem in North America: A preliminary overview. Government Printing Office, Washington, DC and Queens Printer, Ottawa, Canada. 6 pp.

U.S.-Canada Research Consultation Group, 1980, Second Report of the U.S.-Canada Research Consultation Group, Washington, DC and Ottawa, Canada. 9 pp.

U.S. National Academy of Sciences, 1983, Acid Deposition, Atmospheric Processes in Eastern North America. Committee on Atmospheric Transport and Chemical Transformation in Acid Precipitation, National Research Council, National Academy Press, Washington, DC. 375 pp.

Wetstone, G.S. and Rosencranz, A., 1983, Acid Rain in Europe and North America. Environmental Law Institute, Washington, DC. 244 pp.

INDEX

Directions for use:

Keywords were initially indexed according to paragraph numbers in each individual chapter. Since many paragraphs were split to insert illustrations during the final typing of the camera-ready pages, it is possible that keywords might appear in the text one page (not including inserted tables and figures) before or after their citation in the index.

The editor wishes to apologize for the problems associated with the lack of capitalization of proper names and locations. Only the initial letters of main entries in the index were designated to be capitalized when the computer indexing program was developed.